Bioprospecting of Microorganism-Based Industrial Molecules

Bioprospecting of Microorganism-Based Industrial Molecules

Edited by

Sudhir P. Singh

Center of Innovative and Applied Bioprocessing (DBT-CIAB)
Mohali, Punjab, India

Santosh Kumar Upadhyay

Department of Botany
Panjab University
Chandigarh, India

WILEY

This edition first published 2022

Registered Offices
John Wiley & Sons, Inc., 111 River Street, Hoboken, NJ 07030, USA
John Wiley & Sons Ltd, The Atrium, Southern Gate, Chichester, West Sussex, PO19 8SQ, UK

Editorial Office
The Atrium, Southern Gate, Chichester, West Sussex, PO19 8SQ, UK

For details of our global editorial offices, customer services, and more information about Wiley products visit us at www.wiley.com.

Wiley also publishes its books in a variety of electronic formats and by print-on-demand. Some content that appears in standard print versions of this book may not be available in other formats.

Library of Congress Cataloging-in-Publication Data

Names: Singh, Sudhir P., editor. | Upadhyay, Santosh Kumar, editor.
Title: Bioprospecting of microorganism-based industrial molecules / edited by Sudhir P. Singh, Center of Innovative and Applied Bioprocessing (DBT-CIAB), Mohali, India, Santosh Kumar Upadhyay, Department of Botany, Panjab University, Chandigarh, India.
Description: First edition. | Hoboken, NJ : Wiley, 2021. | Includes bibliographical references and index.
Identifiers: LCCN 2021028318 (print) | LCCN 2021028319 (ebook) | ISBN 9781119717249 (cloth) | ISBN 9781119717157 (adobe pdf) | ISBN 9781119717263 (epub)
Subjects: LCSH: Biotechnological microorganisms. | Biomolecules. | Industrial microorganisms.
Classification: LCC TP248.27.M53 B5635 2021 (print) | LCC TP248.27.M53 (ebook) | DDC 660.6–dc23
LC record available at https://lccn.loc.gov/2021028318
LC ebook record available at https://lccn.loc.gov/2021028319

Cover Design: Wiley
Cover Image: © Graphics Master/Shutterstock

Set in 9.5/12.5pt STIXTwoText by Straive, Pondicherry, India
Printed and bound by CPI Group (UK) Ltd, Croydon, CR0 4YY

C9781119717249_221121

Contents

About the Editors

Dr. Sudhir P. Singh is currently Scientist-D at the Center of Innovative and Applied Bioprocessing (DBT-CIAB), Mohali, India. He has been working in the area of molecular biology and biotechnology for more than a decade. Currently, his primary focus of research is gene mining and biocatalyst engineering for the development of approaches for transformation of agro-industrial residues and under- or un-utilized side-stream biomass into value-added bioproducts. He has explored the metagenomic resources from diverse ecological niches and developed enzyme systems for catalytic biosynthesis of functional sugar molecules such as D-allulose, turanose, fructooligosaccharides, glucooligosaccharides, 4-galactosyl-kojibiose, xylooligosaccharides, levan, dextran, resistant starch, and so on. Dr. Singh has published over 60 research articles, 04 review articles, and 06 books (edited). Further, he has 06 patents (granted) to his credit as an inventor. He has been conferred International Bioprocessing Association-Young Scientist Award (2017), School of Biosciences-Madurai Kamraj University (SBS-MKU) Genomics Award (2017), Professor Hira Lal Chakravarty Award, ISCA, DST, (2018), Gandhian Young Technological Innovation Award to team (as a guide) (2019), and the member of the National Academy of Sciences, India (2020).

Dr. Santosh Kumar Upadhyay is currently working as an assistant professor in the Department of Botany, Panjab University, Chandigarh, India. Prior to this, Dr. Upadhyay was DST-INSPIRE faculty at the National Agri-Food Biotechnology Institute, Mohali, Punjab, India. He did his doctoral work at the CSIR-National Botanical Research Institute, Lucknow, and received his Ph.D. in Biotechnology from UP Technical University, Lucknow, India. He has been working in the field of Plant Biotechnology for more than 14 years. His present research focuses in the area of functional genomics. He is involved in the bio-prospecting and characterization of various insect toxic proteins from plant biodiversity, and defense and stress signaling genes in bread wheat. His research group at PU has characterized numerous important gene families and long noncoding RNAs related to the abiotic and biotic stress tolerance and signaling in bread wheat. He has also established the method for genome editing in bread wheat using CRISPR-Cas system and developed a tool SSinder for CRISPR target site prediction. His research contribution led the publication of more than 60 research papers in leading journals of international repute. Further, there are more than five national and international patents, four books to his credit. In recognition of his substantial research record, he has been awarded NAAS Young scientist award (2017–2018) and NAAS-Associate (2018) from the National Academy of Agricultural Sciences, India, INSA Medal for Young Scientist (2013) from the Indian National Science Academy, India, NASI-Young Scientist Platinum Jubilee Award (2012) from the National Academy of Sciences, India, and Altech Young Scientist Award (2011). He has also been the recipient of the prestigious DST-INSPIRE Faculty Fellowship (2012), and SERB-Early Career Research Award, (2016) from the Ministry of Science and Technology, Government of India. Dr. Upadhyay also serves as a member of the editorial board and reviewer of several peer-reviewed international journals.

List of Contributors

Jayanthi Abraham
Microbial Biotechnology Lab, School of Bioscience and Technology, Vellore Institute of Technology, Vellore, Tamil Nadu, India

Maria F. Salazar Affonso
Centro de Ciências da Vida, Universidade do Vale do Taquari – Univates, Lajeado, RS, Brazil

Cristobal N. Aguilar
Bioprocesses and Bioproducts Research Group, Food Research Department, School of Chemistry, Autonomous University of Coahuila, Saltillo, Coahuila, México

Anjana
Center of Innovative and Applied Bioprocessing, Mohali, Punjab, India

Débora Bublitz Anton
Programa de Pós-Graduação em Biotecnologia (PPGBiotec), Universidade do Vale do Taquari – Univates, Lajeado, RS, Brazil

Rwivoo Baruah
Microbiology & Fermentation Technology Department, CSIR-Central Food Technological Research Institute, Mysuru, Karnataka, India

SA Belorkar
Department of Microbiology and Bioinformatics, Atal Bihari Vajpayee University, Bilaspur, Chattisgarh, India

Priyanka Bhat
Department of Chemical Engineering, National Institute of Technology Karnataka, Mangalore, Karnataka, India

José de Jesús Cázares-Marinero
Department of Research & Development, Polioles, S.A. de C.V, Mexico

Jasmita Chauhan
Department of Microbiology, Marwadi University, Rajkot, Gujarat, India

Verônica Contini
Programa de Pós-Graduação em Biotecnologia (PPGBiotec), Universidade do Vale do Taquari – Univates, Lajeado, RS, Brazil
Programa de Pós-Graduação em Ciências Médicas (PPGCM), Universidade do Vale do Taquari – Univates, Lajeado, RS, Brazil

Pritha Chakraborty
Microbial Biotechnology Lab, School of Bioscience and Technology, Vellore Institute of Technology, Vellore, Tamil Nadu, India

Ankita Chatterjee
Microbial Biotechnology Lab, School of Bioscience and Technology, Vellore Institute of Technology, Vellore, Tamil Nadu, India

Jorge Alberto Vieira Costa
Laboratory of Biochemical Engineering, College of Chemistry and Food Engineering, Federal University of Rio Grande (FURG), Rio Grande, RS, Brazil

Bruno Dahmer
Programa de Pós-Graduação em Biotecnologia (PPGBiotec), Universidade do Vale do Taquari – Univates, Lajeado, RS, Brazil

Michele Greque de Morais
Laboratory of Microbiology and Biochemistry, College of Chemistry and Food Engineering, Federal University of Rio Grande (FURG), Rio Grande, RS, Brazil

Claucia F. Volken de Souza
Programa de Pós-Graduação em Biotecnologia (PPGBiotec), Universidade do Vale do Taquari – Univates, Lajeado, RS, Brazil
Programa de Pós-Graduação em Sistemas Ambientais Sustentáveis, Universidade do Vale do Taquari – Univates, Lajeado, RS, Brazil

Rodrigo G. Ducati
Programa de Pós-Graduação em Biotecnologia (PPGBiotec), Universidade do Vale do Taquari – Univates, Lajeado, RS, Brazil

Márcia I. Goettert
Programa de Pós-Graduação em Biotecnologia (PPGBiotec), Universidade do Vale do Taquari – Univates, Lajeado, RS, Brazil
Programa de Pós-Graduação em Ciências Médicas (PPGCM), Universidade do Vale do Taquari – Univates, Lajeado, RS, Brazil

Saswata Goswami
Center of Innovative and Applied Bioprocessing, Mohali, Punjab, India

Chung-Yu Guan
Department of Environmental Engineering, National Ilan University, Yilan City, Taiwan

Prabuddha Gupta
Department of Microbiology, Marwadi University, Rajkot, Gujarat, India

Prakash M. Halami
Microbiology & Fermentation Technology Department, CSIR-Central Food Technological Research Institute, Mysuru, Karnataka, India

Ayerim Hernández-Almanza
School of Biological Science, Autonomous University of Coahuila, Torreón, Coahuila, México

M. Indira
Department of Biotechnology, Vignan's Foundation for Science, Technology & Research, Vadlamudi, Andhra Pradesh, India

Gabrielle Guimarães Izaguirres
Laboratory of Biochemical Engineering, College of Chemistry and Food Engineering, Federal University of Rio Grande (FURG), Rio Grande, RS, Brazil

Jyoti Singh Jadaun
Department of Botany, Dayanand Girls Postgraduate College, Kanpur, Uttar Pradesh, India

Daniel Kuhn
Programa de Pós-Graduação em Biotecnologia (PPGBiotec), Universidade do Vale do Taquari – Univates, Lajeado, RS, Brazil

Prashant Kumar
Department of Civil Engineering, North Eastern Regional Institute of Science and Technology, Nirjuli, Arunachal Pradesh, India

Pradeep Kumar
Department of Forestry, North Eastern Regional Institute of Science and Technology, Nirjuli, Arunachal Pradesh, India

S. Krupanidhi
Department of Biotechnology, Vignan's Foundation for Science, Technology & Research, Vadlamudi, Andhra Pradesh, India

Araceli Loredo-Treviño
Bioprocesses and Bioproducts Research Group, Food Research Department, School of Chemistry, Autonomous University of Coahuila, Saltillo, Coahuila, México

Gloria A. Martínez-Medina
Bioprocesses and Bioproducts Research Group, Food Research Department, School of Chemistry, Autonomous University of Coahuila, Saltillo, Coahuila, México

Pragati Misra
Department of Molecular and Cellular Engineering, Jacob Institute of Biotechnology and Bioengineering, Sam Higginbottom University of Agriculture, Technology and Sciences, Prayagraj, Uttar Pradesh, India

Geetanjali Mishra
Ladybird Research Laboratory, Department of Zoology, University of Lucknow, Lucknow, Uttar Pradesh, India

Juliana Botelho Moreira
Laboratory of Microbiology and Biochemistry, College of Chemistry and Food Engineering, Federal University of Rio Grande (FURG), Rio Grande, RS, Brazil

Tejas Oza
Department of Microbiology, Marwadi University, Rajkot, Gujarat, India

Arun Kumar Pal
Department of Molecular and Cellular Engineering, Jacob Institute of Biotechnology and Bioengineering, Sam Higginbottom University of Agriculture Technology and Sciences, Prayagraj, Uttar Pradesh, India

Ratih Pangestuti
Research and Development Division of Marine Bio Industry, Indonesian Institute of Sciences (LIPI), West Nusa Tenggara, Republic of Indonesia

Lorena Pedraza-Segura
Department of Chemical, Industrial and Food Engineering, Universidad Iberoamericana, Mexico City, Mexico

K. Abraham Peele
Department of Biotechnology, Vignan's Foundation for Science, Technology & Research, Vadlamudi, Andhra Pradesh, India

Yanuariska Putra
Research and Development Division of Marine Bio Industry, Indonesian Institute of Sciences (LIPI), West Nusa Tenggara, Republic of Indonesia

Puji Rahmadi
Research Center for Oceanography, Indonesian Institute of Sciences (LIPI), Jakarta, Republic of Indonesia

Amit K. Rai
Institute of Bioresources and Sustainable Development (DBT-IBSD), Sikkim Centre, Gangtok, Sikkim, India

Mahendrapalsingh Rajput
Department of Microbiology, Marwadi University, Rajkot, Gujarat, India

Nathiely Ramírez-Guzman
Center for Interdisciplinary Studies and Research, Universidad Autónoma de Coahuila, Saltillo, México

Keyur Raval
Department of Chemical Engineering, National Institute of Technology Karnataka, Mangalore, Karnataka, India

Luis V. Rodríguez-Durán
Multidisciplinary Academic Unit Mante, Universidad Autónoma de Tamaulipas, Tamaulipas, México

Gaurav Sanghvi
Department of Microbiology, Marwadi University, Rajkot, Gujarat, India

Suman Sanju
Department of Botany, M.C.M.D.A.V. College, Kangra, Himanchal Pradesh, India

Gerardo Saucedo-Castañeda
Biotechnology Department, Universidad Autónoma Metropolitana-Iztapalapa, Mexico City, Mexico

Karishma Seem
Division of Biochemistry, Indian Agricultural Research Institute, New Delhi, India

Manisha Sharma
Center of Innovative and Applied Bioprocessing (DBT-CIAB), Mohali, Punjab, India

Tomoya Shintani
Department of Nutritional Representative - Supplement Adviser, The Japanese Clinical Nutrition Association, Tokyo, Japan

Pradeep Kumar Shukla
Department of Biological Sciences, Faculty of Science, Sam Higginbottom University of Agriculture, Technology and Sciences, Prayagraj, Uttar Pradesh, India

Evi Amelia Siahaan
Research and Development Division of Marine Bio Industry, Indonesian Institute of Sciences (LIPI), West Nusa Tenggara, Republic of Indonesia

Sudhir P. Singh
Center of Innovative and Applied Bioprocessing (DBT-CIAB), Mohali, Punjab, India

Liliane Martins Teixeira
Laboratory of Microbiology and Biochemistry, College of Chemistry and Food Engineering, Federal University of Rio Grande (FURG), Rio Grande, RS, Brazil

Himani Thakkar
Department of Biochemistry, All India Institute of Medical Sciences, New Delhi, New Delhi, India

Aditi Thakur
Department of Biotechnology, Harlal institute of Management and Technology, Greater Noida, Uttar Pradesh, India

Luís F. Saraiva Macedo Timmers
Programa de Pós-Graduação em Biotecnologia (PPGBiotec), Universidade do Vale do Taquari – Univates, Lajeado, RS, Brazil
Programa de Pós-Graduação em Ciências Médicas (PPGCM), Universidade do Vale do Taquari – Univates, Lajeado, RS, Brazil

Vijay Tripathi
Department of Molecular and Cellular Engineering, Jacob Institute of Biotechnology and Bioengineering, Sam Higginbottom University of Agriculture Technology and Sciences, Prayagraj, Uttar Pradesh, India

Ujwalkumar Trivedi
Department of Microbiology, Marwadi University, Rajkot, Gujarat, India

Santosh Kumar Upadhyay
Department of Botany, Panjab University, Chandigarh, India

T. C. Venkateswarulu
Department of Biotechnology, Vignan's Foundation for Science, Technology & Research, Vadlamudi, Andhra Pradesh, India

Deepak Kumar Verma
Agricultural and Food Engineering Department, Indian Institute of Technology, Kharagpur, West Bengal, India

Vinnyfred Vincent
Department of Biochemistry, All India Institute of Medical Sciences, New Delhi, New Delhi, India

Camile Wünsch
Programa de Pós-Graduação em Biotecnologia (PPGBiotec), Universidade do Vale do Taquari – Univates, Lajeado, RS, Brazil

Tripti Yadav
Ladybird Research Laboratory, Department of Zoology, University of Lucknow, Lucknow, Uttar Pradesh, India

Chang-Ping Yu
Graduate Institute of Environmental Engineering, National Taiwan University, Taipei, Taiwan

Preface

Microorganisms are the most abundant and diverse unicellular organisms on the earth, equipped with distinguishable characteristics and functionality. Bioprospecting offers exploitation of microbial resource for societal benefits. The useful applications of microbial diversity have been established in medical, food, textile, and agro-industrial sectors. Microorganisms are source of biomolecules of health significance, antibiotic, anti-insecticidal compounds, biosurfactants, biofilm, etc. Microbial resources for production of bio-hydrogen, biomethane, biodiesel, and microorganisms have been reported. Microorganisms also contribute to the aspects of food security and bioremediation. Many microbial enzymes have been employed for executing a wide range of chemical interconversions, producing diverse biomolecules of functional properties, and for other industrial bioprocessing aspects such as agri-biomass hydrolysis, and so on. Microbial strains have been identified with crucial implications for the bioremediation of polluted niches. The microbial gums of useful applications in medical, food, cosmetics, and agricultural sectors are produced by a range of microorganisms, e.g. bacteria, fungi, and algae. A wide range of microbial compounds with antioxidant, antiaging, and skin lightening effects have been identified. Microbial bioprocesses have been developed for production of organic acids, which are building blocks of many valuable products such as degradable polymers, polyesters, and bioplastics. Microbes have been demonstrated to be cell factories for production of bioactive compounds and recombinant proteins of biomedical relevance. Many microbial molecules of antimicrobial, antiviral, anti-inflammatory, immunomodulatory, anticancerous, etc., properties have been reported. A diverse types of functional microbial molecules of ultra-reduced calorie, prebiotic, and anti-oxidative effects have been identified. Some phenolic compounds and mycosporine-like amino acids of microbial origin have been reported to be sunscreen and antiaging effects. Several intriguing efforts have been made for developing bioprocesses for the economically viable production of polyhydroxyalkanoates and related copolymers, helpful in addressing the issues related with plastic uses.

This book represents a distinguished repository of knowledge around microbial bioprospecting. The book is useful to academicians and researchers in augmenting their understandings on the aspects mentioned above.

Acknowledgments

We are thankful to the Center of Innovative and Applied Bioprocessing and Panjab University, Chandigarh, for providing facility to complete this book. We are grateful to all the esteemed authors for their exceptional contributions and reviewers for their critical evaluation and suggestions for the quality improvement of the book.

We would like to thank Miss Rebecca Ralf (Commissioning Editor), Miss Kerry Powell (Managing Editor), and Nivetha Udayakumar (Content Refinement Specialist) from John Wiley & Sons, Ltd for their excellent management of this project and anonymous reviewers for their positive recommendations about the book.

We also appreciate the support of our friends and research students, whose discussion and comments were beneficial to shape this book. We thank our numerous colleagues for direct or indirect help in shaping this project.

SPS is grateful to his parents and family for consistent moral support. SPS acknowledges the support from CIAB and the Department of Biotechnology (DBT). SKU wishes to express gratitude to his parents, wife, and daughter for their endless support, patience, and inspiration.

We would like to warmly thank faculties and staffs of the department and university for providing a great working environment.

1

An Introduction to Microbial Biodiversity and Bioprospection

Tomoya Shintani[1], *Santosh Kumar Upadhyay*[2], *and Sudhir P. Singh*[3]

[1] *Department of Nutritional Representative – Supplement Adviser, The Japanese Clinical Nutrition Association, Tokyo, Japan*
[2] *Department of Botany, Panjab University, Chandigarh, India*
[3] *Center of Innovative and Applied Bioprocessing (DBT-CIAB), Mohali, Punjab, India*

1.1 Introduction

1.1.1 Microorganisms

The term "microorganisms" refer to organisms that are so minute that the human eye cannot discern their structure. Almost all the unicellular organisms are included in the category of microorganisms. However, many multicellular organisms are microscopic. The microbes may also exist as cell clusters. Microbes are not limited to only eubacteria and archaea but also includes members from fungi, protozoa, algae, and viruses. Ribosomal RNA is considered a crucial molecule to draw an authentic classification of the life forms. On the basis of comparative rRNA sequence analysis, a basal universal phylogenetic tree had been inferred by Carl Woese [1], representing the overview of organismal history. This phylogenetic tree hypothesized the common origin of all forms of life, emphasizing the importance of microorganisms in the biological diversity at the Earth.

The global compilation of microbiological data has estimated the existence of about 1 trillion microbial species at the Earth [2]. The Earth Microbiome Project had been established in the year 2020, with the main objective to document the uncultured microbial diversity and the functional potential thereof [3]. In the project, sampling of the Earth's microbial communities is being done on an unprecedented scale. Recently, a total of 27 751 samples had been collected from 43 countries to characterize the microbial communities of diverse physicochemical properties and exposed to a wide range of biotic and abiotic factors [4].

1.1.2 Bioprospecting

Bioprospecting is the excavation of useful genetic or biochemical materials from natural resources [5]. The word "bio-" refers to "organism" and "prospecting" refers to the search for precious material. Bioprospecting of the microbial population, residing in diverse

Bioprospecting of Microorganism-Based Industrial Molecules, First Edition.
Edited by Sudhir P. Singh and Santosh Kumar Upadhyay.

ecological niches, is gaining more and more attention with the increasing demand for bioactive molecules and nutraceuticals of pharmaceutical, food, and agro-industrial significance. Microorganisms exist in every corner of this planet, and many of them may not be cultivated. Bioprospecting of biologicals, including microbes, requires interdisciplinary knowledge to exploit the natural resource for societal benefits in a sustainable manner.

A major proportion of Earth contains extreme environmental, where most of the life forms cannot survive. However, extremophiles, which evolve a natural mechanism to thrive the harsh condition, are considered a rich biological resource for extremophiles [6]. The bioactive compounds and the extremozymes have great industrial importance [7]. There is an enormous genetic pool of microbes co-evolved with the higher organisms, including plants and animals. They have manifold promising biomolecules of industrial importance [5]. There are boundless opportunities to explore the treasure box to meet the growing demand for novel bioactive molecules, e.g., food ingredients, agrochemicals, functional biomolecules, antibiotics, enzymes, and so on. Omics is a compelling approach for the exploration of various niches of diverse environmental conditions and the discovery of unusual novel enzymes and other metabolites of human use [8].

1.1.3 Bioprospection of Microorganisms

The microbial resources have potential in the generation of a wide range of high-value compounds. This section has explained a few specific examples, the detail of which can be found in the subsequent chapters.

Biosurfactants are used in many fields, such as water treatment, food processing, health, disinfectants, cosmetics, and medicines [9]. Most biosurfactants are produced employing different strains of bacteria, yeast, and molds [10]. The uses of biosurfactants of microbial origin have been demonstrated in various purposes, for example in food processing, water treatment, cosmetics and pharmaceuticals, and bioremediation.

Microbial gum is a polysaccharide molecule mostly produced by bacteria and fungi [11]. These molecules protect the microbes from harsh environmental conditions [12]. These compounds are of widespread use in foods, pharmaceuticals, and cosmetics. The microbial gums are of prebiotic nature. Its uses have also been demonstrated in wound healing and carrier for drug delivery [13]. The demand for microbial gums is steadily increasing in various industries.

The practicality and potential of microbes in cosmetics have been well established. The use of antiaging products is in the spotlight. A variety of cosmetic compounds are produced from cultured microorganisms such as bacteria, yeasts, fungi, and algae [14]. Many compounds, including antioxidants, peptides, and proteins, hold a promising future in the cosmetic skincare sector. Natural compounds not only have the ability to attract the attention of the market but also have valuable properties of potential medical benefits. In this regard, it is necessary to mine a variety of natural resources for novel compounds. The demand for antiaging and whitening products is expected to steadily increase annually [15].

The microbial population is also useful in cleaning of the environment via bioremediation approach. The metabolic activities of many microorganisms have been found to convert unwanted pollutants into harmless compounds [16]. Many fungal strains have been identified that have nonspecific ability to break down various pollutants. Intensive researches have been conducted to highlight the enzymatic systems involved in bioremediation in order to develop and optimize the bioremediation processes [17]. Microbial interference in addressing the issues related to environmental pollution could offer economically viable and socially acceptable solution. An integration of microorganisms and electrochemical systems has led to developing green technologies that could generate power from wastewater or polluted soils [18, 19]. Microbial strains have been characterized for the development of bioplastic. For example, poly(lactic acid), a bioplastic molecule, can be produced by the microbial fermentation of starch biomass [20]. Development of microbial biotechnology for the production of bioplastic could prevent pollution at a greater extent. Furthermore, microbial communities play critical roles in transforming the abundant agri-biomass into energy products such as biohydrogen, biogas, bio-oil, bioethanol, biodiesel, biofuels, and so on [21, 22]. There is a rapid increase in the interest in using biomethane as a fuel for a variety of applications such as boilers, engines, gas turbines, and fuel cells.

Carbohydrates are one of the most critical nutrients in food. However, in recent years, excessive consumption of glucose and fructose has been linked to obesity and diabetes across the world. Many microbial metabolites of sweet nature and reduced calorie are being explored for their use as a substitute of sugar as food ingredients, e.g. rare sugars – D-allulose, tagatose, turanose, kojibiose, erythritol, and so on [23–25]. These are rare sugars, the monosaccharides that are found in nature only in minute quantities. The microbial resource can be explored for the biotechnological production of more than 50 different rare sugars [26]. Furthermore, microorganisms are a rich bioresource for production of various kinds of oligosaccharide molecules of reduced calorie and prebiotic function [27].

Agriculture today relies on the extensive use of pesticides and fungicides. There is currently a consumer preference to reduce the use of synthetic organic chemicals used in the agricultural treatments. Microbe-derived organic compounds have been experienced very effective and environment friendly. Such compounds have been reported to protect plants from pathogens and provide a better environment for crop growth [28].

Apart from the above, many microbial compounds of medicinal importance have gained attention in the industrial market. Penicillin is a traditional example of microbial importance for the development of therapeutic drug molecules. One of the brilliant examples of bioprospection is the study that received the Nobel Prize in the year 2015. The research was a series of studies by Dr. Omura team, who developed Avermectin and Ivermectin, which are now utilized as drugs and pesticides against parasitic worms, insect pests, and other pathogens [29, 30]. Many microbial proteins, peptides, and metabolites of anticancerous ability have been reported in many studies [31, 32].

1.2 Conclusions and Perspectives

Microbial diversity is a very crucial resource for the execution of exploitable biology. Bioprospection of microbial resource facilitates the identification and isolation of high-value molecules of desirable activity. Researches are being conducted to exploit the microbial diversity, obtaining the biomolecules of pharmaceutical, bioceutical, agricultural, bioremediation, etc., significance. Microorganisms should be exploited as cell factories for the production of biomolecules of health significance, antibiotic, anti-insecticidal compounds, biosurfactants, biofilm, etc. The use of microorganisms should be validated for their meaningful contribution toward food security and bioremediation. Microbial prospecting is required to be intensified with systematic and sustainable approaches.

Acknowledgment

The Department of Biotechnology (DBT), Govt. of India is acknowledged for all support. SKU is grateful to Panjab University, Chandigarh for facility and other support.

References

1 Woese, C.R. (2000). Interpreting the universal phylogenetic tree. *PNAS* 97 (15): 8392–8396.
2 Locey, K.J. and Lennon, J.T. (2016). Scaling laws predict global microbial diversity. *PNAS* 113 (21): 5970–5975.
3 Gilbert, J.A., Jansson, J.K., and Knight, R. (2014). The Earth Microbiome project: successes and aspirations. *BMC Biology* 12: 69.
4 Gilbert, J.A., Jansson, J.K., and Knight, R. (2018). Earth microbiome project and global systems biology. *mSystems* 3: e00217–e00217.
5 Strobel, G. and Daisy, B. (2003). Bioprospecting for microbial endophytes and their natural products. *Microbiology and Molecular Biology Reviews* 67: 491–502.
6 Kaushal, G., Kumar, J., Sangwan, R.S., and Singh, S.P. (2018). Metagenomic analysis of geothermal water reservoir sites exploring carbohydrate-related thermozymes. *International Journal of Biological Macromolecules* 119: 882–895.
7 Sharma, N., Kumar, J., Abedin, M.M. et al. (2020). Metagenomics revealing molecular profiling of community structure and metabolic pathways in natural hot springs of the Sikkim Himalaya. *BMC Microbiology* 20: 246.
8 Zhang, X., Li, L., Butcher, J. et al. (2019). Advancing functional and translational microbiome research using meta-omics approaches. *Microbiome* 7 (1): 154.
9 Santos, D.K., Rufino, R.D., Luna, J.M. et al. (2016). Biosurfactants: multifunctional biomolecules of the 21st century. *International Journal of Molecular Sciences* (3): 401.
10 Mulligan, C.N. (2005). Environmental applications for biosurfactants. *Environmental Pollution* 133 (2): 183–198.
11 Alizadeh-Sani, M., Ehsani, A., Kia, E.M., and Khezerlou, A. (2019). Microbial gums: introducing a novel functional component of edible coatings and packaging. *Applied Microbiology and Biotechnology* 103 (17): 6853–6866.
12 Freitas, F., Torres, C.A., and Reis, M.A. (2017). Engineering aspects of microbial exopolysaccharide production. *Bioresource Technology* 245: 1674–1683.
13 Suresh Kumar, A., Mody, K., and Jha, B. (2007). Bacterial exopolysaccharides–a perception. *Journal of Basic Microbiology* 47 (2): 103–117.
14 Gupta, P.L., Rajput, M., Oza, T. et al. (2019). Eminence of microbial products in cosmetic industry. *Natural Products and Bioprospecting*: 1–12.

15 Łopaciuk, A. and Łoboda, M. (2013). Global beauty industry trends in the 21st century. *Management, Knowledge and Learning International Conference*, 19–21 June 2013, Zadar, Croatia, pp. 19–21.

16 Watanabe, K. (2001). Microorganisms relevant to bioremediation. *Current Opinion in Biotechnology* 12 (3): 237–241.

17 Beattie, A.J., Hay, M., Magnusson, B. et al. (2011). Ecology and bioprospecting. *Austral Ecology* 36 (3): 341–356.

18 Allen, R.M. and Bennetto, H.P. (1993). Microbial fuel-cells: electricity production from carbohydrates. *Applied Biochemistry and Biotechnology* 39: 27–40.

19 Logan, B.E., Hamelers, B., Rozendal, R. et al. (2006). Microbial fuel cells: methodology and technology. *Environmental Science & Technology* 40: 5181–5192.

20 Karamanlioglu, M., Preziosi, R., and Robson, G.D. (2017). Abiotic and biotic environmental degradation of the bioplastic polymer poly (lactic acid): a review. *Polymer Degradation and Stability* 137: 122–130.

21 Khan, M.I., Shin, J.H., and Kim, J.D. (2018). The promising future of microalgae: current status, challenges, and optimization of a sustainable and renewable industry for biofuels, feed, and other products. *Microbial Cell Factories* 17 (1): 36.

22 Nabgan, W., Abdullah, T.A.T., Mat, R. et al. (2017). Renewable hydrogen production from bio-oil derivative via catalytic steam reforming: An overview. *Renewable and Sustainable Energy Reviews* 79: 347–357.

23 Patel, S., Kaushal, G., and Singh, S.P. (2020). A novel D-Allulose 3-epimerase gene from the metagenome of a thermal aquatic habitat and D-Allulose production by *Bacillus subtilis* whole-cell catalysis. *Applied and Environmental Microbiology* 86 (05): e02605–e02619.

24 Agarwal, N., Narnoliya, L.K., and Singh, S.P. (2019). Characterization of a novel amylosucrase gene from the metagenome of a thermal aquatic habitat, and its use in turanose production from sucrose biomass. *Enzyme and Microbial Technology* 131: 109372.

25 Lata, K., Sharma, M., Patel, S.N. et al. (2018). An integrated bio-process for production of functional biomolecules utilizing raw and by-products from dairy and sugarcane industries. *Bioprocess and Biosystems Engineering* 41 (8): 1121–1131.

26 Granström, T.B., Takata, G., Tokuda, M., and Izumori, K. (2004). Izumoring: a novel and complete strategy for bioproduction of rare sugars. *Journal of Bioscience and Bioengineering* 97 (2): 89–94.

27 Singh, S.P., Jadaun, J.S., Narnoliya, L.K., and Pandey, A. (2017). Prebiotic oligosaccharides: special focus on fructooligosaccharides, its biosynthesis and bioactivity. *Applied Biochemistry and Biotechnology* 183 (2): 613–635.

28 Kanchiswamy, C., Malnoy, M., and Maffei, M. (2015). Bioprospecting bacterial and fungal volatiles for sustainable agriculture. *Trends in Plant Science* 20 (4): 206–211.

29 Omura, S. (2008). Ivermectin: 25 years and still going strong. *International Journal of Antimicrobial Agents* 31 (2): 91–98.

30 Omura, S. (2016). A splendid gift from the Earth: the origins and impact of the avermectins (Nobel lecture). *Angewandte Chemie International Edition* 55 (35): 10190–10209.

31 Karpinski, T.M. and Adamczak, A. (2018). Anticancer activity of bacterial proteins and peptides. *Pharmaceutics* 10 (2): 54.

32 Sedighi, M., Zahedi Bialvaei, A., Hamblin, M.R. et al. (2019). Therapeutic bacteria to combat cancer; current advances, challenges, and opportunities. *Cancer Medicine* 8 (6): 3167–3181.

2

Application of Microorganisms in Biosurfactant Production

Lorena Pedraza-Segura[1], Luis V. Rodríguez-Durán[2], Gerardo Saucedo-Castañeda[3], and José de Jesús Cázares-Marinero[4]

[1] *Department of Chemical, Industrial and Food Engineering, Universidad Iberoamericana, Mexico City, Mexico*
[2] *Multidisciplinary Academic Unit Mante, Universidad Autónoma de Tamaulipas, Tamaulipas, México*
[3] *Biotechnology Department, Universidad Autónoma Metropolitana-Iztapalapa, Mexico City, Mexico*
[4] *Department of Research & Development, Polioles, S.A. de C.V, Mexico*

2.1 Biosurfactants Nature and Classification

Biosurfactants (BS) are amphiphilic molecules that, as any surfactant, consist in two contrasting moieties: (i) one hydrophilic part showing affinity to polar substances like water, and (ii) one lipophilic displaying attraction to non-polar media such as oil and fats (Figure 2.1) [1]. The prefix "bio" refers to the fact that these surfactants are exclusively produced through biochemical transformations within microorganisms under certain conditions; this is why BS are also called microbial surfactants [2]. The adjustable binary hydrophilic–lipophilic feature confers surfactants appealing properties for applications in all types of industrial sectors.

Surfactants can be categorized according to their (i) origin, (ii) electrostatic status, and (iii) the ratio of their hydrophilic and lipophilic components (Table 2.1). In the first category, surfactants can be seen as synthetic, semi-synthetic, and BS. Synthetic surfactants are those that have been manufactured on a large scale for many years from raw materials of petrochemical origin. Semi-synthetic surfactants were introduced to the market to offer less hazardous alternatives to conventional surfactants. These are constituted by lipophilic fragments coming from biocompatible motifs like fatty esters, acids, or alcohols extracted from natural oils (*oleum*), ergo these surfactants are also called oleochemical surfactants. Nevertheless, further steps of conventional synthesis are always required to afford finished products (semi-synthetic = half-synthetic). Currently, it is estimated that >27% of consumed surfactants are oleochemicals. Europe and North America have dominated this particular market, and the increased consumption in recent years can be associated with the interest in products for personal care, home care, and agriculture.

In an attempt to solve relevant drawbacks related to low biodegradability profiles of synthetic surfactants, pollution of traditional chemical processes, and sacrifice of natural

Bioprospecting of Microorganism-Based Industrial Molecules, First Edition.
Edited by Sudhir P. Singh and Santosh Kumar Upadhyay.

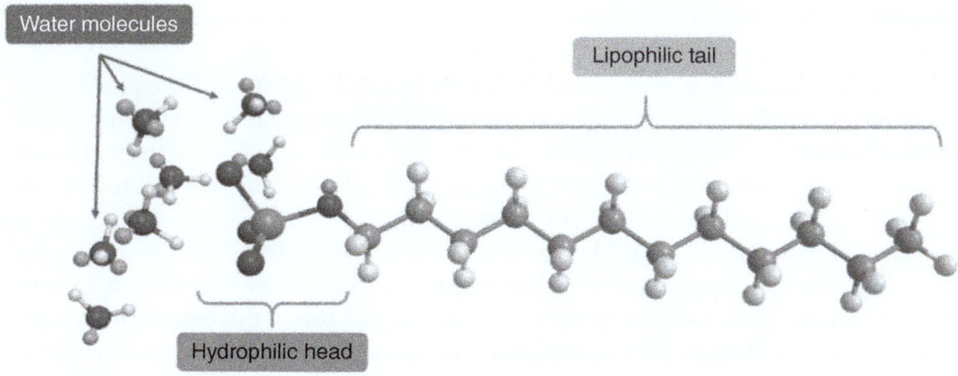

Figure 2.1 3D chemical structure of lauryl sulfate as an example of a surfactant molecule. The hydrophilic head of sulfate (SO_4^{2-}) is surrounded by water molecules, while the lipophilic tail avoids any contact with water. *Source:* Based on [1].

Table 2.1 Classification of surfactants by origin, ionic status, and hydrophilic–lipophilic balance.

Category			Example
I. Origin	Synthetic		Nonylphenol ethoxylates
	Oleochemical		Lauryl alcohol ethoxylates
	Biosurfactants		Sophorolipids
II. Electrostatic status	Oonic	Cationic	Ammonium salts
		Anionic	Lauryl sulfates
		Zwitterionic	Betaines
	Nonionic		Oxirane and 2-methyoxirane copolymers
III. Hydrophilic–lipophilic balance	01–03		Antifoaming agents
	03–08		w/o emulsifiers
	07–10		Wetting agents
	08–16		o/w emulsifiers
	13–16		Detergents
	16–19		Solubilizing agent

resources, new sustainable approaches have recently emerged, implying biotechnological means for surfactant obtention. In this context, BS are known to be secondary metabolites of living organisms just as bacteria and yeast and to be involved in some microbial physiological functions, e.g. cell development, biofilm formation, and osmotic pressure regulation. Furthermore, these play a key role in microbial survival in front of adverse situations such as lack of nutrients and the need to incorporate non-soluble substances like hydrocarbons as the only carbon source [2]. These natural phenomena have been exploited by bioengineering in productive microbial fermentation, and since BS are achieved through complex enzymatic reactions, such biomolecules are endowed of robust chemical architectures, which are tunable from elegant processes of molecular biology, genetic engineering, or careful selection of culture conditions.

In the second category, surfactants are normally grouped according to the presence or absence of formal electrostatic charges in the hydrophilic moiety of the molecule. Thus, cationic surfactants contain formal positive charges (+) on the polar head of the surfactant molecule and are systematically escorted by negative counterions that neutralize the charges. Anionic surfactants have negative formal charges (−), an example of this latter is represented in Figure 2.1. The third class of ionic surfactants is those species comprising both positive and negative charges in the same body (±). These species are known as inner salts or zwitterions (from German *zwitter* = hybrid). The last case of this classification is the absence of formal charges in the surfactant molecule, so these are called nonionic surfactants.

Finally, surfactants can be ranked according to their amphiphilic nature, otherwise by their hydrophilic–lipophilic balance (HLB). Although there are a significant number of theoretical and experimental approaches, the most used HLB system is the Griffin's [3]. In a simplistic way, HLB values are estimated from the division of the mass of the hydrophilic fragment by the mass of the entire molecule, and the resulting quotient is then multiplied by a conventional value of 20. HLB can predict whether a surfactant will behave as an emulsifier, solubilizer, dispersant, or other; therefore, this system has been a useful guide to formulate products containing conventional surfactants for specific applications, and certainly, will be just as convenient for the case of BS. The hydrophilic part of BS is normally constituted by carbohydrates, amino acids, proteins, phosphates, carboxylic acids, or alcohol motifs; and these can be ionic or nonionic. The lipophilic part commonly is long chains of carbon atoms, just as in fatty acids. Both molecular components, hydrophilic and lipophilic, are assembled via linking biochemical functionalities, e.g. ethers (C—O—C), amides (N—C=O), and esters (O—C=O). According to the nature of each moiety hydrophilic and lipophilic, BS are commonly classified in the following groups: (i) glycolipids, (ii) lipopolysaccharides, (iii) lipopeptides, (iv) phospholipids, and (v) fatty acids; each one with specific physicochemical characteristics and physiological roles [4, 5]; for example, Emulsan and other complex lipopolysaccharides are polymeric BS with known emulsification capabilities.

From all the types of BS, glycolipids have the greatest opportunity to be manufactured on a large scale due to the high yield of obtention compared to other BS such as lipoproteins. It is clear that BS produced in higher yields will represent a lower cost for production [6]. This is why glycolipids have captured our attention for this chapter. Glycolipids result from the condensation of aliphatic fatty acids (lipids) and carbohydrates. Their names are taken from the nature of the carbohydrate moiety. Consequently, glycolipids containing sophorose in the hydrophilic segment are called sophorolipids (SL); those containing rhamnose are named rhamnolipids; those with trehalose, trehalose-lipids, and so on. Of all the glycolipid types, SL and rhamnolipids have been among the most studied [7, 8].

SL (Figure 2.2) contain the disaccharide sophorose linked to a fatty acid via an ether function. The fatty acid must be previously hydroxylated somewhere in the carbon chain, usually at the other end of the carboxylic acid function. SL can be found as different chemical structures. The most evident and well known is the divergence between open and cyclic arrangements [9]. Open SL are those that have the chemical functionality of

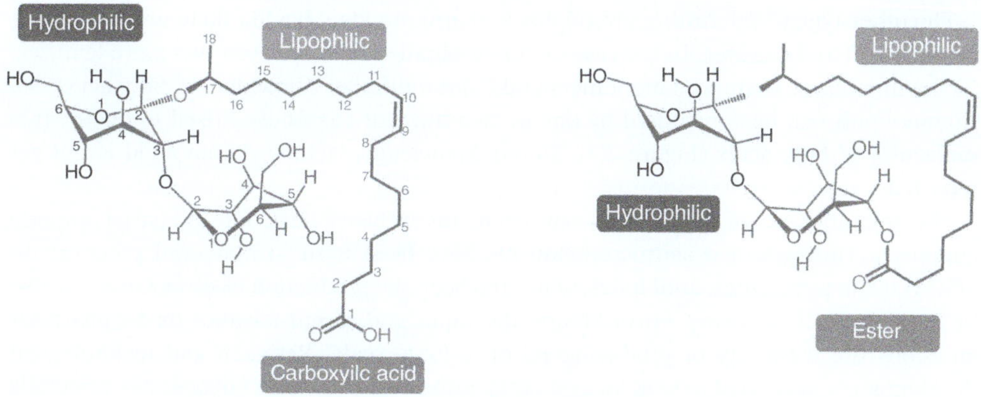

Figure 2.2 Chemical structures for two species of sophorolipids as a result of the condensation of sophorose with oleic acid. On the left side, it is shown the open form of carboxylic acid and on the right side, the cyclic form of the lactone.

carboxylic acid (COOH) at the end of the lipophilic chain, while cyclic arrangements are those having an ester functionality as a result of the condensation between the fatty acid and one of the hydroxyl motifs of the sophorose.

Cyclic esters are called lactones. Hence, there are two types of SL: the acidic forms and the lactone forms. Other less notable molecular variables are (i) the presence or absence of acetyl groups attached to the hydroxyl moieties on the carbohydrate periphery, (ii) the length of the alkyl chain, (iii) the degree of unsaturation (unsaturation = double or triple bonds), (iv) the position of the hydroxyl group in the alkyl chain, (v) the position of the hydroxyl group of the sophorose that serves to build the ether bond with the fatty alcohol, and (vi) the position of the hydroxyl group of sophorose serves to construct the ester bond with the fatty acid in the lactone forms, inter alia.

Figure 2.2 shows the chemical structures of two SL: one in the acid form (left) and one in the lactone form (right). Both species do not contain any acetyl groups but have a fatty acid moiety of 18 carbon atoms with only one unsaturation in position C-9 with *Z* geometry. This fatty acid (oleic acid) must have been previously hydroxylated inside the cell by some biochemical β-oxidation at position C-17. This latter allows its association with the free hydroxyl moiety of the anomeric carbon of sophorose, which produces the ether bridge of sophorose-lipid equally found in both acid and lactone arrangements. Thus, the chemical name for the acid form should be (*S*,*Z*)-17-{[(2*S*,3*R*,4*S*,5*S*,6*R*)-4,5-dihydroxy-6-hydroxymethyl-3-{[(2*S*,3*R*,4*S*,5*S*,6*R*)-3,4,5-trihydroxy-6-hydroxymethyltetrahydro-2*H*-pyran-2-yl]oxy}tetrahydro-2*H*-pyran-2-yl]oxy}octadec-9-enoic acid.

In the case of lactone, the carboxylic acid was condensed with the hydroxyl group in C-5 of the other monosaccharide fragment of sophorose. After imagining the number of possible combinations of structural variations, one would not expect microorganisms to produce unique and pure compounds, but rather a wide variety of many different species. Therefore, the selection of microorganism strain, culture conditions, culture media, and substrates are the first fundamental factors playing a decisive role in the complexity (or simplicity) of obtained mixtures.

The other type of interesting glycolipids is rhamnolipids. Like SL, there are abounding varied possible structures. In the case of rhamnolipids, there are even two more features: (i) the number of rhamnose monomers and (ii) the number of condensed fatty acids. So, rhamnolipids can be constituted by one or two units of rhamnose linked to one or two molecules of fatty acids (Figure 2.3). To our knowledge, lactone rhamnolipid is not yet reported.

The manufacture of surfactants has been an exclusive task for industrial organic chemistry. However, just as microorganisms have been used in industrial processes to afford enzymes, vaccines, antibiotics, wine, and beer, the production of surfactants can also be carried out in this way. Furthermore, the rapid and recent advance on bioprocesses envisions the feasibility of producing BS on a large scale. Research and technological developments have tried to look for cost competitiveness proposing cheaper raw materials and more affordable downstream processes. For example, it has been demonstrated that some agro-industrial wastes such as molasses from sugar industry [10] or whey from dairy industry [11] can be useful. Other aspect on BS is the possibility to improve physicochemical properties or the productivity of microorganisms via biochemical or genetic engineering techniques.

BS have very attractive properties such as low toxicity and high biodegradability (degraded by other microorganisms); which is very well aligned to current needs imposed by international regulations. It now is considered that biocompatibility of a product is a parameter as important as its cost and performance, so the chemical industry is compelled

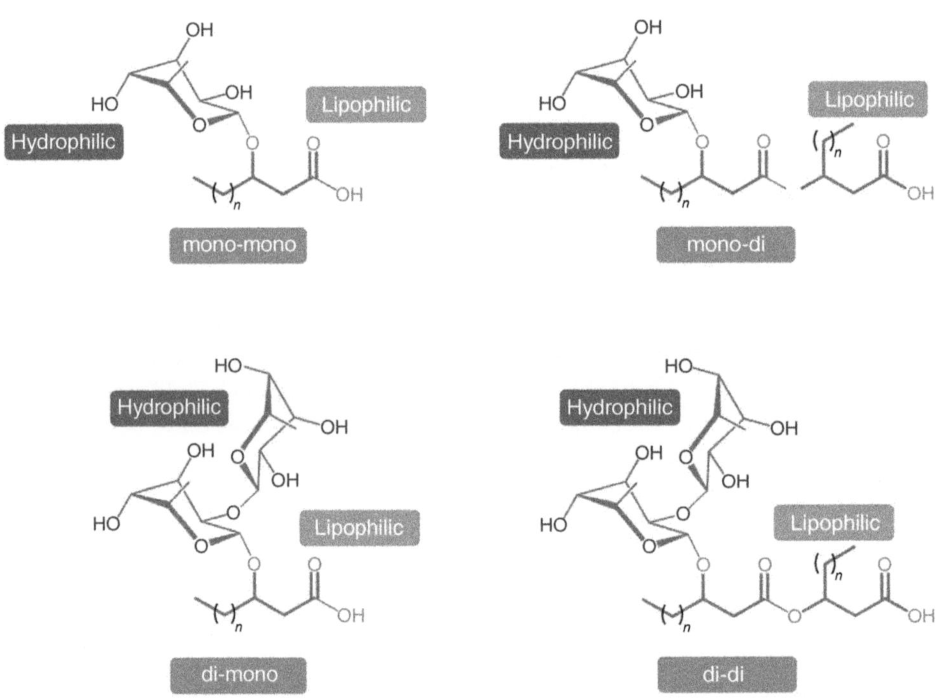

Figure 2.3 General chemical structures for four types of rhamnolipids: monorhamnose-monolipid, monorhamnose-dilipid, dirhamnose-monolipid, and dirhamnose-dilipid.

to implement new strategies for manufacturing more sustainable materials without scarifying efficiency. The numerous advantages of versatile BS compared to synthetic surfactants explain the increasing attention on the topic from the early twenty-first century (Figure 2.4). Although publishable activity on BS dates back 57 years, the boom has emerged for the last 20 years. Only in the last eight years, more than half of the known references on BS have been reported. BS are recognized for resisting a wide range of pH, salt concentration, and temperature [12] and have the ability to reduce surface tension in exactly the same way as chemical and oleochemical surfactants, so these can find the same application niches, i.e. as the key components of countless formulations for almost any sector of the contemporary industry [13]. Hence, BS are excellent candidates for different industrial applications like oil recovery, detergents, cleaning products, degreasers, fertilizers, agrochemicals, textile products, paints, mining, inter alia. Moreover, BS can attend applications where strong eco-friendly features are demanded, e.g. bioremediation of soil and water due to hydrocarbon spills [14, 15], water treatment, food processing [16], health, sanitizers, cosmetics, and pharmaceuticals [17].

As it has been observed, classification for a surfactant molecule comprises all the chemical species that share a binary amphiphilic feature. The production of microbial surfactants involves a strong character of sustainability and circular economy. Its production seems to be the right choice that will revolutionize the way the chemical industry, applications, and markets work. Glycolipids such as SL and rhamnolipids appear to be the species with the greatest potential to be developed at larger scales in the coming years. In addition, there are some synergies with other chemical compounds that can enhance surface activities and performances, which makes ipso facto possible the introduction of BS in the market via innovative formulations.

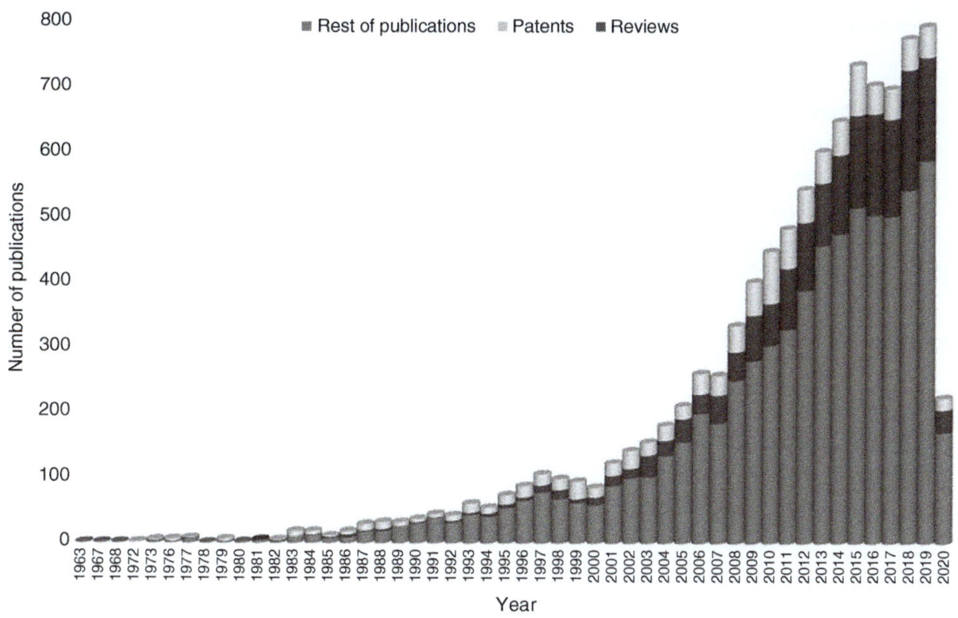

Figure 2.4 Number of publications on biosurfactants from 1963 to April 2020.

Among all primary and secondary metabolites, BS play interesting roles for microbial life. Some authors suggest that their emulsifying properties help microorganisms to adapt to environments, enhancing nutrient availability in soil or water and allowing cell adherence for water-insoluble substrate transport [18]. Since BS can inhibit microbial growth, those microorganisms producing BS can become predominant in their environment [19].

As mentioned above, BS are biodegradable and suitable for different industrial applications; for this reason, there is abundant research about their production and natural sources. Biosynthesis of BS are distributed among archaea, bacteria, yeasts, and molds, but depending on the group, genus, and species of microorganisms, BS structures become significantly different.

2.2 Biosynthesis of BS by Archaea and Bacteria

There are few papers related to Archaea as BS producers, compared to bacteria. Those who have been studied are alkaliphiles from the marine environment [20]. In contrast, bacteria produce mainly glycolipids, such as rhamnolipids, trehalolipids, and glucolipids; biosynthesis is restricted to actinobacteria, proteobacteria, cyanobacteria, and firmicutes (87%), other phyla represent between 1.0 and 2.5% of glycolipid-producing bacteria [21]. On the other hand, lactic acid bacteria (LAB), produce BS showing important roles in LAB colonization due to antimicrobial activities against fungi and bacteria. LAB BS are constituted by polysaccharides, glycolipids, glycolipopeptides, and proteins [22, 23]. Figure 2.5 depicts the distribution of prokaryote-producing BS. A broad review of scientific literature proved that prokaryotes are the predominant group of microorganisms related to BS production (Table 2.2), particularly bacteria.

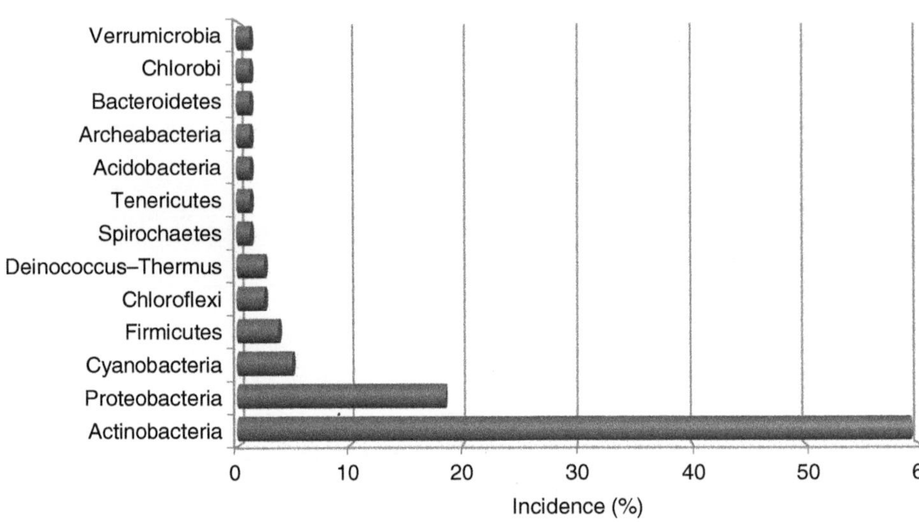

Figure 2.5 Phyla of prokaryote producers of biosurfactants. *Source:* Sharma et al. [22] and Satpute et al. [23].

Table 2.2 Archaea and bacteria as biosurfactant producers.

Biosurfactant	Microorganism	Carbon source
Rhamnolipids	*Pseudomonas aeruginosa*	Oil residues, corn oil, waste frying oil [24–26]
	Pseudomonas fluorescens	Hexadecane, olive oil [26]
	Pseudomonas luteola	Molasses [26]
	Pseudomonas chlororaphis	Glucose [26]
	Pseudomonas putida	Molasses, glucose [26]
	Pseudomonas stutzeri	Palm oil mill effluent [27]
	Pseudomonas pachastrellae	Barley pulp [28]
	Pseudomonas desmolyticum	Hexadecane [29]
	Burkholderia glumae	Canola oil [26]
	Burkholderia plantarii	Glucose [26]
	Burkholderia pseudomallei	Glycerol [26]
	Burkholderia thailandensis	Canola oil, glycerol [26]
	Bacillus pumilus	Crude oil and naphthalene [30]
	Nocardioides sp	n-paraffin [31]
	Serratia rubidaea SNAU02	Mahua oil cake [32]
	Acinetobacter calcoaceticus	Glycerol [26]
	Acinetobacter junii	Soybean oil [33]
	Enterobacter asburiae	Sodium citrate [34]
	Enterobacter hormaechei	Glycerol [26]
	Achromobacter sp.	Lignocellulosic residues [35]
	Pantoea stewartii	Glycerol [26]
	Pseudoxanthomonas sp.	Mannitol [26]
	Renibacterium salmonarium	n-hexadecane [36]
	Streptomyces sp.	Glucose [37]
	Streptomyces coelicoflavus	Olive oil [38]
	Tetragenococcus koreensis	GYP [39]
	Thermus aquaticus	Sunflower oil [26]
Trehalolipids	*Rhodococcus erythropolis*	n-hexadecane [40]
	Rhodococcus sp. PML026	Sunflower oil [41]
	Rhodococcus fascians	n-alkanes [42]
	Rhodococcus opacus	n-alkanes [42]
	Rhodococcus wratislaviensis	n-hexadecane [42]
	Rhodococcus ruber	n-hexadecane [43]
	Corynebacterium matruchotii	[42]
	Arthrobacter parafineus	n-paraffin [42]
	Arthrobacter sp.	Crude oil degrading [42]

Table 2.2 (Continued)

Biosurfactant	Microorganism	Carbon source
	Micrococcus luteus	n-hexadecane [44]
	Tsukamurella spumae	Sunflower oil [45]
	Tsukamurella pseudospumae	Sunflower oil [45]
	Rhodococcus qingshengii	n-hexadecane [46]
Other glycolipids	*Deinococcus caeni PO5*	Jackfruit seed poder [47]
	Brachybacterium paraconglomeratum MSA2	Oil seed cake, wheat bran, treated molasses [47]
	Halomonas sp.	Glucose [48]
	Pseudomonas otitidis	Sodium acetate [49]
Phospholipids	*Corynebacterium lepus*	[50]
Lipopeptides	*Bacillus licheniformes*	Glucose [51]
	Bacillus subtilis	Motor oil [52]
	Bacillus thuringiensis	Glycerol [53]
	Klebsiella sp.	Crude oil [39]
	Aeromonas salmonicida	Gasoline [54]
Polymixin	*Paenibacillus polymixa*	Glucose [55]
Surfactin	*Bacillus subtilis*	Glucose, glycerol [55, 56]
Fengycin	*Bacillus subtilis*	Glucose [57]
Pontifactin	*Pontibacter korlensis*	Glucose [58]
Serrawetin	*S. surfactantfaciens*	[59]
N.A.	*Lactococcus lactis*	Cheese whey [60]
	Streptococcus termophilus	Sucrose and lactose [60]
	Lactobacillus acidophilus	[61]
Protein, polysaccharides	*Lactobacillus pentosus*	Cheese whey [62]

2.3 Biosynthesis of BS by Yeasts and Molds

Yeasts and molds produce BS, mainly glycolipids, as SL and mannosylerythritol-lipids. Table 2.3 shows the most important genera of BS producers, particularly, *Candida, Pseudozyma,* and *Starmerella* (ex *Candida*). Compared to bacteria, yeasts and molds have been less explored as a source of BS, with the exception of SL for oil and cosmetic applications. Candida genus are capable to synthetize both types (SL and MEL) of glycolipids [18]. *Starmerella bombicola* is the most important yeast in SL production and mold *Ustilago maydis* in other glycolipids. As bacteria, yeasts and molds have a highly regulated metabolic pathway for the BS synthesis, since these compounds are secondary metabolites. The carbon source, hydrophobic substrate, and nitrogen limitation are prevailing factors in SL and mannosylerythritol lipids biosynthesis, while for cellobiose lipids nitrogen source is also important [18].

Table 2.3 Yeasts and molds as biosurfactants producers.

Biosurfactant	Microorganism	Carbon source
Mannosylerythritol lipids	*Pseudozyma antarctica*	Glycerol [63]
	Pseudozyma rugulosa	Soybean oil [64]
	Pseudozyma aphidis	Soybean oil, glucose [64]
	Pseudozyma tsukubaensis	Olive oil, arabitol [65]
	Pseudozyma hubiensis	Soybean oil [66]
	Pseudozyma shanxiensis	Soybean oil [66]
	Pseudozyma crassa	Glucose and oleic acid [67]
Cybersan	*Cyberlindnera saturnus*	Glucose [68]
Sophorolipids	*Candida floricola*	Glucose and olive oil [69]
	Candida batistae	Glucose and olive oil [70]
	Candida apicola	Glucose and olive oil [71]
	Candida stellata	Glucose and olive oil [71]
	Candida riodocensis	Glucose and olive oil [71]
	Candida kuoi	Glucose and olive oil [72]
	Meyerozyma guilliermondii	Olive oil [73]
	Starmerella bombicola	Biomass hydrolysate [74], palm oil [75], food waste [76]
	Wickerhamiella domercqiae	Glucose, repeseed oil [77]
	Lachancea thermotolerans	Canola oil [78]
Glycolipids	*Wickerhamomyces anomalus*	Glucose and olive oil [79]
Cellobiose lipids	*Cryptococcus humicola*	Glucose [80]
Glycoprotein	*Curvularia lunata*	PL-2 media [81]
Glycolipoprotein	*Aspergillus ustus*	[82]
Glycolipid	*Ustilago maydis*	[82]

2.4 Screening for BS Producers

As shown above, there are diverse microorganisms able to produce BS, but few of them have the kinetic characteristics for large-scale production since bio-based product market is appealing [83] for new or better-producing microorganisms and is a challenge for biotechnology. Microbial screening has been performed on different environments, such as hydrocarbon and oil-contaminated soil, coastal and offshore [84–86], gas platforms [87], seawater biofilm [88], marine environment [89], mangrove sediments [90], food materials [91], and Amazon rainforest [92]. Many of the isolates are bacteria related to *Pseudomona* and *Bacillus* genera. In the case of yeasts, *Candida*, *Starmerella* (ex *Candida*), and *Pseudozyma* are the most used genera; but more microorganisms are emerging nowadays as BS producers.

Accurate methodologies for isolation and characterization of microorganisms are important to discover specific metabolic capabilities. By traditional methods, microorganisms are isolated from environmental samples and tested for BS production by qualitatively, e.g. CTAB agar assay or hemolysis test as a primary approach. A different strategy limits the screening to oil or hydrocarbon-contaminated sites, resulting in an increment of isolates related to BS production [93]. In both cases, it is necessary to employ quantitative or semi-quantitative methods for BS evaluation: drop collapse test, emulsion index, TLC and superficial tension, or techniques such as HPLC/UPLC, LC-MS, or MS. Nevertheless, this methodology is only for cultivable microorganisms, so there is an undetermined number of potential producers in different environments. At this moment, the metagenomic approach is not a common practice for the screening of BS producers [94].

2.5 A Case Study: SL by Solid-State Fermentation (SSF), Kinetics, and Reactor Size Estimation

SL are BS that are produced commonly by liquid fermentation (LF) using glucose as the hydrophilic carbon source and oleic acid (C18:1) as the hydrophobic source [95]. Glucose can be replaced by sugar cane or sugar beet juices, which have high contents of sugars and nitrogen and avoid the use of the expensive yeast extract and urea for the fermentation media [96]. Some potential hydrophobic sources are solid at fermentation temperature (e.g. stearic acid, m.p. = 69.3 °C), complicating their use as substrates in these liquid processes. In such a case, solid-state fermentation (SSF) is a potentially useful alternative. Furthermore, the SSF avoids the problems that generally occur during the production of SL by liquid fermentation, such as foaming and viscosity increase [96].

BS production in liquid media is carried out with aeration and forced agitation. However, this causes problems when BS production starts because a large amount of foam is generated. Moreover, there is a tendency for microorganisms to accumulate within the foam, eliminating cells from the culture medium. Also, the presence of foam increases the risk of cross-contamination and reduces the efficiency of oxygen transfer between the liquid and gas phase in the bioreactor [97]. The use of antifoams has disadvantages. These can be toxic to microorganisms, have high costs, and are extra-components that must be separated from the BS during the purification processes [98].

Scientific literature about BS shows a clear tendency in the use of liquid fermentations over SSF, being batch processes at laboratory scale the most used methods. Nonetheless, recent studies, including our group in Mexico, point out that SSF is an advantageous alternative for BS production since oxygen transfers are efficient, and no foaming problems are seen during fermentations, especially when production yields are greater than 0.3 $g_{BS}/kg_{dr.}$

Our research group has some experience with the use of non-pathogenic microorganisms for producing SL by SSF both at the laboratory and pilot level. Using glucose, vegetable oils, and solid supports of natural origin, production of BS in our hands has been monitored from respirometry studies using an analytical device patented by our group [99] without disturbing the culture. For example, Figure 2.6 shows the carbon dioxide production rate (CDPR) recorded in real-time during BS production. This measurement is an invaluable tool for the process because it allows us to make decisions in real-time. In this

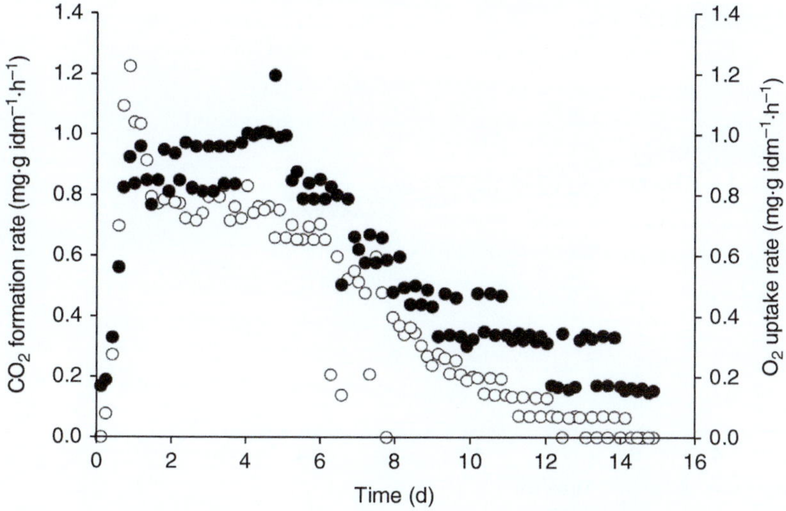

Figure 2.6 CO_2 formation rate (empty symbols) and O_2 uptake rate (full symbols) during the production of SL in SSF.

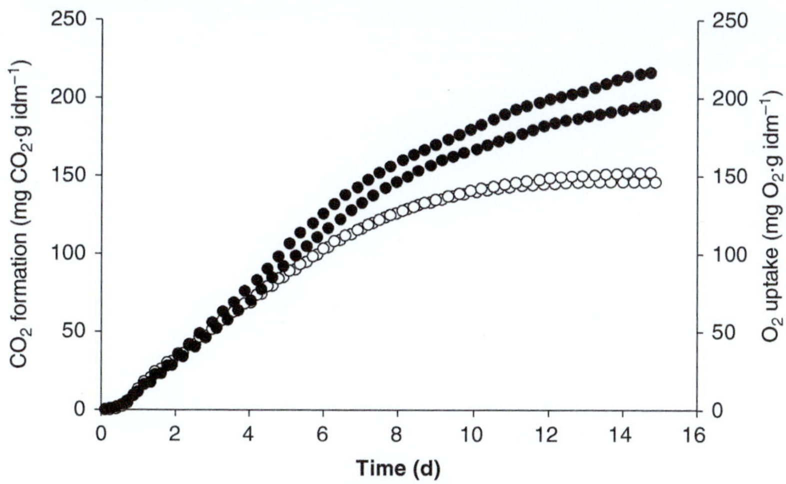

Figure 2.7 Total CO_2 formation (empty symbols) and O_2 uptake (full symbols) during sophorolipid production in SSF.

example, it is seen an imminent increase of CDPR (ca 1.2) followed by a rapid decrease (ca 0.8) that stabilizes in a *plateau* to drop off subsequently. This is the common behavior observed in many of our productive fermentation. In Figure 2.7, it is observed that O_2 consumption rate is similar to CO_2 production rate during the first days of incubation. After that, greater O_2 consumption is observed with respect to CO_2 production (Figure 2.7). This is reflected in a decrease in respiratory quotient (RQ, Figure 2.8).

The RQ (Figure 2.8) is less than 1.0 during the entire cultivation time, and the maximum value of the RQ (0.82 mol of CO_2/mol of O_2) is observed around 36 hours of incubation. These

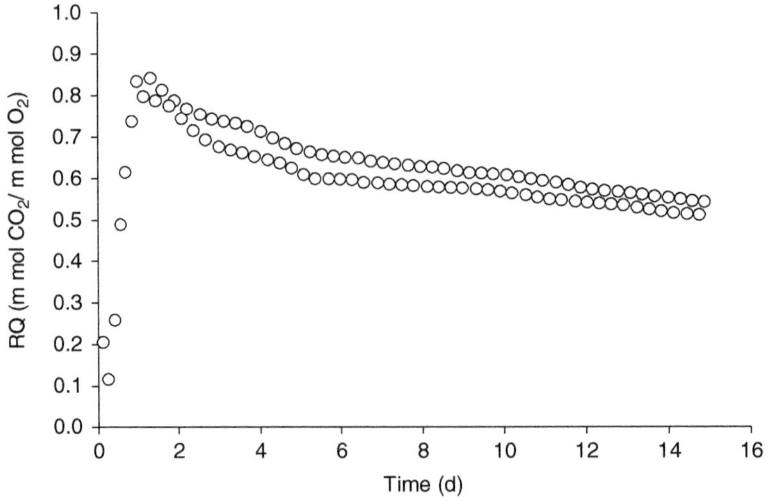

Figure 2.8 Respiratory quotient observed during SL production by SSF.

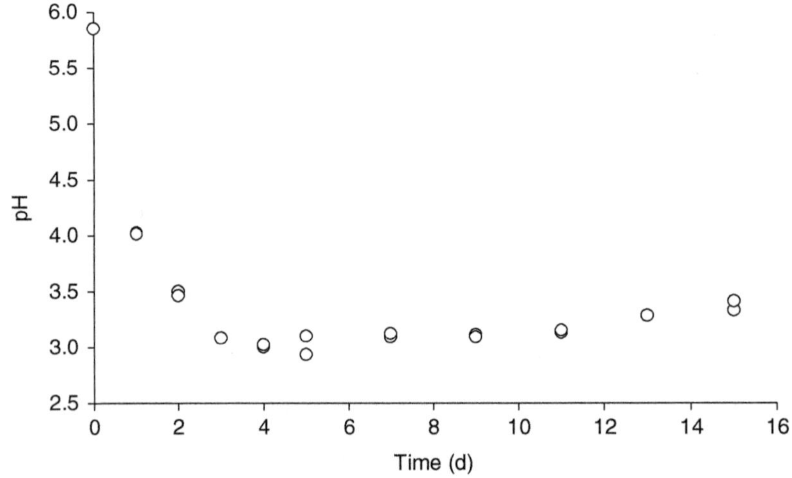

Figure 2.9 Evolution of pH during the production de sophorolipids.

RQ values could be explained by the oxidation of glucose and suggest that, during the first days, fatty acids could be mainly used for synthesizing BS. Literature indicates that fatty acids are hydroxylated and incorporated directly into the synthesis of BS [100], whereas, after glucose depletion, fatty acids can be mainly used as energy source assimilated via β-oxidation.

Another parameter that can be easily measured to follow BS production is pH. The change of pH values over time is shown in Figure 2.9. pH values show a significant decrease during the first days of cultivation (from 6 to 3) and then remain. This behavior is similar in liquid fermentation [100]. The addition of alkali has been proposed to control the pH values around 3.5, which are optimal for sophorolipid production [101]. However, for SSF, it is not feasible due to the lack of homogeneity.

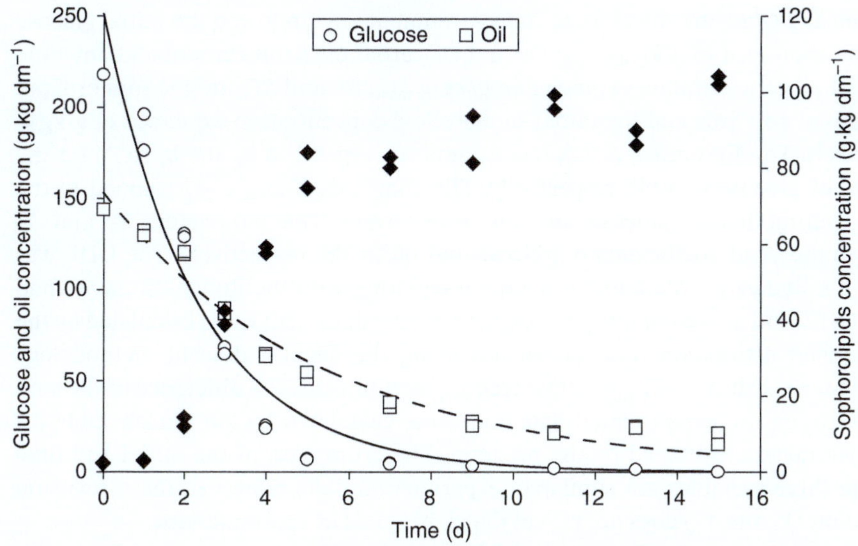

Figure 2.10 Kinetics for sophorolipid production (black) and substrates (○, □) uptake in SSF.

Figure 2.10 shows the consumption of hydrophilic and lipophilic substrates over time. Glucose consumption is recorded around the fifth day of cultivation; at this point, the consumption of the lipophilic substrate is around 65%. However, the lipophilic substrate is still consumed until the end of incubation, where consumption of around 90% was observed. It seems that lipophilic substrate remains as the only carbon source available and is used to obtain energy, which is reflected in the low increase in BS production registered from fifth day of incubation.

It is important to mention that the decrease in the production of BS could be correlated with the decrease in CDPR observed by respirometry (Figure 2.10). In this case, CDPR is a variable of the process determining the time of the maximum production of BS according to the physiological state of the yeasts in the SSF.

From the analysis of the experimental data, three balance equations for the substrate's consumption are proposed based on modified first-order decay kinetics. These three equations are coupled to the formation of BS through the conversion yields, corresponding consumption of glucose, and oil in the formation of BS. The set of ordinary differential equations (ODE) corresponds to initial value problems and are shown below:

$$\frac{dGluc}{dt} = -k_1\left(Gluc - B\right)Gluc = Gluc_0 \,@\, t = t_0 \tag{2.1}$$

$$\frac{dOil}{dt} = -k_2\left(Oil - C\right)Oil = Oil_0 \,@\, t = t_0 \tag{2.2}$$

$$\frac{dSL}{dt} = -Y_1\frac{dGluc}{dt} - Y_2\frac{dOil}{dt}SL = SL_0 \,@\, t = t_0 \tag{2.3}$$

Where *Gluc* and $Gluc_0$ are the glucose concentration at any time, and the initial glucose concentration expressed in g/kg $_{\text{dry mass}}$; *Oil* and Oil_0 are the oil concentration at any time and the initial oil concentration expressed in g/kg$_{\text{dry mass}}$; *SL* and SL_0 are the sophorolipid concentrations at any time and the initial sophorolipid concentration expressed in g/kg$_{\text{dry mass}}$, respectively. The first-order reaction constants correspond to k_1 and k_2 (h^{-1}) for the consumption of glucose and oil, respectively. The constants *B* and *C* correspond to the residual concentrations of glucose and oil, respectively. The parameters Y_1 and Y_2 correspond to the yield coefficients of glucose and oil in BS, respectively. The ODE was solved with the Berkeley – Madonna program, integrating with the Runge–Kutta method to solve the ODE, and also comparing the experimental values with those calculated by the model. Parameter estimation was carried out using the deepest descent methodology proposed by Marquardt in 1963 [102]; this technique minimizes the difference of the sum of squares between the experimental data and those calculated by the model. Table 2.4 shows the parameters estimated by the program. The estimation of the initial and final values of the three variables are similar to experimental data. However, the conversion yield estimation (Y_1 and Y_2) does not reflect the global yield of raw materials.

Figure 2.10 presents the results of the model simulation (lines) compared to the experimental data. Calculated values for substrate consumption fit well to those obtained experimentally. Glucose and oil consumption have a classic descending behavior of first-order kinetics. However, the pattern of the BS formation, the most important variable of the process, does not fit correctly to experimental data, particularly during the first 3 days of culture. The curve calculated by the model is not sigmoidal; this latter is a typical pattern for microbial fermentations. Levenspiel [103] pointed out that the kinetics of the formation of the product of interest (sophorolipid in this case) is the basis for obtaining the design equations of ideal reactors. This is done by plotting the inverse of the product formation rate against the sophorolipid concentration (Figure 2.11).

Using this plot, the size of the batch reactor will be proportional to the area under the curve of the inverse of the reaction rate, evaluated between the limits of product formation

Table 2.4 Estimated parameters to characterize the kinetics of the production of sophorolipids by SSF. The Runge–Kutta method of 4 order was used, in a time interval of 0–360 hours with an increase in time step of 0.05 hours.

Parameters		Units	Parameter value
First-order constant rate	k_1	h^{-1}	0.01784
First-order constant rate	k_2	h^{-1}	0.00771
Yield of glucose to sophorolipids	Y_1	g SL/g Glucose	3.99×10^{-5}
Yield of oil to sophorolipids	Y_2	g SL/g oil	0.69998
Residual glucose constant	*B*	g/kg dry mass	0.06616
Residual oil	*C*	g/kg dry mass	0.15279
Initial glucose	G_0	g/kg dry mass	251.20
Initial oil	Oil_0	g/kg dry mass	153.20
Initial sophorolipid	SL_0	g/kg dry mass	5.92×10^{-14}

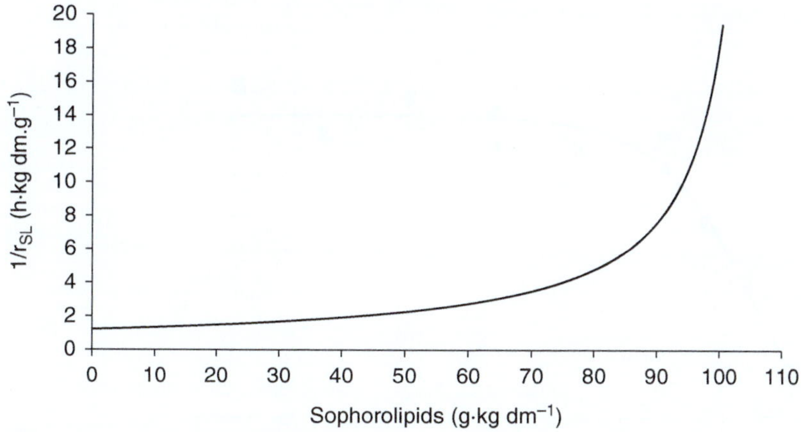

Figure 2.11 Plot of the equation design of ideal bioreactors for the production of sophorolipids using a first-order reaction rate.

(0–100 g/kg$_{dry\ mass}$). On the other hand, in the case of continuous culture of SSF, the size of the bioreactor will be proportional to a rectangle whose base is the increase in product formation (0–100 g/kg$_{dry\ mass}$) and the height of the rectangle is the value of the inverse of the reaction rate ($1/r_{SL}$) evaluated at the point of discharge (~ 100 g/kg$_{dry\ mass}$). A comparison of these two types of bioreactor systems indicates that the batch reactor is much smaller than the continuous culture reactor; therefore, up to this level of analysis, the batch reactor is the most suitable type of reactor to produce the sophorolipid by SSF. However, it is necessary to deepen the analysis of the concavity of the sophorolipid production curve to have a model that better represents the behavior of the production of BS in SSF.

To analyze the change in the concavity of the sophorolipid production curve already indicated, the Gompertz model was used. This model is a sigmoidal equation that has the characteristic of not being symmetric at the inflexion point; as is the case with the logistic equation, it is a very flexible model to simulate behavior such as that described for the production of BS in SSF. Equations (2.4) and (2.5) show the differential and integral form of the model. Equation (2.6) shows the solution of Equation (2.5) for the initial conditions of the fermentation:

$$\frac{dP}{dt} = kPLn\frac{P_{max}}{P} \tag{2.4}$$

$$P = P_{max}\exp\left(-b\exp\left(-kt\right)\right) \tag{2.5}$$

$$t \to 0; P_0 = P_{max}\exp\left(-b\right) \tag{2.6}$$

Where P is the product concentration (sophorolipid), P_{max} and P_0, are the maximum (88.9) and initial (1.61) concentrations of BS in g/kg$_{dry\ mass}$; k is a specific first-order constant (0.032 h^{-1}), and b (8.898) is a parameter associated with the initial concentration of BS. The mentioned parameters were estimated using Excel's Solver subroutine. Figure 2.12 shows the result of the simulation of the experimental data by the Gompertz model. It is worth

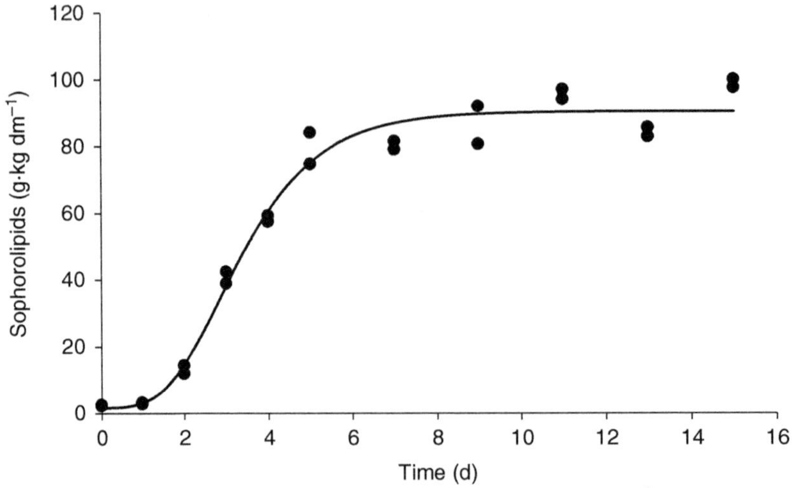

Figure 2.12 Simulation of sophorolipid production using the Gompertz model during the SSF. Symbols are the experimental and data solid line is the data calculated by the model.

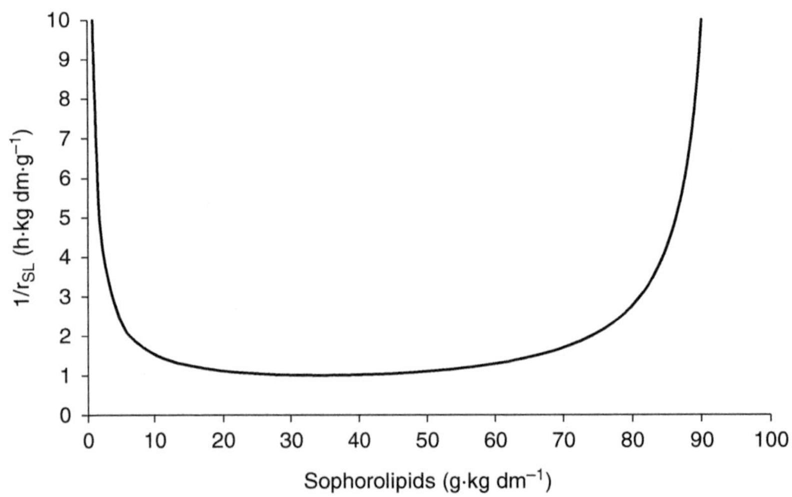

Figure 2.13 The plot of the equation design of ideal bioreactors for the production of sophorolipids using the Gompertz model.

noticing the use of the Gompertz model, allows predicting the change in the concavity of the production rate of BS in SSF. The correlation coefficient between experimental and calculated data was 0.98.

In a similar way to the methodology applied in Figure 2.11, the analysis of Figure 2.13 was carried out. Thus, in the same way, Levenspiel [103] indicates that from the kinetics of the formation of the product of interest (sophorolipid), it can be obtained. The design equations of the ideal reactors are obtained plotting the inverse of the product formation rate against the sophorolipid concentration (Figure 2.13. From this graph, the size of the reactor per batch will be proportional to the area under the curve of the inverse of the reaction rate, evaluated between the limits of the formation of product (~ 0–100 g/$kg_{dry\ mass}$). Compared to Figure 2.11, in this case, the size of the intermittent reactor increases when the level of BS is less than 40 (g/kg dry mass), this is due to the change in concavity modeled by the Gompertz equation, which simulates more precisely the production of BS.

On the other hand, in the case of continuous culture in SSF, the size of the mixed flow bioreactor will be proportional to a rectangle whose base is the increase in product formation (0–100 g/$Kg_{dry\ mass}$) and its height is the value of the inverse of the reaction rate ($1/r_{SL}$) evaluated at the point of discharge (~ 100 g/kg dry mass). In this case, there is no significant change in reactor size, either using the first-order model or the Gompertz model. The comparison of these two types of bioreactor systems indicates that the batch reactor is smaller than the continuous culture reactor. Therefore, up to this point, the batch bioreactor is the most suitable to produce BS by SSF. The analysis of the areas indicates that the continuous reactor is almost 40 times larger than the batch reactor to obtain the same sophorolipid concentration (~ 100 g/kg dry mass). On the other hand, continuous reactors are very attractive due to their high productivity; however, these have the risk of contamination, especially in a relatively long-time process. This possibility should not be ruled out for the future. To make it possible, it is essential to go deeper into the study of SSF bioreactors.

2.6 Conclusions and Perspectives

BS production seems to be the right choice that will change in the future the way the chemical industry works. BS can attend applications in a wide range of sectors as bioremediation, water treatment, food processing, health, sanitizers, cosmetics, and pharmaceuticals. All these compounds are produced by archaea, bacteria, yeasts, and molds. Isolation, screening, and selection of non-pathogenic microorganisms are a critical task for the safety production of BS, as well as culture media design, modeling, and optimization of culture conditions and scale up are indispensable to achieve an economic feasible process. Other tools, such as genetic modification, CRISPR-Cas, and metabolic engineering, are emerging as different strategies to improve the yield and productivity of fermentations. Glycolipids seem to be the species with the highest potential to be developed at larger scales, particularly in the case of BS, which are probably the most promising species because these are produced through non-pathogenic yeasts with the best observable yields for a BS. SSF comes out as an excellent alternative for BS production because it troubleshoots specific problems of traditional liquid fermentations. However, there is still a long way to go to make this emerging technique the preferred one for producing BS at large scale (e.g. improvements of mass and heat transfer, avoiding the accumulation of metabolic heat and carbon dioxide, inter alia). Fortunately, some trends appear to be beneficial to the SSF process, such as the use of single-cell microorganisms tolerant to agitated bioreactors. We have given some examples of our own experience producing BS in SSF. The critical variables are (i) sophorolipid production, (ii) substrates consumption, (iii) respiratory kinetic parameters (CO_2 production, O_2 uptake, respiratory quotient), and (iv) pH. For the simulation of sophorolipid production, it is preferred to use the Gompertz model. Accurate kinetic description is important for estimating the size of the reactors. In our case, it seems that the most convenient regime for BS production is in batch cultures, reducing the risks of contaminations. On the other hand, the analysis of the rate of carbon dioxide formation is a very relevant process monitoring tool, which allows determining the moment in which the maximum production of BS occurs, on-line, in real-time, and without disturbing the fermentation. The progress of the disciplines in the interface of chemistry, engineering, and biology shows an extremely attractive opportunity for the research, manufacturing, and application of BS. BS' potential in the chemical industry is immeasurable because BS cover and exceed the scope of synthetic surfactants, due to more innovative and sustainable chemistry, as well as a strong character of a circular economy.

References

1 Sar, P., Ghosh, A., Scarso, A., and Saha, B. (2019). Surfactant for better tomorrow: applied aspect of surfactant aggregates from laboratory to industry. *Research on Chemical Intermediates* 45: 6021–6041.

2 Santos, D.K.F., Rufino, R.D., Luna, J.M. et al. (2016). Biosurfactants: multifunctional biomolecules of the 21st century. *International Journal of Molecular Sciences* 17: 401.

3 Ontiveros, J.F., Pierlot, C.M., Catté, M. et al. (2014). A simple method to assess the hydrophilic and lipophilic balance of food and cosmetic surfactants using the phase inversion temperature of C10E4/n-octane/water emulsions. *Colloids and Surfaces, A: Physicochemical and Engineering Aspects* 458: 32–39.

4 Henkel, M. and Hausmann, R. (2019). Diversity and classification of microbial surfactants. In: *Biobased Surfactants*, 2e (eds. D. Hayes, D. Solaiman and R. Ashby), 41–63. Elsevier.

5 Vijayakumar, S. and Saravanan, V. (2015). Biosurfactants-types, sources, and applications. *Research Journal of Microbiology* 10: 181–192.

6 Dhanarajan, G. and Sen, R. (2014). Cost analysis of biosurfactant production from a scientist's perspective. In: *Biosurfactants: Production and Utilization-Processes, Technologies, and Economics*, Surfactant Science, vol. 159 (eds. N. Kosaric and F. Vardar-Sujan), 153–160. CRC Press.

7 Konishi, M., Yoshida, Y., and Horiuchi, J.-i. (2015). Efficient production of sophorolipids by *Starmerella bombicola* using a corncob hydrolysate medium. *Journal of Bioscience and Bioengineering* 119: 317–322.

8 Funston, S.J., Tsaousi, K., Rudden, M. et al. (2016). Characterising rhamnolipids production in *Burkholderia thailandesis* E264, a non-pathogenic producer. *Applied Microbiology and Biotechnology* 100: 7945–7956.

9 Delbeke, E.I.P., Movsisyan, M., Van Geem, K.M., and Stevens, C.V. (2016). Chemical and enzymatic modification of sophorolipids. *Green Chemistry* 18: 76–104.

10 Sharma, M., Patel, S.N., Lata, K. et al. (2016). A novel approach of integrated bioprocessing of cane molasses for the production of prebiotic and functional bioproducts. *Bioresource Technology* 219: 311–318.

11 Lata, K., Sharma, M., Patel, S.N. et al. (2018). An integrated bio-process for production of functional biomolecules utilizing raw and by-products from dairy and sugarcane industries. *Bioprocess and Biosystems Engineering* 41: 1121–1131.

12 Perfumo, A., Banat, I.M., and Marchant, R. (2018). Going green and cold: biosurfactants from low-temperature environments to biotechnology applications. *Trends in Biotechnology* 36: 277–289.

13 Fracchia, L., Ceresa, C., Franzetti, A. et al. (2015). Industrial application of biosurfactants. *Surfactant Science Series* 159: 245–267.

14 Patowary, K., Patowary, R., Deka, S., and Kalita, M.C. (2017). Characterization of biosurfactant produced during degradation of hydrocarbons using crude oil as sole source of carbon. *Frotiers in Microbiology* 8: 279.

15 Chaprão, M.J., Ferreira, I.N.S., Correa, P.F. et al. (2015). Application of bacterial and yeast biosurfactants for enhanced removal and biodegradation of motor oil from contaminated sand. *Electronic Journal of Biotechnology* 18: 471–479.

16 Nitschke, M. and Sousa Silva, S. (2018). Recent food applications of microbial surfactants. *Critical Reviews in Food Science and Nutrition* 58: 631–638.

17 Diaz de Rienzo, M.A., Banat, I.M., Dolamn, B. et al. (2015). Sophorolipid biosurfactants: possible uses as antibacterial and antibiofilm agent. *New Biotechology* 32: 720–726.

18 Jezierska, S., Claus, S., and Van Bogaert, I. (2018). Yeast glycolipid biosurfactants. *FEBS Letters* 592 (8): 1312–1329.

19 Mimee, B., Labbé, C., and Bélanger, R.R. (2009). Catabolism of flocculosin, an antimicrobial metabolite produced by *Pseudozyma flocculosa*. *Glycobiology* 19 (9): 995–1001.

20 Charlesworth, J.C. and Burns, B.P. (2015). Untapped resources: biotechnological potential of peptides and secondary metabolites in archaea. *Archaea* 282035: 7.

21 Abdel-Mawgoud, A.M. and Stephanopoulos, G. (2018). Simple glycolipids of microbes: chemistry, biological activity and metabolic engineering. *Synthetic and Systems Biotechnology* 3 (1): 3–19.

22 Sharma, D., Saharan, B.S., and Kapil, S. (2016). *Biosurfactants of Lactic Acid Bacteria*, 86. Switzerland: Springer.

23 Satpute, S.K., Kulkarni, G.R., Banpurkar, A.G. et al. (2016). Biosurfactant/s from *Lactobacilli* species: Properties, challenges and potential biomedical applications. *Journal of Basic Microbiology* 56 (11): 1140–1158.

24 Mata-Sandoval, J.C., Karns, J., and Torrents, A. (2001). Effect of nutritional and environmental conditions on the production and composition of rhamnolipids by *P. aeruginosa* UG2. *Microbiological Research* 155 (4): 249–256.

25 Kaskatepe, B. and Yildiz, S. (2016). Rhamnolipid biosurfactants produced by *Pseudomonas* species. *Brazilian Archives of Biology and Technology* 59.

26 Chong, H. and Li, Q. (2017). Microbial production of rhamnolipids: opportunities, challenges and strategies. *Microbial Cell Factories* 16 (1): 137.

27 Fazli, R.R. and Hertadi, R. (2019). Production and characterization of rhamnolipids from bioconversion of palm oil mill effluent by the halophilic bacterium *Pseudomonas stutzeri* BK-AB12. *Environmental Progress & Sustainable Energy* 38 (3): e13007.

28 Tan, Y.N. and Li, Q. (2018). Microbial production of rhamnolipids using sugars as carbon sources. *Microbial Cell Factories* 17 (1): 89.

29 Jadhav, M., Kalme, S., Tamboli, D., and Govindwar, S. (2011). Rhamnolipid from *Pseudomonas desmolyticum* NCIM-2112 and its role in the degradation of Brown 3REL. *Journal of Basic Microbiology* 51 (4): 385–396.

30 Calvo, C., Toledo, F., and González-López, J. (2004). Surfactant activity of a naphthalene degrading *Bacillus pumilus* strain isolated from oil sludge. *Journal of Biotechnology* 109 (3): 255–262.

31 Vasileva-Tonkova, E. and Gesheva, V. (2005). Glycolipids produced by Antarctic *Nocardioides sp.* during growth on n-paraffin. *Process Biochemistry* 40 (7): 2387–2391.

32 Nalini, S. and Parthasarathi, R. (2018). Optimization of rhamnolipid biosurfactant production from *Serratia rubidaea* SNAU02 under solid-state fermentation and its biocontrol efficacy against *Fusarium* wilt of eggplant. *Annals of Agrarian Science* 16 (2): 108–115.

33 Dong, H., Xia, W., Dong, H. et al. (2016). Rhamnolipids produced by indigenous *Acinetobacter junii* from petroleum reservoir and its potential in enhanced oil recovery. *Frontiers in Microbiology* 7: 1710.

34 Hošková, M., Schreiberová, O., Ježdík, R. et al. (2013). Characterization of rhamnolipids produced by non-pathogenic *Acinetobacter* and *Enterobacter* bacteria. *Bioresource Technology* 130: 510–516.

35 Joy, S., Rahman, P.K., Khare, S.K., and Sharma, S. (2019). Production and characterization of glycolipid biosurfactant from *Achromobacter sp.*(PS1) isolate using one-factor-at-a-time (OFAT) approach with feasible utilization of ammonia-soaked lignocellulosic pretreated residues. *Bioprocess and Biosystems Engineering* 42 (8): 1301–1315.

36 Christova, N., Tuleva, B., Lalchev, Z. et al. (2004). Rhamnolipid biosurfactants produced by *Renibacterium salmoninarum* 27BN during growth on n-hexadecane. *Zeitschrift für Naturforschung. Section C* 59 (1-2): 70–74.

37 Yan, X., Sims, J., Wang, B., and Hamann, M.T. (2014). Marine actinomycete *Streptomyces sp.* ISP2-49E, a new source of rhamnolipid. *Biochemical Systematics and Ecology* 55: 292–295.

38 Kalyani, A.L.T., Naga Sireesha, G., Aditya, A. et al. (2014). Production optimization of rhamnolipid biosurfactant by *Streptomyces coelicoflavus* (NBRC 15399T) using Plackett-Burman design. *European Journal of Biotechnology and Bioscience* 1 (5): 07–13.

39 Lee, S.-C., Lee, S.-J., Kim, S.-H. et al. (2008). *Characterization of new biosurfactant produced by Klebsiella sp.* Y6-1 isolated from waste soybean oil. *Bioresource Technology* 99 (7): 2288–2292.

40 Luong, T.M., Ponamoreva, O.N., Nechaeva, I.A. et al. (2018). Characterization of biosurfactants produced by the oil-degrading bacterium *Rhodococcus erythropolis* S67 at low temperature. *World Journal of Microbiology and Biotechnology* 34 (2): 20.

41 White, D.A., Hird, L.C., and Ali, S.T. (2013). Production and characterization of a trehalolipid biosurfactant produced by the novel marine bacterium *Rhodococcus sp.*, strain PML026. *Journal of Applied Microbiology* 115 (3): 744–755.

42 Shao, Z. (2011). Trehalolipids. In: *Biosurfactants* (ed. G. Soberón-Chávez), 121–143. Berlin, Heidelberg: Springer.

43 Kuyukina, M.S., Ivshina, I.B., Philp, J.C. et al. (2001). Recovery of *Rhodococcus* biosurfactants using methyl tertiary-butyl ether extraction. *Journal of Microbiological Methods* 46 (2): 149–156.

44 Tuleva, B., Christova, N., Cohen, R. et al. (2009). Isolation and characterization of trehalose tetraester biosurfactants from a soil strain *Micrococcus luteus BN56*. *Process Biochemistry* 44 (2): 135–141.

45 Kügler, J.H., Muhle-Goll, C., Kühl, B. et al. (2014). Trehalose lipid biosurfactants produced by the actinomycetes *Tsukamurella spumae* and *T. pseudospumae*. *Applied Microbiology and Biotechnology* 98 (21): 8905–8915.

46 Wang, Y., Nie, M., Diwu, Z. et al. (2019). Characterization of trehalose lipids produced by a unique environmental isolate bacterium *Rhodococcus qingshengii* strain FF. *Journal of Applied Microbiology* 127 (5): 1442–1453.

47 Mnif, I. and Ghribi, D. (2016). Glycolipid biosurfactants: main properties and potential applications in agriculture and food industry. *Journal of the Science of Food and Agriculture* 96 (13): 4310–4320.

48 Dhasayan, A., Kiran, G.S., and Selvin, J. (2014). Production and characterisation of glycolipid biosurfactant by *Halomonas* sp. MB-30 for potential application in enhanced oil recovery. *Applied Biochemistry and Biotechnology* 174 (7): 2571–2584.

49 Singh, P. and Tiwary, B.N. (2016). Isolation and characterization of glycolipid biosurfactant produced by a *Pseudomonas otitidis* strain isolated from Chirimiri coal mines, India. *Bioresources and Bioprocessing* 3 (1): 42.

50 Deleu, M. and Paquot, M. (2004). From renewable vegetables resources to microorganisms: new trends in surfactants. *Comptes Rendus Chimie* 7 (6-7): 641–646.

51 Joshi, S.J., Geetha, S.J., and Desai, A.J. (2015). Characterization and application of biosurfactant produced by *Bacillus licheniformis* R2. *Applied Biochemistry and Biotechnology* 177 (2): 346–361.

52 Bezza, F.A. and Chirwa, E.M.N. (2015). Production and applications of lipopeptide biosurfactant for bioremediation and oil recovery by *Bacillus subtilis* CN2. *Biochemical Engineering Journal* 101: 168–178.

53 Deepak, R. and Jayapradha, R. (2015). Lipopeptide biosurfactant from *Bacillus thuringiensis* pak2310: a potential antagonist against *Fusarium oxysporum*. *Journal de Mycologie Medicale* 25 (1): e15–e24.

54 Kamal, I., Blaghen, M., Lahlou, F.Z., and Hammoumi, A. (2015). Evaluation of biosurfactant production by *Aeromonas salmonicida sp.* degrading gasoline. *International Journal of Applied Microbiology and Biotechnology Research* 3: 89–95.

55 Sajna, K.V., Höfer, R., Sukumaran, R.K. et al. (2015). White Biotechnology in Biosurfactants. In: *Industrial Biorefineries and White Biotechnology* (eds. A. Pandey, R. Höfer, M. Taherzadeh, et al.), 499–521. Amsterdam, Oxford, Waltham: Elsevier.

56 Yi, G., Liu, Q., Lin, J. et al. (2017). Repeated batch fermentation for surfactin production with immobilized *Bacillus subtilis* BS-37: Two-stage pH control and foam fractionation. *Journal of Chemical Technology & Biotechnology* 92 (3): 530–535.

57 Tang, Q., Bie, X., Lu, Z. et al. (2014). Effects of fengycin from *Bacillus subtilis* fmbJ on apoptosis and necrosis in *Rhizopus stolonifer*. *Journal of Microbiology* 52 (8): 675–680.

58 Balan, S.S., Kumar, C.G., and Jayalakshmi, S. (2016). Pontifactin, a new lipopeptide biosurfactant produced by a marine *Pontibacter korlensis* strain SBK-47: purification, characterization and its biological evaluation. *Process Biochemistry* 51 (12): 2198–2207.

59 Clements, T., Ndlovu, T., Khan, S., and Khan, W. (2019). Biosurfactants produced by Serratia species: Classification, biosynthesis, production and application. *Applied Microbiology and Biotechnology* 103 (2): 589–602.

60 Rodrigues, L., Moldes, A., Teixeira, J., and Oliveira, R. (2006). Kinetic study of fermentative biosurfactant production by *Lactobacillus* strains. *Biochemical Engineering Journal* 28 (2): 109–116.

61 Satpute, S.K., Mone, N.S., Das, P. et al. (2019). Inhibition of pathogenic bacterial biofilms on PDMS based implants by *L. acidophilus* derived biosurfactant. *BMC Microbiology* 19 (1): 1–15.

62 Vecino, X., Barbosa-Pereira, L., Devesa-Rey, R. et al. (2015). Optimization of extraction conditions and fatty acid characterization of *Lactobacillus pentosus* cell-bound biosurfactant/bioemulsifier. *Journal of the Science of Food and Agriculture* 95 (2): 313–320.

63 Morita, T., Konishi, M., Fukuoka, T. et al. (2007). Microbial conversion of glycerol into glycolipid biosurfactants, mannosylerythritol lipids, by a basidiomycete yeast, *Pseudozyma antarctica* JCM 10317T. *Journal of Bioscience and Bioengineering* 104 (1): 78–81.

64 Morita, T., Fukuoka, T., Imura, T., and Kitamoto, D. (2009). Production of glycolipid biosurfactants by basidiomycetous yeasts. *Biotechnology and Applied Biochemistry* 53 (1): 39–49.

65 Morita, T., Fukuoka, T., Kosaka, A. et al. (2015). Selective formation of mannosyl-l-arabitol lipid by *Pseudozyma tsukubaensis* JCM16987. *Applied Microbiology and Biotechnology 99* (14): 5833–5841.

66 Saika, A., Koike, H., Fukuoka, T., and Morita, T. (2018). Tailor-made mannosylerythritol lipids: current state and perspectives. *Applied Microbiology and Biotechnology* 102 (16): 6877–6884.

67 Fukuoka, T., Kawamura, M., Morita, T. et al. (2008). A basidiomycetous yeast, *Pseudozyma crassa*, produces novel diastereomers of conventional mannosylerythritol lipids as glycolipid biosurfactants. *Carbohydrate Research* 343 (17): 2947–2955.

68 Balan, S.S., Kumar, C.G., and Jayalakshmi, S. (2019). Physicochemical, structural and biological evaluation of Cybersan (trigalactomargarate), a new glycolipid biosurfactant produced by a marine yeast, *Cyberlindnera saturnus* strain SBPN-27. *Process Biochemistry* 80: 171–180.

69 Konishi, M., Fujita, M., Ishibane, Y. et al. (2016). Isolation of yeast candidates for efficient sophorolipids production: their production potentials associate to their lineage. *Bioscience, Biotechnology, and Biochemistry* 80 (10): 2058–2064.

70 Konishi, M., Fukuoka, T., Morita, T. et al. (2008). Production of new types of sophorolipids by *Candida batistae*. *Journal of Oleo Science* 57 (6): 359–369.

71 Kurtzman, C.P., Price, N.P., Ray, K.J., and Kuo, T.M. (2010). Production of sophorolipid biosurfactants by multiple species of the *Starmerella* (*Candida*) *bombicola* yeast clade. *FEMS Microbiology Letters* 311 (2): 140–146.

72 Kurtzman, C.P. (2012). *Candida kuoi* sp. nov., an anamorphic species of the Starmerella yeast clade that synthesizes sophorolipids. *International Journal of Systematic and Evolutionary Microbiology* 62 (9): 2307–2311.

73 Sharma, P., Sangwan, S., and Kaur, H. (2019). Process parameters for biosurfactant production using yeast *Meyerozyma guilliermondii* YK32. *Environmental Monitoring and Assessment* 191 (9): 531.

74 Samad, A., Zhang, J., Chen, D., and Liang, Y. (2015). Sophorolipid production from biomass hydrolysates. *Applied Biochemistry and Biotechnology* 175 (4): 2246–2257.

75 Shah, M.U.H., Sivapragasam, M., Moniruzzaman, M. et al. (2017). Production of sophorolipids by *Starmerella bombicola* yeast using new hydrophobic substrates. *Biochemical Engineering Journal* 127: 60–67.

76 Wang, H., Roelants, S.L., To, M.H. et al. (2019). *Starmerella bombicola*: recent advances on sophorolipid production and prospects of waste stream utilization. *Journal of Chemical Technology & Biotechnology* 94 (4): 999–1007.

77 Chen, J., Song, X., Zhang, H., and Qu, Y. (2006). Production, structure elucidation and anticancer properties of sophorolipid from *Wickerhamiella domercqiae*. *Enzyme and Microbial Technology* 39 (3): 501–506.

78 Mousavi, F., Beheshti-Maal, K., and Massah, A. (2015). Production of sophorolipid from an identified current yeast, *Lachancea thermotolerans* BBMCZ7FA20, isolated from honey bee. *Current Microbiology* 71 (2): 303–310.

79 Souza, K.S.T., Gudiña, E.J., Azevedo, Z. et al. (2017). New glycolipid biosurfactants produced by the yeast strain *Wickerhamomyces anomalus* CCMA 0358. *Colloids and Surfaces B: Biointerfaces* 154: 373–382.

80 Morita, T., Ishibashi, Y., Fukuoka, T. et al. (2011). Production of glycolipid biosurfactants, cellobiose lipids, by *Cryptococcus humicola* JCM 1461 and their interfacial properties. *Bioscience, Biotechnology, and Biochemistry* 75 (8): 1597–1599.

81 Paraszkiewicz, K., Kanwal, A., and Długoński, J. (2002). Emulsifier production by steroid transforming filamentous fungus *Curvularia lunata*. Growth and product characterization. *Journal of Biotechnology* 92 (3): 287–294.

82 Bhardwaj, G., Cameotra, S.S., and Chopra, H.K. (2013). Biosurfactants from fungi: a review. *Journal of Petroleum & Environmental Biotechnology* 4 (6): 1–6.

83 Ahuja, K. and Singh, S. (2019). Biosurfactant market Size. Delaware, USA. Global Market Insights. https://www.gminsights.com/pressrelease/biosurfactants-market-size (Revised 5 August 2020).

84 Nishanthi, R., Kumaran, S., Palani, P. et al. (2010). Screening of biosurfactants from hydrocarbon degrading bacteria. *Journal of Ecobiotechnology* 2 (5): 47–53.

85 Saravanan, V. and Vijayakumar, S. (2012). Isolation and screening of biosurfactant producing microorganisms from oil contaminated soil. *Journal of Academia and Industrial Research* 1 (5): 264–268.

86 Cai, Q., Zhang, B., Chen, B. et al. (2014). Screening of biosurfactant producers from petroleum hydrocarbon contaminated sources in cold marine environments. *Marine Pollution Bulletin* 86 (1-2): 402–410.

87 Cai, Q., Zhang, B., Chen, B. et al. (2015). Screening of biosurfactant-producing bacteria from offshore oil and gas platforms in North Atlantic Canada. *Environmental Monitoring and Assessment* 187 (5): 284.

88 Hamed, S.B., Smii, L., Ghram, A., and Maaroufi, A. (2012). Screening of potential biosurfactant-producing bacteria isolated from seawater biofilm. *African Journal of Biotechnology* 11 (77): 14153–14158.

89 Banat, I., Thavasi, R., and Jayalakshmi, S. (2011). Biosurfactants from marine bacterial isolates. In: *Current Research, Technology and Education Topics in Applied Microbiology and Microbial Biotechnology* (ed. A. Mendez-Vilas), 1367–1373. Formatex Research Center.

90 Saimmai, A., Tani, A., Sobhon, V., and Maneerat, S. (2012). Mangrove sediment, a new source of potential biosurfactant-producing bacteria. *Annals of Microbiology* 62 (4): 1669–1679.

91 Konishi, M., Maruoka, N., Furuta, Y. et al. (2014). Biosurfactant-producing yeasts widely inhabit various vegetables and fruits. *Bioscience, Biotechnology, and Biochemistry* 78 (3): 516–523.

92 Sena, H.H., Sanches, M.A., Rocha, D.F.S. et al. (2018). Production of biosurfactants by soil fungi isolated from the Amazon forest. *International Journal of Microbiology 2018*: 5684261.

93 Biniarz, P., Łukaszewicz, M., and Janek, T. (2017). Screening concepts, characterization and structural analysis of microbial-derived bioactive lipopeptides: a review. *Critical Reviews in Biotechnology* 37 (3): 393–410.

94 Mnif, I. and Ghribi, D. (2015). Microbial derived surface active compounds: properties and screening concept. *World Journal of Microbiology and Biotechnology* 31 (7): 1001–1020.

95 Koh, A., Wong, A., Quinteros, A. et al. (2017). Influence of sophorolipid structure on interfacial properties of aqueous-arabian light crude and related constituent emulsions. *Journal of the American Oil Chemists' Society* 94 (1): 107–119.

96 Jiménez-Peñalver, P., Castillejos, M., Koh, A. et al. (2018). Production and characterization of sophorolipids from stearic acid by solid-state fermentation, a cleaner alternative to chemical surfactants. *Journal of Cleaner Production* 172: 2735–2747.

97 Yeh, M.S., Wei, Y.H., and Chang, J.S. (2006). Bioreactor design for enhanced carrier-assisted surfactin production with *Bacillus subtilis*. *Process Biochemistry* 41 (8): 1799–1805.

98 Vardar-Sukan, F. (1998). Foaming: Consequences, prevention and destruction. *Biotechnology Advances* 16 (5): 913–948.

99 Saucedo-Castañeda, G., Favela-Torres, E., Viniegra-González, G., Torres-Mancera, M.T., Figueroa-Montero, A., and Rosales Zamora, G. (2016). Sistema de respirometría con administración remota para el monitoreo en línea de la concentración de CO_2 y O_2 y flujo de los flujos de salida de procesos biológicos. Mexican patent 336733, Issued 22 January 2016.

100 Oliveira, M.R., Camilios-Neto, D., Baldo, C. et al. (2014). Biosynthesis and production of sophorolipids. *International Journal of Scientific and Technology Research* 3 (11): 133–146.

101 Daverey, A. and Pakshirajan, K. (2010). Sophorolipids from *Candida bombicola* using mixed hydrophilic substrates: production, purification and characterization. *Colloids and Surfaces B: Biointerfaces* 79 (1): 246–253.

102 Marquardt, D.W. (1963). An algorithm for least-squares estimation of nonlinear parameters. *Journal of the Society for Industrial and Applied Mathematics* 11 (2): 431–441.

103 Levenspiel, O. (1999). *Chemical Reaction Engineering*, 3e. New York: Wiley.

3

Microbial Gums

Current Trends and Applications

Rwivoo Baruah and Prakash M. Halami

Microbiology & Fermentation Technology Department, CSIR-Central Food Technological Research Institute, Mysuru, Karnataka, India

3.1 Introduction

Microbial gums are polysaccharides produced by various microbes like bacteria, fungi, and algae as a metabolite. These microbial gums protect their producers from harsh environmental conditions and sometimes act as secondary carbon sources. These gums have a unique property to form a gel when in contact with water. They are used in several industrial applications. Applications of microbial gums are seen in biotechnology, food, cosmetics, environmental and biomedical uses and, in recent times, have seen tremendous progress in this field of research, such as improving microbial gum behavior by chemical modifications or enhancing microbial gum production by bioprocess engineering techniques. Another reason for the increasing interest in microbial gums is due to the fact that they are biodegradable in nature as opposed to synthetic polymeric gums. Compared to plant-based gums, microbial gums can be produced in large quantities in less time by controlled fermentation [1].

Microbial gums are complex polysaccharides composed of a chain of monosaccharide units joined by glycosidic linkages having high molecular weight. These gums can be classified on the basis of their monomer composition into (i) homopolysaccharides (HoPS) and (ii) heteropolysaccharides (HePS). HoPS are polysaccharides, which are composed of a single type of monomer units like dextran, levan, and pullulan. HePS are polysaccharides, which are composed of more than one type of monomer unit and sometimes composed of noncarbohydrate groups. Microbial gums can be produced as extracellular metabolites such as exopolysaccharide (EPS) or as endo-polysaccharides [2].

A class of microbial gums originating from *Sphingomonas* spp., which includes gellan, welan, and diutan, are collectively called sphingans. They have a fairly similar structure but a few different properties [3]. These polysaccharides, even in small concentrations, can increase the viscosity in a solution. These properties are due to the unique composition,

Bioprospecting of Microorganism-Based Industrial Molecules, First Edition.
Edited by Sudhir P. Singh and Santosh Kumar Upadhyay.

linkages among residues, and molecular weight found in different microbial gums from a wide range of bacteria, fungi, and algae [4].

The viscosifying effects of microbial gums, which are nontoxic in nature, have been especially exploited in the food industry as hydrocolloids, emulsifiers, and stabilizers.

This chapter focuses on various aspects of microbial gums like their biosynthesis, production, structure, and functional properties. In this chapter important examples of microbial gums like xanthan, gellan, and pullulan have been highlighted. Further, various applications of microbial gums in food, cosmetics, medical, and nanotechnology have been discussed.

3.2 Biosynthesis of Microbial Gums

Microbial gums are produced in two different ways: (i) extracellularly by means of glycosyltransferase enzymes secreted from the microbial cells, or (ii) intracellularly, where a cascade of enzymes produced inside the cells is responsible for precursor synthesis, glycosyltransferase, polymerization, and export of the microbial biopolymer (Figure 3.1). HoPS, such as dextran, alternan, and levan, are produced by extracellular glycosylsucrase enzymes that hydrolyses sucrose to produce long-chain biopolymers. Well-known microbial gums such as pullulan, a HoPS of α-(1 → 6)-linked polymer of maltotriose subunits, is produced by fungus *Aureobasidium pullulans* using enzyme glucosyltransferase. UDP-glucose acts as a precursor for pullulan synthesis. The exact biosynthesis pathway for pullulan is still not known [5]. Curdlan, a (1→3)-β-glucan produced by *Alcaligenes faecalis,* is synthesized by the action of an operon containing four genes, namely crdA, crdS, crdC, and *crdR*. *crdS* is curdlan synthase, a glucosyltransferase enzyme, which uses UDP-glucose as a

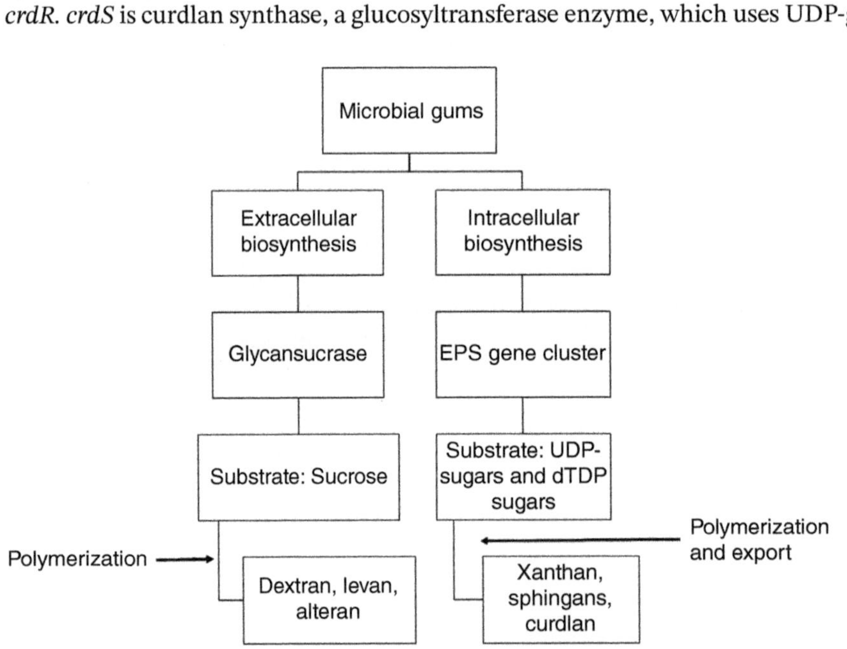

Figure 3.1 Schematics of biosynthesis of microbial gums. *Source:* Based on [3].

precursor. The *crdC* gene is responsible for polymerization of curdlan and subsequent translocation of the biopolymer across the cytoplasmic membrane [6].

HePS are synthesized intracellularly and subsequently exported out from the cells. HePS shows a similar biosynthesis pathway to HoPS-like pullulan. This pathway is controlled by a collection of genes called EPS gene cluster, which is responsible for regulation, chain-length determination, polymerization, and export of HePS. The precursors for HePS biosynthesis are UDP and dTDP sugars, and in rare occurrences, non-sugar groups like acetyl group or pyruvate group could be attached to the biopolymer chain [7, 8].

Biosynthesis of microbial gums is highly strain-dependent, and these variability can be seen as differences in biopolymer yield, linkages present between moieties and molecular weight. These differences also can change the functional properties of these microbial gums. Modification of genes involved in the biosynthesis of microbial gums through metabolic engineering of the strains can yield modified biopolymers or produce a higher quantity of microbial gums [9].

3.3 Production of Microbial Gums

Microbial gums are of great interest across many industries since they can be produced in large quantities, and research into several variations in the methods of production has led to economically viable alternatives. Xanthan gum, a widely used microbial gum, is industrially produced using glucose or invert sugars in batch fermentation mode. The culture conditions for xanthan production include maintaining a temperature between 28 and 30 °C, pH~7.0, and aeration rate higher than 0.3 (v/v), which is carried out for 100 hours [10]. Industrial grade xanthan is obtained by downstream processing that involves sterilization of the fermentation broth to kill the bacterial cells and to deactivate any enzymes. Cells are separated from the broth by centrifugation, and the cell-free supernatant (CFS) is subjected to alcohol precipitation. The precipitated xanthan gum is either reprecipitated or spray-dried and subsequently milled to the desired size [11]. Pullulan, another important microbial gum, is industrially produced similarly to xanthan gum. For the production of pullulan, liquefied starch is used as carbon source, and culture conditions include a temperature of 30 °C, pH~6.5 in aerated conditions for 100 hours to obtain the maximum pullulan yield of around 70% of the substrate used [5]. Other microbial gums like dextran can also be purified from CFS using ethanol precipitation. However, an alternate method for dextran production is by using either CFS-containing enzyme dextransucrase or separately purified dextransucrase. Dextransucrase is added to sucrose in a controlled reaction condition, which results in dextran of high purity [12].

Centrifugal packed-bed reactor (CPBR) was used for the production of xanthan gum, where the bacterial cells were immobilized in the rotating fibrous matrix. The pumping and circulation of the media through the matrix ensured the cells having good liquid and gas transfer. This process resulted in good separation of xanthan and cells, and could lead to less energy-consuming downstream processing [13].

The use of alternative substrates for microbial gum production has seen interest among researchers. Xanthan gum production by using hydrolyzed starch from rice, barley, and corn starch, as well as the use of acid whey, sugarcane molasses, waste sugar beet pulp, and peach pulp, have resulted in competitive yields but less than that of glucose [14]. Coconut milk as a substrate was used to produce pullulan, where 54g/l pullulan was produced in 36 hours of fermentation at 30 °C [15]. Cane molasses have been used to produce welan gum using sulfuric acid hydrolysis as pretreatment, which resulted in a high welan gum yield of 41 g/l. The resulted welan gum was found to have acceptable molecular weight and rheological properties with better thermal stability with control substrate glucose [16].

3.4 Structure and Properties of Microbial Gums

Microbial gums possess the ability to form "gels" or increase the viscosity of a liquid. These properties are due to the specific macromolecular structural organization seen in these gums. Structural properties such as monomer composition, linkages among the residues, branched or unbranched structure, and neutral or charged properties are essential factors that are analyzed during characterization of these gums [2]. Techniques such as gas chromatography (GC) or high-performance anion-exchange chromatography (HPAEC) are used for monomer determination. Fourier-transformed infrared spectroscopy is used for the analysis of functional groups present in microbial gums. Linkages present are determined by nuclear magnetic resonance like ^{1}H and ^{13}C spectroscopy or by methylation analysis using GC [17].

Physicochemical properties of microbial gums are another critical factor, that when determined, can establish possible functional applications of microbial gums in various industries. Important physicochemical properties commonly analyzed include molecular weight determination, microstructure analysis, and determination of thermal properties. High-performance size-exclusion chromatography, accompanied by refractive index (RI) detector or for higher accuracy with multi-angle laser light-scattering detector, is generally used for molecular determination [18]. Microstructure analysis can be determined by different techniques such as scanning electron microscope and atomic force microscope, while the thermal analysis is carried out using thermal gravimetric analyzer and differential scanning calorimetry.

Different combinations of structural and physicochemical properties of these biopolymers can result in desirable properties for industrial applications. EPS produced from *Streptococcus thermophilus* CRL1190 has a high molecular weight (1782 kDa), and a porous microstructure, displaying high water holding capacity and oil holding capacity. EPS CRL1190 also possessed antioxidative, emulsifying, and flocculating activities [19].

Microbial gums have been utilized for the synthesis of hydrogels. Hydrogels can be described as hydrophilic polymeric networks that can imbibe large quantities of liquid. Hydrogels can be used to replace tissue as implants or scaffolds, and those made using microbial gums are nontoxic and biodegradable [1].

3.5 Types of Microbial Gums

In this section, we will be briefly describing different microbial gums that are important in various industries due to their properties. The different types of microbial gums and their producer organisms have been mention in Table 3.1.

Table 3.1 Microbial gums and their producer organisms.

Microbial gum	Microorganism	Monosaccharides present	Linkages present	References
Xanthan gum	*Xanthomonas campestris*	Glucose, mannose, and glucoronic acid (2:2:1)	β-(1,4), β-(1,2), and α-(1,3)	[14]
Gellan gum	*Sphingomonas elodea, Sphingomonas paucimobilis* ATCC 31461	D-glucose, D-glucoronic acid, and l-rhamnose (2:1:1)	β-(1,3), β-(1,4), and α-(1,4)	[20]
Welan gum	*Sphingomonas* spp ATCC 31555	L-mannose, L-rhamnose, D-glucose, and D-glucuronic acid (1.0:4.5:3.1:2.3)	β-(1,3), β-(1,4), and α-(1,4)	[3]
Rhamsan gum	*Sphingomonas* spp. RH-1	D-glucose, D-glucoronic acid and L-rhamnose	β-(1,3), β-(1,4), β-(1,6), and α-(1,4)	[21]
Diutan gum	*Sphingomonas* sp. ATCC 53159	D-glucose, D-glucoronic acid, and L-rhamnose	β-(1,3), β-(1,4), and α-(1,4)	[22]
Pullulan	*Aureobasidium pullulans* *Eurotium chevalieri, Cryphonectria parasitica, Cyttaria darwinii,* and yeast *Rhodotorula baracum*	D-glucose	α-(1,6) and α-(1,4)	[5, 23]
Scleroglucan	*Sclerotium glucanicum*	D-glucose	β-(1,3) and β-(1,6)	[24]
Curdlan	*Alcaligenes faecalis*	D-glucose	β-(1,3) and β-(1,6)	[6]
Levan	*Bacillus subtilis, Bacillus polymyxa, Aerobacter levanicum, Leuconostoc mesenteroides, Streptococcus* sp., *Pseudomonas* sp., and *Corynebacterium laevaniformans.*	D-fructose	β-(2,6) and β-(2,1)	[12]
Dextran	Various LAB genera-like *Leuconostoc, Lactobacillus, Streptococcus, Pediococcus,* and *Weissella*	D-glucose	α-(1,6), α-(1,2), α-(1,3) and α-(1,4)	[25, 26]

3.5.1 Xanthan Gum

Xanthan is a microbial gum produced by *Xanthomonas campestris*, a plant pathogenic bacteria. The molecular weight of xanthan, in general, can be 2000 kDa, but can also reach a higher weight of up to 50 000 kDa. Xanthan is a HePS and is composed of three different monosaccharides: glucose, mannose, and glucoronic acid in the molar ratio of 2:2:1 along with *O*-acetyl and pyruvate groups [14]. Xanthan gum is resistant to degradation by enzymes and is highly pseudoplastic in nature, and it can maintain its viscosity over a wide range of temperatures and pH. In the presence of high shear stress, the xanthan gum viscosity is reduced but is recovered immediately upon the removal of the stress. This phenomenon occurs in xanthan due to the formation of intermolecular aggregates and polymer entanglement when mixed with water [27].

Xanthan gum is a highly efficient food hydrocolloid, emulsifier, and stabilizer, which is commonly used in the food industry. It is nontoxic in nature and is used as a natural food additive. The United States Food and Drug Administration (FDA) and European Economic Community have approved the use of xanthan gum as a safe food additive in 1984 and 1980, respectively [28]. After the approval of xanthan gum from *X. campestris* as a food additive, the demand for it has seen a steady rise at a rate of 5–10% annually [11]. In many scenarios, xanthan gum is mixed with plant-based gum, guar gum, to obtain solutions of even higher viscosity than that from either one gum when used alone [29]. Xanthan gum has been reported to act as a prebiotic for dairy-based lactic acid bacteria by improving bacterial viability under low pH, refrigeration, and the presence of bile salts [30]. Xanthan gum acts as a low-calorie food hydrocolloid, as well as a fat replacer in various fat-rich foods such as salad dressing, mayonnaise, ice creams, and cakes. The mixture of xanthan gum and guar gum had been reported to improve the crip and porous characteristics of deep-fried foods when they were incorporated into the batter mix [29].

3.5.2 Sphingans

Sphingans are a group of EPS produced by *Sphingomonas* spp. They are of different types, namely, gellan gum produced by *Sphingomonas elodea*, welan gum produced by *Sphingomonas* spp. ATCC 31555, diutan gum produced by *Sphingomonas* sp. ATCC 53159, and rhamsan gum produced by *Sphingomonas* sp. RH-1. All these EPS are structurally similar with identical backbone, but differ in the presence of side chain locations (Figure 3.2) [3].

3.5.2.1 Gellan Gum

Gellan gum is produced by *Sphingomonas elodea* and *Sphingomonas paucimobilis* ATCC 31461 under submerged fermentation using glucose as carbon source with a 40–50% yield [31]. Gellan gum is an anionic HePS, which is composed of d-glucose, d-glucoronic acid, and l-rhamnose in a molar ratio of 2:1:1, along with side chains of O-linked acyl, acetyl, and glyceryl groups [20]. The presence of acyl groups in gellan gum affects the rheology of the gel formed, resulting in soft, elastic, and thermos-reversible gels. Chemical deacylation of gellan gum results in a more hard, brittle, and optically clear gels; this process generally used to produce industrial gellan gums (Gelrite or Kelcogel) [31]. Gellan gum generally has a molecular weight of 500 kDa and has the ability to form 3D network

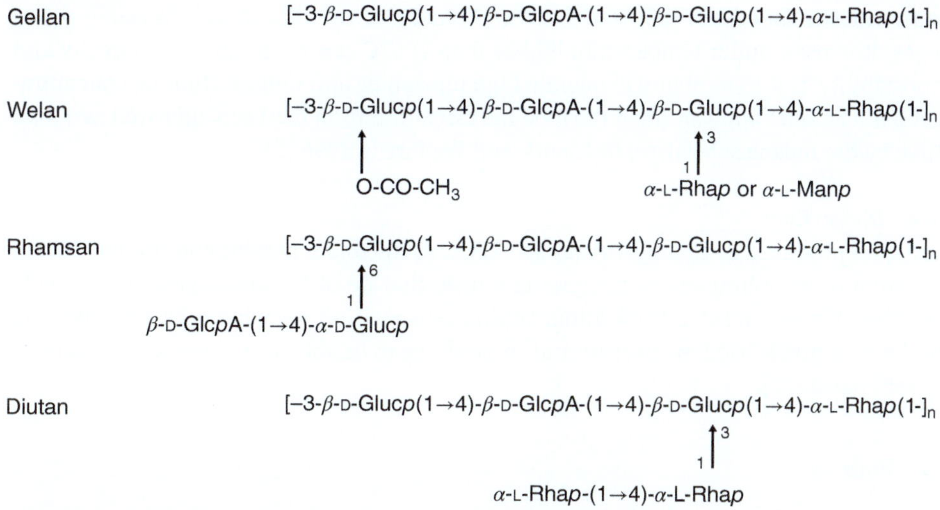

Figure 3.2 Structures of different sphingans. *Source:* Based on [3].

in the presence of monovalent or divalent cations such as Ca^{2+} or Na^{1+}. Gellan gum is used for its gelling properties in food products like dairy, jellies, and confectionery products as a hydrocolloid and can be used to prepare edible films [2]. It has even been approved as a safe food additive by US and EU authorities for their use in food products as a gelling, suspending, and stabilizing agent, either alone or in combination with other food hydrocolloids [31]. Gellan has good compatibility with other gums or biopolymers such as xanthan or locust bean gum. Gellan, when combined with starch, improves the gas and moisture retention in a food product, which results in the reduction of starch used and enhancement in the flavor. Gellan gum is used to form gels in a concentration of 0.04–0.05% (w/v), which is very low compared to other gums [20].

3.5.2.2 Welan Gum

Welan gum produced by *Sphingomonas* spp ATCC 31555 (formerly known as *Alcaligenes* sp. ATCC 31555). Welan gum is also an anionic HePS, which is composed of l-mannose, l-rhamnose, d-glucose, and d-glucuronic acid in the molar ratios 1.0:4.5:3.1:2.3. The side chain present could be either l-rhamnose or l-mannose [3]. Welan gum was previously known as EPS S-130 and commercially produced by CP Kelco, division of J.M. Huber Corporation. The molecular weight of welan gum is about 1000 kDa, and it is non-gelling polysaccharide, unlike gellan explained above [32]. Even so, welan possesses high viscosifying properties due to which it has wide-spread applications in cement technology and oil drilling industries. It has the potential to be used for food application as it can act as a thickening, suspending, binding, and emulsifying agent [3].

3.5.2.3 Rhamsan Gum

Rhamsan gum produced by *Sphingomonas* spp. RH-1 under aerobic conditions using glucose-containing medium. Rhamsan gum has a similar backbone like all sphingans consisting of glucose, glucuronic acid, and rhamnose; and the difference lies with the side chain present, which in the case of rhamsan are a *O*,6-linked glucosyl disaccharide linked by β(1,6) linkages. Rhamsan under temperature higher than 100 °C can maintain high viscosity and thermostability. Due to its ability to tolerate high phosphate and sodium chloride concentration, it has a broader application in the food industry. Rhamsan has been approved as a food additive by the Japanese Ministry of Health and Welfare in 1996 [21].

3.5.2.4 Diutan Gum

Diutan gum produced by *Sphingomonas* sp. ATCC 53159 under aerobic conditions. Diutan differs from other sphingans by possessing a disaccharide of l-rhamnosyl units as a side chain. Diutan gum can act as thickening, binding, suspending, stabilizing, and emulsifying agent. Diutan gum is used in concrete and cement due to its ability to improve the viscosity of cement paste [22].

3.5.3 Pullulan

Pullulan is a HoPS composed of glucose monomers as organized as maltotriose subunits linked by α-(1,6) linkages. It is produced by fungus *Aureobasidium pullulans* intracellularly using UDP-glucose as a precursor followed by export outside the cell [5]. Other fungi have also been reported to produce pulluan, e.g. *Eurotium chevalieri, Cryphonectria parasitica, Cyttaria darwinii,* and yeast *Rhodotorula baracum* [23]. Pullulan is a neutral polysaccharide, and its molecular weight can range between 10 and 3000 kDa, and this variability is dependent on its production conditions. Pullulan shares a similar composition with maltodextrin, but it possesses more α-(1,6) linkages (30%) compared to the later (20%). The higher occurrence of α-(1,6) linkages results in show digestibility of pullulan compared to starch and hence is used for low-calorie food products [12].

Pullulan is a non-gel forming polysaccharide, but it increases the viscosity of a solution, when in the aqueous state even in relatively low concentrations. Pullulan can maintain its viscosity even in high heat, changes in pH, and metal ions [33]. To improve the properties of pullulan, different chemical modifications have been made in pullulan such as small functional groups like carboxymethyl, sulfate, phosphate, acetate, and alkyl esters [34]. Pullulan is used to make capsule shells, tablet coatings, and edible flavored films due to its film-forming properties. Pullulan has been used as a nontoxic food additive in Japan since 1976 and has been regarded as GRAS by US FDA [12]. When consumed, pullulan can alter the composition of intestinal microbiota, specifically stimulating the growth of bifidobacteria and hence acting as a prebiotic [35]. Recently, when phthalyl pullulan nanoparticle was treated with probiotic *Lactobacillus plantarum*, the production of antimicrobial peptide plantaricin increased and showed more significant inhibition of *Escherichia coli* and *Lis. monocytogenes* growth [36].

3.5.4 Other Microbial Gums

Besides the above mentioned microbial gums, there are several other gums produced from microbial sources; a few examples are as follows.

Scleroglucan is an EPS produced by fungus *Sclerotium glucanicum*. It is HoPS of glucose linked by β-(1,3) and β-(1,6) linkages where every third glucose is connected by β-(1,6) linkages. The molecular mass of scleroglucan can vary from strain to strain. The average molecular weight ranges between 320 and 6000 kDa [24]. It is used as thickeners, stabilizers, and gelling agents in food products due to its resistance to temperature, hydrolysis, and electrolytes. It can inhibit syneresis in starch-based foods that occur during refrigeration. It is also used in oil recovery operations in the sea [12].

Curdlan is a β-glucan composed of glucose monomer linked by β-(1,3) linkages. Curdlan is produced by bacterium *Alcaligenes faecalis*. Curdlan is not soluble in water or acidic solution but is only soluble in alkaline pH. Curdlan forms two types of gels at different temperatures: (i) high-set gel (at above 80 °C) and (ii) low-set gel (between 60 and 80 °C) [2]. The high-set gel of curdlan is used in food products, as it is a thermo-irreversible gel and stable during deep-fat frying and freeze-thawing. Curdlan is used in low-calorie foods, as a fat replacer and in food products like freezable tofu noodles [6].

Levan is a HoPS composed of fructose units linked by β-(2,6) linkages in the main chain with β-(2,1) linked branches. Levan has been reported to be produced by a wide range of microbes such as Bacillus subtilis, *Bacillus polymyxa, Aerobacter levanicum, Leuconostoc mesenteroides, Streptococcus* sp., *Pseudomonas* sp., and *Corynebacterium laevaniformans*. Levan has a molecular weight of 2–100 MDa that varies between producer organisms. Levan is used in food, pharmaceutical, and cosmetic industries. It is used as stabilizing, emulsifying, and sweetening agents in food products. In the cosmetic industry, levan is utilized for its moisturizing effect, cell proliferation effect, and anti-inflammatory effects [12].

Dextran is a type of α-glucan, which consists of a linear chain of α-(1,6) linked D-glucose units with various branched linkages at α-(1,2), α-(1-3), and α-(1-4) positions of D-glucose [25]. The degree of branching and molecular weight of dextran varies with dextransucrase enzyme and producers strain. Dextran is commercially available in different molecular weights and is commonly used as molecular weight standards in GPC analysis. Dextran is used in the food industry as a food hydrocolloid and is prebiotic in nature [26].

Alginate, an important gum, is a copolymer of D-mannuronic acid and L-guluronic acid. It is usually produced by brown algae. However, alginate is also produced by bacterial sources like *Azotobacter vinelandii, Azotobacter chroococcum,* and several species of *Pseudomonas*. These are called bacterial alginates. Bacterial alginates are used in pharmaceutical industries due to their high purity, and they possess immunomodulatory activity [37].

Carrageenan is a gel-forming biopolymer produced by edible red seaweed (algae). It is composed of alternating units of D-galactose and 3,6-anhydro-galactose joined by α-1,3 and β-1,4-glycosidic linkage. On the basis of their solubility in potassium chloride, carrageenans are classified into λ, κ, ι, ε, μ–carrageenans. They have average molecular weight of above 100 kDa with 15–40% ester sulfate groups. Carrageenan has been reported to possess immunomodulatory, anticoagulant, antithrombotic, antiviral, and antitumor effects [38].

3.6 Applications of Microbial Gums

By virtue of their various properties, microbial gums are utilized in several types of applications in research and industry. In many cases, one microbial gum is used with another gum of either microbial or nonmicrobial origins that results in the enhancement of the final

Table 3.2 Applications of microbial gums.

Microbial gum	Application	Properties	References
Xanthan	Food Hydrocolloid	Viscosifier, stabilizer and emulsifier	[28]
Xanthan	Fat replacer	Viscosifier, and emulsifier	[29]
Gellan	Food Hydrocolloid	Gelling, suspending and stabilizing agent	[31]
Welan	Cement technology	Viscosifier	[3]
Welan	Oil drilling industries	Viscosifier	[3]
Rhamsan	Food Hydrocolloid	Viscosifier, high salt tolerance	[21]
Diutan	Cement technology	Viscosifier	[22]
Pullulan	Food Additive	Prebiotic	[35, 39]
Pullulan	Edible films, capsule shells and tablet coatings	Non-toxic, film-forming gels	[12]
Curdlan	Fat replacer	Viscosifier, ability to form thermo-irreversible gel	[6]
Levan	Sweeting agent	Sugar containing viscosifier	[12]
Levan	Cosmetic products	Moisturizing, cell proliferation and anti-inflammatory effects	[12]
Dextran	Food Hydrocolloid	Viscosifier, stabilizer and emulsifier	[26]
Xanthan and curdlan	Edible film in food packaging	Excellent miscibility and high tensile strength	[40]
Gellan	Hydrogel for bone regeneration	Bio-compatible, bio-degradable, non-toxic	[41]
Pullulan and hyaluronic acid	Hydrogel for wound healing	Bio-compatible, bio-degradable, non-toxic	[42]
Xanthan and welan	Nanoemulsions for delivery of bioactive compounds	Emulsifier	[43]

desire property. In this section, we will describe the use of different microbial gums in various sectors. The different types of applications of microbial gums have been mention in Table 3.2.

3.6.1 Food Applications

As explained in the previous sections, many microbial gums have properties pertaining to food applications by being used as food hydrocolloids, emulsifiers, stabilizers, and gelling agents. In recent times, the application of microbial gums in food is diversified to a large extent. Prebiotic properties of pullulan have been utilized to make prebiotic bread. The addition of 2% pullulan showed promising result in sensory evaluations and displayed resistance to gastric juice. Pullulan showed higher prebiotic activity score (PAS) than other microbial gums like xanthan and gellan [39].

Studies on the impact of different gums in goat milk were conducted by accessing the textural and microbial properties of yogurt using xanthan, carrageenan, locust bean gum, gaur gum, and modified starches. The yogurt samples containing both xanthan and locust bean gum showed superior properties than the control sample, such as enhanced firmness, consistency, cohesiveness, and viscosity. The gums supplemented also showed bifidogenic activity enhancing the growth of *Bifidobacterium animalis* Bb12 [4].

In recent times, dextran from LAB genus *Weissella* has been utilized as a natural food hydrocolloid to prepare high-fiber containing bread and gluten-free bread [26]. In situ dextran production by *Weissella confusa* A16 in wholegrain sorghum bread containing 50% wheat flour resulted in the improved texture of the bread. In situ production by *Weissella confusa* A16 resulted in 0.56% dextran of bread weight. It helped in masking undesirable flavors of wholegrain bread [44].

Microbial gums have been recently utilized as edible films in food technology resulting in a nontoxic, biocompatible, and biodegradable alternative to synthetic plastic films. The functional properties of the edible films can be enhanced further by the blending of microbial gums with lipids, hydrocolloids, or reinforcement agents [2]. Edible films composed of cassava starch and different microbial gums like xanthan, gellan, and pullulan were obtained by casting, and various properties were evaluated. Gellan gum was found to be most suitable for this purpose with the concentration of 10–20% blend by enhancing the film strength and resistance to breaking and preserving films against starch retrogradation throughout the storage time and reducing the moisture sorption capacity of starch films while reducing water vapor permeability [45]. Edible film blends were prepared using different ratios of xanthan and curdlan. Blends of xanthan and curdlan in a ratio of 5:5 and 4:6 showed excellent miscibility and high tensile strength. It confirmed that the blends of edible film have great potential as food packaging material [40].

Microbeads from whey protein isolate and gellan gum were used as a carrier for bioactive compounds of flaxseed. The encapsulation of flaxseed oil (15%) resulted in a controlled release only in the small intestine and was able to pass the gastric barriers without any modifications. This directed release of the bioactive compound can improve their bioavailability and functionality [46].

3.6.2 Biomedical Applications

Among all the microbial gums used today, dextran is the first example of a microbial biopolymer to be used for pharmaceutical applications. It was used as a plasma volume expander since 1953 [47]. In recent times, hydrogel form of microbial gums has been used for a different biomedical purpose. Hydrogels are 3D scaffolds made from biomaterials, which provide a suitable microenvironment for tissue engineering and regenerative medicine. The basic advantages of hydrogels are that they are biocompatible, biodegradable, nontoxic, and have flexibility similar to human tissue. Microbial gums like xanthan, gellan, and dextran are mostly used for hydrogel preparation. Composite hydrogels or blends using two or more gums are used for biomedical applications with superior properties. Some applications of microbial gums used as hydrogels are seen in tissue engineering used as scaffolds, in wound healing as skin barricade due to their adhesive properties, and in drug delivery [1].

Gellan gum hydrogel with magnesium-enriched calcium carbonate, calcium carbonate, and magnesium carbonate using urease-mediated mineralization was used for bone regeneration application [41]. Bioactive molecule curcumin was grafted on a hyaluronic-acid-modified pullulan polymer. In vitro studies showed no cytotoxicity, enhanced cell proliferation, antioxidative effect, and antimicrobial activity against *E. coli* and *S. aureus*. In vivo studies on Wistar rats showed enhanced wound healing compared with only hyaluronic acid and pullulan polymer [42]. Gellan gum infused with a bioactive glass (BAG) used to encapsulate human adipose stem cells for osteogenic differentiation, displayed increased cell viability, improved mechanical properties, and strong bone mineralization [48]. A chitosan and xanthan gum-based hydrogel was used for drug delivery of the antiviral drug acyclovir. Parameters such as swelling dynamics and entrapment efficiency were studied, and the results concluded the potential for such hydrogel for the efficient antiviral drug delivery [49].

3.6.3 Applications in Nanotechnology

In recent years, microbial gums owing to their unique functional properties have been employed in nanotechnology, which has subsequent applications in food and pharmaceutical industries. Various nanostructure has been developed using microbial gums such as nanoparticles, nanofibers, nanocomplexes, and nanoemulsions [50].

A gellan gum containing silver nanoparticle-based hydrogen sulfide (H_2S) biosensor for the detection of meat spoilage was developed. Using this system, H_2S was detected in chicken breast, and silver carp; sensor present in the packaging of meats changes color from yellow to white in the presence of H_2S. This is a cost-effective, non-destructive, and robust method for the detection of meat spoilage [51]. Xanthan gum-shellac nanoparticles loaded with cinnamon bark extract were assembled by anti-solvent precipitation method. The nanoencapsulation of the bioactive extract resulted in improved thermal stability and pH resistance having antioxidative properties of the extract [52]. The insulin delivery system triggered by pH was developed using acryloyl cross-linked dextran dialdehyde (ACDD) nanoparticles to immobilize glucose oxidase. This system works under gastric conditions (pH 1.2), as well as intestinal conditions (pH 7.4) and proved to be a reliable vector for drug delivery [53]. Nanoemulsions made using microbial gums xanthan and welan in combination with Rosemary essential oil were developed. The nanostructure constructed was less than 150 nm in diameter and showed the best physical stability at 0.4% gum concentration with welan gum containing nanoemulsions showing better properties [43].

3.7 Conclusions and Perspectives

Microbial gums play a key role in many industries and have seen a steady rise in demand over the years. Some gums such as xanthan and gellan are used more often than others due to their unique functional properties. In recent times, there has been an increase in research on microbial gums, especially on increasing their production, understanding their properties and function, creating novel composites of gums with unique properties, and exploring different microbiomes for novel gums. Advancements in molecular biology, structural biology, bioprocess engineering, and metabolic engineering will help in the future research of microbial gums. The focus of obtaining microbial gums through an economical and sustainable manner by the use of cheap substrates or increasing the productivity of a strain by molecular biology would be crucial areas that future research could make great strides.

Acknowledgments

RB acknowledges the research associate fellowship provided by the Council of Scientific and Industrial Research (CSIR). The authors acknowledge the contributions of Dr. SVN Vijayendra, Dept. Microbiology and Fermentation Technology for critical comments and suggestions. Authors acknowledge the kind approval of the Director, CSIR-CFTRI Mysuru, for this work. This manuscript has been approved by planning, monitoring, and coordination (PMC) section, CFTRI-CSIR vide approval number PMC/2020-21/122.

References

1 Ahmad, S., Ahmad, M., Manzoor, K. et al. (2019). A review on latest innovations in natural gums based hydrogels: preparations & applications. *International Journal of Biological Macromolecules* 136: 870–890.

2 Alizadeh-Sani, M., Ehsani, A., Kia, E.M., and Khezerlou, A. (2019). Microbial gums: introducing a novel functional component of edible coatings and packaging. *Applied Microbiology and Biotechnology* 103 (17): 6853–6866.

3 Kaur, V., Bera, M.B., Panesar, P.S. et al. (2014). Welan gum: microbial production, characterization, and applications. *International Journal of Biological Macromolecules* 65: 454–461.

4 Park, Y.W., Oglesby, J., Hayek, S.A. et al. (2019). Impact of Different Gums on Textural and Microbial Properties of Goat Milk Yogurts during Refrigerated Storage. *Food* 8 (5): 169.

5 Prajapati, V.D., Jani, G.K., and Khanda, S.M. (2013). Pullulan: an exopolysaccharide and its various applications. *Carbohydrate Polymers* 95 (1): 540–549.

6 Mcintosh, M., Stone, B.A., and Stanisich, V.A. (2005). Curdlan and other bacterial (1→ 3)-β-D-glucans. *Applied Microbiology and Biotechnology* 68 (2): 163–173.

7 Mozzi, F., Vaningelgem, F., Hébert, E.M. et al. (2006). Diversity of heteropolysaccharide-producing lactic acid bacterium strains and their biopolymers. *Applied and Environmental Microbiology* 72 (6): 4431–4435.

8 Baruah, R., Das, D., and Goyal, A. (2016). Heteropolysaccharides from lactic acid bacteria: current trends and applications. *Journal of Probiotics and Health* 4 (141): 2.

9 Shanmugam, M. and Abirami, R.G. (2019). Microbial polysaccharides-chemistry and applications. *Journal of Biologically Active Products from Nature* 9 (1): 73–78.

10 Flickinger, M.C. and Drew, S.W. (1999). *Encyclopedia of Bioprocess Technology*. Wiley.

11 Rosalam, S. and England, R. (2006). Review of xanthan gum production from unmodified starches by *Xanthomonas comprestris* sp. *Enzyme and Microbial Technology* 39 (2): 197–207.

12 Park, J.K. and Khan, T. (2009). Other microbial polysaccharides: pullulan, scleroglucan, elsinan, levan, alternant, dextran. In: *Handbook of Hydrocolloids* (eds. G.O. Phillips and P.A. Williams), 592–614. Woodhead Publishing.

13 Yang, S.T., Lo, Y.M., and Min, D.B. (1996). Xanthan Gum Fermentation by *Xanthomonas campestris* Immobilized in a Novel Centrifugal Fibrous-Bed Bioreactor. *Biotechnology Progress* 12 (5): 630–637.

14 Becker, A., Katzen, F., Pühler, A., and Ielpi, L. (1998). Xanthan gum biosynthesis and application: a biochemical/genetic perspective. *Applied Microbiology and Biotechnology* 50 (2): 145–152.

15 Thirumavalavan, K., Manikkadan, T.R., and Dhanasekar, R. (2009). Pullulan production from coconut by-products by *Aureobasidium pullulans*. *African Journal of Biotechnology* 8 (2): 254–258.

16 Ai, H., Liu, M., Yu, P. et al. (2015). Improved welan gum production by *Alcaligenes* sp. ATCC31555 from pretreated cane molasses. *Carbohydrate Polymers* 129: 35–43.

17 Wei, D.I., Zhang, Y.C., Hua-Xi, Y.I. et al. (2018). Research methods for structural analysis of lactic acid bacteria induced exopolysaccharides. *Chinese Journal of Analytical Chemistry* 46 (6): 875–882.

18 Ruijssenaars, H.J., Stingele, F., and Hartmans, S. (2000). Biodegradability of food-associated extracellular polysaccharides. *Current Microbiology* 40 (3): 194–199.

19 Lobo, R.E., Gómez, M.I., de Valdez, G.F., and Torino, M.I. (2019). Physicochemical and antioxidant properties of a gastroprotective exopolysaccharide produced by *Streptococcus thermophilus* CRL1190. *Food Hydrocolloids* 96: 625–633.

20 Giavasis, I., Harvey, L.M., and McNeil, B. (2000). Gellan gum. *Critical Reviews in Biotechnology* 20 (3): 177–211.

21 Xu, X.Y., Zhu, P., Li, S. et al. (2014). Rhamsan gum production by *Sphingomonas* sp. CGMCC 6833 using a two-stage agitation speed control strategy. *Biotechnology and Applied Biochemistry* 61 (4): 453–458.

22 Xu, L., Gong, H., Dong, M., and Li, Y. (2015). Rheological properties and thickening mechanism of aqueous diutan gum solution: Effects of temperature and salts. *Carbohydrate Polymers* 132: 620–629.

23 Dailin, D.J., Low, L.Z.M.I., Kumar, K. et al. (2019). Agro-industrial waste: a potential feedstock for pullulan production. *Biosciences, Biotechnology Research Asia* 16 (2): 229–250.

24 Survase, S.A., Saudagar, P.S., Bajaj, I.B., and Singhal, R.S. (2007). Scleroglucan: fermentative production, downstream processing and applications. *Food Technology and Biotechnology* 45 (2): 107–118.

25 Capek, P., Hlavoňová, E., Matulová, M. et al. (2011). Isolation and characterization of an extracellular glucan produced by *Leuconostoc garlicum* PR. *Carbohydrate Polymers* 83 (1): 88–93.

26 Baruah, R., Maina, N.H., Katina, K. et al. (2017). Functional food applications of dextran from *Weissella cibaria* RBA12 from pummelo (*Citrus maxima*). *International Journal of Food Microbiology* 242: 124–131.

27 Sworn, G. (2009). Xanthan gum. In: *Handbook of hydrocolloids* (eds. G.O. Phillips and P.A. Williams), 186–203. Woodhead Publishing.

28 Garcıa-Ochoa, F., Santos, V.E., Casas, J.A., and Gómez, E. (2000). Xanthan gum: production, recovery, and properties. *Biotechnology Advances* 18 (7): 549–579.

29 Chawla, R.P.G.R. and Patil, G.R. (2010). Soluble dietary fiber. *Comprehensive Reviews in Food Science and Food Safety* 9 (2): 178–196.

30 Sadek, Z.I., El-Shafei, K., and Murad, H.A. (2006). Utilization of xanthan gum and inulin as prebiotics for lactic acid bacteria. *Deutsche Lebensmittel-Rundschau* 102 (3): 109–114.

31 Sá-Correia, I., Fialho, A.M., Videira, P. et al. (2002). Gellan gum biosynthesis in *Sphingomonas paucimobilis* ATCC 31461: genes, enzymes and exopolysaccharide production engineering. *Journal of Industrial Microbiology and Biotechnology* 29 (4): 170–176.

32 Sandford, P.A., Cottrell, I.W., and Pettitt, D.J. (1984). Microbial polysaccharides: new products and their commercial applications. *Pure and Applied Chemistry* 56 (7): 879–892.

33 Khan, T., Park, J.K., and Kwon, J.H. (2007). Functional biopolymers produced by biochemical technology considering applications in food engineering. *Korean Journal of Chemical Engineering* 24 (5): 816–826.

34 Tiwari, S., Patil, R., Dubey, S.K., and Bahadur, P. (2019). Derivatization approaches and applications of pullulan. *Advances in Colloid and Interface Science* 269: 296–308.

35 Leathers, T.D. (2003). Biotechnological production and applications of pullulan. *Applied Microbiology and Biotechnology* 62 (5-6): 468–473.

36 Hong, L., Kim, W.S., Lee, S.M. et al. (2019). Pullulan Nanoparticles as prebiotics enhance the antibacterial properties of *Lactobacillus plantarum* Through the Induction of Mild Stress in Probiotics. *Frontiers in Microbiology* 10: 142.

37 Sabra, W., Zeng, A.P., and Deckwer, W.D. (2001). Bacterial alginate: physiology, product quality and process aspects. *Applied Microbiology and Biotechnology* 56 (3-4): 315–325.

38 Necas, J. and Bartosikova, L. (2013). Carrageenan: a review. *Veterinární Medicína* 58 (4).

39 NithyaBalaSundari, S., Nivedita, V., Chakravarthy, M. et al. (2020). Characterization of microbial polysaccharides and prebiotic enrichment of wheat bread with pullulan. *LWT* 122: 109002.

40 Mohsin, A., Zaman, W.Q., Guo, M. et al. (2020). Xanthan-Curdlan nexus for synthesizing edible food packaging films. *International Journal of Biological Macromolecules* https://doi.org/10.1016/j.ijbiomac.2020.06.008.

41 Douglas, T.E., Łapa, A., Samal, S.K. et al. (2017). Enzymatic, urease-mediated mineralization of gellan gum hydrogel with calcium carbonate, magnesium-enriched calcium carbonate and magnesium carbonate for bone regeneration applications. *Journal of Tissue Engineering and Regenerative Medicine* 11 (12): 3556–3566.

42 Duan, Y., Li, K., Wang, H. et al. (2020). Preparation and evaluation of curcumin grafted hyaluronic acid modified pullulan polymers as a functional wound dressing material. *Carbohydrate Polymers* 238: 116195.

43 Martin-Piñero, M.J., García, M.C., Santos, J. et al. (2020). Characterization of novel nanoemulsions, with improved properties, based on rosemary essential oil and biopolymers. *Journal of the Science of Food and Agriculture* https://doi.org/10.1002/jsfa.10430.

44 Wang, Y., Trani, A., Knaapila, A. et al. (2020). The effect of in situ produced dextran on flavour and texture perception of wholegrain sorghum bread. *Food Hydrocolloids* 106: 105913.

45 Sapper, M., Talens, P., and Chiralt, A. (2019). Improving functional properties of cassava starch-based films by incorporating xanthan, gellan, or pullulan gums. *International Journal of Polymer Science* 2019 https://doi.org/10.1155/2019/5367164.

46 Kuhn, K.R., e Silva, F.G.D., Netto, F.M., and da Cunha, R.L. (2019). Production of whey protein isolate–gellan microbeads for encapsulation and release of flaxseed bioactive compounds. *Journal of Food Engineering* 247: 104–114.

47 Nwodo, U.U., Green, E., and Okoh, A.I. (2012). Bacterial exopolysaccharides: functionality and prospects. *International Journal of Molecular Sciences* 13 (11): 14002–14015.

48 Vuornos, K., Ojansivu, M., Koivisto, J.T. et al. (2019). Bioactive glass ions induce efficient osteogenic differentiation of human adipose stem cells encapsulated in gellan gum and collagen type I hydrogels. *Materials Science and Engineering: C* 99: 905–918.

49 Malik, N.S., Ahmad, M., Minhas, M.U. et al. (2020). Chitosan/Xanthan Gum based hydrogels as potential carrier for an antiviral drug: fabrication, characterization, and safety evaluation. *Frontiers in Chemistry* 8: 50. https://doi.org/10.3389/fchem.2020.00050.

50 Taheri, A. and Jafari, S.M. (2019). Nanostructures of gums for encapsulation of food ingredients. In: *Biopolymer Nanostructures for Food Encapsulation Purposes* (ed. S.M. Jafari), 521–578. Academic Press.

51 Zhai, X., Li, Z., Shi, J. et al. (2019). A colorimetric hydrogen sulfide sensor based on gellan gum-silver nanoparticles bionanocomposite for monitoring of meat spoilage in intelligent packaging. *Food Chemistry* 290: 135–143.

52 Muhammad, D.R.A., Doost, A.S., Gupta, V. et al. (2020). Stability and functionality of xanthan gum–shellac nanoparticles for the encapsulation of cinnamon bark extract. *Food Hydrocolloids* 100: 105377.

53 Jamwal, S., Ram, B., Ranote, S. et al. (2019). New glucose oxidase-immobilized stimuli-responsive dextran nanoparticles for insulin delivery. *International Journal of Biological Macromolecules* 123: 968–978.

4

Antiaging and Skin Lightening Microbial Products

Prabuddha Gupta, Ujwalkumar Trivedi, Mahendrapalsingh Rajput, Tejas Oza, Jasmita Chauhan, and Gaurav Sanghvi

Department of Microbiology, Marwadi University, Rajkot, Gujarat, India

4.1 Introduction

Cosmetics are products aimed to be applied to the human body for beautifying, promoting attractiveness, or as a cleansing agent without affecting the structure and function of the body [1]. Globally, the demand for cosmetic products has increased significantly. Five main segments, skincare, haircare, makeup, fragrances, and toiletries, make up the global cosmetic market. The diversity of these five segments works as a complementation to suffice the global consumer demand. It is estimated that internationally, the revenue from the cosmetic market will rise to $429.88 billion by 2022 with 4.3% of annual growth [2]. Globally, the cosmetics and personal care industry will receive significant profits. The segment of antiaging and skin lightening products has received the highest market value due to consumer awareness regarding environmental exposure and preserving a youthful appearance.

The prevalent use of skincare and antiaging is getting associated with social prestige and environmental justice [3]. The belief and to follow the lighter skin has been rooted since ancient times. In South Asia, during colonial legacy times, it is believed that the white race people ruled over the dark [4]. In the context of skin-related products, the developing nations are proving to be best-emerging profit-making markets as white skin is desirable and have more perspective and potential of social advantages for marital and even career prospects [5, 6]. Due to these factors, multinational companies are forming strong foothold in countries like India, China, Japan, and Russia. According to the CAGR reports, the market of antiaging cream and skin lightening cream is expected to grow at 4.64% annually to its value and 4.31% annually to its volume by the period 2018–2025.

The contribution of the advertising agencies is argued to play a pivotal role in reinforcing and taking the advantages of diversity in developing nations due to caste, race, and age. Antiaging and skin lightening products of multinational companies can be availed from local convenience stores to commercial internet sites. A content analysis of advertisements

Bioprospecting of Microorganism-Based Industrial Molecules, First Edition.
Edited by Sudhir P. Singh and Santosh Kumar Upadhyay.

conducted by Li and group (2008) for skin-related products in India, Korea, Japan, and Hong Kong depicted that "smooth, young, transparent, fair, wrinkle-free" is considered as "good skin." Further, this analysis also indicates that these narratives enhanced finances by using them in advertising the brand of cosmetics.

India, due to its ancient cultural practices in beauty correlated with the strong bond among caste and community biases, seems to outsize the market share in antiaging and skin fairness cream [4]. Due to liberalization in the twentieth century, currently, massive diversity of creams, lotions, scrubs, and of multinational companies for men and women are sold in retail outlets in India. It is estimated that approximately 50% of India's entire cosmetic market is held by the skincare segment. The worth of skincare products in India varies between US $450 and 535 million [7]. Based on the cultural practices and caste system, in India, it is believed that lighter/fairer skin will avail common people higher status in society [7]. Phillips A. [8] has discussed that skin color has a significant impact on the personal and societal life within the Indian social context as it is correlated with moral and behavioral qualities. Moreover, one use of higher skin lightening creams was prominently showcased in the survey study carried out among undergraduate students from 26 countries [9]. According to that survey, approximately 19% of 799 students are reported for the usage of skin lightening products, and the use was associated with depression, low social support, and even risky sexual behaviors. With this context, it is very important to understand the biology of aging and the factors affecting the skin aging so that appropriate products can be purchased having minimum side reactions and health hazards.

4.2 Aging

Aging is a consequence of gradual and cumulative loss in the function of an organism's life processes with time. Such changes are ipso facto to the progressive deterioration of all the tissues of the body [10]. Aging may also be described as an evolutionary attributed whereby the lifespan (in years) of an individual from a species is predecided [11]. Although an individual starts to age right after birth, the real signs of aging are apparent only after a certain period of age, which is broadly known as "biological maturity" [12]. These signs are accompanied by the reduction of organ efficiency and increased mortality. Aging is believed to be ubiquitous to the living organisms. Nevertheless, in certain species, aging does not imply the criterion of reduced fertility and increased mortality [13]. Physiological, anatomical, and biological changes that an organism undergoes during aging are summarized in (Figure 4.1).

4.2.1 Structure of Skin

Skin is the foremost and most vital organs of the human body spanning about 20 sq. ft. in the area and contributing about 15% of body weight. It serves as an anatomical barrier of the immune system protecting the human body from external threats of infections and toxic compounds [14]. Skin is also associated with a fine network of sensory neurons, ensuring balanced sensitivity and reflexes against hot and cold stimuli. It also plays a pivotal role in thermal homeostasis of the human body [15]. Anatomy of skin comprises three layers, namely, epidermis, dermis, and subcutaneous tissue layer. The epidermis is predominantly clustered with keratin producing keratinocytes while basal dermis layer

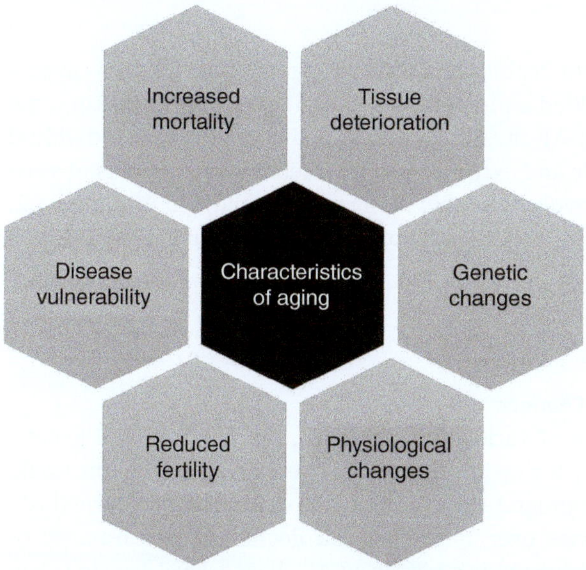

Figure 4.1 Summary of various characteristics of aging.

homes melanin-producing melanocytes [16, 17]. The dermal–epidermal junction comprises the extracellular matrix, which is secreted by epidermal fibroblasts. The extracellular matrix is composed of complex polysaccharides like chondroitin sulfate and hyaluronic acid and structural proteins such as collagen, elastin, and fibrin [15]. Together, these biomolecules provide strength and elasticity to the skin. Aging of the skin is a result of abrogation in the structure or function of any of the above mentioned histological or biochemical components of the skin.

It is evident that skin aging is a consequence of a variety of factors encountered by an individual during his lifespan. A broad classification of various aging theories is summarized in Table 4.1.

Table 4.1 Classification of aging theories.

Classification / theories	Genetic mutation	Wear and tear	Cellular waste accumulation
Antagonistic pleiotropy	■		
Waste accumulation			■
Free radical	■	■	■
Insulin resistance			■
Advanced Glycosylation			■
Telomeres shortening	■		
The error catastrophe	■		
Autoimmunity		■	
Circadian regulation		■	
Evolution	■		

4.2.2 Skin Aging Factors

The cosmetically relevant and noticeable changes in aging skin are (i) hyperpigmentation or age spots, (ii) wrinkled and saggy skin, (iii) dryness and dull skin tone, (iv) thin and injury-prone skin [18]. All the above-mentioned changes in an individual arise due to the interplay of intrinsic and extrinsic factors. Mechanistic insights of various intrinsic and extrinsic factors involved in skin aging are summarized in Figure 4.2. In the present section, we will explore the details of these factors, and we will try to touch their mechanisms.

4.2.3 Intrinsic Skin Aging Factors

4.2.3.1 Anatomical and Histological Changes

Intrinsic factors are the combination of various physiological, biochemical, and genetic changes that an individual undergoes during its lifespan. An overall reduction in the thickness of aging skin is attributed to the progressive decline in the dermal and epidermal cellular ability to divide and proliferate over time. Thus, a decline in the number of keratinocytes and fibroblasts will reduce skin's elastin and collagen level causing it to become thin and saggy [19].

4.2.3.2 Telomere Shortening

The fact that skin cells lose their regenerative ability after several divisions are associated with a phenomenon known as "Hayflick's Limit," and was proposed by Leonard Hayflick and colleagues [11]. According to him, somatic cells have a finite proliferative ability, which can be extrapolated to predict the lifespan of any species. Mechanistic insights into Hayflick's limit theory was described independently by Watson and Olovnikov in the 1970s. According to them, "telomere" being the terminal portion of linear eukaryotic genome gets

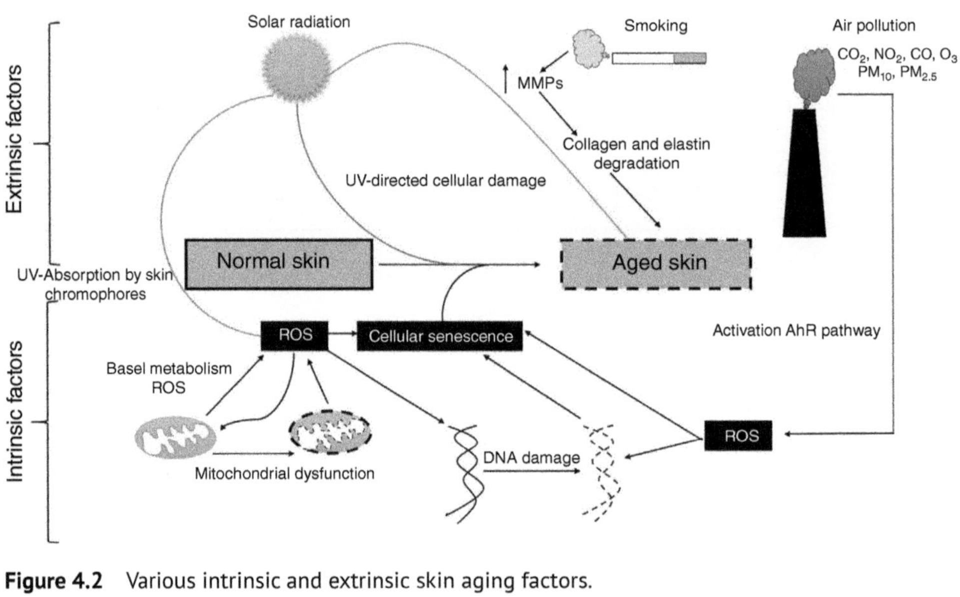

Figure 4.2 Various intrinsic and extrinsic skin aging factors.

shortened progressively upon successive cellular division, and cell ceases to divide further upon reaching a threshold [20, 21].

4.2.3.3 Metabolic ROS Production

Reactive oxygen species (ROS) are oxygen-containing reactive molecules that are normally produced in cells as a metabolic by-product. The major source of metabolic ROS is mitochondria [22]. Young cells have sufficient levels of antioxidant enzymes like superoxide dismutase and peroxidase to counter the damaging effects of oxidative stress generated by ROS [23]. It is, however, known that aged cells lose the ability to express the sufficient levels of antioxidant enzymes, and hence remedial ROS can progressively damage essential cellular components leading to senescence [24]. Such damaging implications are also true for aging skin cells resulting in cell reduction. ROS can also degrade the components (collagen, elastin, etc.) of the epidermal extracellular matrix, causing an overall decrease in strength and elasticity of skin [25].

4.2.3.4 Upregulation of Matrix Metalloproteinases

Matrix metalloproteinases (MMPs) belong to the family of endopeptidases whose function is to maintain the proteinaceous content of the extracellular matrix in the connective tissues. In the skin, MMPs degrade the excessive matrix collagen, thereby maintaining the flexibility of the skin [26]. Aged cells are found to produce suboptimal levels of MMPs, which enhances the collagen degradation in aging skin. Elevated ROS levels are also known to increase MMPs expression, thereby enhancing skin damage [19, 25].

4.2.3.5 Mitochondrial Dysfunction

Aging, in many ways, affects organellar efficiency. Mitochondria, which is regarded as the powerhouse of the cell, also maintains a fine balance of ROS and antioxidants in the cell. A dysfunctional mitochondria arising due to mutations or ROS damage increases the propensity of oxidative stress in the cell, thereby increasing the damage [27]. Mitochondrial dysfunction associated senescence (MiDAS) is one of the major intrinsic factors involved in skin aging [28, 29].

4.2.3.6 Mutations and Oncogenesis

Nuclear and mitochondrial DNA mutate either spontaneously or due to exposure to mutagens. Eukaryotic cells have efficient repair systems to neutralize these mutations. Aging cells (and also young cells in specific genetic variants) lack efficient DNA repair systems, which may lead to activation of oncogenes (like *ras*), causing oncogene-induced senescence [30]. Activation of oncogenes in the skin can increase the levels of proteases known as senescence-associated secretory phenotypes, which can induce early senescence in fibroblasts and keratinocytes [31]. Double-stranded breaks, mutations, and telomere shortening can affect the overall structure of chromatin in aging cells [32]. Abrogation in the chromatin remodeling mechanism can arrest the cells at G1 checkpoint and can lead to early senescence. A nonfunctional or inhibited histone deacetylase severely affects the chromatin remodeling process and causes early senescence in fibroblasts [33].

4.3 Extrinsic Skin Aging Factors

Skin is an anatomical protective barrier of the human body towards the harmful biotic and abiotic agents of the environment. Younger skin can efficiently repair and recover against the damaging effects of these harmful factors. However, overexposure to any of these extrinsic factors has a detrimental effect on skin and triggers aging.

4.3.1 Photoaging

Photoaging refers to an accelerated skin aging process due to overexposure with the UV component of sunlight [34]. For an individual with light skin, prolonged exposure to the sun will have histological, biochemical, and genetic consequences. UV-exposed skin display thickened epidermis with an increased quantity of melanocytes and abnormal elastic fibers [35]. Conversely, the dermis layer reduces in density with a decrease in the number of fibroblasts, keratinocytes, and reduced levels of extracellular matrix collagen [10]. Sustained UV exposure to skin is also known to induce higher expression of MMPs, leading to the degradation of elastin and collagen, causing fibroblast senescence [36]. In the normal state, tissue inhibitor of metalloproteinase (TIMP) keeps a check on the levels of MMPs, thereby maintaining the homeostasis. However, during photoaging, expression of TIMP reduces reinforcing the MMPs-mediated collagen and elastin degradation [15]. Solar UV component is thoroughly studied for causing direct DNA damage and indirect cellular damage by producing ROS. Trans-urocanic acid is a subcutaneous chromophore known to absorb the UV-A radiation. Upon excitation, it generates many ROSs causing oxidative stress and senescence in fibroblasts and associated cells. Dermal collagen also contains significant composition UV-B absorbing amino acids such as cysteine, histidine, tyrosine, tryptophan, and phenylalanine. They collectively contribute to the generation of ROS beyond a certain threshold [37]. UV-induced ROS can also damage nuclear and mitochondrial DNA causing an alteration in gene expression, dysfunctional mitochondria, oncogenesis, and senescence.

4.3.2 Tobacco Smoking

Tobacco smoking is highly addictive due to nicotine dependence and is considered to be one of the leading cause of several types of cancers and premature deaths. The nicotine content of tobacco triggers the neural reward pathway reinforcing its dependence and addiction [38]. Tobacco smoke elicits skin aging in two ways. First, the tobacco smoke gets absorbed on the exposed skin depositing numerous toxic compounds. These compounds act as ROS stressors elevating the oxidative stress in skin cells. Such conditions will enhance skin wrinkles via mechanisms already mentioned in previous sections. Recurrent skin exposure to tobacco smoke causes irreversible damage to the skin's texture and its regenerative ability [39]. Second, the inhaled smoke can enhance the expression of MMPs, disturbing the extracellular matrix components. The toxic compounds present in tobacco smoke also increase the plasma neutrophil elastase activity, causing the degradation of elastin fibers from the extracellular matrix. Additionally, tobacco smoke can significantly downregulate TGF-β1 receptor causing the subsequent reduction in procollagen gene expression [40].

4.3.3 Air Pollution

With the advancement in technology and industries, our environment is flooded with pollutants having a broad range of pathophysiological implications. Any substance released in the environment with harmful consequences can be regarded as a pollutant. Skin is exposed to a variety of chemical and biological pollutants present in the indoor and outdoor air [41]. The major categories of air pollutants consist of particulate matter ($PM_{2.5}$ and PM_{10}), gases (ozone, carbon dioxide, carbon monoxide, sulfur dioxide, and nitrogen dioxide) and volatile organic compounds (VOCs) [42]. These pollutants upon skin absorption can induce direct oxidative damage to skin cells and extracellular matrix elements. Conversely, they may also display indirect toxic effects by stimulating the aryl hydrocarbon receptor (AhR) transduction pathway [43]. AhR is present in skin cells whose natural ligand is a xenobiotic compound dioxin. AhR activation induces the expression of cytochrome P450 enzymes, which are involved in xenobiotic compound degradation. Certain by-products/intermediates of this pathway are toxic to the cell in the sense that they may generate oxidative stress [43]. AhR pathway can be activated by several VOCs and gaseous pollutants with variable efficacies. Recurrent exposer of skin to such air pollutants will accelerate the skin aging process.

Overall, our skin needs to be carefully looked in before applying any cream, lotion, or ointment on the skin. Normally, creams are composed of either synthetic or natural ingredients. To date, due to high demand and consumption under the skincare segment, much of the creams are combinations of synthetic chemicals. However, drawbacks like side reactions, allergy are shifting the interest of the market on organic, herbal, or natural cosmetics. Also, to fulfil current market demand, most of the companies are working on finding a biological source for the production of important cosmetic compounds instead of going to generate new synthetic compounds. This trend will lead to the sustainable production of cosmetics using biological origin to satisfy the consumers' demand [44].

4.4 Why Microbes

The microbial products are of utmost importance because of their versatility, and there are multiple reasons behind it, viz., easy culturing, infinite production, the end product being eco-friendly, and nontoxic [44]. When it comes to cosmetics, any product with applicable effectiveness remains in high demand and usage on one of the sensitive and environmentally exposed human organ skin, which is also of a socio and psychological concern for the human being that makes it even more effective. However, the skin is a multilayer organ consisting of differentiated cells playing different functions associated with defense, skin color, sensation, prevention of water loss, and structural integrity; therefore, it is crucial to have formulations that do not interfere with other structures and mechanisms. Pharmaceutical ingredients that are practiced currently in the market are being utilized beyond its capacity, and hence at multiple levels, it reveals certain side effects because of synthetic and chemical nature, while microbial products have got the advantage over it having nontoxic nature [45]. As we know, culturing microbes is conventional and easy on a large scale nowadays due to automated specifically designed fermenter machines, which keeps optimized conditions for the entire production cycle. Therefore, biological products have got the ease of adaptation to the human body and environment as they possess organic nature and comparatively fewer side effects. Another parameter associated with the benefit of using microbial products is they can be modified according to physicochemical requirements and optimized according to physicochemical needs, for example, usage of several enzymes into textile and detergent industries [45]. However, it can be applicable for enzymatic activities, but similar studies support the modification of microbial products and enhancements of physicochemical properties of active compounds under different conditions. For example, the production of keratinase enzyme by *Bacillus subtilis* and production of rhamnolipids by *Pseudomonas aeruginosa* where keratinase enzyme remains active molecule produced by microbes to utilize keratin from surrounding, and rhamnolipid remains secondary metabolite secreted to enhance surface activity to enhance adherence and proliferation/differentiation of skin cells [46, 47]. Cosmetics require multiple formulations for specific applications like powder, ointment, face wash, skin lightening, cleansing, hair removing, smoothening, moistening, etc., so suitable selection of microbial product and its effectiveness under chemical formulation is a crucial task to project future of the formulated product and its long-lasting effectiveness.

4.4.1 Bacterial Compounds

Bacteria are known to produce diverse and novel compounds with antiaging and antioxidant properties. Bacterial cultures have some definite advantages over fungal cultures, like fast growth, short life cycle, ease of genetic manipulation, and broader range pH and temperature tolerance. These types of bacterial compounds are spread in different domains of biomolecules ranging from peptides, carbohydrate derivatives, organic acids, lipid and derivatives, amino acids and secondary metabolites, etc. The class and range of these commercial important biologically active compounds produced by bacteria include polysaccharides, oligosaccharides, exopolysaccharides (EPS), biosurfactants, proteins, peptides, vitamins, etc. Although the yield not only depends upon the type of product but also upon the kind of bacteria, its metabolic profile, nutritional source, and physiological conditions, advances in genetic engineering, fermentation technology, and advanced purification methodologies have helped in enhancing the bacterial product yields to a greater level.

4.4.2 Polysaccharides and Oligosaccharides

4.4.2.1 Hyaluronic Acid

Hyaluronic acid (HA) is a high-molecular-weight glycosaminoglycan heteropolysaccharide made of repeated disaccharide units of N-acetyl-D-glucosamine (GlcNAc) and D-glucuronic acid (GlcA) linked by Β-(1→3) and Β-(1→4) glycosidic bonds. Many reports have shown numerous commercial applications of HA and its salts in cosmetic products like makeup kits, moisturizers, lotions, sprays, lipsticks, etc. [48]. HA can be derived from several natural sources, yet HA derived from bacterial fermentation and rooster combs are only used in cosmetics. Its major role in cosmetic formulations is for skin-conditioning agents and viscosity-enhancing agents. This is due to its ability to its large water retention capacity that provides hydration to the skin. Antiaging creams are another product using a large amount of HA [49]. The anti-wrinkle effect of HA has also been reported, which was due to effective moisturization and improved skin elasticity [50]. The emergence of HA as a dermal filler in recent years has swiftly replaced collagen in nonsurgical cosmetic procedures [49].

HA is a major component of the dermal skin matrix present in every tissue and body fluid. As an effect of the aging process, there is a gradual decrease in the glycosaminoglycan HA content in the dermal layer resulting in skin dryness glycosaminoglycan (GAG) hyaluronic acid. Protection from ultraviolet radiation, tissue regeneration, and water retention properties of HA has been reported [51, 52].

Cyclodextrins are a class of cyclic oligosaccharides that are made of glucose molecules linked with α-(1→4) linkage to form a close ring structure. Cyclodextrins find large applications in cosmetic products [53]. Bacterial cyclodextrinase enzyme helps to obtain a cyclic form of dextrin. Production of cyclodextrins is reported in *Brevibacillus brevis* strain CD 162 and *Mycobacterium terrae* KNR 9 [54, 55]. Cyclodextrins obtained from bacterial enzymatic transformations find applications in skin-cleaning products due to its potential to increase the solubility of substances that are secreted by the skin, including fats. The addition of cyclodextrin complex with tanning agent dihydroxyacetone helps to vanish the odor of later. In a similar case, glutathione used to reduce melanin pigmentation, skin whitening, and improved skin texture. But the odor of glutathione prevents its extensive use in cosmetic products. But if the glutathione is mixed with cyclodextrins, it prevents the bad odor of glutathione yet holding the same effect on the skin [56]. Cyclodextrin is used to decrease the rate of volatilization of esters, sustained release, and longevity of aroma in perfumes, deodorants, and room freshener products and detergents. Fine cyclodextrin powders are also used in talcum, diapers, sanitary pads, diaper pads, napkins, etc., to control the odor.

4.4.2.2 Bacterial Cellulose

Bacterial cellulose (BC) is a natural homopolysaccharide made of repeating with B-(1→4) linkages [57]. Bacterial cellulose shares all the properties shown by plant cellulose. The plant cellulose complexes with other natural polymers like hemicellulose and lignin, but BC is produced and obtained in relatively pure form. BC can be from several groups of bacteria like *Gluconacetobacter xylinus, Gluconacetobacter hansenii, Glucoacetobacter pasteurianus* [58, 59]. BC is used to make biomembranes that can be used in medicine and cosmetics and to its high degree of hydration has enhanced its uses in moisturizing creams [60]. BC shows skin regeneration activity, and its biopolymer has been projected for the development of skin moisturizers and skin repair [61]. In a study carried out on 30 volunteers, BC masks have been checked for the treatment of dry skin and improving skin hydration, and its in vivo efficacy was checked after single applications [61].

4.4.2.3 Astaxanthin and Equol

Astaxanthin is a terpene of carotenoids family and is obtained from many algal and bacterial species. Astaxanthin is known to have antioxidant properties. A recent study focused on the comparison of Astaxanthin and Equol was carried out by gene expression analysis to analyze the properties like antioxidants, extracellular matrix integrity, growth factors, and inflammatory biomarkers. These findings indicate that Equol's efficiency is greater than that of Astaxanthin [62]. Equol is also added in topical and oral applications to increase skin health and decrease photoaging.

Equol is a comparatively novel phytochemical used as a constituent for human skincare. Equol is an isoflavan, which has a polyphenolic structure produced in plant and food sources [63]. Equol [7-hydroxy-3-(4′-hydroxyphenyl)-chroman] belongs to the group of compounds known as nonsteroidal phytoestrogens. The heterocyclic structure contains two reactive hydroxyls and one relatively inert and unreactive oxygen in the central furan ring. It has selective estrogen receptor modulator (SERM) property that gives an improved/constant delivery into the dermal layers of the skin [63, 64], thus inhibiting dermal aging and improves facial appeal [63, 65]. Equol has been studied extensively for more than two decades [66]. But it has gained interest in its new application in skincare, where several reports using topical applications of Equol have demonstrated improved human dermal parameters, such as in vitro, in ex vivo, and in vivo studies and applications in cosmetics [63–66]. In effect, Equol is a unique cosmetic ingredient for which in vitro gene expression biomarkers have been shown to relate to protein expression [65].

Equol is produced in the human intestine metabolized from daidzein by bacterial flora in the intestines. Daidzein is a type of isoflavone present in soybeans and other plant sources. In recent years, many bacterial isolates have been found from human fecal samples, which can produce Equol. These bacteria can prove to be a unique and novel source of Equol production by fermentation for commercial cosmetic products.

4.4.3 Fungi Compounds

4.4.3.1 Tyrosinase Inhibition

Melanin is the major pigment accountable for skin tone and color. Due to continuous exposure, there is a sequential change in the function and structure of the skin. Due to overexposure to UV radiation, the melanin in the skin is increased due to the augmented level of tyrosinase activity. Tyrosinase is a rate-limiting enzyme converting tyrosine to dihydroxyphenylalanine (DOPA) [67]. Further dopaquinone is converted to dopachrome with final product eumelanin leading to skin hyperpigmentation. Additionally, factors like hormonal changes signal transduction pathways also play a major role in skin pigmentation. Studies reporting the antityrosinase inhibitory activity are described using the *Pleurotus* species. Fruiting bodies of *P. citrinopileatus* species were reported for its tyrosinase activity and melanin production inhibition. Kojic acid, a low-molecular-weight organic acid, is widely known for its application in cosmetic and nutraceutical industries. Kojic acid is reported to be found in many fungal species with different types of substrates [68]. Due to its tyrosinase inhibition activity, kojic acid is used as an antioxidant in the cosmetic industry substituent to hydroquinone as the skin whitening agent.

4.4.3.2 Hyaluronidase Inhibition

Hyaluronic acid (HA) is a natural glucose-based polymer, which holds an important role in skin revitalization and moisture-holding capacity [69]. In the aging process, the level of hyaluronic acid decreases with age, leading to moisture loss in the skin. Tropical lotions and creams provide alternatives, but such lotions are facing challenges because of their inflammatory response [69]. Dudek-Makuch and Studzińska-Sroka [70] reported the anti-hyaluronidase activity from *Aesculus hippocastanum* species, and their reports suggest that inhibition of HA will work as a protector to skin and compounds like saponins.

4.4.3.3 Collagenase and Elastase Inhibition

The epidermis of the skin is attached firmly by connective tissue beneath dermis. The fibroblast (dermis) produces two proteins: collagen and elastin, which serves as a protective role in the skin [71]. Elastin is an important protein providing elasticity to the connective tissues, viz., cartilage, lungs, and skin. In aging phenomena, degradation of elastin is occurring due to the elevated level of elastase enzyme. Collagen is an important part of skin responsible for maintaining elasticity and flexibility of the skin. During the aging process, a significant decrease in the level of collagen, elastin, and HA results in loss of flexibility, moisture, and strength of skin, resulting in wrinkle formation. [72]. Additionally, MMPs are endopeptidases that degrade the extracellular matrix associated with pathological and physiological conditions like inflammatory response and also carcinogenesis [73]. Kim et al. [71] has reported the mycelial extracts of *Tricholoma* and *Grifola* sp. possess the anti-collagenase and anti-elastase activity. Further two-hybrid compounds, ascorbyl-3-pcoumarate and the ascorbyl-2-p-coumarate mixture of ascorbic acid and p-coumaric acid serve as necessary cosmeceutical agents for increasing the collagen synthesis from dermal fibroblast and also decreasing the MMP protein expression [74].

4.4.4 Algae Compounds

Due to increasing awareness of natural compounds, attention has been shifted toward the use of novel bioactive compounds from microbial sources. Among these microbial communities, the importance of algae is a prominent one due to the presence of numerous biologically active compounds [75]. Algae are eukaryotic organisms with a primary photosynthetic pigment. Most of them have an autotrophic mode of nutrition and are related to various polyphyletic groups [76]. Algae can be broadly classified based on the chlorophyll pigments and type of reserve food materials. According to the pigments, a few major types of class are Chlorophyceae (green algae), Phaeophyceae (brown algae), Xanthophyceae (yellow green), Rhodophyceae (red algae), Chrysophyceae (golden algae), Bacillariophyceae (diatoms), and Dinoflagellates [77]. Based upon the cell size and morphology, they can be further grouped as microalgae and macroalgae (seaweeds). Furthermore, depending upon salinity, algae can occur in freshwater, marine, or brackish water habitats [78].

Algae are known to produce abundant bioactive molecules associated with antiaging and skin lightening products, which have potential applications in the cosmetic industry [79]. Algal products widely used in the cosmetic industry include bioactive compounds such as antioxidants, sunscreens lotions, UV protectants, anti-photoaging compounds, skin smoothing, thickening, gelling, moisturizing, skin lightening, and de-pigmenting agents to enrich the competence of skin against abrasions, tanning, sunburn, aging, and increase the overall radiance of skin [80–82]. These products comprise complex and simple compounds such as carbohydrates, proteins (enzymes, peptides, and amino acids), lipids (PUFA, MUFA, omega-3-fatty acids, sterols), pigments, and vitamins, antioxidants, and secondary metabolites [83]. These wide arrays of compounds can be extracted, solubilized, isolated, and purified by several different methods based on their physicochemical properties, molecular size, and solubility for utilization in cosmetic products. To use these potential bioactive molecules in the cosmetic application, the downstream recovery, purification of these ingredients of the whole process must be streamlined as per GRAS (Generally Recognized as Safe) practice, under the sections 201(s) and 409 of the Federal Food, Drug, and Cosmetic Act.

4.4.4.1 Carbohydrates from Algae

Algae are very well known to produce a wide array of polysaccharides. These polysaccharides play a crucial role in imparting the structure of cell walls and act as energy storage units during unfavorable stressful conditions. Some of the major types of biologically active polysaccharides in algal tissues are fucoidan, lamnarian, alginates, agar, carrageenan, galactan, porphyran, glucan, and ulvan, which demonstrate typical structural and functional characteristics [84]. These polysaccharides are nontoxic, cost-effective, easily extracted, and abundant in the algal biomass, which can be used as a substitute against synthetically derived gelling agents, or silicone-derived ingredients. These compounds provide moisturizing and have antioxidant properties; hence they are used as gelling agents and stabilizers in emulsions [85, 86] (Table 4.2).

Table 4.2 Algal biomolecules, sources, origins, and applications in cosmetics.

Group of compound	Active compound	Algal species	Cosmetic properties	References
Polysaccharides	Agar	*Gelidium sp.* *Gracilaria sp.*	Emulsion stabilizers, gelling agents, thickeners	[87–89]
	Alginates	*Laminaria sp.* *Lessonia sp.* *Macrocystis sp.*	Emulsion stabilizers, chelating agents, gelling, moisturizing, photo-protective	[84, 90–92]
	Fucoidan	*Saccharina sculpera* *Fucus vesiculosus,* *Turbinariaconoides,* *Undaria pinnatifida*	Antiaging, anti-wrinkle, antioxidative properties, UV protectant, skin lightening, spot reduction, elastase, tyrosinase inhibitors, skin fibroblast stimulation	[93–97]
	Laminarans	*Laminaria japonica,* *Ecklonia kurome,* *Eiseniabicyclis,* *Saccharina,* *Ascophyllum, Fucus,* *Sargassum,* and *Undaria*	Antioxidant, anti-cellulite agent	[81, 98, 99]

Table 4.2 (Continued)

Group of compound	Active compound	Algal species	Cosmetic properties	References
	Porphyran	*Porphyra sp.*	Antioxidant and antiaging	[78, 88, 93, 100]
	Carrageenan	*Acanthophora muscoides* *Chondrus crispus*	Gelling agents, protective colloids, thickeners	[101–103]
	Ulvan	*Ulva armoricana, Ulva lactuca, Ulva rotundata,* and *Ulva rigida*	Antioxidative, chelating, gelling, moisturizing, and protective agents	[88, 104, 105]
Pigments	Astaxanthin	*Haematococcus pluvialis*	Antiaging	[106]
	β-carotene	*Chlorella, Spirulina, Dunaliella salina*	Antiaging, antioxidant, radical scavengers, anti-photoaging agents, and colorants	[88, 107–111]
	Chlorophylls (a, b, c, d)	*Chlorella, Nannochloropsis*	Coloring agents, UV-protectant, reduce inflammation, Reducing wrinkles, inhibiting matrix metalloproteinases (MMPs)	[88, 92, 112–114]
	Fucoxanthin	*Fucus, Dictyota, Saragassum,* and *Laminaria*	Anti-tyrosinase activity in guinea pig UVB irradiated, and melanogenesis in mice UVB irradiated	[88, 115–117]
	Phycobiliproteins (APC, PE, PC)	*Cyanobacteria*	Antioxidant, anti-inflammatory, colorants, radical scavenging agents	[88, 97, 118–121]
Secondary metabolites	Phlorotannins	*Ecklonia cava, Eiseniabicyclis*	Tyrosinase and hyaluronidase inhibitors, antioxidants, antiaging, UV-photoprotectant, chelating agents	[79, 88, 122]
Mycosporine-like amino acids (MAAs)	Scytonemin, Asterina-330, Shinorine, Palythine Porphyra-334	*Scytonema, Lyngbya, Porphyra umbilicalis*	Photo-protectant, sunscreen against UV-A	[79, 111, 122–124]

Source: Wang et al. [85] and Wang et al. [86].

4.4.4.2 Fucoidan

Fucoidans are heteropolysaccharides containing L-fucose sugar and sulfate ester groups [10]. They are mostly derived from the brown seaweed. In some seaweeds, along with fucose sugar, other sugars such as glucose, galactose, mannose, xylose, and glucuronic acid (GlcA) have also been reported [125]. The application of fucoidans in the cosmetic industry has been widely investigated. Numerous study has been carried on the inhibitory aspects of topical application of fucoidans on the skin against aging and photo-damaged effects [93, 94, 115]. Fucoidans are known to enhance multiplication skin fibroblast and augment the retention of native collagen [95, 126]. Fucoidans are known to protect the elastic fiber network in human dermal cells, and hence they are used as a natural whitening agent in lotion and skin creams to reduce age spot and increased radiance, increase smooth texture skin [96, 127]. In human skin, melanocytes cells are known to impart the skin complexion. The pigment melanin is produced and stored inside the melanocytes and is primarily obtained from amino acid L-tyrosine, which is converted to dopaquinone by the action of the enzyme tyrosinase. Hence, suppression of tyrosinase-related proteins plays an important role in anti-melanogenesis and skin whitening [128]. In this regard, several studies have demonstrated the inhibition of enzyme tyrosinase, antioxidant activity, and anti-melanogenesis of fucoidan fraction from various macroalgae [96, 128, 129]. Furthermore, the property of fucoidans to readily solubilize in water makes it attractive to combine into several ranges of cosmetic formulations such as lotions, creams, ointments, and other beauty products.

4.4.4.3 Laminaran

Laminaran is also widely known as laminarin or leucosin. Laminarin is a glucan polysaccharide made up of $\beta(1\rightarrow3)$-glucan with $\beta(1\rightarrow6)$-branches with a $\beta(1\rightarrow3)$:$\beta(1\rightarrow6)$ in the ratio of 3:1. They are obtained from brown seaweeds *Laminaria japonica, E. kurome, Eiseniabicyclis, Saccharina, Ascophyllum, Fucus, Sargassum,* and *Undaria* species. Several reports have confirmed to have anticancerous, anti-inflammatory, anticoagulant, and antioxidant properties in laminarin extracts of seaweeds [87]. Laminarin is widely used as filler composition in cosmetic products for the treatment of wrinkles, folds, and as anti-cellulite agents [98, 99].

4.4.4.4 Ulvans

Ulvans are cell wall polysaccharide derived from the genus *Ulva* (*Ulvaarmoricana, Ulvalactuca, Ulvarotundata,* and *Ulvarigida*) of green macroalgae belonging to the Chlorophyta group. Ulvan roughly contributes from 9 to 36% dry weight of the biomass of *Ulva* and primarily consist of sulfated rhamnose, uronic acids (glucuronic acid and iduronic acid), and xylose units [88, 104, 105]. The ulvans are widely known for the antioxidative, chelating, gelling, moisturizing, and protective properties. The antioxidant capacity of ulvan is modulated by ROS scavenging and inhibition of lipid peroxidation. Besides, ulvans are also known to suppress antiaging by enhancing the endogenous enzymatic antioxidant systems such as superoxide dismutase, catalase, and glutathione peroxidase activities owing to the abundant sulfated polysaccharides [88, 90, 104]. Moreover, the gelling property of ulvan is not significantly affected in the presence of various cations (Ca^{2+}, Cu^{2+}, and Zn^{2+}), in the pH range of 7.5–8.0 and temperature up to 180 °C makes it a rheological suitable in cosmetic formulations [88, 91].

4.4.4.5 Porphyran

Porphyrans are exclusively derived from *Porphyra* species belonging to genus *Rhodophyta*. Porphyran is a sulfated polysaccharide consisting of D-galactose, 3,6-anhydro-L-galactose, 6-*O*-methyl-D-galactose, and L-galactose 6-sulfate units [151]. Porphyran demonstrates high solubility in an aqueous solvent and safe for the skin when it is formulated in cosmetic products as a solvent in liquid or emulsion form. The bio-functional activity of porphyrans includes skin whitening, anti-inflammatory, and moisturizing properties and hence used in topical or systemic skin cleansers, cosmetic wipes, cosmetic cotton, sanitary cotton, wet tissue, skin lotion, skin cream, ointments, gel, face pack, shaving agents, facial cleansing agents, etc. [78, 88, 93, 100].

4.4.4.6 Carrageenan

Carrageenan is naturally derived from the *Rhodophyta* genus (carrageenophytes). Carrageenan is a sulfated polysaccharide with galactose units with alternating α (1–3) and β (1–4) bonds. There are three main classes of carrageenan, kappa (κ), iota (I), and lambda (λ) owing to differences in the degree of sulfation. K-carrageenan has one sulfate moiety per disaccharide, I-carrageenan has two sulfate moieties, while λ -carrageenan has three. Furthermore, I-carrageenan and K-carrageenan are used as gelling agents, and λ -carrageenan are used as thickening and viscosifier agent in cosmetic application due to its unique rheological properties [130, 131]. Carrageenan is known to decrease the effect of photoaging, sunburn, wrinkles, and skin cancer induced by harmful ultraviolet-B rays (UVB) (290–320 nm) [132]. Several studies have demonstrated significant photo-protection properties of carrageenan against detrimental effects of UVB-induced cell damage and ROS, suggesting its crucial role as a photo-protective cosmetic excipient in skin lightening products [101–103].

4.4.4.7 Agar

Agar is commonly known as agar or agarose, and is fundamentally a combination of two polysaccharides, agaropectin, and agarose obtained from several different genus of *Rhodophyta*. Agarose, the chief component of agar, consists of straight-chain polymeric units of agarobiose. Agarobiose is made up of disaccharide units of D-galactose and 3,6-anhydro-L-galactopyranose. While the agaropectin consists of β-1,3-linked D-galactose units altered with sulfate and pyruvate moieties [78, 88]. As mentioned earlier, the gelling property of these polysaccharides exclusively depends on the degree of sulfation and the number of 3, 6-anhydrogalactose molecules present.

The distinctive property of agar to easily melt at 85 °C and solidify at 32–40 °C makes it a prominent candidate in various industrial, laboratory, cosmetic, and culinary applications. Furthermore, agar is a biocompatible and inert substance and hence can be readily formulated with various other biochemical compounds used in cosmetic products such as hand lotions, deodorants, antiaging treatment creams, facial and acne treatment, and so on [78, 82, 88, 92, 122].

Other agarose-derived polysaccharides (agar oligosaccharides) also have antiaging, anti-melanogenesis, skin brightening, and ROS scavenging properties [92, 133, 134]. There are two forms of agar oligosaccharides, namely, neo-form and agaro-form. Neo-form AOSs are called neoagarooligosaccharides (NAOSs) and have repeating neoagarobiose units composed of d-galactose at the nonreducing end and 3,6-anhydro-L-galactose [92].

4.4.4.8 Alginic Acids

Alginic acids are obtained from brown seaweed, i.e. *Pheaeophyta* genus. Alginic acid is made up of linear polymers of α-l-guluronic acid and β-d-mannuronic acid units connected by 1 $\rightarrow$ 4 glycosidic bonds. Alginates can rapidly form hydrogels in the existence of divalent cations (Mg^{2+}, Ca^{2+}etc) at room temperature. Moreover, Na^{+}-alginate and Ca^{2+}-alginate are used in microencapsulation and immobilization of various compounds, enzymes, cells, etc. owing due to its gelling, biocompatibility, and biodegradability properties [135]. The chelating and gelling properties of alginates have widely attracted its use in cosmetic products as thickeners, emulsion stabilizers, hand jellies, ointment bases, skin lightening face packs, wound dressing for skin recovery, etc. The alginate masks are known to restore type I and type III collagen synthesis in the skin, and also helps in moisturizing by enhancing the water holding capacity of the skin. Furthermore, alginate is known to provide vital minerals and microelements to deeper dermal layers in the skin [136]. Several commercially available formulation is exploiting the benefit of alginate in the treatment of ulcers, wounds, and skin burns anti-cellulite programs, face sculpturing, and skin regeneration therapy [136–138] (Figure 4.3).

4.4.5 Pigments from Algae

Algae contain three major classes of photosynthetic pigments categorized as chlorophylls, carotenoids (carotenes and xanthophylls), and phycobilins [139]. The chlorophylls and carotenes are fat-soluble molecules and can be readily extracted by using organic solvents (acetone, methanol, or DMSO). However, the phycobilins being water-soluble can be extracted by using polar and nonpolar solvent mixtures. Most of these pigments can counteract against the UV light exposure during growth and development of algae and are known to involve in protecting the cell against harmful oxidative ROS species [140, 141]. Hence, these pigments are used in cosmetic formulations not only to reduce the oxidation of oils, occurring due to rancidity, but also to reduce skin aging, owing due to antioxidant nature. Moreover, these pigments are a natural source of dyes and colorants, making it useful in the various cosmetic industry as the replacement against synthetic dyes (Table 4.2). The various pigments and their cosmeceutical application with respect to antiaging and skin lightening are discussed as follows.

4.4.5.1 Phycobiliproteins

Phycobiliproteins obtained from algae and cyanobacteria are categorized into three main classes depending on absorption properties: phycoerythrins (PE, $\lambda = 540$–570 nm), phycocyanins (PC, $\lambda = 610$–620 nm), and allophycocyanins (APC, $\lambda = 650$–655 nm) [118]. PEs are of utmost interest owing to their fluorescent and antioxidant properties. These pigments are red and are known to absorb light in the green wavelength ($\lambda = 498$ and $\lambda = 565$ nm) and emit in the yellow wavelength ($\lambda = 573$ nm). In vitro, the antioxidant activity of PE is a well-studied phenomenon in *Caenorhabditis elegans*, which resembles the human aging pathways, wherein PE is known to upregulate enzymes such as superoxide dismutase, catalase activities, and reduce oxidative stress damage [142]. The PC and APC are also known to express antioxidant activities and radical-scavenging properties in several studies [97, 119,

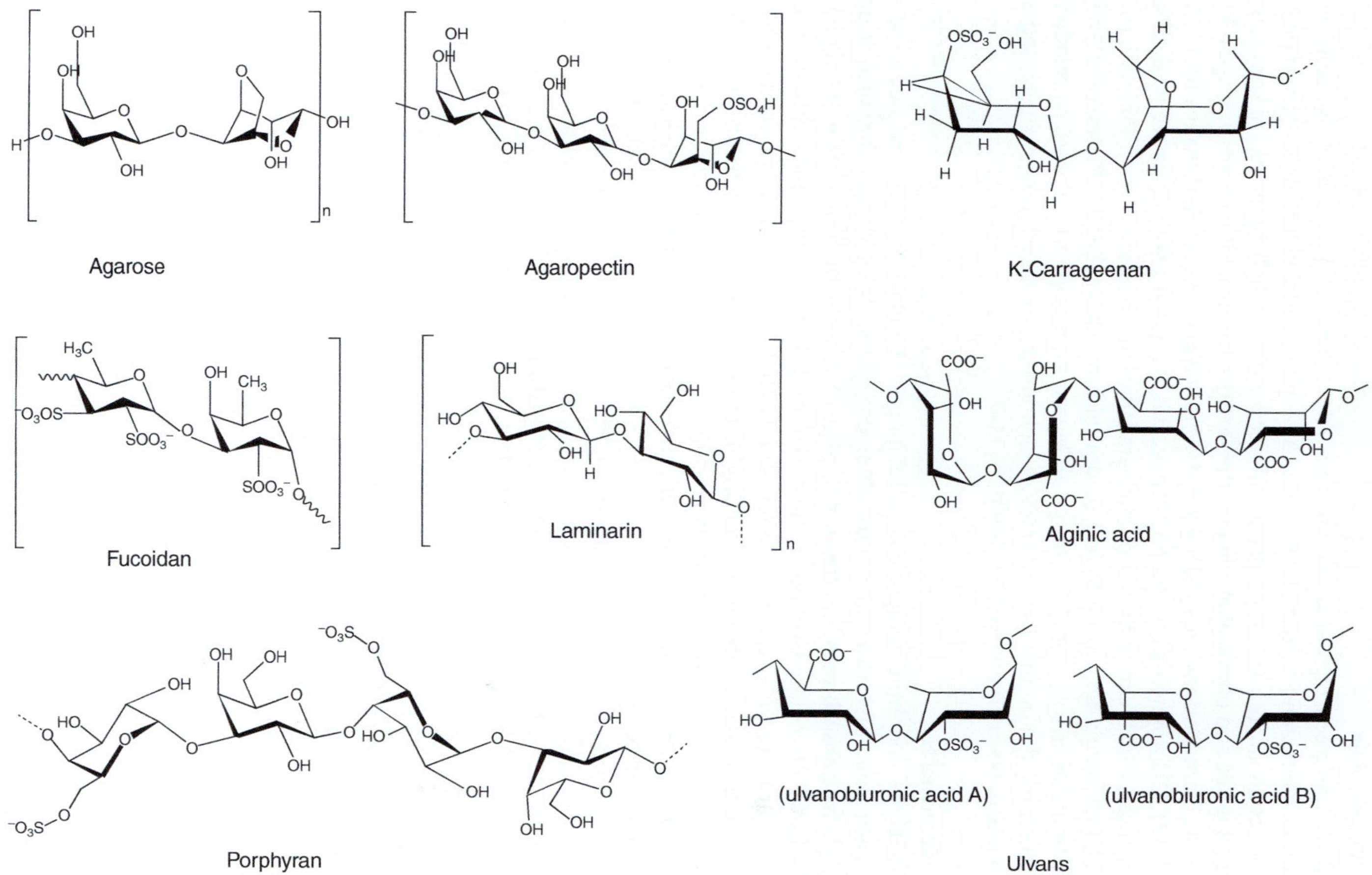

Figure 4.3 Important carbohydrates from algae used in cosmetic applications. *Source:* Aderibigbe et al. [136], Dantas et al. [137] and Huang anf Fu [138].

120]. PC is involved in inducing expression of heme-oxygenase-1 (HO-1), thereby reducing the UV-B-induced cell death in skin cells, demonstrating the photo-protective role [121].

4.4.5.2 Chlorophylls

Chlorophylls are green pigments and are fat-soluble. They are predominantly found in algae, cyanobacteria, and higher plants. Both chlorophyll-a and chlorophyll-b are known for their exceptional antioxidant properties. Moreover, chlorophyll pigments are known to absorb light in red and blue regions; hence, they are used as an excellent coloring agent in cosmeceutical formulations and food industry (identified as E140 and E141 in the European Union) [112]. Chlorophyll-a and its breakdown product (pheophorbide-a) are used as a photosensitizer, which can carry out photodegradation of harmful UV rays. Both chlorophyll and pheophorbide-a are effective in diminishing skin-related inflammation and reducing wrinkles arising from UV rays [113]. Moreover, pheophorbide-a has a profound role in inhibiting the production of MMPs, which are induced by UV rays, suggesting a potential role in the cosmetic formulation [113, 114].

4.4.5.3 Carotenoids

Carotenoids are lipophilic compounds, belonging to the class of tetraterpenoids, containing 8 isoprene units with at least 40 carbon atoms [106, 143]. The peculiar feature of carotenoids is to absorb wavelengths in the range of 400–550 nm (violet to green light), resulting in deeply colored yellow, orange, red, or colorless compounds subjected to presence or absence of conjugated double bonds [107]. Carotenoids can be classified into two groups based upon the availability of oxygen. Xanthophyll group shows the presence of oxygen (e.g. astaxanthin, fucoxanthin, zeaxanthin, violaxanthin, loroxanthin, and lutein), while carotenes groups are devoid of oxygen and contain pure hydrocarbons (α, β, γ-carotene, and lycopene) [106]. Carotenoids are present in the stratum corneum of human skin, which imparts the normal skin color and UV protection [108]. The worldwide carotenoids production is expected to reach $1.53 billion by 2021, primarily dominated by β-carotene, lycopene, astaxanthin, zeaxanthin, and lutein owing to their well-established applications in cosmetics, food, confectionary, and aquaculture (Figure 4.4).

4.4.5.4 β-carotene

β-carotene are known to be precursors of vitamin-A and are known to impart orange color. β-carotene are widely explored for providing good vision, healthy skin texture, and functioning of the mucous membrane [79]. The antioxidant potential of β-carotene is widely demonstrated by numerous studies, and hence they are used as radical scavenging as an antioxidant in cosmetic creams, formulations, sunscreen lotions to enhance the appearance of dry or damaged skin, reducing skin flake, restoring moisture, and suppleness of the skin. β-carotene is very popularly used in sunscreen lotions as a photo-protective ingredient because of its ability to protect the skin against early aging damaged by UV rays [109]. Furthermore, β-carotene is a pro-oxidant involved in the peroxidation of lipids [110]. Numerous microalgae such as *Chlorella*, *Spirulina*, and *Dunaliella salina* are cultivated for the production of β-carotene.

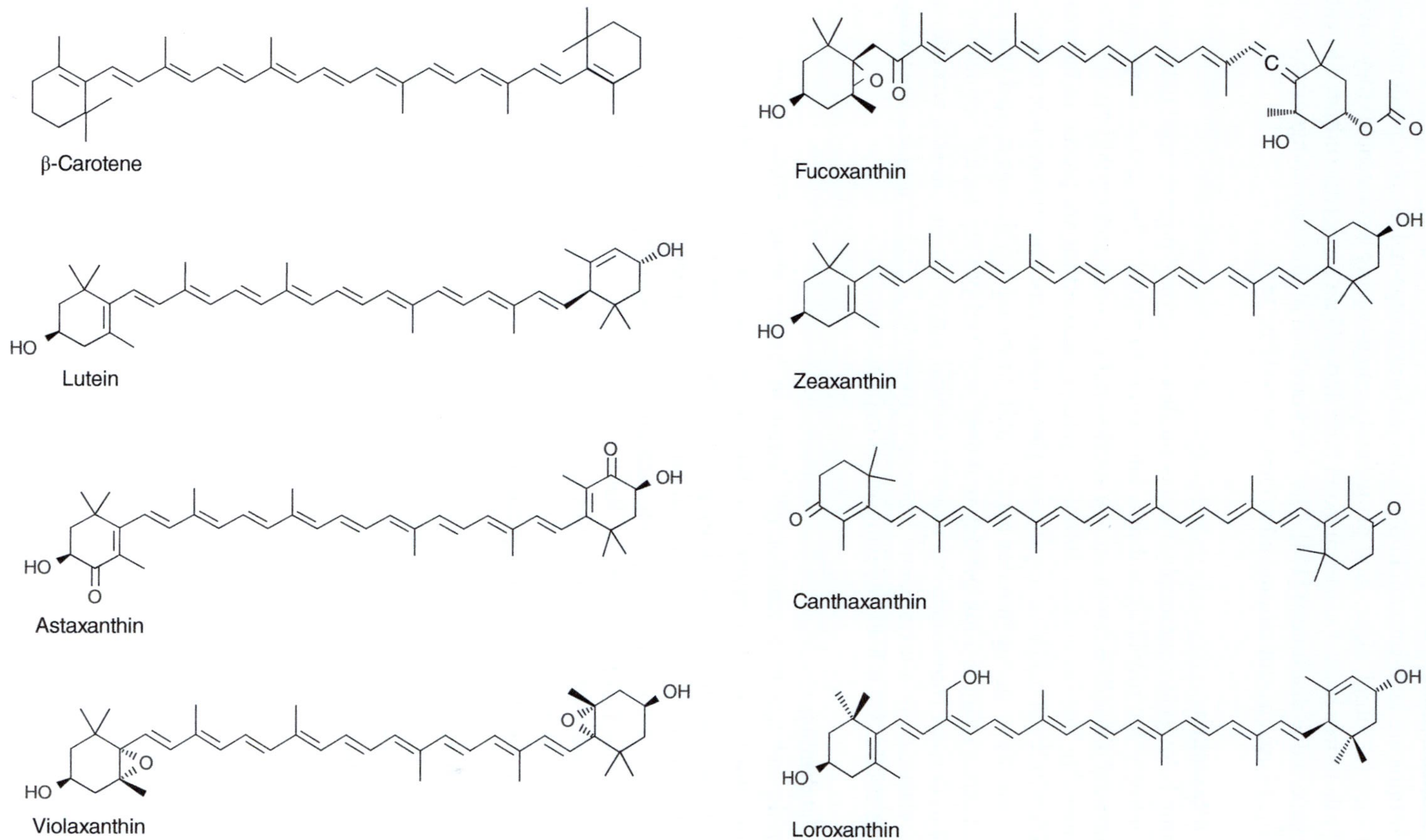

Figure 4.4 Important carotenoids pigments from algae used in cosmetic applications.

4.4.5.5 Canthaxanthins

Canthaxanthins are keto-carotenoid pigments and are orange-red in color, predominantly found in bacteria, fungi, algae, land plants, and animals. Being a carotenoid, they have a potential role in free radical scavenging and have antioxidant properties (initiated by catalase and superoxide dismutase) [144]. They are known to produce an orange-brown color in the skin are used as sun tanning pills (Orobronze) [108, 145].

4.4.5.6 Astaxanthin

Astaxanthin belongs to a xanthophyll and is another recognized pigment used in the aquaculture industry for providing red and pink color to aquatic fish and shrimps [106]. *Haematococcus pluvialis* is a well-known species that produces astaxanthin up to 3–7% of dry cell mass. In human dermal fibroblasts (HDF), astaxanthin are proven to be potent against UVA-induced oxidative impairment [146]. Astaxanthin neutralizes free radicals or other oxidants by involving in redox reaction and slowing down the age-related skin process. In another study, a 6-mg per day oral dose and 2 ml per day topical application of astaxanthin-derived *Haematococcus pluvialis* showed improvements in skin wrinkle, age spot size, elasticity, skin texture, and increase in the moisture content of corneocyte layer suggesting its effective role in antiaging and skin protection [147].

4.4.5.7 Fucoxanthin

Fucoxanthin belongs to a xanthophyll found in brown algae (*Fucus, Dictyota*, and *Laminaria*) and diatoms. Fucoxanthin is known for its high commercial value in the global market for its various biological properties such as anti-obesity, antioxidant, anti-inflammatory, anticancer, and skin photo-protection [115]. Fucoxanthin suppress ROS formation, reduce cell DNA damage and are known to delay the apoptosis induced by hydrogen peroxide [116]. Fucoxanthin has been reported to reduce the enzyme tyrosinase in UVB-irradiated dermal cells of guinea pig and inhibit melanogenesis in UVB-irradiated mice. Oral treatment of fucoxanthin has been found to downregulate the skin m-RNA expression associated with melanin production [117].

4.4.5.8 Zeaxanthin

The xanthophyll cycle is involved in the removal and addition of epoxy group from xanthophyll. The zeaxanthin is the product of a de-epoxidized xanthophylls and has known to be involved in a photoprotective role and as an antioxidant. Although zeaxanthin was primarily known to be isolated from *Zea maize*, they are also found to be present in algae as well. Zeaxanthin is known to inhibit tyrosinase enzyme averting hyperskin pigmentation, thereby acting as a skin whitening agent. The pure extract of *Nannochloropsis oculata*, which contains zeaxanthin, as an antityrosinase substance [148]. Lutein- and zeaxanthin-containing creams (Lutemax® 2020) are known to promote skin lightening and improve skin health conditions [149].

4.4.5.9 Violaxanthin

Violaxanthin is an orange-colored, natural xanthophyll pigment occurring in a variety of algae and land plants. Violaxanthin pigments are biosynthesized from its precursor zeaxanthin by undergoing epoxidation carried out by de-epoxidase enzyme. It is approved as a

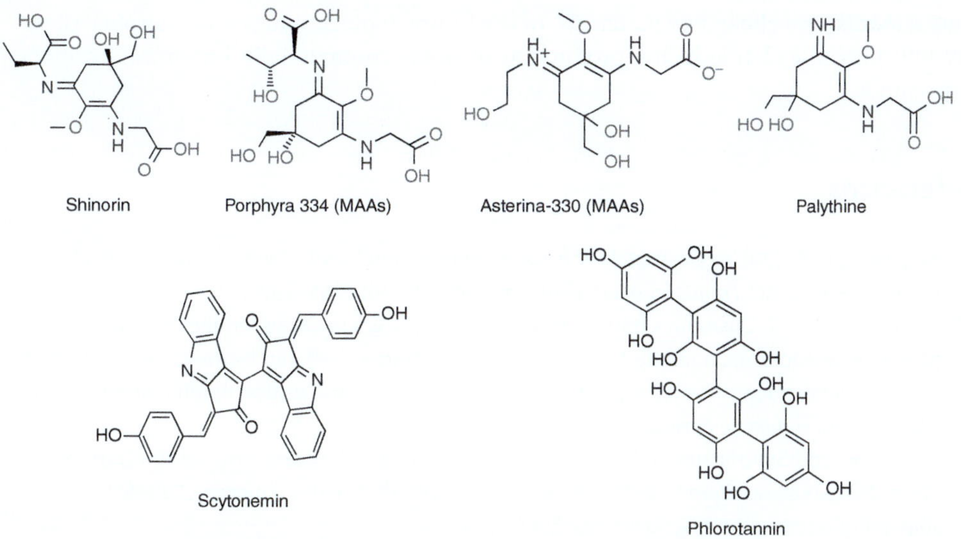

Figure 4.5 Important secondary metabolites and mycosporine-like amino acids from algae used in cosmetic applications.

food additive coloring agent and only in countries like Australia and New Zealand (INS number 161e). Violaxanthin extract from microalgae *Nannochloropsis oceanica* has shown to reduce skin photoaging, induced by ultraviolet B (UVB) radiation in HDF [150].

4.4.6 Secondary Metabolites

Various secondary metabolites are known to obtain from algae and are formulated in cosmetic products. Phlorotanins derived from algae *Eisenia bicyclis*, *Ecklonia* cava are proved to benefit the skin by reducing the elastase activity significantly. Mycosporine-like amino acids (MAAs) such as scytonemin, asterina-330, shinorine, and palythine are well-known UV-protective compounds, found in most of the algal and cyanobacterial species and widely used as photoaging protectant [79, 122–124, 141] (Figure 4.5).

4.5 Conclusions and Perspectives

Skin aging is linked with the many complex factors like geographical, environmental, physical, and molecular factors. Skin aging is a natural process connected with the aging of any life forms. However, factors like environmental, molecular, and geographical have a significant impact on the process of skin aging. Most of the scientific reports suggest the role of compounds like antioxidants, vitamins, and enzymes in keeping the skin to rejuvenate for a longer period. The increasing awareness about the toxic effects of chemical compounds has augmented a sharp rise in demand for biological and natural compounds. With the current demand, most cosmetic companies are focusing on biological compounds due to their easy extraction, higher yields, low space and maintenance requirements, and cheap raw materials. In the future, biotechnology and microbial diversity will play a lead role in the production of novel commercially important cosmetic compounds.

References

1 de Castro, M.L. (2011). Cosmetobolomics as an incipient '-omics' with high analytical involvement. *TrAC Trends in Analytical Chemistry* 30 (9): 1365–1371.

2 Rajput, N. (2016). Cosmetics market by category (skin & sun care products, hair care products, deodorants, makeup & colour cosmetics, fragrances) and by distribution channel (general departmental store, supermarkets, drug stores, brand outlets)-global opportunity analysis and industry. *Forecast*: 2014–2022.

3 Zota, A.R. and Shamasunder, B. (2017). The environmental injustice of beauty: framing chemical exposures from beauty products as a health disparities concern. *American Journal of Ostetrics and Gynaecology* 217 (4): 418–421.

4 Shankar, P.R. and Subish, P. (2016). Fair skin in South Asia: an obsession? *Journal of Pakistan Association of Dermatology* 17 (2): 100–104.

5 Dadzie, O.E. and Petit, A. (2009). Skin bleaching: highlighting the misuse of cutaneous depigmenting agents. *Journal of the European Academy of Dermatology and Venereology* 23 (7): 741–750.

6 Traore, A., Kadeba, J.C., Niamba, P. et al. (2005). Use of cutaneous depigmenting products by women in two towns in Burkina Faso: epidemiologic data, motivations, products and side effects. *International Journal of Dermatology* 44: 30–32.

7 Karnani, A. (2007). Doing well by doing good—case study:'Fair & Lovely'whitening cream. *Strategic Management Journal* 28 (13): 1351–1357.

8 Philips, A. (2004). Gendering colour: Identity, femininity and marriage in Kerala. *Anthropologica*: 253–272.

9 Peltzer, K., Pengpid, S., and James, C. (2016). The globalization of whitening: prevalence of skin lighteners (or bleachers) use and it's social correlates among university students in 26 countries. *International Journal of Dermatology* 55 (2): 165–172.

10 Landau, M. (2007). Exogenous factors in skin ageing. In *Environmental Factors in Skin Diseases* (Vol. 35, pp. 1-13). Karger Publishers.

11 Hayflick, L. and Moorhead, P.S. (1961). The serial cultivation of human diploid cell strains. *Experimental Cell Research* 25 (3): 585–621.

12 Libertini, G. (2019). Ageing definition. In: *Encyclopedia of Gerontology and Population Aging* (eds. D. Gu and M.E. Dupre), 1–10. Cham: Springer International Publishing.

13 Kirkwood, T.B. and Austad, S.N. (2000). Why do we age? *Nature* 408 (6809): 233–238.

14 Jabłońska-Trypuć, A., Krętowski, R., Kalinowska, M. et al. (2018). Possible mechanisms of the prevention of doxorubicin toxicity by cichoric acid—antioxidant nutrient. *Nutrients* 10 (1): 44.

15 Gu, Y., Han, J., Jiang, C., and Zhang, Y. (2020). Biomarkers, oxidative stress and autophagy in skin ageing. *Ageing Research Reviews* 101036.

16 Arda, O., Göksügür, N., and Tüzün, Y. (2014). Basic histological structure and functions of facial skin. *Clinics in Dermatology* 32 (1): 3–13.

17 Eckhart, L. and Zeeuwen, P.L. (2018). The skin barrier: Epidermis vs environment. *Experimental Dermatology* 27 (8): 805–806.

18 Parrado, C., Mercado-Saenz, S., Perez-Davo, A. et al. (2019). Environmental stressors on skin ageing. Mechanistic Insights. *Frontiers in Pharmacology 10*.

19 Rittié, L. and Fisher, G.J. (2015). Natural and sun-induced ageing of human skin. *Cold Spring Harbour Perspectives in Medicine* 5 (1): a015370.

20 Olovnikov, A.M. (1973). A theory of marginotomy: the incomplete copying of template margin in enzymic synthesis of polynucleotides and biological significance of the phenomenon. *Journal of Theoretical Biology* 41 (1): 181–190.

21 Watson, J.D. (1972). Origin of concatemeric T7DNA. *Nature: New Biology* 239 (94): 197–201.

22 Murphy, M.P. (2009). How mitochondria produce reactive oxygen species. *Biochemical Journal* 417 (1): 1–13.

23 Hayyan, M., Hashim, M.A., and AlNashef, I.M. (2016). Superoxide ion: generation and chemical implications. *Chemical Reviews* 116 (5): 3029–3085.

24 Devasagayam, T.P.A., Tilak, J.C., Boloor, K.K. et al. (2004). Free radicals and antioxidants in human health: current status and prospects. *Japi 52* (794804): 4.

25 Fisher, G.J., Quan, T., Purohit, T. et al. (2009). Collagen fragmentation promotes oxidative stress and elevates matrix metalloproteinase-1 in fibroblasts in aged human skin. *The American Journal of Pathology* 174 (1): 101–114.

26 Birkedal-Hansen, H.W.G.I., Moore, W.G.I., Bodden, M.K. et al. (1993). Matrix metalloproteinases: a review. *Critical Reviews in Oral Biology and Medicine* 4 (2): 197–250.

27 Forrester, S.J., Kikuchi, D.S., Hernandes, M.S. et al. (2018). Reactive oxygen species in metabolic and inflammatory signaling. *Circulation Research* 122 (6): 877–902.

28 Haas, R.H. (2019). Mitochondrial dysfunction in aging and diseases of aging. *Biology* 8: 48.

29 Wiley, C.D., Velarde, M.C., Lecot, P. et al. (2016). Mitochondrial dysfunction induces senescence with a distinct secretory phenotype. *Cell Metabolism* 23 (2): 303–314.

30 Serrano, M., Lin, A.W., McCurrach, M.E. et al. (1997). Oncogenic ras provokes premature cell senescence associated with accumulation of p53 and p16INK4a. *Cell* 88 (5): 593–602.

31 Ghosh, K. and Capell, B.C. (2016). The senescence-associated secretory phenotype: critical effector in skin cancer and ageing. *Journal of Investigative Dermatology* 136 (11): 2133–2139.

32 Smogorzewska, A. and de Lange, T. (2002). Different telomere damage signalling pathways in human and mouse cells. *The EMBO Journal* 21 (16): 4338–4348.

33 Munro, J., Barr, N.I., Ireland, H. et al. (2004). Histone deacetylase inhibitors induce a senescence-like state in human cells by a p16-dependent mechanism that is independent of a mitotic clock. *Experimental Cell Research* 295 (2): 525–538.

34 Kligman, L.H. (1986). Photoaging: manifestations, prevention, and treatment. *Dermatologic Clinics* 4 (3): 517–528.

35 Rhodes, A.R., Albert, L.S., Barnhill, R.L., and Weinstock, M.A. (1991). Sun-induced freckles in children and young adults. A correlation of clinical and histopathologic features. *Cancer* 67 (7): 1990–2001.

36 Fisher, G.J., Datta, S.C., Talwar, H.S. et al. (1996). Molecular basis of sun-induced premature skin ageing and retinoid antagonism. *Nature* 379 (6563): 335–339.

37 Kim, H.H., Lee, M.J., Lee, S.R. et al. (2005). Augmentation of UV-induced skin wrinkling by infrared irradiation in hairless mice. *Mechanisms of Ageing and Development* 126 (11): 1170–1177.

38 Hatsukami, D.K., Stead, L.F., and Gupta, P.C. (2008). Tobacco addiction. *The Lancet* 371 (9629): 2027–2038.

39 Ernster, V.L., Grady, D., Miike, R. et al. (1995). Facial wrinkling in men and women, by smoking status. *American Journal of Public Health* 85 (1): 78–82.

40 Yin, L., Morita, A., and Tsuji, T. (2003). Tobacco smoke extract induces age-related changes due to modulation of TGF-β. *Experimental Dermatology* 12: 51–56.

41 Krutmann, J., Bouloc, A., Sore, G. et al. (2017). The skin ageing exposome. *Journal of Dermatological Science* 85 (3): 152–161.

42 Schikowski, T. and Hüls, A. (2020). Air pollution and skin aging. *Current Environmental Health Reports* 7 (1).

43 Brinkmann, V., Ale-Agha, N., Haendeler, J., & Ventura, N. (2019). The Aryl Hydrocarbon Receptor (AhR) in the Aging Process: Another Puzzling Role for This Highly Conserved Transcription Factor. *Frontiers in Physiology, 10.*

44 Singh, J., Sharma, D., Kumar, G., and Sharma, N.R. (eds.) (2018). *Microbial Bioprospecting for Sustainable Development.* Springer.

45 Waites, M.J., Morgan, N.L., Rockey, J.S., and Higton, G. (2009). *Industrial Microbiology: An Introduction.* Wiley.

46 Sanghvi, G., Patel, H., Vaishnav, D. et al. (2016). A novel alkaline keratinase from Bacillus subtilis DP1 with potential utility in cosmetic formulation. *International Journal of Biological Macromolecules* 87: 256–262.

47 Maier, R.M. and Soberon-Chavez, G. (2000). Pseudomonas aeruginosa rhamnolipids: biosynthesis and potential applications. *Applied Microbiology and Biotechnology* 54 (5): 625–633.

48 Becker, L.C., Bergfeld, W.F., Belsito, D.V. et al. (2009). Final report of the safety assessment of hyaluronic acid, potassium hyaluronate, and sodium hyaluronate. *International Journal of Toxicology* 28: 5–67.

49 Marcellin, E., Steen, J.A., and Nielsen, L.K. (2014). Insight into hyaluronic acid molecular weight control. *Applied Microbiology and Biotechnology* 98 (16): 6947–6956.

50 Allemann, I.B. and Baumann, L. (2008). Hyaluronic acid gel (Juvéderm™) preparations in the treatment of facial wrinkles and folds. *Clinical Interventions in Ageing* 3 (4): 629.

51 Tzellos, T.G., Klagas, I., Vahtsevanos, K. et al. (2009). Extrinsic ageing in the human skin is associated with alterations in the expression of hyaluronic acid and its metabolizing enzymes. *Experimental Dermatology* 18 (12): 1028–1035.

52 Ganceviciene, R., Liakou, A.I., Theodoridis, A. et al. (2012). Skin anti-ageing strategies. *Dermato-endocrinology* 4 (3): 308–319.

53 Del Valle, E.M. (2004). Cyclodextrins and their uses: a review. *Process Biochemistry* 39 (9): 1033–1046.

54 Rajput, K.N., Patel, K.C., and Trivedi, U.B. (2016). β-Cyclodextrin production by cyclodextrin glucanotransferase from an alkaliphile Microbacterium terrae KNR 9 using different starch substrates. *Biotechnology Research International 2016.*

55 Kim, M.H., Sohn, C.B., and Oh, T.K. (1998). Cloning and sequencing of a cyclodextrin glycosyltransferase gene from Brevibacillus brevis CD162 and its expression in Escherichia coli. *FEMS Microbiology Letters* 164 (2): 411–418.

56 Matsuda, H., Ito, K., Taki, A. et al. (1995). *U.S. Patent No. 5,447,920*. Washington, DC: U.S. Patent and Trademark Office.

57 Chawla, P.R., Bajaj, I.B., Survase, S.A., and Singhal, R.S. (2009). Microbial cellulose: fermentative production and applications. *Food Technology and Biotechnology* 47 (2): 107–124.

58 Çoban, E.P. and Biyik, H. (2011). Evaluation of different pH and temperatures for bacterial cellulose production in HS (Hestrin-Scharmm) medium and beet molasses medium. *African Journal of Microbiology Research* 5 (9): 1037–1045.

59 Mohite, B.V., Salunke, B.K., and Patil, S.V. (2013). Enhanced production of bacterial cellulose by using Gluconacetobacter hansenii NCIM 2529 strain under shaking conditions. *Applied Biochemistry and Biotechnology* 169: 1497–1511.

60 Ioelovich, M. (2008). Cellulose as a nanostructured polymer: a short review. *BioResources* 3 (4): 1403–1418.

61 Amnuaikit, T., Chusuit, T., Raknam, P., and Boonme, P. (2011). Effects of a cellulose mask synthesized by a bacterium on facial skin characteristics and user satisfaction. *Medical Devices (Auckland, NZ) 4*: 77.

62 Lephart, E.D. (2019). Equol's efficacy is greater than astaxanthin for antioxidants, extracellular matrix integrity & breakdown, growth factors and inflammatory biomarkers via human skin gene expression analysis. *Journal of Functional Foods* 59: 380–393.

63 Lephart, E.D. (2017). Resveratrol, 4′ acetoxy resveratrol, R-equol, racemic equol or S-equol as cosmeceuticals to improve dermal health. *International Journal of Molecular Sciences* 18 (6): 1193.

64 Gopaul, R., Knaggs, H.E., and Lephart, E.D. (2012). Biochemical investigation and gene analysis of equol: a plant and soy-derived isoflavonoid with antiaging and antioxidant properties with potential human skin applications. *BioFactors* 38 (1): 44–52.

65 Oyama, A., Ueno, T., Uchiyama, S. et al. (2012). The effects of natural S-equol supplementation on skin ageing in postmenopausal women: a pilot randomized placebo-controlled trial. *Menopause* 19 (2): 202–210.

66 Setchell, K.D., Brown, N.M., and Lydeking-Olsen, E. (2002). The clinical importance of the metabolite equol – a clue to the effectiveness of soy and its isoflavones. *The Journal of Nutrition* 132 (12): 3577–3584.

67 Meng, T.X., Zhang, C.F., Miyamoto, T. et al. (2012). The melanin biosynthesis stimulating compounds isolated from the fruiting bodies of Pleurotus citrinopileatus. *Journal of Cosmetics. Dermatological Sciences and Applications* 2 (03): 151.

68 Oh, M.J., Hamid, M.A., Ngadiran, S. et al. (2011). Ficus deltoidea (Mas cotek) extract exerted anti-melanogenic activity by preventing tyrosinase activity in vitro and by suppressing tyrosinase gene expression in B16F1 melanoma cells. *Archives of Dermatological Research* 303 (3): 161–170.

69 Saranraj, P. and Naidu, M.A. (2013). Hyaluronic acid production and its applications a review. *International Journal of Pharmaceutical and Biological Archiv* 4 (5): 853–859.

70 Dudek-Makuch, M. and Studzińska-Sroka, E. (2015). Horse chestnut–efficacy and safety in chronic venous insufficiency: an overview. *Revista Brasileira de Farmacognosia* 25 (5): 533–541.

71 Kim, S.Y., Go, K.C., Song, Y.S. et al. (2014). Extract of the mycelium of T. matsutake inhibits elastase activity and TPA-induced MMP-1 expression in human fibroblasts. *International Journal of Molecular Medicine* 34 (6): 1613–1621.

72 Ndlovu, G., Fouche, G., Tselanyane, M. et al. (2013). in vitro determination of the anti-ageing potential of four southern African medicinal plants. *BMC Complementary and Alternative Medicine* 13 (1): 304.

73 Thomas, N.V., Manivasagan, P., and Kim, S.K. (2014). Potential matrix metalloproteinase inhibitors from edible marine algae: a review. *Environmental Toxicology and Pharmacology* 37 (3): 1090–1100.

74 Kwak, J.Y., Park, S., Seok, J.K. et al. (2015). Ascorbyl curates as multifunctional cosmeceutical agents that inhibit melanogenesis and enhance collagen synthesis. *Archives of Dermatological Research* 307 (7): 635–643.

75 Pimentel, F.B., Alves, R.C., Rodrigues, F. et al. (2018). Macroalgae-derived ingredients for cosmetic industry – an update. *Cosmetics* 5 (1).

76 Andersen, R.A. (1992). Diversity of eukaryotic algae. *Biodiversity and Conservation* 1 (4): 267–292.

77 Sahoo, D. and Seckbach, J. (2015). *The Algae World*. Springer.

78 Kim, S.-K. and Chojnacka, K. (2015). *Marine Algae Extracts: Processes, Products, and Applications*. Wiley.

79 Gupta, P.L., Rajput, M., Oza, T. et al. (2019). The eminence of microbial products in the cosmetic industry. *Natural Products and Bioprospecting* 9 (4): 267–278.

80 de Jesus Raposo, M.F., de Morais, A.M.B., and de Morais, R.M.S.C. (2015). Marine polysaccharides from algae with potential biomedical applications. *Marine Drugs* 13 (5): 2967–3028.

81 Saewan, N. and Jimtaisong, A. (2015). Natural products as photoprotection. *Journal of Cosmetic Dermatology* 14 (1): 47–63.

82 Wang, H.-M.D., Chen, C.-C., Huynh, P., and Chang, J.-S. (2015). Exploring the potential of using algae in cosmetics. *Bioresource Technology* 184: 355–362.

83 Sá, A.G.A., de Meneses, A.C., de Araújo, P.H.H., and Oliveira, D.d. (2017). A review on the enzymatic synthesis of aromatic esters used as flavour ingredients for food, cosmetics and pharmaceuticals industries. *Trends in Food Science & Technology* 69: 95–105.

84 Priyan Shanura Fernando, I., Kim, K.-N., Kim, D., and Jeon, Y.-J. (2018). Algal polysaccharides: potential bioactive substances for cosmeceutical applications. *Critical Reviews in Biotechnology*: 1–15.

85 Wang, H.-M., Chou, Y.-T., Wen, Z.-H. et al. (2013). Novel biodegradable porous scaffold applied to skin regeneration. *PLoS One* 8 (6): e56330.

86 Wang, J., Jin, W., Hou, Y. et al. (2013). Chemical composition and moisture-absorption/retention ability of polysaccharides extracted from five algae. *International Journal of Biological Macromolecules* 57: 26–29.

87 Fabrowska, J., Łęska, B., Schroeder, G., Messyasz, B., & Pikosz, M. (2015). Biomass and extracts of algae as material for cosmetics. In Marine Algae Extracts (eds S.-K. Kim and K. Chojnacka) (pp. 681–706). Wiley. https://doi.org/10.1002/9783527679577.ch38

88 Pereira, L. (2018). Seaweeds as source of bioactive substances and skin care therapy – cosmeceuticals, algotheraphy, and thalassotherapy. *Cosmetics* 5 (4): 68. https://doi.org/10.3390/cosmetics5040068.

89 Qin, Y. (2018). 1 – seaweed bioresources. In: *Bioactive Seaweeds for Food Applications* (ed. Y. Qin), 3–24. Academic Press.

90 Wijesekara, I., Pangestuti, R., and Kim, S.-K. (2011). Biological activities and potential health benefits of sulfated polysaccharides derived from marine algae. *Carbohydrate Polymers* 84 (1): 14–21.

91 El Gamal, A.A. (2010). Biological importance of marine algae. *Saudi Pharmaceutical Journal* 18 (1): 1–25.

92 Kim, J.H., Lee, J.-E., Kim, K.H., and Kang, N.J. (2018). Beneficial effects of marine algae-derived carbohydrates for skin health. *Marine Drugs* 16 (11).

93 Pallela, R., Na-Young, Y., and Kim, S.-K. (2010). Anti-photoaging and photoprotective compounds derived from marine organisms. *Marine Drugs* 8 (4): 1189–1202.

94 Teas, J. and Irhimeh, M.R. (2017). Melanoma and brown seaweed: an integrative hypothesis. *Journal of Applied Phycology* 29 (2): 941–948.

95 Song, Y.S., Li, H., Balcos, M.C. et al. (2014). Fucoidan promotes the reconstruction of skin equivalents. *The Korean Journal of Physiology & Pharmacology: Official Journal of the Korean Physiological Society and the Korean Society of Pharmacology* 18 (4): 327–331.

96 Fitton, J.H., Dell'Acqua, G., Gardiner, V.-A. et al. (2015). Topical benefits of two fucoidan-rich extracts from marine macroalgae. *Cosmetics* 2 (2): 66–81.

97 Li, Y.-J., Han, Z., Ge, L. et al. (2016). C-phycocyanin protects against low fertility by inhibiting reactive oxygen species in ageing mice. *Oncotarget* 7 (14): 17393–17409.

98 Barg, H. (2013). *Filler composition comprising beta-glucans* (United States Patent No. US20130196944A1). https://patents.google.com/patent/US20130196944A1/en

99 Yvin, J.-C., Levasseur, F., and Hud'Homme, F. (1999). *Use of laminarin and oligosaccharides derived therefrom in cosmetics and for preparing a skin treatment drug*

100 Saito, M. (2005). *Porphyran-containing cosmetic* (Patent No. JP2005336148A). https://patents.google.com/patent/JP2005336148A/en

101 Cheong, K.-L., Qiu, H.-M., Du, H. et al. (2018). Oligosaccharides derived from red seaweed: production, properties, and potential health and cosmetic applications. *Molecules* 23 (10): 2451.

102 Laurent, L., & Bebot, C. (2018). *Cosmetic composition comprising at least one lambda-carrageenan polysaccharide in combination with at least one specific polyol, and process for the cosmetic treatment of keratin fibers with the composition, and use of the composition for hair care* (United States Patent No. US20180296458A1).

103 Pangestuti, R. and Kim, S.-K. (2011). Biological activities and health benefit effects of natural pigments derived from marine algae. *Journal of Functional Foods* 3 (4): 255–266.

104 Kidgell, J.T., Magnusson, M., de Nys, R., and Glasson, C.R.K. (2019). Ulvan: a systematic review of extraction, composition and function. *Algal Research* 39: 101422.

105 Ray, B. and Lahaye, M. (1995). Cell-wall polysaccharides from the marine green alga Ulva "rigida" (ulvales, Chlorophyta). Extraction and chemical composition. *Carbohydrate Research* 274: 251–261.

106 Gong, M. and Bassi, A. (2016). Carotenoids from microalgae: a review of recent developments. *Biotechnology Advances* 34 (8): 1396–1412.

107 Novoveská, L., Ross, M.E., Stanley, M.S. et al. (2019). Microalgal carotenoids: a review of production, current markets, regulations, and future direction. *Marine Drugs* 17 (11).

108 Fitzpatrick, J.E., High, W.A., and Kyle, W.L. (2018). Discolorations of the skin. In: *Urgent Care Dermatology: Symptom-Based Diagnosis (pp. 441–460)* (eds. J.E. Fitzpatrick, W.A. High and W.L. Kyle). Elsevier.

109 Sharif, H.R., Goff, H.D., Majeed, H. et al. (2017). Physicochemical stability of β-carotene and α-tocopherol enriched nanoemulsions: Influence of carrier oil, emulsifier and antioxidant. *Colloids and Surfaces A: Physicochemical and Engineering Aspects* 529: 550–559.

110 Polyakov, N.E., Leshina, T.V., Konovalova, T.A., and Kispert, L.D. (2001). Carotenoids as scavengers of free radicals in a Fenton reaction: Antioxidants or pro-oxidants? *Free Radical Biology & Medicine* 31 (3): 398–404.

111 Çelik, S.E., Bekdeser, B., Tufan, A.N., and Apak, R. (2017). Modified radical scavenging and antioxidant activity measurement of β-Carotene with β-Cyclodextrins complexation in aqueous medium. *Analytical Sciences* 33 (3): 299–305.

112 Solymosi, K. and Mysliwa-Kurdziel, B. (2017). Chlorophylls and their derivatives used in food industry and medicine. *Mini Reviews in Medicinal Chemistry* 17 (13): 1194–1222.

113 Busch, T., Cengel, K.A., and Finlay, J. (2009). Pheophorbide a as a photosensitizer in photodynamic therapy: in vivo considerations. *Cancer Biology & Therapy* 8 (6): 540–542.

114 Xodo, L.E., Rapozzi, V., Zacchigna, M. et al. (2012). The chlorophyll catabolite pheophorbide as a photosensitizer for photodynamic therapy. *Current Medicinal Chemistry* 19 (6): 799–807.

115 Pangestuti, R., Siahaan, E.A., and Kim, S.-K. (2018). Photoprotective substances derived from marine algae. *Marine Drugs 16* (11) https://doi.org/10.3390/md16110399.

116 Peng, J., Yuan, J.-P., Wu, C.-F., and Wang, J.-H. (2011). Fucoxanthin, a marine carotenoid present in brown seaweeds and diatoms: metabolism and bioactivities relevant to human health. *Marine Drugs* 9 (10): 1806–1828.

117 Shimoda, H., Tanaka, J., Shan, S.-J., and Maoka, T. (2010). The anti-pigmentary activity of fucoxanthin and its influence on skin mRNA expression of melanogenic molecules. *The Journal of Pharmacy and Pharmacology* 62 (9): 1137–1145.

118 Bermejo Román, R., Alvárez-Pez, J.M., Acién Fernández, F.G., and Molina Grima, E. (2002). Recovery of pure B-phycoerythrin from the microalga Porphyridium cruentum. *Journal of Biotechnology* 93 (1): 73–85.

119 Romay, C. and Gonzalez, R. (2000). Phycocyanin is an antioxidant protector of human erythrocytes against lysis by peroxyl radicals. *Journal of Pharmacy and Pharmacology* 52 (4): 367–368.

120 Singh, N.K., Sonani, R.R., Awasthi, A. et al. (2016). Phycocyanin moderates ageing and proteotoxicity in Caenorhabditis elegans. *Journal of Applied Phycology* 28 (4): 2407–2417.

121 Kim, K.M., Lee, J.Y., I'm, A.-R., and Chae, S. (2018). Phycocyanin protects against UVB-induced apoptosis through the PKC α/βII-Nrf-2/HO-1 dependent pathway in human primary skin cells. *Molecules* 23 (2): 478.

122 Bedoux, G., Hardouin, K., Burlot, A.S., and Bourgougnon, N. (2014). Bioactive components from seaweeds: cosmetic applications and future development. In: *Advances in Botanical Research*, vol. 71 (ed. N. Bourgougnon), 345–378. Academic Press.

123 Babitha, S., & Kim, E.-K. (2011). Effect of Marine Cosmeceuticals on the Pigmentation of Skin. In S.-K. Kim, *Marine Cosmeceuticals* (pp. 63–66). CRC Press.

124 Chaves-Peña, P., de la Coba, F., Figueroa, F.L., and Korbee, N. (2020). Quantitative and qualitative HPLC analysis of mycosporine-like amino acids extracted in distilled water for cosmetical uses in four rhodophyta. *Marine Drugs* 18 (1): 27.

125 Li, B., Lu, F., Wei, X., and Zhao, R. (2008). Fucoidan: structure and bioactivity. *Molecules* 13 (8): 1671–1695.

126 Pielesz, A. and Paluch, J. (2014). Fucoidan as an inhibitor of thermally induced collagen glycation examined by acetate electrophoresis. *Electrophoresis* 35 (15): 2237–2244.

127 Fujimura, T., Tsukahara, K., Moriwaki, S. et al. (2000). Effects of natural product extract on contraction and mechanical properties of fibroblast populated collagen gel. *Biological & Pharmaceutical Bulletin* 23 (3): 291–297.

128 Smit, N., Vicanova, J., and Pavel, S. (2009). The hunt for natural skin whitening agents. *International Journal of Molecular Sciences* 10 (12): 5326–5349.

129 Jesumani, V., Du, H., Pei, P. et al. (2020). Comparative study on skin protection activity of polyphenol-rich extract and polysaccharide-rich extract from Sargassum vachellianum. *PLoS One* 15 (1): e0227308.

130 Bagal Kestwal, D.R., Pan, M.H., and Chiang, B.-H. (2019). Properties and applications of gelatin, pectin, and carrageenan gels. In: *Bio Monomers for Green Polymeric Composite Materials* (eds. P. Visakh, O. Bayraktar and G. Menon), 117–140. Wiley. https://doi.org/10.1002/9781119301714.ch6.

131 Kozlowska, J., Pauter, K., and Sionkowska, A. (2018). Carrageenan-based hydrogels: effect of sorbitol and glycerin on the stability, swelling and mechanical properties. *Polymer Testing* 67: 7–11.

132 Thevanayagam, H., Mohamed, S.M., and Chu, W.-L. (2014). Assessment of UVB-photoprotective and antioxidative activities of carrageenan in keratinocytes. *Journal of Applied Phycology* 26 (4): 1813–1821.

133 Ariga, O., Okamoto, N., Harimoto, N., and Nakasaki, K. (2014). Purification and characterization of α-neoagarooligosaccharide hydrolase from Cellvibrio sp. OA-2007. *Journal of Microbiology and Biotechnology* 24 (1): 48–51. https://doi.org/10.4014/jmb.1307.07018.

134 Chen, H.-M. and Yan, X.-J. (2005). Antioxidant activities of agaro-oligosaccharides with different degrees of polymerization in the cell-based system. *Biochimica et Biophysica Acta* 1722 (1): 103–111.

135 Orive, G., Hernández, R.M., Gascón, A.R., and Pedraz, J.L. (2006). Encapsulation of cells in alginate gels. In: *Immobilization of Enzymes and Cells* (ed. J.M. Guisan), 345–355. Humana Press.

136 Aderibigbe, B.A. and Buyana, B. (2018). Alginate in wound dressings. *Pharmaceutics* 10 (2).

137 Dantas, M.D.M., Cavalcante, D.R.R., Araújo, F.E.N. et al. (2011). Improvement of dermal burn healing by combining sodium alginate/chitosan-based films and low level laser therapy. *Journal of Photochemistry and Photobiology. B, Biology* 105 (1): 51–59.

138 Huang, S. and Fu, X. (2010). Naturally derived materials-based cell and drug delivery systems in skin regeneration. *Journal of Controlled Release: Official Journal of the Controlled Release Society* 142 (2): 149–159.

139 Sathasivam, R. and Ki, J.-S. (2018). A review of the biological activities of microalgal carotenoids and their potential use in healthcare and cosmetic industries. *Marine Drugs* 16 (1).

140 Benedetti, S., Benvenuti, F., Pagliarani, S. et al. (2004). Antioxidant properties of a novel phycocyanin extract from the blue-green alga Aphanizomenon flos-aquae. *Life Sciences* 75 (19): 2353–2362.

141 Cikoš, A.-M., Jerković, I., Molnar, M. et al. (2019). New trends for macroalgal natural products applications. *Natural Product Research* 0 (0): 1–12.

142 Sonani, R.R., Rastogi, R.P., Singh, N.K. et al. (2017). Phycoerythrin averts intracellular ROS generation and physiological functional decline in eukaryotes under oxidative stress. *Protoplasma* 254 (2): 849–862.

143 Varela, J.C., Pereira, H., Vila, M., and León, R. (2015). Production of carotenoids by microalgae: achievements and challenges. *Photosynthesis Research* 125 (3): 423–436.

144 Esatbeyoglu, T. and Rimbach, G. (2017). Canthaxanthin: from molecule to function. *Molecular Nutrition & Food Research* 61 (6).

145 Koller, M., Muhr, A., and Braunegg, G. (2014). Microalgae as versatile cellular factories for valued products. *Algal Research* 6: 52–63.

146 Camera, E., Mastrofrancesco, A., Fabbri, C. et al. (2009). Astaxanthin, canthaxanthin and beta-carotene differently affect UVA-induced oxidative damage and expression of oxidative stress-responsive enzymes. *Experimental Dermatology* 18 (3): 222–231.

147 Tominaga, K., Hongo, N., Karato, M., and Yamashita, E. (2012). Cosmetic benefits of astaxanthin on humans subjects. *Acta Biochimica Polonica* 59 (1): 43–47.

148 Shen, C.-T., Chen, P.-Y., Wu, J.-J. et al. (2011). Purification of algal anti-tyrosinase zeaxanthin from Nannochloropsis oculata using supercritical anti-solvent precipitation. *The Journal of Supercritical Fluids* 55 (3): 955–962.

149 Juturu, V., Bowman, J.P., and Deshpande, J. (2016). Overall skin tone and skin-lightening-improving effects with oral supplementation of lutein and zeaxanthin isomers: a double-blind, placebo-controlled clinical trial. *Clinical, Cosmetic and Investigational Dermatology* 9: 325–332.

150 Kim, H.-M., Jung, J.H., Kim, J.Y. et al. (2019). The protective effect of Violaxanthin from *Nannochloropsis oceanica* against ultraviolet b-induced damage in normal human dermal fibroblasts. *Photochemistry and Photobiology* 95 (2): 595–604.

151 Inoue, N., Yamano, N., Sakata, K. et al. The Sulfated Polysaccharide Porphyran Reduces Apolipoprotein B100 Secretion and Lipid Synthesis in HepG2 Cells. *Bioscience, Biotechnology, and Biochemistry* 73 (2): 447–449. https://doi.org/10.1271/bbb.80688 . (23 February 2009).

5

Application of Microorganisms in Bioremediation

Himani Thakkar and Vinnyfred Vincent

Department of Biochemistry, All India Institute of Medical Sciences, New Delhi, New Delhi, India

5.1 Introduction

The microorganisms are known for their diversity and adaptability in a wide range of environmental conditions. Their versatility to utilize and adapt to any available substrate in order to gain energy kept the balance in the ecosystem until humans become dominant species. With the advent of industrial revolution, persistent contamination of soil and water by industrial waste and several other human activities have become a threat to a large number of ecosystems. The advancement of the technology and the adaptability of microorganisms had led to the development of a process known as bioremediation.

The idea of utilizing microorganisms in waste management came from ancient Rome wastewater treatment facilities. Ancient Romans developed a sophisticated sewage system to fulfil their water needs, where drained water was cleaned by the natural biodegradation process. Later in 1960s, George M. Robinson for the first time introduced the concept of bioremediation, where he started treating petroleum waste with bacterial culture. This initiative led to the implementation of bioremediation for the management of pollutants.

Bioremediation is the technique of removing pollutants from the environment by exploiting the metabolic potential of microorganisms to transform harmful pollutants into harmless products. Bioremediation provides an eco-friendly, economically viable, and socially acceptable offer to remove contaminants from the environment (in soil or water). The microbes using the redox reaction (carbon or metal) convert/modify the pollutant into harmless products like carbon dioxide or methane or water or cell biomass. Bacteria, archaea, fungi, and microalgae play a key role in the process of bioremediation.

Bioremediation involves biotransformation and biodegradation of toxic pollutants to nontoxic compounds. Biotransformation is the structural modification of toxic compounds, while biodegradation is the breakdown of organic to inorganic compounds by microorganisms in a way that it reduces the persistence and toxicity of the compound [1, 2].

For the bioremediation process to be successful, favorable environmental conditions and substrate need to be available for the microbes for degradation of pollutants [3], since

Bioprospecting of Microorganism-Based Industrial Molecules, First Edition.
Edited by Sudhir P. Singh and Santosh Kumar Upadhyay.

bioremediation depends on the interaction between the substrate (pollutant), microorganisms, and environmental factors (pH, temperature, nutrients, etc.) [4]. Biostimulation, bioaugmentation, bioaccumulation, and biosorption are widely used bioremediation techniques. Biostimulation involves the modification of the environment to stimulate the bioremediation activity of the naturally occurring microbial community. Bioaugmentation is a practice of increasing the microbial load at the site of contamination by adding cultured microorganisms. Bioaccumulation is a natural process where microorganisms uptake organic or metal pollutants from the contaminated area (soil or water) and accumulates them in the intracellular space for degradation [5, 6], and biosorption is a process of extraction of elements from the site of contamination through adsorption to make it available for microbes [7].

5.2 Microbial Bioremediation

Microorganisms are the most abundant and diverse group of organisms in the biosphere [8]. Their universal presence and capability to survive in diverse environmental conditions along with the advent of molecular microbiology promoted scientists to isolate and characterize the microbial population at the site of contamination that may have role in bioremediation [9]. The microbes using their metabolic or catalytic potentials degrade or transform toxic compounds like polyaromatic hydrocarbons, heavy metals, and heterocyclic compounds [9].

Depending on the environmental conditions of the site where the biodegradation of a pollutant occurs, the metabolic processes are broadly categorized as aerobic and anaerobic. In aerobic bioremediation, microorganisms use oxygen as the final electron acceptor for the oxidation of organic contaminants (petroleum, polyaromatic hydrocarbons, and phenols). Oxygen is also utilized in the metabolic reactions for degradation of contaminants that serve as a source of energy for microbes. The degradation efficiency of low to moderate molecular weight compounds (aliphatic, alicyclic, and aromatic) is high compared to one with high molecular weight.

In anaerobic bioremediation, degradation reaction by microorganisms is performed in the absence of oxygen. In anaerobic bioremediation, oxidized iron, manganese, sulfate, nitrate, and carbon dioxide act as electron acceptors in place of oxygen in these reactions. Anaerobic reactions are used to treat highly halogenated contaminants like tetrachloroethane, polychlorinated biphenyls (PCBs), and dichloroethene.

Additionally, some of the contaminants are degraded by co-metabolism, where microorganisms degrade a variety of contaminants like polycyclic aromatic hydrocarbons (PAHs), methyl tertiary-butyl ether, vinyl chloride. In this type of reaction, the degradation of contaminant occurs by an enzyme or cofactor that is produced during the metabolism of a compound that is obligate for microbial growth [10]. Co-metabolic bioremediation can be aerobic or anaerobic. The electron donors in co-metabolic aerobic oxidation reactions included methane, butane, propane, ammonia, while methanol, glucose, lactate, and pyruvate serve as substrates for co-metabolic anaerobic reduction reaction. The microorganisms do not receive benefits in terms of energy or biomass derived from the co-metabolic degradation process [11].

The aim of bioremediation is to enhance the degradation capacity of microbes by providing optimum levels of nutrients and other biochemicals and maintaining the environmental factors required for their metabolism. Microorganisms possess characteristic metabolic pathways that are mediated by enzymes specific to a contaminant. The microbial enzymes in bioremediation belong to the groups of oxidoreductases, hydrolases, peroxidases, lyases, transferases, isomerases, ligases, and cellulases [12]. Based on the nature of the pollutant, microbial bioremediation can be classified into bioremediation of organic pollutants and bioremediation of inorganic pollutants.

5.3 Microbial Bioremediation of Organic Pollutants

Industrialization and widespread use of organic compounds such as petroleum oil, BTEX (benzene, toluene, ethylbenzene, and xylenes), PAHs, chlorinated hydrocarbons (PCBs, trichloroethylene), and pesticides accumulate and persist in the ecosystem. Anthropogenic activities are the major source of organic pollutants, which is of global concern now. Some of the organic compounds are highly toxic, and seepage or runoff of these pollutants due to their mobile nature and continuous cycling even pollute rain, fog, and snow. They are also associated with a variety of health issues around the world. Aliphatic and aromatic hydrocarbons can degrade under both anaerobic and aerobic conditions.

5.3.1 Bioremediation of Alkanes

Biodegradation of alkanes initiates with the incorporation of oxygen atoms by alkane oxygenases (mono-oxygenases and dioxygenases introduce one and two oxygens, respectively) or is carried out by either a multi-meric monooxygenase or a cytochrome P450 monooxygenase [13]. Microbes degrade short-chain length alkanes (C_2–C_4) using methane monooxygenases while medium-chain length alkanes (C_5–C_{17}) are degraded using soluble cytochrome P450s and integral membrane non-heme iron monooxygenases, such as AlkB [14, 15].

Aerobic degradation of n-alkanes containing two or more carbon atoms starts with the oxidation of a terminal methyl group to a primary alcohol. The alcohol is further oxidized to the corresponding aldehyde, and finally converted into fatty acid. Fatty acids are oxidized to generate acetyl-CoA [16–18].

The alkane degradation genes on OCT plasmid carried by *Pseudomonas putida* Gpo1 are broadly characterized to study the alkane degradation pathway. The first enzyme of this pathway is an integral-membrane non-heme diiron monooxygenase, named AlkB, that hydroxylates alkanes at the terminal position. This enzyme complex consists of two soluble electron transfer proteins named rubredoxin (AlkG) and rubredoxin reductase (AlkT).

AlkB alkane hydroxylase oxidizes propane, n-butane, and C_5–C_{13} alkanes. The *alkB* like genes are also present in genera of *Burkholderia*, *Acinetobacter*, *Pseudomonas*, *Alcanivorax*, *Oleiphilus*, *Mycobacterium*, *Rhodococcus*, *Nocardia*, and *Prauserella*. The structural and catalytic motifs of the enzyme complex are conserved across the species but differ in terms of substrate range and specificity [19]. *Acinetobacter baylyi* ADP1 mediates the degradation of C_{12}–C_{36} n-alkane by the AlkB-like alkane monooxygenase AlkM. *Acinetobacter* sp. M1 possesses n-alkane hydroxylase-encoding genes (*alkMa* and *alkMb*) for oxidation of C_{20}–C_{44} alkanes.

Besides alkane monooxygenases, two additional hydroxylation systems, methane monooxygenases (MMOs) and cytochrome P450 protein superfamily are involved in a few alkane-degrading microorganisms. There are two forms of methane monooxygenases (MMOs), one is cytoplasmic soluble (sMMOs) present in *Methylococcus*, *Methylosinus*, *Methylocystis*, *Methylomonas*, *Methylomicrobium*, and *Methylocella* genera, another form is membrane-bound (pMMO) present in all methanotrophs except in *Methylocella* [20]. The sMMOs possess a [2Fe-2S] catalytic center and utilize FAD and NADH as cofactors and oxidize C_1–C_8 chain-length alkanes, halogenated alkanes, alkenes, and cycloalkanes. The pMMOs have a mono- or di-copper center and hydroxylate only C_1–C_5 substrates such as alkanes, halogenated alkanes, and alkenes [19]. Cytochrome P450 system is present in Bacteria and Archaea. The soluble cytochrome P450 involved in alkane oxidation are present in Proteobacteria and Actinobacteria [21].

Sulfate-reducing, Deltaproteobacteria carry enzymes for anaerobic alkane degradation. In *Desulfococcus oleovorans* Hxd3, alkane activation occurs in conjugation with fumarate at the subterminal (C-2) carbon atom by alkyl succinate synthase (Ass), yielding (1-methylalkyl) succinates, although other metabolites were also identified indicating activation at the C-3 atom, and the terminal (C-1) carbon atom. The alkylsuccinates undergo carbon rearrangement and decarboxylation, yielding a methylated fatty acid that is two carbons larger than the original n-alkane [22]. The resulting fatty acid is conjugated to CoA and further processed by β-oxidation to generate acetyl-CoA. Three denitrifying isolates (strains HxN1, OcN1, HdN1) oxidize alkane to carbon dioxide coupled with a reduction of NO^{3-} to N_2 under anoxic conditions. Betaproteobacteria strains HxN1 and OcN1 utilize n-alkanes from C_6 to C_8 and C_8 to C_{12}, respectively. Gammaproteobacteria strain HdN1 is more versatile as it can utilizing alkane from C_6 to C_{30} [23, 24].

5.3.2 Bioremediation of Benzene, Toluene, Ethylbenzene, and Xylenes (BTEX)

Benzene, toluene, ethylbenzene, and xylenes (BTEX) are the main components of the volatile organic fraction from crude oil. For toluene degradation, five aerobic pathways and one anaerobic route have been reported [25]. The degradation of toluene by *P. putida* mt-2 has been extensively studied. The enzymes for toluene and xylene degradation are encoded by the *xyl* genes from pWWO plasmid carried by *P. putida* mt-2 [26]. The upper operon of the *xyl* gene encodes the enzymes, xylene monooxygenase that oxidizes the methyl group of toluene, m-xylene, and p-xylene into the corresponding (alkyl)benzoate. Catechol-2,3-dioxygenase catalyzes the transformation of (alkyl)benzoate and the meta-cleavage of the resulting (methyl)catechol, which then enters the Krebs cycle. The anaerobic pathway for toluene degradation from *Thauera aromatica* K172 and *Azoarcus* sp. involves an initial addition of fumarate by benzylsuccinate synthase to the methyl group of toluene.

5.3.3 Bioremediation of Polyaromatic Hydrocarbons

Polyaromatic hydrocarbons (PAHs) are ubiquitous in nature and are of great environmental concern due to their persistence, toxicity, mutagenicity, and carcinogenicity in nature. PAHs are categorized into low-molecular-weight PAHs (LMW PAHs) (compounds with two or three aromatic rings) and high-molecular-weight PAHs (HMW PAHs) (compounds with four or more aromatic rings). LMW PAHs are soluble in water; hence, more susceptible to biodegradation compared to HMW PAHs, LMW PAHs include naphthalene, acenaphthene, fluorene, and phenanthrene [27]. The high hydrophobicity of HMW PAHs increases with the number of aromatic rings, and this property of HMW PAHs contributes to their persistence in the environment [28]. HMW PAHs are pyrene, fluoranthene, benz(a)anthracene, chrysene, etc. Depending on the presence of oxygen, polyaromatic hydrocarbons are degraded by aerobic and anaerobic catabolic reactions. Microorganisms play a major role in the degradation of aromatic hydrocarbons by recycling carbon and the development of a sustainable ecosystem.

Microbial degradation of PAHs under aerobic conditions initiates via incorporation of molecular oxygen into the aromatic ring by dioxygenase enzyme system resulting in the formation of cis-dihydrodiol. These compounds are rearomatized to form dihydroxylated intermediates by the action of dehydrogenase. Dihydrodiols are then cleaved by intradiol or extradiol ring-cleaving dioxygenases forming catechol that are converted to TCA-cycle intermediates [29, 30]. Dioxygenases are multicomponent enzyme that introduces molecular oxygen into their substrate, consists of reductase, ferredoxin, and terminal oxygenase subunits.

Naphthalene is a bicyclic aromatic hydrocarbon ubiquitously present in the environment and has been used as a model compound of PAH degradation. Microorganisms that degrade naphthalene includes *Alcaligenes denitrificans*, *Mycobacterium* sp., *P. putida*, *P. fluorescens*, *P. paucimobilis*, *P. vesicularis*, *P. cepacia*, *P. testosteroni*, *Rhodococcus* sp., *Corynebacterium venale*, *Bacillus cereus*, *Moraxella* sp., *Streptomyces* sp., *Vibrio* sp. and *Cyclotrophicus* sp. The genes involved in the production of enzymes for the degradation process are well studied in *P. putida* strain G7. Plasmid NAH7 in *Pseudomonas* harbors naphthalene catabolic genes (*nah*) that are organized in three operons. One operon (nal) codes for the genes for the upper pathway enzymes that are involved in the conversion of naphthalene to salicylate and the second operon carries *sal* genes that code for enzymes involved in the conversion of salicylate to TCA cycle intermediates. The third is *NahR* that codes for the regulatory protein LysR, which is a positive transcriptional regulator of the *nah* genes in bacteria [31, 32].

Similar to NAH, there are other naphthalene-catabolizing plasmids that show more than 90% nucleotide sequence identity of the genes coding for upper catabolic enzymes [33–36]. *Ralstonia* sp. U2 isolated from oil-contaminated soil [37] carries *nag* genes for naphthalene degradation. The *nag* genes in *Ralstonia* sp U2 are arranged in a similar order as the genes in *nah*-like operon in *Pseudomomas* strain, except for few insertions that include subunits of enzyme salicylate-5-hydroxylase that catalyzes the conversion of naphthalene to gentisate [38] and location of regulatory gene *nagR* and chemotaxis gene *nagY*.

Furthermore, *phn* genes of *Burkholderia* sp. strain RP007 and *Alcaligenes faecalis* strain AFK2 codes for PAH catabolizing enzymes [39, 40]. Strain RP007 can degrade low-molecular-weight PAHs like naphthalene and anthracene, while strain AFK2 can only degrade phenanthrene. The *phn* gene in both strains is different in sequence and organization from the previously described genes. The *phd* genes of *Comamonas teststeroni* strains GZ38A, GZ39, and GZ42 utilize phenanthrene as a sole carbon source [41, 42]. The genes present in these strains were not similar to *nah* like in pseudomonas strains.

The members of *Sphingomonadaceae* (*Sphingomonas, Sphingobium, Novosphingobium*) are known to degrade a wide range of mono- and polyaromatic compounds [43–45]. *Sphingomonas* carry catabolic genes on large plasmids that are arranged in a complex manner and are largely scattered through several clusters but are strongly conserved between different strains [46, 47]. The catabolic versatility of Sphingomonad is due to the presence of different combinations of conserved gene clusters that allows them to adapt and degrade various aromatic compounds proficiently [43].

In addition, a *Sphingobium* sp. strain PNB co-metabolize monoaromatic (fluoranthene, acenaphthene) and polyaromatic compounds (benz[a]anthracene, pyrene, and benzo[a]pyrene), indicating the metabolic versatility of the strain. The strain carries seven sets of ring hydroxylating oxygenase with different substrate specificities. PNB strain is a candidate organism in the bioremediation of PAHs-contaminated sites [48, 49].

Among PAHs degrading bacteria, the genus *Rhodococcus* is a diverse group of gram-positive bacteria in soil that carry three structural genes (*nar*Aa, *nar*Ab, and *nar*B) required for the catabolism of naphthalene [50, 51]. The arrangement of *nar* genes in the transcription unit (operon) differs in different *Rhodococcus* strains. Due to the genetic exchange or reshuffling of genes in Rhodococcus strains during the evolution [50], *nar* genes are present on both plasmids and chromosomes. The nucleotide sequence of *nar*Aa and *nar*Ab genes codes for the components (α- and β- subunit) of naphthalene dioxygenase (NDO). The subunits of NDO in *Rhodococcus* sp. have 31–39% amino acid identity to the NDO subunits in *P. putida*. The *nar*A and *nar*B genes are transcribed as a single unit in the presence of naphthalene, while the naphthalene catabolic genes are induced by salicylate in *Pseudomonas* species. *nar*R1 and *nar*R2 genes code for two regulatory proteins that belong to GntR family of transcription factors [52].

The other most studied gram-positive bacterial species is *Nocardioides* sp. strain KP7. The strain exhibit *phd* genes that code for the different subunits of phenanthrene dioxygenase for the degradation of phenanthrene. Like *nar* genes, *phd*A and *phd*B genes encode for α- and β- subunit of the phenanthrene dioxygenase. The dioxygenase in strain KP7 belongs to a different class of PAH catabolic genes because of the differences in genetic arrangement and sequence of the *phd* genes [53].

A variety of microbes including *Aeromonas* sp., *Alcaligenes faecalis, A. denitrificans, Arthrobacter polychromogenes, Beijerinckia* sp., *Micrococcus* sp., *Mycobacterium* sp., *P. putida, P. paucimobilis, Rhodococcus* sp., *Vibrio* sp., *Nocardia* sp., *Flavobacterium* sp., *Streptomyces* sp. and *Bacillus* sp. can degrade phenanthrene [54, 55]. Three pathways convert phenanthrene dihydrodiols to naphthalene-1,2 diols, generally phenanthrene degradation is initiated by 3,4-dioxygenation to yield cis-3,4-dihydroxy-3,4-dihydrophenanthrene, which undergoes enzymatic dehydrogenation to 3,4-dihydroxyphenanthrene. Subsequently, the diol formed is catabolized to naphthalene-1,2-diol through ortho-cleavage to form 2-(2 -carboxy-vinyl)-naphthalene-1-carboxylic acid and by meta-cleavage to 4-(1-hydroxy-naph thalen-2-yl)-2-oxo-but-3-enoic acid [56].

Fluorene, a PAH, is a major constituent of fossil fuels and coal derivatives. Several bacteria able to use fluorene as their sole source of carbon and energy belong to the genera of *Arthrobacter, Brevibacterium, Burkholderia, Mycobacterium, Pseudomonas,* and *Sphingomonas* [57].

5.3.3.1 Degradation of High-Molecular-Weight Polyaromatic Hydrocarbons

Pyrene is the commonly used model to study the degradation of high-molecular-weight aromatic compounds. *Mycobacterium* has been extensively studied and is a well-known genus to mineralize high-molecular-weight PAHs such as fluoranthene, pyrene, and benzo[a]pyrene [58]. *Mycobacterium vanbaalenii* PYR-1 strain exhibits genes coding for dioxygenases and monooxygenases for catabolism of pyrene [59, 60]. The mycolic acid in the cell walls of these bacteria facilitates the uptake of polyaromatic hydrocarbons [33, 61]. Other high-molecular-weight PAH-degrading bacterial species include *Mycobacterium* sp., *Gardona* sp. *Rhodococcus sp., P. putida, Pseudomonas aeruginosa, P. fluorescence, Burkholderia cepacia, S. yanoikuyae, Flavobacterium* sp. *and Cycloclasticus* sp.

Benzo[a]pyrene (BaP), a five-ring molecule, is abundantly present as an active component of coal tar.

The low water solubility and high oxidation state of the carbons in BaP make it relatively resistant to microbial degradation. The large amount of energy needed to break carbon–carbon bonds makes it less preferable to be used as a source of energy by microorganisms [62].

Bacterial strains isolated from a contaminated site and human skin degrade BaP via co-metabolic mechanism [63, 64]. Some of these bacterial species mineralize BaP-producing nontoxic CO_2 as a by-product with the addition of external carbon source. The pathway utilized by bacteria consists of the rate-limiting initial oxidation by dioxygenase to cis-dihydrodiol [28], the formation of catechol by cis-dihydrodiol dehydrogenase, which becomes a substrate for other dioxygenases that form ring cleavage products. These products could then be either excreted or shuttled into pathways that metabolize smaller PAHs, which are eventually converted into TCA intermediates. Bacteria utilizing benzo[a]pyrene as the sole source of carbon and energy were isolated from enriched mangrove sediments and were identified as *Ochrobactrum anthropi*, *Stenotrophomonas acidaminiphila*, and *Aeromonas salmonicida ss salmonicida* [65].

Mud generated from oil and natural gas drilling when treated with naturally occurring microorganisms from the contaminated site reduced the levels of the petroleum hydrocarbon from 32.2 to 1.5 g kg/1 (95% degradation) within six months [66]. The isotopic labeling and DNA approaches revealed the presence of the microbial community at the site of Deepwater Horizon oil spell that contributed to the degradation of the petroleum hydrocarbons at the site [67]

5.3.4 Fungal Degradation of Polyaromatic Hydrocarbons

Several studies have identified numerous fungal species that catabolize different PAHs in the environment [68]. Polyaromatic hydrocarbons are not utilized directly by fungi instead are co-metabolized to yield either oxidized products or carbon dioxide. Based on the type of degradation mechanism, fungi are classified into two types – ligninolytic fungi or white-rot fungi (uses extracellular enzymes, lignin peroxidase (LiP), manganese peroxidase (MnP), and laccases) and non-ligninolytic fungi (cytochrome P450 monooxygenase like enzymes) [69, 70]. Ligninolytic fungi belong to the group basidiomycetes and non-ligninolytic fungi belong to the group zygomycetes, ascomycetes, and hyphomycetes.

White-rot fungi, *Phanerochaete chrysosporium* and *Pleurotus ostreatus* and *Trametes* utilize both ligninolytic and non-ligninolytic enzymes for the degradation of the PAHs [71].

5.3.4.1 Bioremediation of PAHs by Ligninolytic Fungi

Ligninolytic fungi (white-rot fungi) produce mainly two types of enzymes, peroxidases and laccases. Peroxidases are LiP and MnP [72]. Laccases are copper-containing oxidases that oxidize phenolic and aromatic compounds. LiP and MnP catalyze the oxidization of PAHs to PAH quinones by donating one electron produced by hydroxylation of free radicals in the presence of hydrogen peroxide with heme as cofactor. Wood-rotting fungi, like *P. ostreatus* and *Antrodia vaillantii*, degrade PAH in soil [73]. Rate of degradation by ligninolytic fungal strains varies with a variation of lignolytic enzymes [74]. Unlike bacterial intracellular enzyme system, filamentous nature of the fungi allows them to penetrate the soil and secrete ligninolytic enzymes extracellularly that diffuse into the pores of the soil contaminated with PAH. Ligninolytic enzymes are non-specific; therefore, they oxidize a wide range of aromatic compounds. The oxidation of PAH by ligninolytic enzymes is influenced by the ionization potential of the aromatic compound; MnPs are able to degrade PAH with ionization potential of up to 8.2 eV, while LiP can oxidize PAH with ionization potential less than 7.55 eV [75].

5.3.4.2 Catabolism of PAHs by Non-Ligninolytic Fungi

In non-ligninolytic fungi, cytochrome P450 monooxygenase enzymes catalyze the oxidation of PAHs to arene oxides, which are initial products of PAH metabolism. Arene oxides being unstable either undergo rearrangement to form phenol or are hydrated to trans-dihydrodiol by an epoxide hydrolase. The phenol and trans-dihydrodiol formed are conjugated with sulfate, glucose, glucuronic acid, or xylose. *Cunninghamella elegans* degrade benzo[a]pyrene to 10-hydroxy-3-benzo[a]pyrenesulfate, fluoranthene to fluoranthene trans-2,3-dihydrodiol, 8-and 9-hydroxyfluoranthene trans-2,3-dihydrodiol [76]. Cytochrome P450 monooxygenases contain heme as cofactor and are present in the endoplasmic reticulum. Genome-wide studies have shown that fungal P450 monooxygenases catalyze the oxidation of varying ring-size PAH compounds, namely, phenanthrene, pyrene, and BaP [77]. Soil heavily contaminated with polyaromatic hydrocarbons when treated with fungi *Phanerochaete velutina* for three months showed 94% degradation of PAH both at the laboratory and field scale. Based on the toxicity tests, soil quality was significantly improved after field-scale treatment [78]. The PAH degrading potential of *Anthracophyllum discolor*, a white-rot fungus was evaluated in soil containing a mixture of PAH. The high degradation capacity of different PAH (62% phenanthrene, 73% anthracene, 54% fluoranthene, 60% pyrene, and 75% benzo(a)pyrene) by the fungus was associated with the production of MnP [79]. An overview of the pathways involved in bioremediation of organic pollutants is described in Figure 5.1.

5.3.5 Bioremediation of Pesticides by Microbes

Microorganisms with pesticide degrading capacity have been isolated from a wide variety of pesticide-contaminated sites. The soil is the medium that mostly gets these chemicals when they are applied to agricultural crops; additionally, the pesticide industry's effluent,

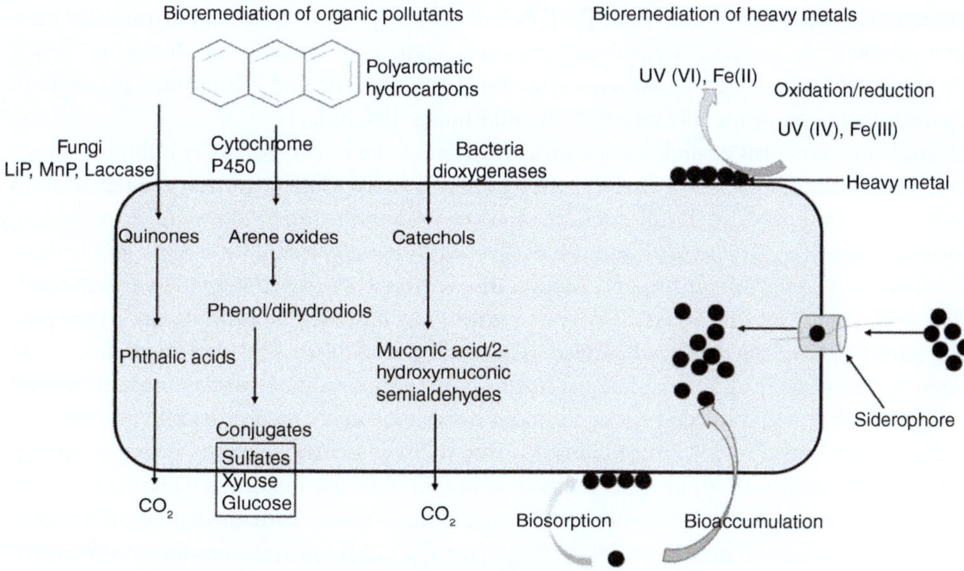

Figure 5.1 Metabolic pathways for bioremediation of organic pollutants and heavy metals.

sewage sludge, activated sludge, wastewater, natural waters, sediments, and areas surrounding the manufacture of pesticides are also a rich source of pesticide degrader.

Dieldrin, aldrin, alpha-BHC, beta-BHC, deltaBHC, gamma-BHC (Lindane), heptachlor, endosulfan, and dichloro diphenyl trichloro ethane (DDT) are some of the most widely used organochlorine pesticides. Organochlorine pesticides are more toxic and recalcitrant to degradation compared to organophosphorus. Among the organophosphorus compounds, glyphosate, chlorpyrifos, parathion, methyl parathion, diazinon, coumaphos, monocrotophos, fenamiphos, and phorate have been used extensively.

The complete breakdown of pesticides into inorganic components is termed as biomineralization. In some cases, the degradation leads to the formation of less complex and less toxic organic compounds, referred to as partial biodegradation, while in some conditions, partial degradation results in formation and accumulation of more toxic and less soluble metabolites compared to the parent compound. In such cases, the intermediate metabolites inhibit the growth and proliferation of other microorganisms as well.

For example, a strain of *Pseudomonas stutzeri* utilizes parathion as a carbon source releasing diethylphosphorothioanate but could not metabolize it further even as an alternate source of phosphorus or sulfur. Also, degradation of endosulfan in soil by some fungi and bacteria leads to the formation of toxic endosulfan sulfate and less toxic endosulfan diol [80]. Other example for partial degradation is DDT, which undergoes degradation to form metabolites like dichlorodiphenyldichloroethylene and [dichloro-2,2-bis(p-chlorophenylethane)] that are toxic and more persistent than the parent compound.

Pesticide-degrading bacterial species belong to the genera *Flavobacterium*, *Arthrobacter*, *Azotobacter*, *Burkholderia*, and *Pseudomonas*. The degradation potential varies among species and with the nature of compound. Some species of *Bacillus* and *Pseudomonas* cannot degrade chlorpyrifos completely, while *Staphylococcus* sp. ES-2 and *Streptomyces* sp. HP-11 have metabolic pathways for complete degradation of chlorpyrifos and its hydrolysis product TCP into polar metabolites [81]. *Pseudomonas* sp. and *Klebsiella pneumoniae* have been shown to possess hydrolytic enzymes that are capable of breaking down s-triazineherbicides, such as atrazine [82, 83]. *Pseudomonas* and *Alcaligenes* sp. degrade organochlorine pesticide like endosulfan and lindane [84, 85].

Fungi bio-transform pesticides by converting them to an intermediate by inducing minor structural changes and is released to soil, where it is susceptible to further biodegradation by bacteria [86]. Several fungi such as *Agrocybe semiorbicularis, Auricularia auric-ula, Coriolus versicolor, Dichomitus squalens, Flam-mulina velupites, Hypholoma fasciculare, P. ostreatus, Stereum hirsutum*, and *Avatha discolor* have ability to degrade various pesticides [87]. White-rot fungi (e.g., *P. chrysosporium*) has high degradation because they possess lignin peroxidase and manganese-dependent peroxidases that can degrade a wide range of pesticides [88]. *Phanerochaete chrysosporium* mineralize bound chloroaniline and 2, 4-dichlorophenol, such that these compounds become available for degradation [89].

Microorganisms transform pesticides by using them as a carbon source, nitrogen source, any other mineral source, or a final electron acceptor in the respiratory chain. Pesticide degradation occurs in the metabolic and co-metabolic process. Pesticide degradation occurs by different metabolic pathways depending upon the nature of the microbe and the environmental conditions. The principal reactions involved are oxidative, reduction, hydroxylation, and conjugation.

Oxygenation is generally the first step in the biotransformation of pesticides. The oxidative transformation is mediated by oxidative enzymes (cytochrome p450, peroxidases, and polyphenol oxidases, peroxidase, laccase, and tyrosinase).

Hydrolytic Transformations: Hydrolytic enzymes cleave the bonds of a substrate by adding hydrogen or hydroxyl group from water molecules to the product. Esterase and few lipases and proteases are involved in ester hydrolysis. Four types of esterases have been characterized in *Pseudomonas fluorescens*, each differing in protein structure, cellular localization, and substrate specificity. *Pseudomonas* sp. strain ADP, carries three *atz* genes on a self-transmissible plasmid pADP-1 that can completely mineralize atrazine [90].

Aromatic Nitroreductive Processes: Reductive metabolism of nitroaromatic compounds is mediated by nitroreductase enzymes that are found in aerobic and anaerobic bacteria, and several genera of fungi [91]. The anions like chloride, sulfate when strongly bond to the hydrocarbon ring prevents the microbes from attacking the ring structure, reductive transformation facilitates the removal of the anions in the ring structure such that the compound becomes more susceptible for degradation by microbes [92].

In bacteria, three pathways of reductive metabolism of nitroaromatics have been characterized: aromatic nitroreduction, partial nitroreduction, and hydrogenation [91]. Nitroreductases are flavoproteins that use NAD(P)H as reducing equivalents, require FMN/FAD as cofactors, and have varying sensitivities to O_2 concentrations. A gene cluster designated *phn*, consisting of 17 genes from *Escherichia coli*, is responsible for the degradation of a wide range of phosphonates as it encodes for C–P lyase[93].

Pesticide Conjugation Reactions: The pesticide conjugation is termed as the metabolic process, whereby an exogenous or endogenous natural compound is complexed to a pesticide or its metabolite(s) facilitating detoxification, compartmentalization, sequestration, and mineralization. This process includes xyloxylation, alkylation, acylation, and nitrosylation. This type of biotransformation is common in the fungal biodegradation of pesticides. The bio-transformed pesticide is released into the soil, where it is susceptible to further metabolism by bacteria. Both fungi and bacteria use methylation as a common conjugation reaction to detoxify xenobiotics [94]. For example, the formation of O-methylated pentachlorophenol by *Trichoderma virgatum* and bacterial cultures, which is less toxic, but more recalcitrant [95].

5.4 Microbial Degradation of Heavy Metals

The rise of industrialization has led to the increased release of heavy metals in the environment, which has become a cause of concern for human and environmental health. Microorganisms such as bacteria, fungi, or algae have the ability to transform toxic metals into less toxic forms by altering the ionic state of the metals. Microorganisms adopt different strategies to adjust to and transform heavy metals in the contaminated areas. Microorganisms have adopted different strategies to interact and survive in heavy metal contaminant areas. Microbes use various approaches to transform the element into a less harmful form [96]. The adsorption of metals by the microbes is influenced by the composition of the cell envelope and its adsorption capacity. The presence of groups carrying negative charge (carboxyl, amino, phosphoryl, and sulfo) on the cell wall of microbes helps them to attach to the heavy metals by the mechanism of proton exchange. The heavy metal bioremediation initiates with the process of mobilization or immobilization by redox reactions, ion exchange, and complexation [97].

In the process of immobilization, heavy metals like Fe, As, Cr, and Hg are converted into their insoluble or stationary form in sediments by either enzymatic oxidation, reduction, complexation, or siderophores formation. Mobilization helps to transform the toxic metals to mineral or organic acids and increases their bioavailability such that heavy metals can reach microbial metabolic systems [98]. In this process, some heavy metals lose electrons and get transformed into a useful or less toxic state. One of the examples is the oxidation of uranium by *Acidothiobacillus ferrooxidans* and *Acidothibacillus thiooxidans* from its ore where oxidation of UV (IV) to UV (VI) is coupled with the redox reduction of Fe (III) to Fe (II). Denison Mines in Ontario, Canada in 1984, produced 10–15% of the total uranium using the above approach, and the extraction efficiency was observed to be within the range of 70–80% [99]. Microbes also enhance the solubility of ions like Fe(III) and As(V) by reducing them to Fe(II) and As(III), respectively, thereby promoting leaching from soil [100]. The idea that bacteria can oxidize arsenite to arsenate was exploited to detoxify drinking water in Bangladesh. In the process, ferrous ions are oxidized abiotically to ferric [Fe(III)] iron, which then precipitates and immobilizes arsenic [101].

Other detoxification mechanisms of heavy metals include complexation where a complex of the inorganic metals is formed by the addition of the ligand that is easily removed from the solid waste. Complexation is achieved through metal binding, vacuole compartmentalization, and volatilization. The major microbial complexing agents are citric acid, tricarboxylic acids, and alcohols, siderophores, and toxic metal-binding compounds [97].

Siderophores are low-molecular-weight high-affinity specific iron chelators that are excreted by microbes under iron starvation conditions. The microbes acquire iron by complexing the ferric ions with excreted siderophores and forming a ferri-siderophore complex. The complex binds to specific receptor present on the microbial cell surface. The complex gets translocated by active transport and is released inside the cell. Some of the siderophore-producing bacteria are *Micrococcus luteus* and *Bacillus silvestris* [102]. Siderophores are classified into hydroxamate, catecholate, and carboxylate types based on their chemical properties [103]. Siderophores are specific for Fe (III), but *P. aeruginosa* chelate a wide range of metals, i.e. Ag^{+}, Al^{3+}, Cd^{2+}, Co^{2+}, Cr^{2+}, Cu^{2+}, Eu^{3+}, Ga^{3+}, Hg^{2+}, Mn^{2+}, Ni^{2+}, Pb^{2+}, Sn^{2+}, Tb^{3+}, Tl^{+} and Zn^{2+} [104]. The siderophores produced by *Fusarium solani* complex with Cu and Zn ions [105].

Siderophores are classified into hydroxamate, catecholate, and carboxylate types based on their chemical properties [103]. The hydroxamate siderophores are the most common group of siderophores that consists of C (=O) N–(OH) R, where R is an amino acid, and its derivatives are mainly produced by bacteria and fungi. The catecholate siderophores bind iron through the formation of the hexadentate octahedral complex. Several well-known bacteria, such as *E. coli* and *Salmonella typhimurium* produce these types of siderophores [106]. The carboxylate siderophores bind to iron ions through the carboxyl and hydroxyl groups and are produced by rhizosphere bacteria such as *Rhizobium*.

Microorganisms bind to and concentrate the heavy metals from the environment by the process of biosorption and bioaccumulation [107]. Biosorption is the physiochemical process during which the heavy metals bind to the cellular surface of microorganisms. Biosorption is a passive and faster process, while bioaccumulation is an active metabolic process.

Biosorption is a cost-efficient method for reducing heavy metal contamination from industrial waste. The capacity of biomass to recover the metal ions depends on physical, chemical, and biological properties. Since the cell wall of microbes is the first site that makes contact with the metals, the nature of the cellular components plays an important role in binding with the metals. In common, increasing the pH increases the overall negative charge on the cell surface, which favors the electrochemical attraction and adsorption of cations. Anions interact more strongly with cells with increasing concentration of positive charges. The anionic functional groups present in the peptidoglycan, teichoic acids, and teichuronic acids of Gram-positive bacteria, and the peptidoglycan, phospholipids, and lipopolysaccharides of Gram-negative bacteria are the components primarily responsible for the anionic character and metal-binding capability of the cell wall [108]. The carboxyl groups of the cell wall peptidoglycan of *Streptomyces pilosus* are responsible for the binding of copper; similarly, amine groups of *Corynebacterium glutamicum* are responsible for the binding to anions via electrostatic attraction. Functional groups like carboxyl, phosphonate, amine, and hydroxyl groups on the cell wall facilitate these interactions.

Micrococcus luteus possess the capacity to remove copper (408 mg/g) and for lead (1965 mg/g) using the biosorption technique [109]. The biomass of *C. glutamicum* obtained after the lysine fermentation has higher lead uptake capacity (up to 2.74 mmol g/1 at pH 5) compared to other commercially available expensive sorbents like natural zeolite, activated carbon, and synthetic ion exchange resin [110].

Metal-degrading microbes include bacterial species like *Pseudomonas*, *Desulfovibrio*, *Bacillus*, and *Geobacter*, fungi such as *Rhizopus arrhizus*, yeast such as *Saccharomyces cerevisiae*, and algae such as *Chaetomorpha linum* and marine microalgae (seaweed) [108, 111]. Compared with other microbes, bacteria are considered to be better biosorbent due to their high surface-to-volume ratios, as well as the presence of several potential active adsorption sites in their cell wall.

Bioaccumulation is an energy-dependent process where microorganisms take up metals into their intracellular space using translocation complexes through the lipid bilayer.

Inside the intracellular space, these metals are sequestered by peptide ligands. The bioaccumulation of metals such as mercury, lead, silver, cadmium, nickel, caesium, cobalt, chromium, and uranium is carried out by microbes using ion pumps, ion channels, endocytosis, and lipid permeation [112, 113]. During bioaccumulation, the heavy metal ions make an inactive complex with other high-affinity ligands. Some bacteria produce metal-binding proteins such as metallothioneins (MTs) to increase the metal binding capacity as a response to increased metal exposure [114, 115]. Recombinant *E. coli* strain was designed to improve Cd^{2+} binding capability by inserting a gene coding for a metal-binding peptide (phytochelatins) with a repetitive metal-binding motif (Glu-Cys) nGly [116].

Biosurfactants are surface-active agents produced by the microbes that alter the conditions of surfaces through adsorption, thereby lowering surface tension between liquids or between a liquid and a solid [117, 118]. The amphipathic nature of biosurfactants (hydrophilic and hydrophobic moieties) help them to interact with any type of surface and modify these surfaces. Biosurfactants are either high-molecular-weight polymers that bind tightly to solid surfaces or are low-molecular-weight molecules that lower the surface and interfacial tensions [117]. Based on the nature of the lipid molecules, they are categorized as glycolipid, lipopeptide, or protein–lipid–polysaccharide complex.

Bacillus sp. isolated from a marine sponge produced anionic lipopeptide biosurfactant that was named as MSI 54 lipopeptide. The efficacy of the biosurfactant for removal of heavy metal was analyzed using atomic absorption spectroscopy and energy-dispersive X-ray spectroscopy on the surface of fresh vegetables and wastewater; it showed high affinity against cationic heavy metals such as Pb, Cd, Hg, and Mn [119].

Rhamnolipids are a major class of biosurfactants that are produced by *P. aeruginosa* [120]. Rhamnolipids are glycolipid moieties associated with rhamnose (glycan) that capture heavy metal ions through electrostatic or complexation techniques. The anionic nature of the rhamnolipids complexes with metals in the order $K^+ < Mg^{2+} < Mn^{2+} < Ni^{2+} < Co^{2+} < Ca^{2+} < Hg^{2+} < Fe^{3+} < Zn^{2+} < Cd^{2+} < Pb^{2+} < Cu^{2+} < Al^{3+}$. Anionic biosurfactants like rhamnolipid, sophorolipid, and surfactin chelate cations like metals [121]. Rhamnolipids are also effective in the removal of hydrocarbon. Rhamnolipid when added to artificially contaminated soil, removed 80–100% cadmium and lead in the soil [122]. There are several other biosurfactants produced by a wide range of bacterial and yeast species such as sophorolipid, cellobiose lipids, trehalose lipids, surfactin, and subtilin [123].

5.5 Factors Affecting Bioremediation

It was not until the early 1980s that the bioremediation trials were performed under natural environmental conditions, and it was realized that the environmental factors have a pronounced effect on the remediation process. The environmental factors include physicochemical nature of the contaminated environment, nutrient availability, temperature, moisture/water content, pressure, and pH. Besides environmental factors, various biotic factors affect the behavior and the growth of microbial cells. Insufficient information on the factors influencing the process often reduces the efficacy of the process. As discussed before, bioremediation is an outcome of three-way interaction that is a pollutant, the environmental matrix, and microbial characteristics.

5.5.1 Abiotic Factors

The metabolic reactions in the microbial system are involved in the degradation of chemical pollutants; these reactions are carried in the presence of enzyme as the catalyst, enzyme performs optimally only over a very narrow range of physicochemical parameters. Deviation from the optimal conditions reduces the efficiency of the degradation reaction. Few microorganisms can adapt in adverse conditions and carry out optimal degradation over a range of physicochemical parameters. Biosorption, a key step in toxic metal removal by microbes, is a pH-dependent phenomenon, whereby pH influences the isoelectric point in a solution and the net negative charge on the microbial cell surface. For example, the optimum value of pH for the solubility of phosphorus is 6.5, and hence below and above this, the solubility declines.

Pollutant bioavailability is also one of the important factors that is affected by the physiochemical parameters. pH values affect the solubility of metal ions and the organic pollutants in the environment, thereby affecting the adsorption by microbes. Cations, such as NH_4^+, Mg^{2+}, Ca^{2+} are more soluble at acidic pH, whereas anionic forms (NO^{3-}, NO^{2-}, PO_4^{3-}, Cl^-) are better solubilized at alkaline pH. The optimum pH for most bacteria is between 6 and 8, for filamentous fungi between three and seven, and between four and seven for yeast [124].

Depending on the type of microbes (mesophilic, thermophilic, or psychrotrophics), degradation is carried out at different temperature ranges, like for mesophilic microbes, temperature range from 15 to 45°. Temperature affects the viscosity and solubilization of contaminants as well the rate of transport of nutrients through the cell membrane. The metabolic activity involved in bioremediation is more efficient during the spring and summer seasons as for many microbes, the enzymes denature at high temperature (>40°). High pressure and low temperature present in deep oceans are not favorable for microbial activity.

The presence of electron acceptors in the contaminated area also affects the process of bioremediation. As explained earlier, the primary step for the degradation of pollutants is the oxygenation of the substrate by the addition of molecular oxygen by oxygenase. The aerobic bioremediation largely depends on the oxygen availability at the site of contamination. Anaerobic degradation is a slow process and is normally important for catabolic biodegradation pathways [125].

Furthermore, nutrients like nitrogen, calcium, sulfur, magnesium, and other trace elements are required for microbial growth and cellular metabolism. Enough nutrients are required for efficient microbial degradation. An appropriate ratio of carbon to nitrogen and phosphorus is also essential; ammonia and phosphate serve as a source of nitrogen and phosphorus, respectively. Nitrogen is needed for the synthesis of amino acids and proteins, while phosphorus is used in the process of synthesis and degradation of cellular components. Water helps in the transportation of nutrients and other components to the microbial system. The quantity of water in the contaminated area has a great influence on the metabolic process as it is associated with oxygen exchange. Excessive water content has a negative impact as it clogs the soil and prevents the oxygen exchange across the soil, thereby alleviating the process of bioremediation.

5.5.2 Biotic Factors

The microbial population and its inherent degradation machinery (plasmid) within the contaminated environment are some of the most significant determinants for the outcome of the bioremediation process. Plasmid-encoding genes for specific enzymes that are specific for the substrate is one of the factors that affect substrate degradation. Degradation of some of the pollutants is performed by a mixed population of microbes known as microbial consortia. The consortia have high biodegrading potential as the microbes carry a large repertoire of enzymes that help them to adapt to the containment area. There exits division of labor inside the microbial consortium, where one species initiates the reaction while others finish the process. The species that initiate the process of degradation are known as primary consumers, whereas those degrading intermediate products are considered as secondary consumers. Secondary consumers facilitate the growth of primary consumers by secreting products that act as growth factors.

Microbes adapt to the contaminant environment for utilization of a specific contaminant by three different mechanisms that include induction or repression of specific enzymes, genetic changes that result in the appearance of new metabolic potentials capabilities, and the selective enrichment of a specific sub-population with the ability to transform the target compound.

5.6 Advances in Bioremediation

With the recent advancements in the field of biotechnology (gene editing, bioinformatics) and microbiology, microorganisms, their degradation pathways, and corresponding enzymes are engineered in a way to enhance the in-situ degradation of target pollutants. The advent of these techniques provided an economically feasible and environmentally friendly way to mineralize the pollutants compared to the costly chemical-based approaches.

The combination of system biology and metabolic engineering facilitates the designing of microbes with significant potential for the degradation of pollutants. The system biology approach provides a quantitative understanding of the microbial system and its behavior under different environmental conditions at the molecular, cellular, population, community, and ecosystem levels [126, 127].

Metabolic engineering uses systems biology and "omics" techniques (genomics, transcriptomics, metabolomics, and proteomics) to provide deeper insight into the genetics and physiology of the complex microbial metabolic pathways. The aim of the omics tools is to manipulate the metabolic pathways or alter enzymatic reactions to improve the synthesis of products that are relevant for biodegradation.

The steps involved in engineering a metabolic degradative pathway using system biology and metabolomics tools are (i) first is to know the pollutant to be removed from the contaminated environment and the pathway involved, (ii) to find a suitable host, its genetics, and the factors responsible for microbial growth, (iii) last is to modify the existing pathway or build and optimize a new pathway in the selected host system.

A wide spectrum of computational tools like databases, algorithms, pathway prediction systems (PPDs) have been designed that provide information on various aspects like physicochemical properties of the pollutant, enzymes, and metabolites involved in the catabolic pathway [128]. The chemical and biodegradation databases are useful for evaluating the current biological pathways [129]. The software tools developed grant basic knowledge for bioengineering a metabolic pathway.

The chemical toxicity predicting software tools utilize the quantitative structure regulatory activity relationship models to calculate the toxicity of the pollutants. The models include *Sarah Nexus*, *VirtualToxLab*, *Toxicity Estimation Software Tool*, and *ToxiPred*. [130].

Biodegradative databases comprise the University of Minnesota Biocatalysis/Biodegradation Database (EAWAG-BBD), Biodegradation Network-Molecular Biology database (Bionemo), a database of biodegradative oxygenases (oxDBase), Metacyc, BioCyc. EAWAG-BBD was initially developed by the University of Minnesota, USA, and later hosted by Eaway (Swiss Federal Institute of Aquatic Science and Technology), which is a remarkable tool that contains information related to microbes, enzymes, genes, and reactions involved in the microbial degradation pathways of toxic chemicals. EAWAG-BBD database comprises 219 microbial degradation pathways, 1503 chemical reactions, 993 enzymes, 534 microorganisms, 249 biotransformation rules, 50 organic functional groups, 76 reactions of naphthalene 1,2-dioxygenase, and 109 reactions of toluene dioxygenase.

MetaCyc and BioCyc databases are developed by SRI International. MetaCyc is a curated database of experimentally elucidated metabolic pathways from all domains of life; it contains 2801 pathways from 3123 different organisms. BioCyc is a collection of 16817 Pathway/Genome Databases, plus software tools for exploring them. Key aspects of BioCyc database are that the availability of computationally predicted metabolic pathways and operons and have data integrated from other databases, including gene essentiality, regulatory networks, protein features, and GO annotations. Besides providing the relevant information, their application assists in identifying a reaction or the enzyme that can be modified [130]. These databases are interconnected with more specific protein databases such as RCSB Protein Data Bank or BRENDA or ExPASy or Enzyme or NCBI to provide information describing genes and enzymes involved in the degradation reactions.

Another database, OxDBase: A database of biodegradative oxygenases, is developed by the CSIR-Institute of Microbial Technology, Chandigarh, India, stores information on oxygenases derived from published literature and databases [131]. Oxygenases, as discussed before, are one of the most important enzymes involved in the aerobic degradation of aromatic compounds. OxDBase contains information on over 240 oxygenases, including monooxygenases and dioxygenases. All enzyme entries contain information about the reactions in which enzymes are involved, their family and subfamily, structures and gene links, and literature citations. The information stored in this database is linked to other databases like UM-BBD, BRENDA, ENZYME, and LIGAND.

Several PPDs have been developed to predict biodegradation pathways for the unexplored toxic compounds. The UM-BBD- PPS is part of EAWAG-BBD Pathway Prediction system that predicts plausible pathways for microbial degradation of chemical compounds based on the biotransformation rules and reactions found in the EAWAG-BBD database or the scientific literature [132]. It can predict the biodegradation pathways under different conditions (like aerobic or anaerobic).

Other PPDs include PathPred, Biochemical Network Integrated Computational Explorer, From Metabolite to Metabolite (FMM), Metabolic Tinker [133, 134]. These pathways prediction tools supply information on the metabolic pathways of different organisms using the data derived from the Kyoto Encyclopedia of Genes and Genomes (KEGG) in the form of the KEGG REACTION database and KEGG repair database [135]. FMM provides information on post-translational modification (PTM) of enzymes from numerous species.

The second crucial step is the selection of hosts. Numerous microbial species have been studied over the years for the biodegradation of toxic compounds, but no single species possess all characteristics required for optimal degradation. *P. putida* KT2440, a saprophytic pseudomonad, has the ability to survive and thrive in a wide range of environmental conditions. The multi-omics approach has made it feasible to study the genes encoding proteins for metabolic degradation pathways within a model organism. 16S rRNA-phylogenetic microarrays, Phylochip, have been used to study microbial communities. Next-generation sequencing provided the whole genomes of several pollutant degrading bacteria. Whole-genome sequencing of *Arthrobacter* sp. strains LS16 and YC-RL1 has revealed that they contain metabolic networks that are involved in the bioremediation of several aromatic compounds like naphthalene, fluorene, 4-nitrophenol, bisphenol A, biphenyl, and p-xylene [136].

Complete genome sequence combined with transcriptomics and proteomics allows the users to identify the status of RNA and proteins involved in the pathway. Furthermore, metabolomics and flux analyses are effective in determining the rate of metabolic reactions on a real-time basis and to study cellular metabolites produced under specific environmental conditions. Genomic sequences of microorganisms with biodegradation capabilities are available through comprehensive databases such as NCBI genome database or MicroScope, a microbial genome annotation, and analysis platform [137]. Flux balance analysis and metabolic flux analysis are two major approaches to use metabolic models to predict intracellular fluxes.

The third step is to build the pathway and optimize its performance in the context of host metabolism. The information obtained from the databases and omics tools assist in constructing a metabolic biochemical pathway in a host cell by assembling various enzymes from different organisms. One of the examples of the application of computational tools for engineering a metabolic pathway is for 1,2,3-trichloropropane (TCP), which is also highly toxic and recalcitrant to biodegradation. The synthetic pathway for aerobic degradation of TCP was integrated into the chromosome of *P. putida* KT2440 by functionally arranging various TCP-degrading enzymes from different bacteria. Figure 5.2 represents the workflow for the development of genetically engineered microbes with biodegradation pathways for improving the process of bioremediation. The introduction of new codons into bacteria host might result in the accumulation of intermediate metabolites that inhibit the growth, or the new enzymes introduced might crosstalk with the native metabolic pathways. The adverse effects of introducing new genes in the host cells are taken into considered and quantitatively characterized by computational modeling. Scientists developing *E. coli* BL21(DE3) strain overexpressing the artificial pathway for degradation of TCP created a unique mathematical model to predict the effects of metabolite toxicity and burden to assess the behavior of *E. coli* under different conditions [138]. Similarly, the degradation of TCP by *P. putida* was improved by optimizing different engineering approaches in a way

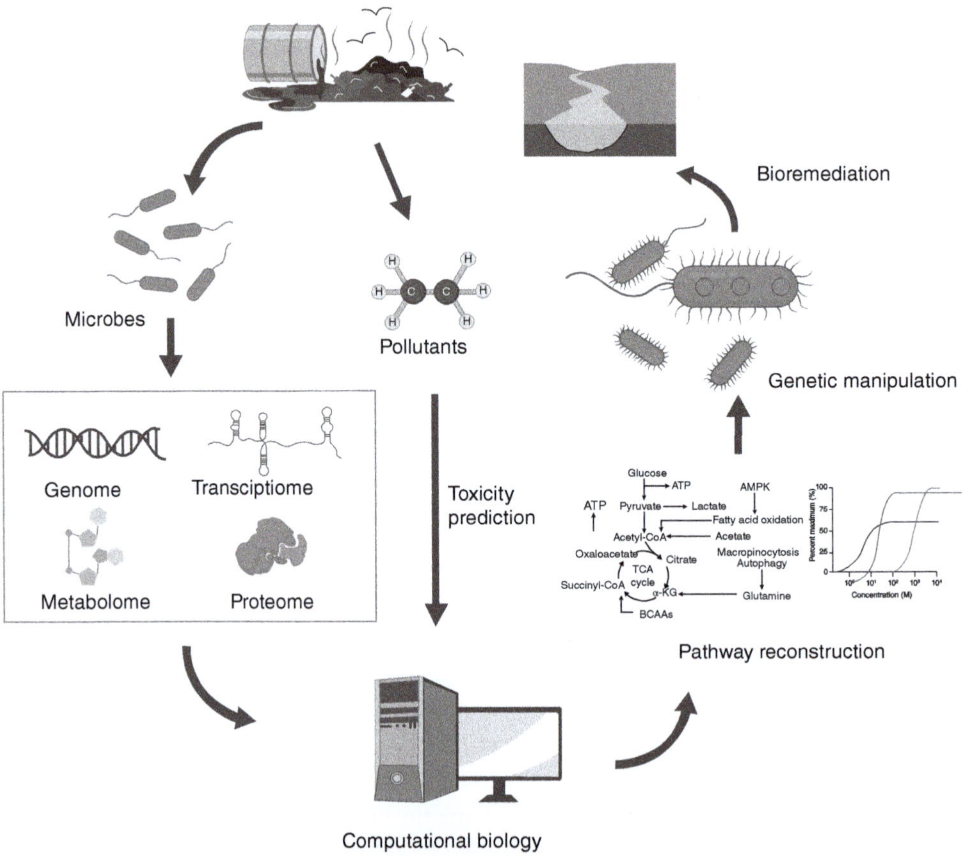

Figure 5.2 Workflow for the development of genetically engineered microorganisms for bioremediation. *Source:* Based on Biorender software.

that P. putida was able to utilize TCP efficiently as the sole source of carbon for growth [139]. An immobilized synthetic metabolic pathway composed of enzymes from three different microorganisms (haloalkane dehalogenase from *Rhodococcus rhodochrous*, and haloalcohol dehalogenase and epoxide hydrolase from *Agrobacterium radiobacter*) has been designed for completed degradation of TCP to glycerol. The immobilized pathway was observed to convert a high percentage of TCP to glycerol when tested using a packed bed reactor, also removed TCP from water with high efficiency [140].

5.7 Conclusions and Perspectives

Biodegradation is an eco-friendly and cost-effective method for the removal of pollutants from the environment through microbial activity. With the increasing number and type of pollutants in different habitats, bioremediation will be an effective and reliable method for the recovery of contaminated sites. As there are several factors that limit the efficiency of microbial degradation, it is critical to understand the structure and functions of microbial communities in the impacted area and how they respond and adapt to different conditions. Developing recombinant microbial strains and overcoming the rate-limiting factors for the degradation process would be beneficial to the bioremediation industry, as well as minimize the risks of further spreading of pollutants in the ecosystem.

References

1 Kedderis, G.L. (2010). 1.07 – biotransformation of toxicants. In: *Comprehensive Toxicology*, 2e (ed. C.A. McQueen), 137–151. Oxford: Elsevier.

2 Jørgensen, S.E. (2008). Biodegradation. In: *Encyclopedia of Ecology* (eds. S.E. Jørgensen and B.D. Fath), 366–367. Oxford: Academic Press.

3 Riccardi, C. (2005). Characterization of bacterial population coming from a soil contaminated by polycyclic aromatic hydrocarbons (PAHs) able to degrade pyrene in slurry phase. *Annals of Microbiology* 55 (2): 85–90.

4 National Research Council (1993). *In Situ Bioremediation: When Does it Work?* Washington, DC: The National Academies Press https://doi.org/10.17226/2131.

5 Hlihor, R.-M., Apostol, L.-C., and Gavrilescu, M. (2017). Environmental bioremediation by biosorption and bioaccumulation: principles and applications. In: *Enhancing Cleanup of Environmental Pollutants: Volume 1: Biological Approaches* (eds. N.A. Anjum, S.S. Gill and N. Tuteja), 289–315. Cham: Springer International Publishing.

6 Tyagi, M., da Fonseca, M.M.R., and de Carvalho, C.C.C.R. (2011). Bioaugmentation and biostimulation strategies to improve the effectiveness of bioremediation processes. *Biodegradation* 22 (2): 231–241.

7 Chojnacka, K. (2010). Biosorption and bioaccumulation – the prospects for practical applications. *Environment International* 36 (3): 299–307.

8 Curtis, T.P., Sloan, W.T., and Scannell, J.W. (2002). Estimating prokaryotic diversity and its limits. *Proceedings. National Academy of Sciences. United States of America* 99 (16): 10494–10499.

9 Watanabe, K. (2001). Microorganisms relevant to bioremediation. *Current Opinion in Biotechnology* 12 (3): 237–241.

10 Hazen, T.C. (2010). Cometabolic Bioremediation. In: *Handbook of Hydrocarbon and Lipid Microbiology* (ed. K.N. Timmis), 2505–2514. Berlin, Heidelberg: Springer.

11 Guimarães, B.C.M., Arends, J.B.A., van der Ha, D. et al. (2010). Microbial services and their management: recent progresses in soil bioremediation technology. *Applied Soil Ecology* 46 (2): 157–167.

12 Karigar, C.S. and Rao, S.S. (2011). Role of microbial enzymes in the bioremediation of pollutants: a review. *Enzyme Research* 2011: 805187.

13 van Beilen, J.B., Wubbolts, M.G., and Witholt, B. (1994). Genetics of alkane oxidation by Pseudomonas oleovorans. *Biodegradation* 5 (3–4): 161–174.

14 Rojo, F. (2009). Degradation of alkanes by bacteria. *Environmental Microbiology* 11 (10): 2477–2490.

15 Austin, R.N. and Groves, J.T. (2011). Alkane-oxidizing metalloenzymes in the carbon cycle. *Metallomics* **3** (8): 775–787.

16 Hamme, J.D.V., Singh, A., and Ward, O.P. (2003). Recent advances in petroleum microbiology. *Microbiology and Molecular Biology Reviews* 67 (4): 503–549.

17 Wentzel, A., Ellingsen, T.E., Kotlar, H.-K. et al. (2007). Bacterial metabolism of long-chain n-alkanes. *Applied Microbiology and Biotechnology* 76 (6): 1209–1221.

18 Watkinson, R.J. and Morgan, P. (1990). Physiology of aliphatic hydrocarbon-degrading microorganisms. *Biodegradation* 1 (2): 79–92.

19 van Beilen, J.B. and Funhoff, E.G. (2007). Alkane hydroxylases involved in microbial alkane degradation. *Applied Microbiology and Biotechnology* 74 (1): 13–21.

20 McDonald, I.R., Miguez, C.B., Rogge, G. et al. (2006). Diversity of soluble methane monooxygenase-containing methanotrophs isolated from polluted environments. *FEMS Microbiology Letters* 255 (2): 225–232.

21 Funhoff, E.G., Bauer, U., García-Rubio, I. et al. (2006). CYP153A6, a Soluble P450 Oxygenase catalyzing Terminal-Alkane Hydroxylation. *Journal of Bacteriology* 188 (14): 5220–5227.

22 Callaghan, A.V., Morris, B.E.L., Pereira, I.a.C. et al. (2012). The genome sequence of Desulfatibacillum alkenivorans AK-01: a blueprint for anaerobic alkane oxidation. *Environmental Microbiology* 14 (1): 101–113.

23 Zedelius, J., Rabus, R., Grundmann, O. et al. (2011). Alkane degradation under anoxic conditions by a nitrate-reducing bacterium with possible involvement of the electron acceptor in substrate activation. *Environmental Microbiology Reports* 3 (1): 125–135.

24 Grundmann, O., Behrends, A., Rabus, R. et al. (2008). Genes encoding the candidate enzyme for anaerobic activation of n-alkanes in the denitrifying bacterium, strain HxN1. *Environmental Microbiology* 10 (2): 376–385.

25 Shinoda, Y., Sakai, Y., Uenishi, H. et al. (2004). Aerobic and anaerobic toluene degradation by a newly isolated denitrifying bacterium, Thauera sp. strain DNT-1. *Applied and Environmental Microbiology* 70 (3): 1385–1392.

26 Domínguez-Cuevas, P., González-Pastor, J.-E., Marqués, S. et al. (2006). Transcriptional tradeoff between metabolic and stress-response programs in pseudomonas putida KT2440 cells exposed to toluene. *The Journal of Biological Chemistry* 281 (17): 11981–11991.

27 Pannu, J.K., Singh, A., and Ward, O.P. (2003). Influence of peanut oil on microbial degradation of polycyclic aromatic hydrocarbons. *Canadian Journal of Microbiology* 49 (8): 508–513.

28 Cerniglia, C.E. (1993). Biodegradation of polycyclic aromatic hydrocarbons. *Current Opinion in Biotechnology* 4 (3): 331–338.

29 Kanaly, R.A. and Harayama, S. (2000). Biodegradation of high-molecular-weight polycyclic aromatic hydrocarbons by bacteria. *Journal of Bacteriology* 182 (8): 2059–2067.

30 Mallick, S., Chakraborty, J., and Dutta, T.K. (2011). Role of oxygenases in guiding diverse metabolic pathways in the bacterial degradation of low-molecular-weight polycyclic aromatic hydrocarbons: a review. *Critical Reviews in Microbiology* 37 (1): 64–90.

31 Boronin, A.M. and Kosheleva, I.A. (2010). Diversity of naphthalene biodegradation systems in soil bacteria. In: *Handbook of Hydrocarbon and Lipid Microbiology* (ed. K.N. Timmis), 1155–1163. Berlin, Heidelberg: Springer.

32 Titok, M.A. and Al-Shammri, F.J.H. (2012). Efficiency of naphalene degradation by pseudomonas bearing Nah-plasmids in model soll system. *IRAQI Academic Scientific Journals*, Article 1 10 (1): 121–130.

33 Peng, R.-H., Xiong, A.-S., Xue, Y. et al. (2008). Microbial biodegradation of polyaromatic hydrocarbons. *FEMS Microbiology Reviews* 32 (6): 927–955.

34 Menn, F.M., Applegate, B.M., and Sayler, G.S. (1993). NAH plasmid-mediated catabolism of anthracene and phenanthrene to naphthoic acids. *Applied and Environmental Microbiology* 59 (6): 1938–1942.

35 Yang, Y., Chen, R.F., and Shiaris, M.P. (1994). Metabolism of naphthalene, fluorene, and phenanthrene: preliminary characterization of a cloned gene cluster from Pseudomonas putida NCIB 9816. *Journal of Bacteriology* 176 (8): 2158–2164.

36 Simon, M.J., Osslund, T.D., Saunders, R. et al. (1993). Sequences of genes encoding naphthalene dioxygenase in Pseudomonas putida strains G7 and NCIB 9816-4. *Gene* 127 (1): 31–37.

37 Habe, H. and Omori, T. (2003). Genetics of polycyclic aromatic hydrocarbon metabolism in diverse aerobic bacteria. *Bioscience, Biotechnology, and Biochemistry* 67 (2): 225–243.

38 Zhou, N.Y., Fuenmayor, S.L., and Williams, P.A. (2001). nag genes of Ralstonia (formerly Pseudomonas) sp. strain U2 encoding enzymes for gentisate catabolism. *Journal of Bacteriology* 183 (2): 700–708.

39 Laurie, A.D. and Lloyd-Jones, G. (1999). The phn genes of Burkholderia sp. strain RP007 constitute a divergent gene cluster for polycyclic aromatic hydrocarbon catabolism. *Journal of Bacteriology* 181 (2): 531–540.

40 Kiyohara, H., Takizawa, N., Date, H. et al. (1990). Characterization of a phenanthrene degradation plasmid from Alcaligenes faecalis AFK2. *Journal of Fermentation and Bioengineering* 69 (1): 54–56.

41 Goyal, A.K. and Zylstra, G.J. (1996). Molecular cloning of novel genes for polycyclic aromatic hydrocarbon degradation from Comamonas testosteroni GZ39. *Applied and Environmental Microbiology* 62 (1): 230–236.

42 Goyal, A.K. and Zylstra, G.J. (1997). Genetics of naphthalene and phenanthrene degradation by Comamonas testosteroni. *Journal of Industrial Microbiology & Biotechnology* 19 (5–6): 401–407.

43 Basta, T., Buerger, S., and Stolz, A. (2005). Structural and replicative diversity of large plasmids from sphingomonads that degrade polycyclic aromatic compounds and xenobiotics. *Microbiology (Reading, England)* 151 (Pt 6): 2025–2037.

44 Stolz, A. (2009). Molecular characteristics of xenobiotic-degrading sphingomonads. *Applied Microbiology and Biotechnology* 81 (5): 793–811.

45 Vila, J., Tauler, M., and Grifoll, M. (2015). Bacterial PAH degradation in marine and terrestrial habitats. *Current Opinion in Biotechnology* 33: 95–102.

46 Romine, M.F., Fredrickson, J.K., and Li, S.-M.W. (1999). Induction of aromatic catabolic activity in Sphingomonas aromaticivorans strain F199. *Journal of Industrial Microbiology & Biotechnology* 23 (4–5): 303–313.

47 Kertesz*, M.A. and Kawasaki, A. (2010). Hydrocarbon-Degrading Sphingomonads: Sphingomonas, Sphingobium, Novosphingobium, and Sphingopyxis. In: *Handbook of Hydrocarbon and Lipid Microbiology* (ed. K.N. Timmis), 1693–1705. Berlin, Heidelberg: Springer.

48 Roy, M. (2013). Catabolic versatility of Sphingobium sp. Strain PNB capable of degrading structurally diverse aromatic compounds. *Journal of Bioremediation & Biodegradation* 04 (01).

49 Khara, P., Roy, M., Chakraborty, J. et al. (2014). Functional characterization of diverse ring-hydroxylating oxygenases and induction of complex aromatic catabolic gene clusters in Sphingobium sp. PNB. *FEBS Open Bio* 4: 290–300.

50 Kulakov, L.A., Chen, S., Allen, C.C.R., and Larkin, M.J. (2005). Web-type evolution of rhodococcus gene clusters associated with utilization of naphthalene. *Applied and Environmental Microbiology* 71 (4): 1754–1764.

51 Larkin, M.J., Kulakov, L.A., and Allen, C.C.R. (2010). Rhodococcus: genetics and functional genomics. In: *Handbook of Hydrocarbon and Lipid Microbiology* (ed. K.N. Timmis), 1345–1353. Berlin, Heidelberg: Springer.

52 Di Gennaro, P., Terreni, P., Masi, G. et al. (2010). Identification and characterization of genes involved in naphthalene degradation in Rhodococcus opacus R7. *Applied Microbiology and Biotechnology* 87 (1): 297–308.

53 Saito, A., Iwabuchi, T., and Harayama, S. (2000). A novel phenanthrene dioxygenase from Nocardioides sp. Strain KP7: expression in Escherichia coli. *Journal of Bacteriology* 182 (8): 2134–2141.

54 Balashova, N.V., Kosheleva, I.A., Golovchenko, N.P., and Boronin, A.M. (1999). Phenanthrene metabolism by Pseudomonas and Burkholderia strains. *Process Biochemistry* 35 (3): 291–296.

55 Seo, J.-S., Keum, Y.-S., Hu, Y. et al. (2006). Phenanthrene degradation in Arthrobacter sp. P1-1: Initial 1,2-, 3,4- and 9,10-dioxygenation, and meta- and ortho-cleavages of naphthalene-1,2-diol after its formation from naphthalene-1,2-dicarboxylic acid and hydroxyl naphthoic acids. *Chemosphere* 65 (11): 2388–2394.

56 Moser, R. and Stahl, U. (2001). Insights into the genetic diversity of initial dioxygenases from PAH-degrading bacteria. *Applied Microbiology and Biotechnology* 55 (5): 609–618.

57 Ghosal, D., Dutta, A., Chakraborty, J. et al. (2013). Characterization of the metabolic pathway involved in assimilation of acenaphthene in Acinetobacter sp. strain AGAT-W. *Research in Microbiology* 164 (2): 155–163.

58 Hennessee, C.T., Seo, J.-S., Alvarez, A.M., and Li, Q.X. (2009). Polycyclic aromatic hydrocarbon-degrading species isolated from Hawaiian soils: Mycobacterium crocinum sp. nov., Mycobacterium pallens sp. nov., Mycobacterium rutilum sp. nov., Mycobacterium rufum sp. nov. and Mycobacterium aromaticivorans sp. nov. *International Journal of Systematic and Evolutionary Microbiology* 59 (2): 378–387.

59 Sho, M., Hamel, C., and Greer, C.W. (2004). Two distinct gene clusters encode pyrene degradation in Mycobacterium sp. strain S65. *FEMS Microbiology Ecology* 48 (2): 209–220.

60 Churchill, P.F., Morgan, A.C., and Kitchens, E. (2008). Characterization of a pyrene-degrading Mycobacterium sp. strain CH-2. *Journal of Environmental Science and Health, Part B* 43 (8): 698–706.

61 Kim, S.-J., Kweon, O., and Cerniglia, C.E. (2010). Degradation of polycyclic aromatic hydrocarbons by mycobacterium strains. In: *Handbook of Hydrocarbon and Lipid Microbiology* (ed. K.N. Timmis), 1865–1879. Berlin, Heidelberg: Springer.

62 Ostrem Loss, E.M. and Yu, J.-H. (2018). Bioremediation and microbial metabolism of benzo(a)pyrene. *Molecular Microbiology* 109 (4): 433–444.

63 Lily, M.K., Bahuguna, A., Dangwal, K., and Garg, V. (2009). Degradation of Benzo [a] Pyrene by a novel strain Bacillus subtilis BMT4i (MTCC 9447). *Brazilian Journal of Microbiology* 40 (4): 884–892.

64 Sowada, J., Schmalenberger, A., Ebner, I. et al. (2014). Degradation of benzo[a]pyrene by bacterial isolates from human skin. *FEMS Microbiology Ecology* 88 (1): 129–139.

65 Aziz, A., Agamuthu, P., Alaribe, F.O., and Fauziah, S.H. (2018). Biodegradation of benzo[a]pyrene by bacterial consortium isolated from mangrove sediment. *Environmental Technology* 39 (4): 527–535.

66 Avdalović, J., Đurić, A., Miletić, S. et al. (2016). Treatment of a mud pit by bioremediation. *Waste Management & Research* 34 (8): 734–739.

67 Mahmoudi, N., Porter, T.M., Zimmerman, A.R. et al. (2013). Rapid degradation of Deepwater horizon spilled oil by indigenous microbial communities in Louisiana saltmarsh sediments. *Environmental Science & Technology* 47 (23): 13303–13312.

68 Cerniglia, C.E. and Sutherland, J.B. (2010). Degradation of polycyclic aromatic hydrocarbons by fungi. In: *Handbook of Hydrocarbon and Lipid Microbiology* (ed. K.N. Timmis), 2079–2110. Berlin, Heidelberg: Springer.

69 Hofrichter, M. (2002). Review: lignin conversion by manganese peroxidase (MnP). *Enzyme and Microbial Technology* 30 (4): 454–466.

70 Tortella, G.R., Diez, M.C., and Duran, N. (2005). Fungal diversity and use in decomposition of environmental pollutants. *Critical Reviews in Microbiology* 31 (4): 197–212.

71 Bezalel, L., Hadar, Y., and Cerniglia, C.E. (1997). Enzymatic mechanisms involved in Phenanthrene degradation by the white rot fungus Pleurotus ostreatus. *Applied and Environmental Microbiology* 63 (7): 2495–2501.

72 Chowdhary, P., Shukla, G., Raj, G. et al. (2018). Microbial manganese peroxidase: a ligninolytic enzyme and its ample opportunities in research. *SN Applied Sciences* 1 (1): 45.

73 Chun, S.C., Muthu, M., Hasan, N. et al. (2019). Mycoremediation of PCBs by Pleurotus ostreatus: possibilities and prospects. *Applied Sciences* 9 (19): 4185.

74 Clemente, A.R., Anazawa, T.A., and Durrant, L.R. (2001). Biodegradation of polycyclic aromatic hydrocarbons by soil fungi. *Brazilian Journal of Microbiology* 32 (4): 255–261.

75 Vazquez-Duhalt, R., Westlake, D.W., and Fedorak, P.M. (1994). Lignin peroxidase oxidation of aromatic compounds in systems containing organic solvents. *Applied and Environmental Microbiology* 60 (2): 459–466.

76 Singh, D. and Chen, S. (2008). The white-rot fungus Phanerochaete chrysosporium: conditions for the production of lignin-degrading enzymes. *Applied Microbiology and Biotechnology* 81 (3): 399–417.

77 Syed, K., Doddapaneni, H., Subramanian, V. et al. (2010). Genome-to-function characterization of novel fungal P450 monooxygenases oxidizing polycyclic aromatic hydrocarbons (PAHs). *Biochemical and Biophysical Research Communications* 399 (4): 492–497.

78 Winquist, E., Björklöf, K., Schultz, E. et al. (2014). Bioremediation of PAH-contaminated soil with fungi – from laboratory to field scale. *International Biodeterioration & Biodegradation* 86: 238–247.

79 Acevedo, F., Pizzul, L., Castillo, M.d.P. et al. (2011). Degradation of polycyclic aromatic hydrocarbons by the Chilean white-rot fungus Anthracophyllum discolor. *Journal of Hazardous Materials* 185 (1): 212–219.

80 Cerniglia, C.E. and Sutherland, J.B. (2001). Bioremediation of polycyclic aromatic hydrocarbons by ligninolytic and non-ligninolytic fungi. In: *Fungi in Bioremediation* (ed. G.M. Gadd), 136–187. Cambridge: Cambridge University Press.

81 Supreeth, M., Chandrashekar, M.A., Sachin, N., and Raju, N.S. (2016). Effect of chlorpyrifos on soil microbial diversity and its biotransformation by Streptomyces sp. HP-11. *3 Biotech* 6 (2).

82 de Souza, M.L., Sadowsky, M.J., and Wackett, L.P. (1996). Atrazine chlorohydrolase from Pseudomonas sp. strain ADP: gene sequence, enzyme purification, and protein characterization. *Journal of Bacteriology* 178 (16): 4894–4900.

83 Karns, J.S. and Eaton, R.W. (1997). Genes encoding s-Triazine degradation are plasmid-borne in Klebsiella pneumoniae Strain 99. *Journal of Agricultural and Food Chemistry* 45 (3): 1017–1022.

84 Jayashree, R. and Vasudevan, N. (2007). Persistence and distribution of endosulfan under field condition. *Environmental Monitoring and Assessment* 131 (1–3): 475–487.

85 Kumar, D. and Pannu, R. (2018). Perspectives of lindane (γ-hexachlorocyclohexane) biodegradation from the environment: a review. *Bioresour. Bioprocess* 5 (1): 29.

86 Gianfreda, L. and Rao, M.A. (2004). Potential of extra cellular enzymes in remediation of polluted soils: a review. *Enzyme and Microbial Technology* 35 (4): 339–354.

87 Bending, G.D., Friloux, M., and Walker, A. (2002). Degradation of contrasting pesticides by white rot fungi and its relationship with ligninolytic potential. *FEMS Microbiology Letters* 212 (1): 59–63.

88 Quintero, J.C., Lú-Chau, T.A., Moreira, M.T. et al. (2007). Bioremediation of HCH present in soil by the white-rot fungus Bjerkandera adusta in a slurry batch bioreactor. *International Biodeterioration & Biodegradation* 60 (4): 319–326.

89 Sandermann, H., Heller, W., Hertkorn, N. et al. (1998). A new intermediate in the Mineralization of 3,4-Dichloroaniline by the white rot fungus Phanerochaete chrysosporium. *Applied and Environmental Microbiology* 64 (9): 3305–3312.

90 Martinez, B., Tomkins, J., Wackett, L.P. et al. (2001). Complete nucleotide sequence and organization of the atrazine catabolic plasmid pADP-1 from Pseudomonas sp. strain ADP. *Journal of Bacteriology* 183 (19): 5684–5697.

91 Zablotowicz, R.M., Hoagland, R.E., Lee, H. et al. (2000). Transformation of Nitroaromatic pesticides and related xenobiotics by microorganisms and plants. In: *Pesticide Biotransformation in Plants and Microorganisms*, vol. 777 (eds. J.C. Hall, R.E. Hoagland and R.M. Zablotowicz), 194–216. American Chemical Society.

92 Foght, J., April, T., Biggar, K., and Aislabie, J. (2001). Bioremediation of DDT-contaminated soils: a review. *Bioremediation Journal* 5 (3): 225–246.

93 Stosiek, N., Talma, M., and Klimek-Ochab, M. (2020). Carbon-Phosphorus Lyase – the state of the art. *Applied Biochemistry and Biotechnology* 190 (4): 1525–1552.

94 Hall, J.C., Wickenden, J.S., and Yau, K.Y.F. (2000). Biochemical conjugation of pesticides in plants and microorganisms: an overview of similarities and divergences. In: *Pesticide Biotransformation in Plants and Microorganisms*, vol. 777 (eds. J.C. Hall, R.E. Hoagland and R.M. Zablotowicza), 89–118. American Chemical Society.

95 Cserjesi, A.J. and Johnson, E.L. (1972). Methylation of pentachlorophenol by Trichoderma virgatum. *Canadian Journal of Microbiology* 18 (1): 45–49.

96 Monachese, M., Burton, J.P., and Reid, G. (2012). Bioremediation and tolerance of humans to heavy metals through microbial processes: a potential role for probiotics? *Applied and Environmental Microbiology* 78 (18): 6397–6404.

97 Ayangbenro, A.S. and Babalola, O.O. (2017). A new strategy for heavy metal polluted environments: a review of microbial Biosorbents. *International Journal of Environmental Research and Public Health* 14 (1).

98 Fomina, M. and Gadd, G.M. (2014). Biosorption: current perspectives on concept, definition and application. *Bioresource Technology* 160: 3–14.

99 Schippers, A., Hedrich, S., Vasters, J. et al. (2014). Biomining: metal recovery from ores with microorganisms. *Advances in Biochemical Engineering/Biotechnology* 141: 1–47.

100 Bachate, S.P., Khapare, R.M., and Kodam, K.M. (2012). Oxidation of arsenite by two β-proteobacteria isolated from soil. *Applied Microbiology and Biotechnology* 93 (5): 2135–2145.

101 Hassan, Z., Sultana, M., Khan, S.I. et al. (2019). Ample Arsenite bio-oxidation activity in Bangladesh drinking water wells: a bonanza for bioremediation? *Microorganisms* 7 (8).

102 Cabaj, A. and Kosakowska, A. (2009). Iron-dependent growth of and siderophore production by two heterotrophic bacteria isolated from brackish water of the southern Baltic Sea. *Microbiological Research* 164 (5): 570–577.

103 Khan, A., Singh, P., and Srivastava, A. (2018). Synthesis, nature and utility of universal iron chelator – Siderophore: a review. *Microbiological Research* 212–213: 103–111.

104 Braud, A., Hoegy, F., Jezequel, K. et al. (2009). New insights into the metal specificity of the Pseudomonas aeruginosa pyoverdine-iron uptake pathway. *Environmental Microbiology* 11 (5): 1079–1091.

105 Hong, J.W., Park, J.Y., and Gadd, G.M. (2010). Pyrene degradation and copper and zinc uptake by Fusarium solani and Hypocrea lixii isolated from petrol station soil. *Journal of Applied Microbiology* 108 (6): 2030–2040.

106 Searle, L.J., Méric, G., Porcelli, I. et al. (2015). Variation in siderophore biosynthetic gene distribution and production across environmental and faecal populations of Escherichia coli. *PLoS One* 10 (3): e0117906.

107 Joutey, N.T., Sayel, H., Bahafid, W., and El Ghachtouli, N. (2015). Mechanisms of hexavalent chromium resistance and removal by microorganisms. *Reviews of Environmental Contamination and Toxicology* 233: 45–69.

108 Vijayaraghavan, K. and Yun, Y.-S. (2008). Bacterial biosorbents and biosorption. *Biotechnology Advances* 26 (3): 266–291.

109 Puyen, Z.M., Villagrasa, E., Maldonado, J. et al. (2012). Biosorption of lead and copper by heavy-metal tolerant Micrococcus luteus DE2008. *Bioresource Technology* 126: 233–237.

110 Choi, S.B. and Yun, Y.-S. (2004). Lead biosorption by waste biomass of Corynebacterium glutamicum generated from lysine fermentation process. *Biotechnology Letters* 26 (4): 331–336.

111 Wang, Y., Ho, S.-H., Cheng, C.-L. et al. (2016). Perspectives on the feasibility of using microalgae for industrial wastewater treatment. *Bioresource Technology* 222: 485–497.

112 Satyapal, G.K., Mishra, S.K., Srivastava, A. et al. (2018). Possible bioremediation of arsenic toxicity by isolating indigenous bacteria from the middle Gangetic plain of Bihar, India. *Biotechnology Reports (Amsterdam, Netherlands)* 17: 117–125.

113 Satyapal, G.K. and Rani, S. (2016). Potential role of arsenic resistant bacteria in bioremediation: current status and future prospects. *J Microb Biochem Technol* 8 (3).

114 Wernérus, H., Lehtiö, J., Teeri, T. et al. (2001). Generation of metal-binding staphylococci through surface display of combinatorially engineered cellulose-binding domains. *Applied and Environmental Microbiology* 67 (10): 4678–4684.

115 Samuelson, P., Wernérus, H., Svedberg, M., and Ståhl, S. (2000). Staphylococcal surface display of metal-binding polyhistidyl peptides. *Applied and Environmental Microbiology* 66 (3): 1243–1248.

116 Bae, W., Chen, W., Mulchandani, A., and Mehra, R.K. (2000). Enhanced bioaccumulation of heavy metals by bacterial cells displaying synthetic phytochelatins. *Biotechnology and Bioengineering* 70 (5): 518–524.

117 Ron, E.Z. and Rosenberg, E. (2001). Natural roles of biosurfactants. *Environmental Microbiology* 3 (4): 229–236.

118 Reznik, G.O., Vishwanath, P., Pynn, M.A. et al. (2010). Use of sustainable chemistry to produce an acyl amino acid surfactant. *Applied Microbiology and Biotechnology* 86 (5): 1387–1397.

119 Ravindran, A., Sajayan, A., Priyadharshini, G.B. et al. (2020). Revealing the efficacy of thermostable Biosurfactant in heavy metal bioremediation and surface treatment in vegetables. *Frontiers in Microbiology* 11.

120 Maier, R.M. and Soberón-Chávez, G. (2000). Pseudomonas aeruginosa rhamnolipids: biosynthesis and potential applications. *Applied Microbiology and Biotechnology* 54 (5): 625–633.

121 Santos, D.K.F., Rufino, R.D., Luna, J.M. et al. (2016). Biosurfactants: multifunctional biomolecules of the 21st century. *International Journal of Molecular Sciences* 17 (3).

122 Mulligan, C.N. (2005). Environmental applications for biosurfactants. *Environmental Pollution* 133 (2): 183–198.

123 Pradhan, A.K. and Pradhan, N. (2015). Microbial Biosurfactant for hydrocarbons and heavy metals bioremediation. In: *Environmental Microbial Biotechnology* (eds. L.B. Sukla, N. Pradhan, S. Panda and B.K. Mishra), 91–104. Cham: Springer International Publishing.

124 Boopathy, R. (2000). Factors limiting bioremediation technologies. *Bioresource Technology* 74 (1): 63–67.

125 Srivastava, J., Naraian, R., Kalra, S.J.S., and Chandra, H. (2014). Advances in microbial bioremediation and the factors influencing the process. *International journal of Environmental Science and Technology* 11 (6): 1787–1800.

126 Dangi, A.K., Sharma, B., Hill, R.T., and Shukla, P. (2018). Bioremediation through microbes: systems biology and metabolic engineering approach. *Critical Reviews in Biotechnology*: 1–20.

127 Hall, E.K., Bernhardt, E.S., Bier, R.L. et al. (2018). Understanding how microbiomes influence the systems they inhabit. *Nature Microbiology* 3 (9): 977–982.

128 Medema, M.H., van Raaphorst, R., Takano, E., and Breitling, R. (2012). Computational tools for the synthetic design of biochemical pathways. *Nature Reviews Microbiology* 10 (3): 191–202.

129 Long, M.R., Ong, W.K., and Reed, J.L. (2015). Computational methods in metabolic engineering for strain design. *Current Opinion in Biotechnology* 34: 135–141.

130 Arora, P.K. and Bae, H. (2014). Integration of bioinformatics to biodegradation. *Biological Procedures Online* 16 (1): 8.

131 Arora, P.K., Kumar, M., Chauhan, A. et al. (2009). OxDBase: a database of oxygenases involved in biodegradation. *BMC Research Notes* 2: 67.

132 Gao, J., Ellis, L.B.M., and Wackett, L.P. (2011). The University of Minnesota pathway prediction system: multi-level prediction and visualization. *Nucleic Acids Research* 39 (Web Server issue): W406–W411.

133 Chou, C.-H., Chang, W.-C., Chiu, C.-M. et al. (2009). FMM: a web server for metabolic pathway reconstruction and comparative analysis. *Nucleic Acids Research* 37 (Web Server issue): W129–W134.

134 McClymont, K. and Soyer, O.S. (2013). Metabolic tinker: an online tool for guiding the design of synthetic metabolic pathways. *Nucleic Acids Research* 41 (11): e113.

135 Moriya, Y., Shigemizu, D., Hattori, M. et al. (2010). PathPred: an enzyme-catalyzed metabolic pathway prediction server. *Nucleic Acids Research* 38 (Web Server): W138–W143.

136 Hassan, I., Eastman, A.W., Weselowski, B. et al. (2016). Complete Genome sequence of Arthrobacter sp. Strain LS16, isolated from agricultural soils with potential for applications in bioremediation and bioproducts. *Genome Announcements* 4 (1).

137 Vallenet, D., Calteau, A., Cruveiller, S. et al. (2017). MicroScope in 2017: an expanding and evolving integrated resource for community expertise of microbial genomes. *Nucleic Acids Research* 45 (D1): D517–D528.

138 Demko, M., Chrást, L., Dvořák, P. et al. (2019). Computational modelling of metabolic burden and substrate toxicity in Escherichia coli carrying a synthetic metabolic pathway. *Microorganisms* 7 (11).

139 Gong, T., Xu, X., Che, Y. et al. (2017). Combinatorial metabolic engineering of Pseudomonas putida KT2440 for efficient mineralization of 1,2,3-trichloropropane. *Scientific Reports* 7 (1): 7064.

140 Dvorak, P., Bidmanova, S., Damborsky, J., and Prokop, Z. (2014). Immobilized synthetic pathway for biodegradation of toxic recalcitrant pollutant 1,2,3-trichloropropane. *Environmental Science & Technology* 48 (12): 6859–6866.

6

Microbial Applications in Organic Acid Production

Jyoti Singh Jadaun[1], Amit K. Rai[2], and Sudhir P. Singh[3]

[1] *Department of Botany, Dayanand Girls Postgraduate College, Kanpur, Uttar Pradesh, India*
[2] *Institute of Bioresources and Sustainable Development (DBT-IBSD), Sikkim Centre, Gangtok, Sikkim, India*
[3] *Center of Innovative and Applied Bioprocessing (DBT-CIAB), Mohali, Punjab, India*

6.1 Introduction

Organic acids are low-molecular-weight compounds having one or more functional groups such as carboxylic (−COOH), alcohol (−OH), sulfonic group (R−S(=O)2−OH), and thiol group (R-SH). These are also called building block chemicals due to the presence of these functional groups. Recently, market value of organic acids has increased due to their diverse industrial applications in food, cosmetics and perfumery, textile, plastic, rubber, and pharmaceutical industries [1]. These organic acids can be used as precursor or intermediate compounds to synthesize other valuable products; therefore, they are gaining more attention in the chemical industry. Traditionally, organic acids are produced from a petroleum refinery. However, currently, the microbial fermentation process is used for industrial production of organic acids, such as citric acid (CA), lactic acid (LA), succinic acid (SA), acetic acid (AA), and so on [2]. Naturally, these are widely distributed in nature, and almost every organism produces a number of organic acids in more or less amount [3]. Many organic acids, such as α-ketoglutaric acid, CA, SA, fumaric acid, L-malic acid, and oxaloacetates, are intermediates of tricarboxylic acid (TCA) cycle, and these are dominating the organic acid industry [1]. According to global market analysis, the organic acid market is predicted to reach up to US $36.86 billion by 2026. Commercial applications of organic acids and their derivatives are expanding due to their biodegradable nature as they are utilized to synthesize bioplastics. Microbial fermentation has several advantages over petroleum refinery, such as a sustainable, cost-effective, and ecofriendly process [4]. In the near future, microbial production of organic acids can replace the petroleum-derived chemicals if we succeed in developing an effective process for each trading organic acid. Therefore, there is a need to develop more robust, cost-effective, and less time-consuming microbial technology for production of organic acids. Development of novel microbial screening techniques, upgradation of genome or transcriptome sequencing technologies,

Bioprospecting of Microorganism-Based Industrial Molecules, First Edition.
Edited by Sudhir P. Singh and Santosh Kumar Upadhyay.

and modernization in genetic engineering approaches can provide pace to microbial biotechnology. Technologies for producing organic acids through microbes are also evolving progressively as genomic information of host cells enhances the chances for integration or deletion of target genes at correct loci without affecting basal cell metabolism [5].

Genetic modification of microorganisms enhances the production rate up to multifold [6]. Researchers across the world are trying to manipulate the microbes in such a manner to get the maximum yield of targeted organic acids. Enhancement in yield of a particular organic acid depends on the extensive knowledge available on metabolic pathways for concerned organic acid production. After genetic analysis of microbes, we can transfer genes to appropriate hosts for maximizing the yield [3]. Although it is not compulsory to always get the increased yield in recombinant strain, it may be due to a lack of proper functioning of the host's cellular metabolism [7, 8]. This can be further rectified by using genes from different sources, alteration in gene and promoter combination, and by adding a transcription factor to upregulate the gene expression and codon optimization [4]. Natural producers are not known for a few organic acids, such as adipic acid. In this case, the heterologous expression of genes from different sources worked successfully as production of adipic acid was achieved in *Escherichia coli* [9]. This chapter explores the recent updates on the microbial production of different organic acids.

6.2 Glycolic acid (2C)

Glycolic acid (hydroxyacetic acid) is the smallest α-hydroxy acid (Figure 6.1). Hlasiwetz and Habermann discovered glycolic acid, and later several methods have been developed for its production at commercial level. It is colorless, odorless, and hygroscopic crystalline solid, which is highly soluble in water. It acts as a platform molecule for the generation of polymers, especially polygylcolate (PGA). PGA is a crystalline polyester and insoluble in most organic solvents. It is biodegradable in nature, and it is a preferable material for biomedical applications as it has high mechanical strength and fine gas permeability. Its melting temperature range is 224–227 °C, while the glass transition temperature range is 35–40 °C. In 2019, PGA's global market size was approximately US $4.1 billion, and up to the year 2024, it is predicted to grow US $6.5 billion at CAGR of 9.5% (https://www.marketresearchfuture.com). In 1954, a ring-opening polymerization (through the synthesis of cyclic di-ester) method was introduced by DuPont company for the production of high-molecular-weight PGA. Later in 1960s, an American company developed bioabsorbable suture thread by using PGA [10]. Different medical regenerative devices are made using a copolymer of PGA and polylactic acid [11]. Recently, biological processes have been developed to manufacture glycolic acid in genetically engineered organisms like *E. coli* and yeasts (*Saccharomyces cerevisiae* and *Kluyveromyces lactis*). Koivistoinen et al. [12] reported that engineered *S. cerevisiae* and *K. lactis* are suitable organisms for glycolic acid production as they can survive in 50 g/l glycolic acid. The glyoxylate cycle is engineered in yeasts by introducing the glyoxylate reductase gene from *Arabidopsis thaliana* (Figure 6.2). Further, the production titer can be enhanced by deleting the malate synthase and isocitrate dehydrogenase (cytosolic forms) genes. By following this practice, it was noticed that *S. cerevisiae* and *K. lactis* produced 1 g/l and 15 g/l glycolic acid with D-xylose and ethanol,

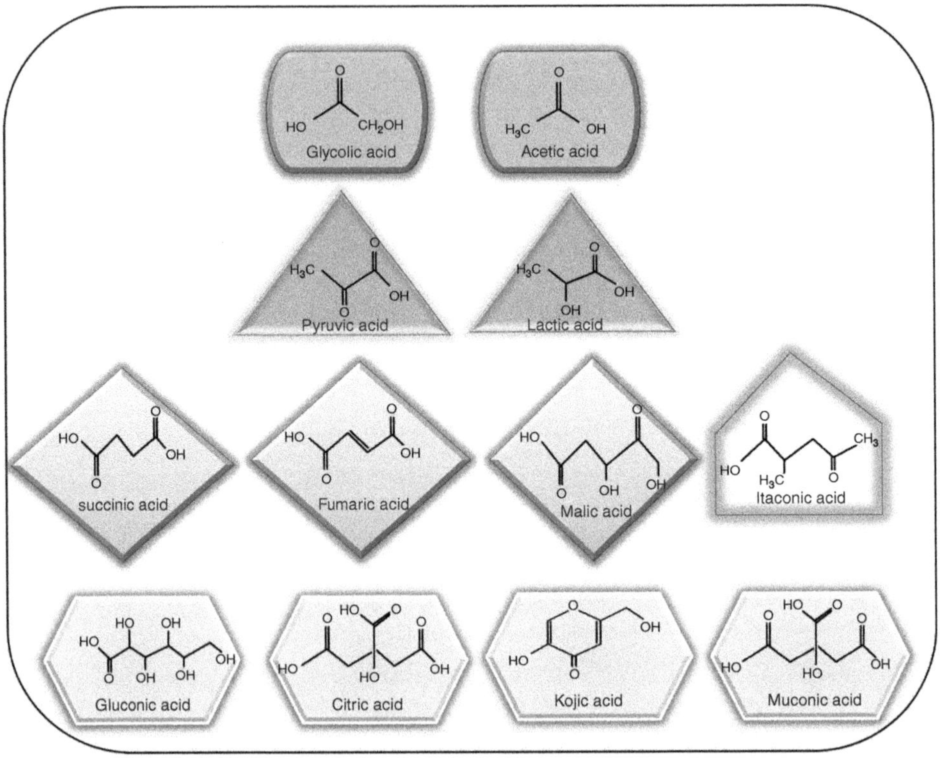

Figure 6.1 Representation of chemical structure of different organic acids.

respectively. In another study, *E. coli* is genetically engineered for glycolic acid production by using a modified pathway named as "glycoptimus." It includes the overexpression of three genes (D-arabinose-5-P isomerase (*kdsD*), class 1 aldolase (*fsaA*), and aldehyde dehydrogenase (*aldA*)), and this pathway is combined with the reshaping of carbon metabolism. Ultimately, the engineered strain produced 2.5 and 3 moles of glycolic acid from hexose and pentose sugars, respectively [13]. Cabulong et al. [14] reported that metabolic engineering of *E. coli* by using the Dahms pathway, the glyoxylate bypass pathway, and the partial reverse glyoxylate pathway (RGP) leads to the production of glycolic acid with a yield of 0.46 g from one-gram D-xylose. In another report, metabolic engineering of *E. coli* was carried out by incorporating glyoxylate bypass pathway by introducing the isocitrate lyase and isocitrate dehydrogenase kinase/phosphatase gene, along with minimizing the flow of glyoxylate in other pathways through inactivation of malate synthase, glyoxylate carboligase, and glycolate oxidase. This engineered *E. coli* produced 1.78 g/l glycolate from 3.23 g/l acetate [15]. A high molar yield (0.99 mol per mol of xylose) of glycolic acid is achieved by utilizing the genetically engineered organism *Corynebacterium glutamicum* [16]. A novel *Enterobacter cloacae* strain was identified, which produced ethylene glycol and glycolic acid from xylonic acid. This bacterium has a metabolic pathway for converting xylonic acid into ethylene glycol [17]. Shifting in carbon source medium alters products' nature as it produces 2,3-butanediol, acetoin, or AA type when glycerol is supplied as the sole carbon source [17].

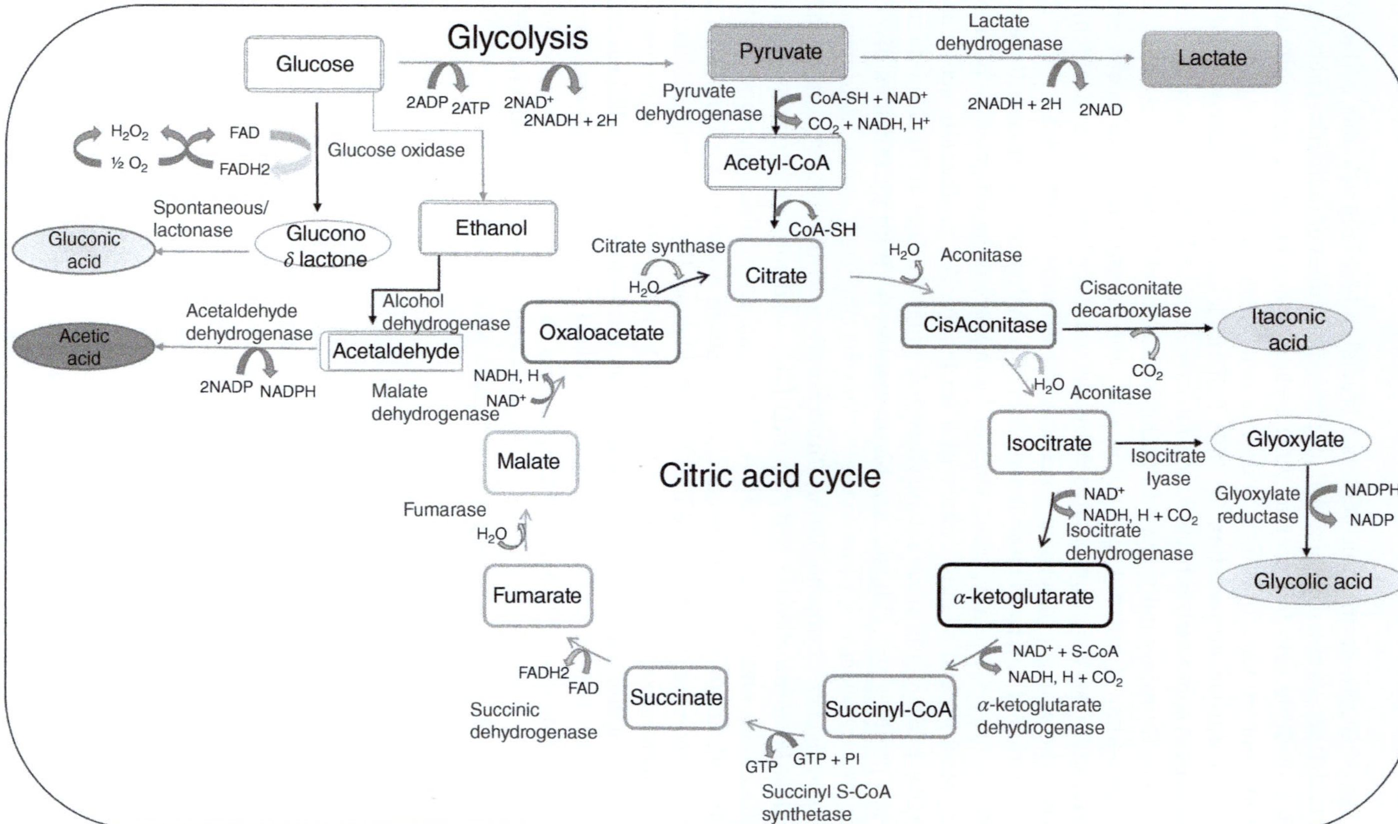

Figure 6.2 Illustration of different metabolic pathways involved in biosynthesis of different organic acids. *Sources:* Based on [36, 46, 60, 77, 107, 108].

6.3 Acetic Acid (2C)

Acetic acid (CH3COOH), or ethanoic acid, is a crucial industrially valuable carboxylic acid (Figure 6.1). It serves as a base material for producing different industrial chemicals, such as acetic anhydride, cellulose acetate, vinyl acetate monomer, and terephthalic acid. Microbial production of AA is one of the oldest known biological processes, e.g., in 1862 Pasteur proposed vinegar fermentation. In 1898 Beijernick and in 1924 Kluyver and de Leeuw discovered AA producing bacteria *Acetobacter aceti* and *Gluconobacter suboxydans*, respectively. Conventionally, the production of acetic acid was developed through oxidative fermentation of ethanol into acetic acid. For the production of acetic acid, sequential oxidation reactions are catalyzed by alcohol dehydrogenase and aldehyde dehydrogenase; therefore, it is termed "oxidative fermentation" (Figure 6.2). For the industrial production of AA, two methods are generally employed: (i) surface culture and (ii) submerged culture. Generally, "seed vinegar," a mixture of heterogeneous microorganisms, is applied instead of pure culture as a starter. For AA production, acetic acid bacteria (AAB) are commonly used that are classified under the Acetobacteraceae family. AAB are gram-positive and obligate aerobic in nature. AAB group includes diverse genera such as *Acetobacter*, *Gluconobacter*, *Gluconacetobacter*, *Asaia*, *Acidomonas*, *Kozakia*, *Saccharibacter*, *Swaminathania*, *Tanticharoenia*, *Granulibacter Ameyamaea*, *Neokomagataea*, *Komagataeibacter*, and *Neoasaia* [18]. AAB commonly occurs in naturally fermented foods and beverages [19]. AAB can rapidly oxidize sugar and alcohol molecules [18, 20]. Entani et al. [21] isolated a novel strain *Acetobacter polyoxogenes* sp. Nov from fermented vinegar, which produced AA of high acidity. Industrial production of AA from lignocellulosic material is also attempted by using a *Clostridium thermoaceticum* that has the ability to convert sugars present in cellulosic and hemicellulosic fractions [22]. The whole-cell of AAB or immobilized cells of AAB can be used for ethanol production. According to a study, *S. cerevisiae* produces 8% ethanol from sugarcane juice, and this ethanol is converted into vinegar by entrapped (calcium alginate) and adsorbed (baggase, wood-shaving, and corn cob) cells of *A. aceti* NRRL 746 [23].

6.4 Pyruvic Acid (3C)

Pyruvic acid (2-oxopropanoic acid, α ketopropionic acid, or acetylformic acid) is one of the most important organic acids as it is an intermediate between glycolysis and TCA cycle (Figures 6.1 and 6.2). It is exploited in different industries, such as food, pharmaceutical, cosmetic, and polymer [24]. Microbial production offers more advantages over the chemical synthesis of pyruvic acid (PA). Bacteria, fungi, and yeast are prospected for the production of PA [25]. Many types of yeast are explored for the production of PA, but majorly *Torulopsis glabrata* species are employed for this purpose as *T. glabrata* IFO 0005 is reported to accumulate PA at a concentration of 67.8 g/l [26]. L-lactic acid is converted into pyruvic acid by methylotropic yeast transformants, *Hansenula polymorpha* and *Pichia pastoris*, which expressed glycolate oxidase (S)-2-hydroxyacid oxidase and endogenous catalase [27]. Further, thiamine-auxotrophic yeast *Yarrowia lipolytica* 374/4 is noted for the production of PA, and the study discloses that the yield of PA is dependent on the thiamine concentration in the medium [28]. PA is one of the central metabolites. Its biosynthesis has limitation of involving various anabolic and catabolic steps, so by ceasing one or more pathway steps its production may be enhanced. By applying this concept, metabolic engineering leads to the development of pyruvate decarboxylase-negative (Pdc^-) strains of *S. cerevisiae*. Another mutant of Pdc^- strain (TAM) yielded PA in the concentration of 135 g/l [29]. Metabolic engineering of *E. coli* has been done by mutating the genes for pyruvate dehydrogenase complex (aceEF), pyruvate formate lyase (pfl), pyruvate oxidase (poxB), phosphoenolpyruvate synthase (pps), lactate dehydrogenase (ldhA), and F1-ATPase activity (atpFH). The constructed strain, ALS1059, could produce 90 g/l pyruvate [30]. In another study, engineered *E. coli* with manipulated fatty acid biosynthesis pathway was able to produce PA at 26.1 g/l [31].

6.5 Lactic Acid (3C)

Lactic acid (2-hydroxypropionic acid) (Figure 6.1) used in various industries such as pharmaceutical (in manufacturing of surgical sutures and prosthetic devices), food (as a flavoring agent, preservatives, buffering agent, and as acidulant), cosmetics (in the formulation of moisturizer due to water-retaining capacity), and textile industries [32, 33]. LA can be converted into other useful chemicals, such as esters' production as acrylate ester, lactate ester, pyruvic acid, propylene glycol, acrylic acid, and propylene oxide [34]. One of LA's most crucial exploitation is in the production of polylactic acid (PLA), a hydrophobic polymer with myriad applications in the medical field due to biocompatibility, processability, and slow degradation rate [35]. As the chemical process of synthesis of LA produces the racemic mixture of D and L forms of LA, the fermentative production of LA is preferable. Pyruvic acid is used as a precursor for the production of LA by using microbes. Pyruvic acid is converted into LA due to the catalytic activity of enzymes, and stereospecificity is dependent on the type of enzyme involved in the reaction, e.g., (i) NAD-dependent L-lactate dehydrogenase, (ii) NAD-dependent D-lactate dehydrogenase. Many bacterial species are reported for the production of optically pure LA (Figure 6.2) [36]. There are certain limitations in LA's production through fermentation technology, one of them is a costly substrate. To overcome this, different studies have been performed by using renewable materials such as starchy, cellulosic, and other agro-industrial residues [37]. Whey liquid spent after separation of protein and fat from milk is a suitable feedstock for LA production due to the presence of a significant concentration of lactose [38]. *Streptococcus bovis* 148 produces LA from starch at the concentration of 14.2 g/l [39].

6.6 Succinic Acid (4C)

Succinic Acid (SA) is also known as 1,4-butanedioic acid or amber acid (Figure 6.1). It is becoming the foremost industrial chemical. It acts as a platform molecule for manufacturing high-value chemicals such as adipic acid, tetrahydrofuran, 2-pyrrolidinone, γ-butyrolactone, *N*-methyl pyrrolidone, and biopolymers such as polybutylene succinate [40]. It also has applications in the agriculture sector (growth regulator), food industry (flavor enhancer, and bread softening agent), and pharmaceutical industry (anticarcinogenic and insulinotropic agent) [41]. Traditionally, SA is produced from the petrochemical process, but currently fermentative production of SA is gaining more attention. The biosynthetic route for the production of SA includes both the aerobic and anaerobic pathways, although in both pathways, glucose is converted into pyruvate, and further enzymatic reactions lead to the production of fumarate, and finally fumarate is converted into succinate by glycolysis. SA is an intermediate of TCA cycle and the end product of anaerobic metabolism; therefore, it is synthesized by microbes, animals, and plants (Figure 6.2). Bacteria and fungi are found as suitable hosts for the production of SA. Another interesting fact is that microbes use CO_2 for the production of SA. Accumulation of SA in a large quantity in microbes is not a common phenomenon, although some bacterial and fungal species are reported as natural producers of SA. Bacterial species include facultative anaerobes such as *Actinobacillus succinogenes*, *Klebsiella pneumonia*, *Enterococcus faecalis*, and strict anaerobes, *Anaerobiospirillum succiniciproducens*, *Bacteroides fragilis*, and *Succinivibrio dextrinosolvens* [42]. *Mannheimia succiniciproducens* MBEL55E is a good producer of SA. It yields 13.5 g/l of SA using glucose as a carbon source under anaerobic condition [43]. Fungal strains screened for production of SA are *Penicillium viniferum, Lentinus degener Byssochlamys nivea, Aspergillus niger, Paecilomyces varioti*, and *Penicillium simplicissimum* [40]. To increase SA yield, metabolic engineering of SA producing and utilizing pathways is carried out in different microbes. The first trial has been attempted in *E. coli* as overexpression of phosphoenolpyruvate carboxylase leading to a 3.5-fold increase in the yield of SA [44].

6.7 Fumaric Acid (4C)

Fumaric acid (2-butenedioic acid, trans isomer) is an intermediate of TCA cycle, so it occurs widely in nature (Figure 6.1). Firstly, it was isolated from plant *Fumaria officinalis* (European herb). It is colorless, non-hygroscopic, crystalline powder having fruit-like taste, with low solubility in water. It has different industrial applications, such as it is used in synthesis of other chemicals (for example L-malic acid and biodegradable polymer), in food industries (as an acidulant), and in pharmaceutical industries (in synthesis of alternative drug such as dimethyl fumarate in treatment of psoriasis) [45]. Different chemical processes are available for the synthesis of FA, but due to unsustainability of petrochemicals, the biochemical process for FA production got attention. It is hypothesized and confirmed by experiments that reductive part of TCA cycle is responsible for the production of FA (Figure 6.2) [46]. Besides TCA pathway, other metabolic pathways (amino acid metabolism and fatty acid metabolism) are also suggested to be involved in FA production in *Rhizopus oryzae* [47]. FA synthesis through TCA involves three steps that are catalyzed by different enzymes: (i) pyruvate decarboxylase, (ii) malate dehydrogenase, and (iii) fumarase. These enzymes from different sources are utilized for the metabolic engineering of organisms for enhancing the yield of FA. A genetically engineered *E. coli* with thermostable fumarase of *Thermus thermophilus* is reported to produce fumaric acid [48]. Engineering of *S. cerevisiae* for production of FA has a limitation that both forms (cytosolic and mitrochondrial) of fumarases cannot perform reversible reaction for the synthesis of fumaric acid from malic acid. They are able to convert only fumaric to malic acid. Metabolically engineered *S. cerevisiae* was able to produce 610 ± 31 mg/l of FA after deletion of FUM1 (fumarase) gene, and further overexpression of pyruvate carboxylase from *R. oryzae* leads to the production of FA in the concentration of 1134 ± 48 mg/l [49]. Further, high titer (5.64 g/l) of FA is achieved with genetically engineered strain of *S. cerevisiae* CEN.PK2-1CDTHI2, having manipulation in reductive and oxidative pathways [50]. Maleate isomerase is an alternative enzyme for the production of FA. The microbial sources for maleate isomerases are *Pseudomonas* [51] and *Arthrobacter* [52]. However, these enzymes have limitations in commercial use due to instability at moderate temperature. Thermostable maleate isomerase is produced by *Bacillus* species like *Bacillus stearothermophilus* and *Bacillus brevis,* which are more useful in the production of FA [53]. For industrial production of FA, generally fermentation process is used, and commonly used organisms are *Rhizopus* species (*R. nigricans*, *R. formosa*, *R. arrhizus*, and *R. oryzae*). *R. arrhizus* and *R. oryzae* are preferably utilized among these species due to their less complicated nutritional requirements and fermentability under aerobic and anaerobic conditions [54]. Although glucose is used as the primary carbon source but many other carbon sources, such as xylose, starch, lignocellulosic biomass, glycerol, and liquid food waste may be used during fermentation process [45, 55].

6.8 Malic Acid (4C)

Malic acid (MA, 2-hydroxybutanedioic acid, $C_4H_6O_5$) is a white, crystalline, and hygroscopic solid (Figure 6.1). It occurs mostly in fruits and vegetables such as grapes, cherries, melon, peach, plum, blueberries, carrot, and broccoli. MA is frequently used in the food and beverage industry (as an acidulant and taste enhancer) and cosmetics industry to reset the pH of cream and gels [56]. It is used in the synthesis of polymalic acid (PMA), a biodegradable polymer, which has several medical applications such as drug carriers and surgical suture [57]. Chemically, it is produced through hydration of maleic acid or fumaric acid, but it is produced in a racemic mixture. Enantioselective hydration of fumaric acid by fumarase enzyme produces L-malic acid (L-MA) only [58]. In 1924, MA was reported in the fermentative product of yeast, [59]. Later, many microbes have been identified, which produce MA by one-step fermentation process. *Aspergillus flavus* is identified as a natural producer of MA, which accumulates L-MA under aerobic conditions in high glucose concentration (Figure 6.2) [60]. This organism was not proceeded further for the manufacturing of food-grade L-MA due to the contamination of aflatoxins, which are naturally produced by this fungus. Other species of *Aspergillus* are also explored to produce MA using glucose as carbon source such as *A. niger* ATCC 9029, *A. niger* ATCC 9142, *A. niger* ATCC 10577, *A. oryzae* NRRL3488, *A. oryzae* DSM 1863 [61]. A mutant strain of *A. niger* MTCC 281 is adapted to produce MA (77.38 ± 0.51 g/l) using crude glycerol as a substrate [62]. *Zygosaccharomyces rouxii* V19, a sugar-tolerant yeast, produced MA (74.9 g/l) after the addition of 0.5% glutamic acid [63]. Different species of *Penicillium* like *Penicillium rubens* 67, *Penicillium clerotiorum* K304, and *Penicillium viticola* 152 are identified as a producer of calcium malate [61]. Further metabolic engineering of *A. niger* strain S575 leads to the production of high titer of MA without citric acid (CA) accumulation. Final engineered strain S1149 produced up to 201.13 g/l of MA in fed-batch fermentation [64]. Yeast is considered as an appropriate host for the production of industrially valuable chemicals, but for MA, yeast could not be considered as a suitable host because of its lower production. Although genetically modified yeast was able to produce MA and for this experiment, *S. cerevisiae* strain, which was glucose-tolerant, C_2-independent pyruvate decarboxylase-negative was utilized for the incorporation of three genetic modifications simultaneously. The resulted strain produced MA titer up to 59 g/l [65]. In another study, *Aureobasidium pullulans* var. melanogenum GXZ6 is reported to produce PMA [66]. Researchers attempted to produce MA through acid hydrolysis of PMA [67].

6.9 Itaconic Acid (5C)

Itaconic acid (IA, 2-methylenesuccinic acid or methylene butanedioic acid) is a white, crystalline, unsaturated dicarboxylic acid, which is highly soluble in water and in alcohols also [68]. The unique property of IA is the presence of multifunctional groups especially conjugation of methylene group with carboxylic acid (Figure 6.1). It has wide industrial applications such as it is utilized for the synthesis of resins, fibers, polymers, as co-builder in detergents and as anti-scaling agent in oilfields. In the pharmaceutical industry, it is used in the manufacturing of bioactive compounds, and in the food industry, it is used as a food additive [69]. Conversion of IA into methyl methacrylate (Plexiglass) has massive potential for increasing IA's global market value. IA was first identified as a product of pyrolytic distillation of citric acid. Kinoshita (1932) reported that a green color species of *Aspergillus,* isolated from salted plum, was able to produce IA; thereafter, this fungus was named as *Aspergillus itaconicus* [70]. Later in 1939, *Aspergillus terreus* was identified as a producer of IA by Calam and his team. PFIZER company initiated the fermentation process for the microbial production of IA [71]. Different strains of *A. terreus* have been screened for IA production and *A. terreus* NRRL 1960 (ATCC 10020) was found more suitable for the production of a good yield of IA [70]. Other species of *Aspergillus,* such as *A. niger* and *A. oryzae* are also explored as IA's potential producer [72]. Tabuchi et al. [73] reported that *Candida* sp. produced IA up to 35% yield under phosphate-limited conditions. It has been observed that basidiomycete *Ustilago maydis* can produce 44.5 g/l˙ IA using 200 g/l glucose and 75 mM ammonium [74]. Another organism *Pseudozyma antarctica* NRRL Y-7808 is able to produce IA from glucose and other sugars also under a controlled supply of nitrogen [75]. *Pseudozyma* was earlier considered as yeast having basidiomycetes affinity, but now it is renamed as Moesziomyces and placed under Ustilaginales order [76]. Several factors affect the yield during microbial production of IA such as strain, nutrient media, oxygen concentration, nitrogen presence, and mode of fermentation process [70]. To enhance the yield of IA, mutant, and genetically modified strains are examined. CA, a TCA cycle intermediate, acts as a precursor molecule for IA's biosynthesis, and it involves aconitase and *cis*-aconitate decarboxylase enzymes (Figure 6.2) [77]. Genetic engineering of *E. coli* by addition and deletion of genes have the potential to increase the yield of IA. A strain was developed by overexpressing the cis-aconitate decarboxylase (*cad*) and aconitase B gene (*acnB*) gene in isocitrate dehydrogenase gene (*icd*) deleted host cell, and produced 4.34 g/l IA. [78]. To further increase the IA production, a model-based approach was followed by Harder et al. [79], and they successfully showed that *E. coli* strain (ita23: MG1655 Δ*aceA* Δ*sucCD* Δ*pykA* Δ*pykF* Δ*pta* Δ*Picd::cam_BBa_J23115* pCadCS) produced 32 g/l IA during fed-batch cultivation. In their study, Noh et al. [80] genetically modified *E. coli* to produce IA by using cheap carbon sources such as acetate. The genetically modified strain of *E. coli* WCIAG4 was able to produce 3.57 g/l itaconic acid by using acetate [80]. Genetic engineering of *A. niger* with cis-aconitate decarboxylase gene from *A. terreus* enables it to produce IA, but quantity happens to be very low. Further modification with the addition of cytosolic citrate synthase citB leads to higher titer (26.2 g/l) of IA [81, 82]. Genetic manipulation in *S. cerevisiae* leads to IA production (168 mg/l) [83]. Recently, a nonconventional salt-tolerant yeast, *Pichia kudriavzevii,* was genetically modified with the addition of aconitic acid decarboxylase gene (*cad*) from *A. terreus,* and achieved the titer 135 ± 5 mg/l of IA. Further, the production is doubled on addition of mitochondrial tricarboxylate transporter (*Pk_mttA*). A higher titer (505 ± 17.7 mg/l) in shake flask condition is obtained after knocking out of isocitrate dehydrogenase (*icd*) through CRISPR-Cas9 system [84].

6.10 Gluconic Acid (6C)

Gluconic acid (GA, 2,3,4,5,6-pentahydroxy caproic acid) is a mild and noncorrosive organic acid (Figure 6.1). It is soluble in water, and it acts as a good chelator at high pH. It is used in food, pharmaceutical, and construction industries. In 1880, Boutroux discovered the microbial production of GA in the culture of AAB. In 1992, firstly, it was isolated from a fungal source, *A. niger* [85]. Other than *A. niger*, many filamentous fungi are reported as GA producers, such as *Penicillium, Scopulariopsis, Gliocadium,* and *Gonatobotrys* [86]. Bacterial genus as a producer of GA are *Pseudomonas, Gluconobacter, Acetobacter, Micrococcus, Enterobactor, Zymomonas,* and *Azospirillum* [85, 87]. At the cellular level, gluconic acid is synthesized outside the cell membrane by the enzyme glucose oxidase, localized in the cell wall of GA-producing fungi (Figure 6.2). Two-step reaction is followed to synthesize GA from glucose. GA catalyzes the conversion of glucose (β-D glucopyranose) into D-glucono-1, 5 lactone with the release of H_2O_2 (converted into O_2 and H_2O by enzyme catalase). Further, D-glucono-1, 5 lactone is either spontaneously, or through gluconolactonase enzyme, is processed into GA [88]. This process involving oxidase/catalase is generally present in fungi, while bacteria follow different mechanisms for GA production. A membrane-bound PQQ (quinoprotein)-dependent glucose dehydrogenase (bound to the cytoplasmic membrane) is responsible for GA synthesis in *Acinetobacter calcoaceticus*. Similar enzyme is also found in other bacterial species such as *Klebsiella aerogenes, Pseudomonas aeruginosa*, and *Rhodobacter sphaeroides* [89]. *Gluconobacter oxydans* is exceptional with a NADP-dependent glucose dehydrogenase and membrane-bound glucose dehydrogenase; therefore, it directly converts glucose into gluconic acid [90]. At the industrial level, gluconic acid is produced through the fermentation of *A. niger*. Different parameters affected GA's final yield, and most important are dissolved oxygen and pH [86]. In a study, different fermentations (submerged, semisolid-state, surface, and solid-state surface) are compared with varying substrate concentrations by using *A. niger*, and results revealed that the highest yield of GA (94.7%) is obtained under solid-state fermentation [91].

6.11 Citric Acid (6C)

Citric acid (2-hydroxy-propane-1, 2, 3-TCA) is an industrially crucial organic acid as it is mostly used in the food, pharmaceutical, and cosmetic industries (Figure 6.1). Citric acid (CA) is water-soluble and has a pleasant acidic taste. It can act as a chelating or buffering agent [92]. CA's estimated annual demand is two million tons, leading to demand of less expensive and renewable sources [93]. In 1893, the fermentation of CA was firstly observed by Wehmer through the culture of *Penicillium glaucum* using sugar medium. In 1917, the pioneering work for CA production was carried out by Currievia through submerged fermentation of *A. niger* in starch or sugar-based medium. This method is well adapted for CA's industrial production due to its advantages over the chemical method [94, 95]. Now several species of bacteria, yeast, and fungi are reported for CA production [96]. Some bacterial species reported for CA production are *Arthrobacter paraffinens*, *Corynebacterium* sp., and *Bacillus licheniformis*. Filamentous fungi are well-known producer of CA, and common genera are *Aspergillus* and *Penicillium* [92, 97]. Other than *A. niger*, many species of *Aspergillus* such as *Aspergillus fonsecaeus*, *Aspergillus awamori*, *Aspergillus saitoi*, *Aspergillus wentii*, *Aspergillus fenicis* and *Aspergillus clavatus* are reported for the production of CA. Different species of *Candida* such as *Candida lipolytica* [98], *Candida tropicalis* [99], *Candida guillermondii* [100], *Candida parapsilosis* [101], and *Candida oleophila* are also examined for CA production. Since last decade, *Y. lipolytica,* a non-conventional yeast, is main attraction for CA production [92, 102]. Most strains of *Y. lipolytica* excrete CA as a result in disturbance in metabolism generally in the nitrogen-deficient medium [103]. However, optimization of CA yield has been performed with wild, as well as mutant strains and findings suggest that mutant strains can produce multifold CA than wild types [104]. Different carbon sources such as n-paraffins, hexadecanes, edible oils, glucose, molasses, starch hydrolysates, and glycerol have been tried to maximize the CA yield [102, 105, 106]. Further, the metabolic engineering of *Y. lipolytica* has been established for achieving the high titer of CA, as increment in CA synthesis (overexpression of pyruvate carboxylase gene), supplement of precursor molecules, or removal of feedback inhibition (Figure 6.2) [107, 108].

6.12 Kojic Acid (6C)

Kojic acid (KA, 5-hydroxy-2-hydroxymethylγ-pyrone) is a colorless, crystalline (prismatic needles) weak acid having a solubility in polar substances like water and alcohols (Figure 6.1). KA has numerous industrial applications in the medical industry (as antibacterial, antifungal, antitumor, antidiabetic agent, and as a pain killer), food industry (as a flavor enhancer and antioxidant), agriculture (as anti-melanosis and insecticide activator), and analytical chemistry (as iron determining reagent, iron chelators and in the synthesis of kojic acid-chitosan conjugates) [109]. In the cosmetic industry, it is used to prepare skin whitening creams, soaps, and lotions due to its properties as it acts as UV protector and suppressor of hyperpigmentation. It inhibits melanin formation due to tyrosinase inhibitory activity [110]. For microbial production of KA, different species of *Aspergillus* are explored such as *Aspergillus tamarii*, *Aspergillus oryzae*, *Aspergillus parasiticus*, *Aspergillus flavus*, and *A. candidus* [109]. *Aspergillus flavus* Link 44-1 strain is used by various refreshers to study the optimum conditions for KA production, and studies revealed that different factors such as aeration [111], carbon and nitrogen sources, and pH of media [112] could be crucial factors for maximum production of KA. However, there is always a risk of aflatoxin production by *A. flavus,* although a study done by Madihah et al. [113] showed that aflatoxin production can be inhibited by manipulating the medium composition and culture conditions. For improvement in yield of KA, monokaryotic culture of *A. flavus* was developed, it produced 19.2 g/l of KA, which is higher than heterokaryons (10.5 g/l) [109]. Immobilized cells (entrapped in calcium alginate) of *A. oryzae* was able to accumulate KA up to 83 g/l [114]. Entrapment of *A. oryzae* in plastic composite support (PCS) bioreactor also produced KA up to 3.09 g/l/d [115]. A mutant strain of *A. oryzae* MK107-39 was able to produce multifold KA than parent strain [116]. For more understanding of KA biosynthesis in *A. oryzae,* 14 genes involved in KA biosynthesis have been identified, some essential genes are *kojR* (a transcription factor), *kojA* (an enzyme), *kojT* (a transporter gene) [117]. For enhancement of KA, *kojR* was overexpressed in *A. oryzae* and transformed strain was able to produce d 26.4 g/l kojic acid from 80 g/l glucose. For utilizing the cellulose as substrate, this strain was further transformed with three cellulase genes and final strain successfully used phosphoric acid swollen cellulose as substrate and produced 0.18 g/l of kojic in six days of fermentation [118].

6.13 Muconic and Adipic Acid (C6)

Adipic acid (ADA, hexanedioic acid) is a colorless crystal, and it is one of the most important aliphatic dicarboxylic acid (Figure 6.1). It is mainly used in the synthesis of nylon-6,6 polyamide and other chemicals such as polyurethanes, as lubricant, and polyester polyols. Other than the chemical industry, it is also having application in food industry as acidulant. Chemical synthesis of adipic acid involves benzene, a carcinogen, which derived from petrochemicals and major drawback of this process is emission of green-house gases, so it becomes mandatory to search for biological route for synthesis of adipic acid. Several efforts have been performed to produce ADA and its precursor molecules such as muconic acid and glucaric acid through microbial fermentation. Earlier attempts have been made to produce ADA by combining the biological and chemical processes. Due to unavailability of microbes for production of ADA, recombinant organisms were developed with heterologous expression of genes (Figure 6.3). In this study *E. coli* is metabolically engineered to produce cis, cis-muconic acid (CCM) from glucose through fermentation and further CCM is converted in to adipic acid through chemical hydrogenation process [119]. Metabolic pathway was reconstructed with three genes, *aroZ* (DHS dehydratase from *Klebsiella pneumonia*), *aroY* (PCA decarboxylase from *K. pneumoniae*), and *catA* (catechol 1,2-dioxygenase from *A. calcoaceticus*). The final recombinant strain of *E. coli* showed production of CCM up to 36.8 g/l [120]. In another study, *S. cerevisiae* is genetically engineered to produce CCM from the 3-dehydroshikimate, an intermediate of aromatic amino acid biosynthesis [121]. The engineered pathway involves three enzymes: 3-dehydroshikimate dehydratase, protocatechuic acid decarboxylase, and catechol 1,2-dioxygenase. The final constructed strain successfully produced CCM (1.56 mg/l), which can be chemically hydrogenated for

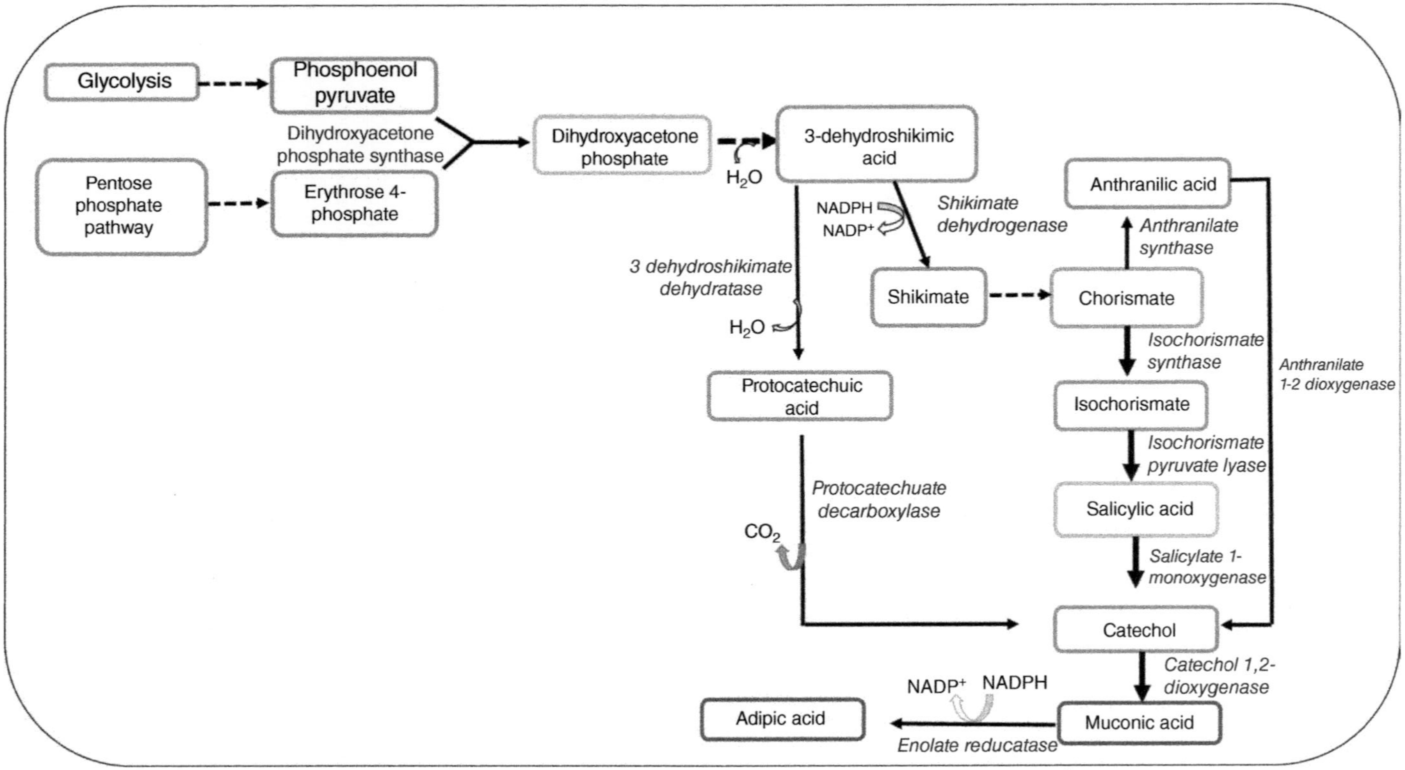

Figure 6.3 Schematic diagram representing the proposed metabolic pathways used for bio-based production of muconic and adipic acid.

production of ADA. Curran et al. reported that recombinant yeast can produce up to 140 mg/l of CCM from glucose [122]. Extensive research for developing the complete pathway for production of ADA leads to development of such pathway in engineered *E. coli*, which is able to produce ADA directly from simple carbon sources [9]. They added one more gene, enoate reductase, from *Clostridium acetobutylicum* (*CaER*) along with genes required for biosynthesis of CCM. This enoate reductase is oxygen-sensitive, and it can convert CCM to ADA under microaerobic conditions. The recombinant *E. coli* produced ADA up to 27.6 ± 1.3 mg/l [9]. Complete biological route for production of ADA was also transferred in *S. cerevisiae* by using the enoate reductase from *Bacillus coagulans* in a *S. cerevisiae* strain having muconic acid biosynthesis pathway. After three-stage fermentation process, transformed yeast cells produced ~284 mg/l CCM and 2.6 mg/l ADA [123].

6.14 Conclusions and Perspectives

Organic acids are the building blocks for the generation of a plethora of chemicals, valuable for industrial applications. Recently, the market of organic acids is increasing continuously due to its tremendous applications in various sectors. To avoid the harmful effects of the chemical process for the production of organic acids, current trend is shifting toward the production of these acids through microbial fermentation processes. Different natural producers of these acids have been isolated, and it is observed that filamentous fungus are dominating producers over the bacteria and yeasts as fungi show higher acid-tolerant capacity. Inception of synthetic biology tools and metabolic engineering leads to development of recombinant organism having the capacity to generate these targeted molecules in desirable amount, which may aid to lower the global prices of organic acids, if they successfully operate at large scale. *E. coli, S. cerevisiae*, *C. glutamicum,* and *Y. lipolytica* are the examples of host cells that are successfully utilized as microbial cell factories for production of different organic acids. However, the main obstacle in microbial production of organic acid is the purification steps, which need to be handled very carefully. Purity of chemicals is directly correlated with market value of the process, so focus should be on the downstream processing along with production process.

Acknowledgments

SPS and AKR acknowledge the Department of Biotechnology (DBT) for all support.

References

1 Sauer, M., Porro, D., Mattanovich, D., and Branduardi, P. (2008). Microbial production of organic acids: expanding the markets. *Trends in Biotechnology* 26 (2): 100–108.

2 Alonso, S., Rendueles, M., and Díaz, M. (2015). Microbial production of specialty organic acids from renewable and waste materials. *Critical Reviews in Biotechnology* 35 (4): 497–513.

3 Becker, J., Lange, A., Fabarius, J., and Wittmann, C. (2015). Top value platform chemicals: bio-based production of organic acids. *Current Opinion in Biotechnology* 36: 168–175.

4 Chen, Y. and Nielsen, J. (2016). Biobased organic acids production by metabolically engineered microorganisms. *Current Opinion in Biotechnology* 37: 165–172.

5 Yin, X., Li, J., Shin, H.D. et al. (2015). Metabolic engineering in the biotechnological production of organic acids in the tricarboxylic acid cycle of microorganisms: advances and prospects. *Biotechnology Advances* 33 (6): 830–841.

6 Förster, A., Aurich, A., Mauersberger, S., and Barth, G. (2007). Citric acid production from sucrose using a recombinant strain of the yeast *Yarrowia lipolytica*. *Applied Microbiology and Biotechnology* 75 (6): 1409–1417.

7 Warnecke, T. and Gill, R.T. (2005). Organic acid toxicity, tolerance, and production in *Escherichia coli* biorefining applications. *Microbial Cell Factories* 4 (1): 1–8.

8 Buschke, N., Schäfer, R., Becker, J., and Wittmann, C. (2013). Metabolic engineering of industrial platform microorganisms for biorefinery applications–optimization of substrate spectrum and process robustness by rational and evolutive strategies. *Bioresource Technology* 135: 544–554.

9 Sun, J., Raza, M., Sun, X., and Yuan, Q. (2018). Biosynthesis of adipic acid via microaerobic hydrogenation of cis, cis-muconic acid by oxygen-sensitive enoate reductase. *Journal of Biotechnology* 280: 49–54.

10 Schmitt, E.E. and Polistina, R.A. (1969). U.S. Patent No. 3,463,158. Washington, DC: U.S. Patent and Trademark Office.

11 Makadia, H.K. and Siegel, S.J. (2011). Poly lactic-co-glycolic acid (PLGA) as biodegradable controlled drug delivery carrier. *Polymers* 3 (3): 1377–1397.

12 Koivistoinen, O.M., Kuivanen, J., Barth, D. et al. (2013). Glycolic acid production in the engineered yeasts *Saccharomyces cerevisiae* and *Kluyveromyces lactis*. *Microbial Cell Factories* 12 (1): 82.

13 Lachaux, C., Frazao, C.J., Krauβer, F. et al. (2019). A new synthetic pathway for the bioproduction of glycolic acid from lignocellulosic sugars aimed at maximal carbon conservation. *Frontiers in Bioengineering and Biotechnology* 7: 359.

14 Cabulong, R.B., Lee, W.K., Bañares, A.B. et al. (2018). Engineering *Escherichia coli* for glycolic acid production from D-xylose through the Dahms pathway and glyoxylate bypass. *Applied Microbiology and Biotechnology* 102 (5): 2179–2189.

15 Li, W., Chen, J., Liu, C.X. et al. (2019). Microbial production of glycolate from acetate by metabolically engineered *Escherichia coli*. *Journal of Biotechnology* 291: 41–45.

16 Lee, S.S., Choi, J.I., and Woo, H.M. (2019). Bioconversion of xylose to ethylene glycol and glycolate in engineered *Corynebacterium glutamicum*. *ACS Omega* 4 (25): 21279–21287.

17 Zhang, Z., Yang, Y., Wang, Y. et al. (2020). Ethylene glycol and glycolic acid production from xylonic acid by *Enterobacter cloacae*. *Microbial Cell Factories* 19: 1–16.

18 Saichana, N., Matsushita, K., Adachi, O. et al. (2015). Acetic acid bacteria: a group of bacteria with versatile biotechnological applications. *Biotechnology Advances* 33 (6): 1260–1271.

19 De Roos, J. and De Vuyst, L. (2018). Acetic acid bacteria in fermented foods and beverages. *Current Opinion in Biotechnology* 49: 115–119.

20 Mamlouk, D. and Gullo, M. (2013). Acetic acid bacteria: physiology and carbon sources oxidation. *Indian Journal of Microbiology* 53 (4): 377–384.

21 Entani, E., Ohmori, S., Masai, H., and Suzuki, K.I. (1985). *Acetobacter polyoxogenes* sp. nov., a new species of an acetic acid bacterium useful for producing vinegar with high acidity. *The Journal of General and Applied Microbiology* 31 (5): 475–490.

22 Brownell, J.E. and Nakas, J.P. (1991). Bioconversion of acid-hydrolyzed poplar hemicellulose to acetic acid by *Clostridium thermoaceticum*. *Journal of Industrial Microbiology* 7 (1): 1–6.

23 Kocher, G.S., Kalra, K.L., and Phutela, R.P. (2006). Comparative production of sugarcane vinegar by different immobilization techniques. *Journal of the Institute of Brewing* 112 (3): 264–266.

24 Xu, P., Qiu, J., Gao, C., and Ma, C. (2008). Biotechnological routes to pyruvate production. *Journal of Bioscience and Bioengineering* 105 (3): 169–175.

25 Maleki, N. and Eiteman, M.A. (2017). Recent progress in the microbial production of pyruvic acid. *Fermentation* 3 (1): 8.

26 Yonehara, T. and Miyata, R. (1994). Fermentative production of pyruvate from glucose by *Torulopsis glabrata*. *Journal of Fermentation and Bioengineering* 78 (2): 155–159.

27 Eisenberg, A., Seip, J.E., Gavagan, J.E. et al. (1997). Pyruvic acid production using methylotrophic yeast transformants as catalyst. *Journal of Molecular Catalysis B: Enzymatic* 2 (4-5): 223–232.

28 Morgunov, I.G., Kamzolova, S.V., Perevoznikova, O.A. et al. (2004). Pyruvic acid production by a thiamine auxotroph of *Yarrowia lipolytica*. *Process Biochemistry* 39 (11): 1469–1474.

29 Van Maris, A.J., Geertman, J.M.A., Vermeulen, A. et al. (2004). Directed evolution of pyruvate decarboxylase-negative *Saccharomyces cerevisiae*, yielding a C2-independent, glucose-tolerant, and pyruvate-hyperproducing yeast. *Applied and Environmental Microbiology* 70 (1): 159–166.

30 Zhu, Y., Eiteman, M.A., Altman, R., and Altman, E. (2008). High glycolytic flux improves pyruvate production by a metabolically engineered *Escherichia coli* strain. *Applied and Environmental Microbiology* 74 (21): 6649–6655.

31 Akita, H., Nakashima, N., and Hoshino, T. (2016). Pyruvate production using engineered *Escherichia coli*. *AMB Express* 6 (1): 1–8.

32 Wee, Y.J., Kim, J.N., and Ryu, H.W. (2006). Biotechnological production of lactic acid and its recent applications. *Food Technology and Biotechnology* 44: 163–172.

33 Cubas-Cano, E., González-Fernández, C., Ballesteros, M., and Tomás-Pejó, E. (2018). Biotechnological advances in lactic acid production by lactic acid bacteria: lignocellulose as novel substrate. *Biofuels, Bioproducts and Biorefining* 12 (2): 290–303.

34 Fan, Y., Zhou, C., and Zhu, X. (2009). Selective catalysis of lactic acid to produce commodity chemicals. *Catalysis Reviews* 51 (3): 293–324.

35 Casalini, T., Rossi, F., Castrovinci, A., and Perale, G. (2019). A perspective on polylactic acid-based polymers use for nanoparticles synthesis and applications. *Frontiers in Bioengineering and Biotechnology* 7: 259.

36 Gao, C., Ma, C., and Xu, P. (2011). Biotechnological routes based on lactic acid production from biomass. *Biotechnology Advances* 29 (6): 930–939.

37 Okano, K., Tanaka, T., Ogino, C. et al. (2010). Biotechnological production of enantiomeric pure lactic acid from renewable resources: recent achievements, perspectives, and limits. *Applied Microbiology and Biotechnology* 85 (3): 413–423.

38 Panesar, P.S., Kennedy, J.F., Gandhi, D.N., and Bunko, K. (2007). Bioutilisation of whey for lactic acid production. *Food Chemistry* 105 (1): 1–14.

39 Narita, J., Nakahara, S., Fukuda, H., and Kondo, A. (2004). Efficient production of L-(+)-lactic acid from raw starch by *Streptococcus bovis 148*. *Journal of Bioscience and Bioengineering* 97 (6): 423–425.

40 Bechthold, I., Bretz, K., Kabasci, S. et al. (2008). Succinic acid: a new platform chemical for biobased polymers from renewable resources. *Chemical Engineering & Technology: Industrial Chemistry-Plant Equipment-Process Engineering-Biotechnology* 31 (5): 647–654.

41 Saxena, R.K., Saran, S., Isar, J., and Kaushik, R. (2017). Production and applications of succinic acid. In: *Current Developments in Biotechnology and Bioengineering* (eds. A. Pandey, S. Negi and C.R. Soccol), 601–630. Elsevier.

42 Cao, Y., Zhang, R., Sun, C. et al. (2013). Fermentative succinate production: an emerging technology to replace the traditional petrochemical processes. *BioMed Research International* https://doi.org/10.1155/2013/723412.

43 Lee, P., Lee, S., Hong, S., and Chang, H. (2002). Isolation and characterization of a new succinic acid-producing bacterium, *Mannheimia succiniciproducens* MBEL55E, from bovine rumen. *Applied Microbiology and Biotechnology* 58 (5): 663–668.

44 Millard, C.S., Chao, Y.P., Liao, J.C., and Donnelly, M.I. (1996). Enhanced production of succinic acid by overexpression of phosphoenolpyruvate carboxylase in *Escherichia coli*. *Applied and Environmental Microbiology* 62 (5): 1808–1810.

45 Ilica, R.A., Kloetzer, L., Galaction, A.I., and Caşcaval, D. (2019). Fumaric acid: production and separation. *Biotechnology Letters* 41 (1): 47–57.

46 Peleg, Y., Battat, E., Scrutton, M.C., and Goldberg, I. (1989). Isoenzyme pattern and subcellular localisation of enzymes involved in fumaric acid accumulation by *Rhizopus oryzae*. *Applied Microbiology and Biotechnology* 32 (3): 334–339.

47 Yu, S.Z., Huang, D., Wen, J.P. et al. (2012). Metabolic profiling analysis for fumaric acid production improvement in *Rhizopus oryzae* by femtosecond laser irradiation. *Bioresource Technology* 114: 610–615.

48 Liu, Y., Song, J., Tan, T., and Liu, L. (2015). Production of fumaric acid from l-malic acid by solvent engineering using a recombinant thermostable fumarase from *Thermus thermophilus* HB8. *Applied Biochemistry and Biotechnology* 175 (6): 2823–2831.

49 Xu, G., Zou, W., Chen, X. et al. (2012). Fumaric acid production in *Saccharomyces cerevisiae* by *in silico* aided metabolic engineering. *PLoS One* 7 (12): e52086.

50 Xu, G., Chen, X., Liu, L., and Jiang, L. (2013). Fumaric acid production in *Saccharomyces cerevisiae* by simultaneous use of oxidative and reductive routes. *Bioresource Technology* 148: 91–96.

51 Nakajima-Kambe, T., Nozue, T., Mukouyama, M., and Nakahara, T. (1997). Bioconversion of maleic acid to fumaric acid by *Pseudomonas alcaligenes* strain XD-1. *Journal of Fermentation and Bioengineering* 84 (2): 165–168.

52 Kato, Y., Yamagishi, J., and Asano, Y. (1995). Maleate cis-trans isomerase from *Arthrobacter* sp. TPU 5446. *Journal of Fermentation and Bioengineering* 80 (6): 610–612.

53 Hatakeyama, K., Goto, M., Uchida, Y. et al. (2000). Molecular analysis of maleate cis-trans isomerase from thermophilic bacteria. *Bioscience, Biotechnology, and Biochemistry* 64 (3): 569–576.

54 Liu, H., Zhao, S., Jin, Y. et al. (2017). Production of fumaric acid by immobilized *Rhizopus arrhizus* RH 7-13-9# on loofah fiber in a stirred-tank reactor. *Bioresource Technology* 244: 929–933.

55 Liu, H., Ma, J., Wang, M. et al. (2016). Food waste fermentation to fumaric acid by *Rhizopus arrhizus* RH7-13. *Applied Biochemistry and Biotechnology* 180 (8): 1524–1533.

56 Goldberg, I., Rokem, J.S., and Pines, O. (2006). Organic acids: old metabolites, new themes. *Journal of Chemical Technology & Biotechnology: International Research in Process, Environmental & Clean Technology* 81 (10): 1601–1611.

57 Ding, H., Helguera, G., Rodríguez, J.A. et al. (2013). Polymalic acid nanobioconjugate for simultaneous immunostimulation and inhibition of tumor growth in HER2/neu-positive breast cancer. *Journal of Controlled Release* 171 (3): 322–329.

58 Chibata, I., Tosa, T., and Takata, I. (1983). Continuous production of L-malic acid by immobilized cells. *Trends in Biotechnology* 1 (1): 9–11.

59 Dakin, H.D. (1924). The formation of l-malic acid as a product of alcoholic fermentation by yeast. *Journal of Biological Chemistry* 61 (1): 139–145.

60 Peleg, Y., Stieglitz, B., and Goldberg, I. (1988). Malic acid accumulation by *Aspergillus flavus*. *Applied Microbiology and Biotechnology* 28 (1): 69–75.

61 Chi, Z., Wang, Z.P., Wang, G.Y. et al. (2016). Microbial biosynthesis and secretion of l-malic acid and its applications. *Critical Reviews in Biotechnology* 36 (1): 99–107.

62 Iyyappan, J., Bharathiraja, B., Baskar, G. et al. (2018). Malic acid production by chemically induced *Aspergillus niger* MTCC 281 mutant from crude glycerol. *Bioresource Technology* 251: 264–267.

63 Taing, O. and Taing, K. (2007). Production of malic and succinic acids by sugar-tolerant yeast *Zygosaccharomyces rouxii*. *European Food Research and Technology* 224 (3): 343–347.

64 Xu, Y., Zhou, Y., Cao, W., and Liu, H. (2020). Improved production of malic acid in *Aspergillus niger* by abolishing citric acid accumulation and enhancing glycolytic flux. *ACS Synthetic Biology* 9 (6): 1418–1425.

65 Zelle, R.M., de Hulster, E., van Winden, W.A. et al. (2008). Malic acid production by *Saccharomyces cerevisiae*: engineering of pyruvate carboxylation, oxaloacetate reduction, and malate export. *Applied and Environmental Microbiology* 74 (9): 2766–2777.

66 Zeng, W., Zhang, B., Chen, G. et al. (2019). Efficient production of polymalic acid by a novel isolated *Aureobasidium pullulans* using metabolic intermediates and inhibitors. *Applied Biochemistry and Biotechnology* 187 (2): 612–627.

67 Zou, X., Zhou, Y., and Yang, S.T. (2013). Production of polymalic acid and malic acid by *Aureobasidium pullulans* fermentation and acid hydrolysis. *Biotechnology and Bioengineering* 110 (8): 2105–2113.

68 Yang, W., Hu, Y., Chen, Z. et al. (2012). Solubility of itaconic acid in different organic solvents: experimental measurement and thermodynamic modeling. *Fluid Phase Equilibria* 314: 180–184.

69 Kuenz, A., Gallenmüller, Y., Willke, T., and Vorlop, K.D. (2012). Microbial production of itaconic acid: developing a stable platform for high product concentrations. *Applied Microbiology and Biotechnology* 96 (5): 1209–1216.

70 da Cruz, J.C., Sérvulo, E.F.C., and de Castro, A.M. (2017). Microbial production of itaconic acid. In: *Microbial Production of Food Ingredients and Additives* (eds. A.M. Holban and A.M. Grumezescu), 291–316. Academic Press.

71 Pfeifer, V.F., Vojnovich, C., and Heger, E.N. (1952). Itaconic acid by fermentation with *Aspergillus terreus*. *Industrial and Engineering Chemistry* 44 (12): 2975–2980.

72 Li, A., Pfelzer, N., Zuijderwijk, R., and Punt, P. (2012). Enhanced itaconic acid production in *Aspergillus niger* using genetic modification and medium optimization. *BMC Biotechnology* 12 (1): 57.

73 Tabuchi, T., Sugisawa, T., Ishidori, T. et al. (1981). Itaconic acid fermentation by a yeast belonging to the genus *Candida*. *Agricultural and Biological Chemistry* 45 (2): 475–479.

74 Maassen, N., Panakova, M., Wierckx, N. et al. (2014). Influence of carbon and nitrogen concentration on itaconic acid production by the smut fungus *Ustilago maydis*. *Engineering in Life Sciences* 14 (2): 129–134.

75 Levinson, W.E., Kurtzman, C.P., and Kuo, T.M. (2006). Production of itaconic acid by *Pseudozyma antarctica* NRRL Y-7808 under nitrogen-limited growth conditions. *Enzyme and Microbial Technology* 39 (4): 824–827.

76 Shoff, C.J. and Perfect, J.R. (2021). Uncommon yeasts and molds causing human disease. In: *Encyclopedia of Mycology*, vol. 1 (eds. Ó. Zaragoza and A. Casadevall), 813–834. Elsevier https://doi.org/10.1016/B978-0-12-809633-8.21023-6.

77 Steiger, M.G., Blumhoff, M.L., Mattanovich, D., and Sauer, M. (2013). Biochemistry of microbial itaconic acid production. *Frontiers in Microbiology* 4: 23.

78 Okamoto, S., Chin, T., Hiratsuka, K. et al. (2014). Production of itaconic acid using metabolically engineered *Escherichia coli*. *The Journal of General and Applied Microbiology* 60 (5): 191–197.

79 Harder, B.J., Bettenbrock, K., and Klamt, S. (2016). Model-based metabolic engineering enables high yield itaconic acid production by *Escherichia coli*. *Metabolic Engineering* 38: 29–37.

80 Noh, M.H., Lim, H.G., Woo, S.H. et al. (2018). Production of itaconic acid from acetate by engineering acid-tolerant *Escherichia coli* W. *Biotechnology and Bioengineering* 115 (3): 729–738.

81 van der Straat, L., Vernooij, M., Lammers, M. et al. (2014). Expression of the *Aspergillus terreus* itaconic acid biosynthesis cluster in *Aspergillus niger*. *Microbial Cell Factories* 13 (1): 11.

82 Hossain, A.H., Li, A., Brickwedde, A. et al. (2016). Rewiring a secondary metabolite pathway towards itaconic acid production in *Aspergillus niger*. *Microbial Cell Factories* 15 (1): 1–15.

83 Blazeck, J., Miller, J., Pan, A. et al. (2014). Metabolic engineering of *Saccharomyces cerevisiae* for itaconic acid production. *Applied Microbiology and Biotechnology* 98 (19): 8155–8164.

84 Sun, W., Vila-Santa, A., Liu, N. et al. (2020). Metabolic engineering of an acid-tolerant yeast strain *Pichia kudriavzevii* for itaconic acid production. *Metabolic Engineering Communications*: e00124.

85 Ramachandran, S., Fontanille, P., Pandey, A., and Larroche, C. (2006). Gluconic acid: properties, applications and microbial production. *Food Technology and Biotechnology* 44 (2).

86 Singh, O.V. and Kumar, R. (2007). Biotechnological production of gluconic acid: future implications. *Applied Microbiology and Biotechnology* 75 (4): 713–722.

87 Rodriguez, H., Gonzalez, T., Goire, I., and Bashan, Y. (2004). Gluconic acid production and phosphate solubilization by the plant growth-promoting bacterium *Azospirillum* spp. *Naturwissenschaften* 91 (11): 552–555.

88 Anastassiadis, S. and Morgunov, I.G. (2007). Gluconic acid production. *Recent Patents on Biotechnology* 1 (2): 167–180.

89 Duine, J.A., Frank Jzn, J., and Jongejan, J.A. (1986). PQQ and quinoprotein enzymes in microbial oxidations. *FEMS Microbiology Reviews* 1 (3-4): 165–178.

90 Pronk, J.T., Levering, P.R., Olijve, W., and Van Dijken, J.P. (1989). Role of NADP-dependent and quinoprotein glucose dehydrogenases in gluconic acid production by *Gluconobacter oxydans*. *Enzyme and Microbial Technology* 11 (3): 160–164.

91 Singh, O.V., Jain, R.K., and Singh, R.P. (2003). Gluconic acid production under varying fermentation conditions by *Aspergillus niger*. *Journal of Chemical Technology & Biotechnology: International Research in Process, Environmental & Clean Technology* 78 (2-3): 208–212.

92 Soccol, C.R., Vandenberghe, L.P., Rodrigues, C., and Pandey, A. (2006). New perspectives for citric acid production and application. *Food Technology and Biotechnology* 44 (2).

93 Carsanba, E., Papanikolaou, S., Fickers, P. et al. (2019). Citric acid production by *Yarrowia lipolytica*. In: *Non-conventional Yeasts: From Basic Research to Application* (ed. A. Sibirny), 91–117. Cham: Springer.

94 Papagianni, M. (2007). Advances in citric acid fermentation by *Aspergillus niger*: biochemical aspects, membrane transport and modeling. *Biotechnology Advances* 25 (3): 244–263.

95 Angumeenal, A.R. and Venkappayya, D. (2013). An overview of citric acid production. *LWT- Food Science and Technology* 50 (2): 367–370.

96 Max, B., Salgado, J.M., Rodríguez, N. et al. (2010). Biotechnological production of citric acid. *Brazilian Journal of Microbiology* 41 (4): 862–875.

97 Cunningham, J.E. and Kuiack, C. (1992). Production of citric and oxalic acids and solubilization of calcium phosphate by *Penicillium bilaii*. *Applied and Environmental Microbiology* 58 (5): 1451–1458.

98 Crolla, A. and Kennedy, K.J. (2001). Optimization of citric acid production from *Candida lipolytica* Y-1095 using n-paraffin. *Journal of Biotechnology* 89 (1): 27–40.

99 Okoshi, H., Sato, S., Mukataka, S., and Takahashi, J. (1987). Citric acid production by *Candida tropicalis* under high dissolved oxygen concentrations. *Agricultural and Biological Chemistry* 51 (1): 257–258.

100 Tisnadjaja, D., Gutierrez, N.A., and Maddox, I.S. (1996). Citric acid production in a bubble-column reactor using cells of the yeast *Candida guilliermondii* immobilized by adsorption onto sawdust. *Enzyme and Microbial Technology* 19 (5): 343–347.

101 Omar, S.H. and Rehm, H.J. (1980). Physiology and metabolism of two alkane oxidizing and citric acid producing strains of *Candida parapsilosis*. *European Journal of Applied Microbiology and Biotechnology* 11 (1): 42–49.

102 Papanikolaou, S., Muniglia, L., Chevalot, I. et al. (2002). *Yarrowia lipolytica* as a potential producer of citric acid from raw glycerol. *Journal of Applied Microbiology* 92 (4): 737–744.

103 Levinson, W.E., Kurtzman, C.P., and Kuo, T.M. (2007). Characterization of *Yarrowia lipolytica* and related species for citric acid production from glycerol. *Enzyme and Microbial Technology* 41 (3): 292–295.

104 Abghari, A. and Chen, S. (2017). Engineering *Yarrowia lipolytica* for enhanced production of lipid and citric acid. *Fermentation* 3 (3): 34.

105 Wojtatowicz, M., Rymowicz, W., and Kautola, H. (1991). Comparison of different strains of the yeast *Yarrowia lipolytica* for citric acid production from glucose hydrol. *Applied Biochemistry and Biotechnology* 31 (2): 165–174.

106 Rane, K. and Sims, K. (1993). Production of citric acid by *Candida lipolytica* Y 1095: effect of glucose concentration on yield and productivity. *Enzyme and Microbial Technology* 15: 646–651.

107 Liu, X.Y., Chi, Z., Liu, G.L. et al. (2013). Both decrease in *ACL1* gene expression and increase in *ICL1* gene expression in marine-derived yeast *Yarrowia lipolytica* expressing

INU1 gene enhance citric acid production from inulin. *Marine Biotechnology* 15 (1): 26–36.

108 Tan, M.J., Chen, X., Wang, Y.K. et al. (2016). Enhanced citric acid production by a yeast *Yarrowia lipolytica* over-expressing a pyruvate carboxylase gene. *Bioprocess and Biosystems Engineering* 39 (8): 1289–1296.

109 Mohamad, R., Mohamed, M.S., Suhaili, N. et al. (2010). Kojic acid: applications and development of fermentation process for production. *Biotechnology and Molecular Biology Reviews* 5 (2): 24–37.

110 Saeedi, M., Eslamifar, M., and Khezri, K. (2019). Kojic acid applications in cosmetic and pharmaceutical preparations. *Biomedicine & Pharmacotherapy* 110: 582–593.

111 Ariff, A.B., Salleh, M.S., Ghani, B. et al. (1996). Aeration and yeast extract requirements for kojic acid production by *Aspergillus flavus* link. *Enzyme and Microbial Technology* 19 (7): 545–550.

112 Rosfarizan, M., Arbakariya, A., Hassan, M.A. et al. (2002). Importance of carbon source feeding and pH control strategies for maximum kojic acid production from sago starch by *Aspergillus flavus*. *Journal of Bioscience and Bioengineering* 94 (2): 99–105.

113 Madihah, M.S., Ariff, A.B., Hassan, M.A. et al. (1996). Enhanced kojic acid production by *Aspergillus flavus* link in growth medium containing methanol. *ASEAN Food Journal* 11: 158–162.

114 Kwak, M.Y. and Rhee, J.S. (1992). Cultivation characteristics of immobilized *Aspergillus oryzae* for kojic acid production. *Biotechnology and Bioengineering* 39 (9): 903–906.

115 Liu, J.M., Yu, T.C., Lin, S.P. et al. (2016). Evaluation of kojic acid production in a repeated-batch PCS biofilm reactor. *Journal of Biotechnology* 218: 41–48.

116 Futamura, T., Okabe, M., Tamura, T. et al. (2001). Improvement of production of kojic acid by a mutant strain *Aspergillus oryzae*, MK107-39. *Journal of Bioscience and Bioengineering* 91 (3): 272–276.

117 Terabayashi, Y., Sano, M., Yamane, N. et al. (2010). Identification and characterization of genes responsible for biosynthesis of kojic acid, an industrially important compound from *Aspergillus oryzae*. *Fungal Genetics and Biology* 47 (12): 953–961.

118 Yamada, R., Yoshie, T., Wakai, S. et al. (2014). *Aspergillus oryzae*-based cell factory for direct kojic acid production from cellulose. *Microbial Cell Factories* 13 (1): 71.

119 Draths, K.M. and Frost, J.W. (1994). Environmentally compatible synthesis of adipic acid from D-glucose. *Journal of the American Chemical Society* 116 (1): 399–400.

120 Niu, W., Draths, K.M., and Frost, J.W. (2002). Benzene-free synthesis of adipic acid. *Biotechnology Progress* 18 (2): 201–211.

121 Weber, C., Brückner, C., Weinreb, S. et al. (2012). Biosynthesis of cis, cis-muconic acid and its aromatic precursors, catechol and protocatechuic acid, from renewable feedstocks by *Saccharomyces cerevisiae*. *Applied and Environmental Microbiology* 78 (23): 8421–8430.

122 Curran, K.A., Leavitt, J.M., Karim, A.S., and Alper, H.S. (2013). Metabolic engineering of muconic acid production in *Saccharomyces cerevisiae*. *Metabolic Engineering* 15: 55–66.

123 Raj, K., Partow, S., Correia, K. et al. (2018). Biocatalytic production of adipic acid from glucose using engineered *Saccharomyces cerevisiae*. *Metabolic Engineering Communications* 6: 28–32.

7

Production of Bioactive Compounds vs. Recombinant Proteins

Maria F. Salazar Affonso[1], *Débora Bublitz Anton*[2], *Daniel Kuhn*[2], *Bruno Dahmer*[2], *Camile Wünsch*[2], *Verônica Contini*[2,3], *Luís F. Saraiva Macedo Timmers*[2,3], *Claucia F. Volken de Souza*[2,4], *Márcia I. Goettert*[2,3], *and Rodrigo G. Ducati*[2]

[1] *Centro de Ciências da Vida, Universidade do Vale do Taquari – Univates, Lajeado, RS, Brazil*
[2] *Programa de Pós-Graduação em Biotecnologia (PPGBiotec), Universidade do Vale do Taquari – Univates, Lajeado, RS, Brazil*
[3] *Programa de Pós-Graduação em Ciências Médicas (PPGCM), Universidade do Vale do Taquari – Univates, Lajeado, RS, Brazil*
[4] *Programa de Pós-Graduação em Sistemas Ambientais Sustentáveis, Universidade do Vale do Taquari – Univates, Lajeado, RS, Brazil*

7.1 Introduction

Developing alternative strategies to battle new and/or resistant microorganisms such as bacteria, viruses, or parasites should be considered essential. Infectious diseases are considered a serious public health problem across the globe, being responsible for a large number of deaths on a daily basis, according to the World Health Organization (WHO). Establishing mechanisms to efficiently control these numbers continues to be a colossal challenge to date [1].

The production and use of bioactive compounds, especially for therapeutic purposes, is attractive for the pharmaceutical industry. The significant advances in the design and synthesis of bioactive compounds are slowly turning the table on the combat against multiple diseases [2]. Drug development is a complex process that takes a long path, deeply associated with biochemical and technological innovations. Such innovations have made possible the discovery of new therapies and have helped improve the quality of life of many people worldwide. The starting point for drug discovery is based on the comprehension of the disease's process, followed by compound manufacturing and testing to validate targets. A drug may be produced synthetically or isolated from microorganisms, plants, or animals.

A central part of developing new chemotherapies lies on outlining and anticipating potential pitfalls, where different strategies can be employed to ease difficulty along the way. When it comes to drug development, for example, many measures can be taken to

Bioprospecting of Microorganism-Based Industrial Molecules, First Edition.
Edited by Sudhir P. Singh and Santosh Kumar Upadhyay.

establish a compound and predict drug resistance. The use of recombinant proteins could also be strategic, as targeting enzymes of biological importance may elude microorganism resistance by the use of metabolites with potential for therapeutic application. Promising approaches such as using antimicrobial peptides synthesized by lactic acid bacteria (LAB) have also shown potential for biomedical use. Genetic engineering made possible the use of LAB as a source for recombinant protein, secrete metabolites with properties against different cancer types, and selectively inhibit microbial growth. Computational methods are becoming a vital component within the process of selective drug design, being a fast and efficient application with reduced costs taking part in this manner. Once a bioactive compound is found, subsequent stages such as *in vitro* and preclinical trials will probe its effectiveness, mechanism of action, and cytotoxicity. These stages will access the safety and adverse effects of the chosen compound as well.

This chapter brings to knowledge how strategies involving the production, evaluation, and use of bioactive compounds can lead to a significant improvement in human health.

7.2 *In vitro* Cell-Based Assays

The prospection of bioactive compounds with therapeutic properties – antibiotic, antitumoral, or anti-inflammatory – involves many steps in order to establish a compound demonstrating effectiveness with no significant toxicity. In the process of bioprospection, it is essential to have experiments carried out *in vitro*, often being performed to determine the effectiveness and mechanism of action of its top hits [3]. *In vitro* cytotoxicity assays are the first step to assess the biocompatibility of any compound.

Determining the cytotoxic potential of compounds in a study is important for two reasons: (i) if designed as anticancer agents, cytotoxicity is important for their action [4]; on the other hand, (ii) if intended for use as pharmaceuticals or cosmetics, they must be nontoxic to cells [5]. *In vitro* cytotoxicity assays with cultured cells are widely used in sensitivity testing of potential pharmaceuticals or cosmetic compounds [6, 7].

Cell culturing is a fast and inexpensive method to test compounds when compared to the cost-intensive animal models. *In vitro* assays may, however, present some limitations. The compound's cytotoxic concentration will not always be the same *in vitro* and *in vivo* [3]. Moreover, differences between *in vitro* and *in vivo* experiments regarding the compound's time of exposure, metabolism, tissue penetration, and excretion can exist [5]. Thereby, the choice of biologically representative cell culture model and type of *in vitro* assay is essential for providing relevant results *in vitro* and for the reproducibility of these results *in vivo*.

Several *in vitro* methods used to assess the cytotoxicity of compounds have been developed using cell cultures. Most of them have been concentrated on evaluating cell growth or survival after exposure to a compound, which can induce cellular changes by different mechanisms [8, 9]. *In vitro* assays can be classified into five major classes according to the nature of the assay: viability, metabolic, survival, transformation, and irritancy [5]. Determining the ideal one to be used will depend on factors such as expected outcome, the cost for reagents, and instrumentation availability, among others.

7.3 Cell Viability Assays

The most used parameter to assess cytotoxicity is cell viability. This type of assay gives an immediate to short-term response for measuring viable cells after exposure to a compound, in many cases taking advantage of the loss of membrane integrity that changes permeability and allows the uptake of certain dyes, thereby staining the cells [9]. According to the measure of endpoint, these can be separated into dye exclusion assays or fluorometric assays [10].

Dye exclusion assays are used to determine the number of viable cells present in a cell suspension. They are based on the principle that living cells have intact membranes that exclude certain dyes, such as trypan blue, eosin, congo red, and erythrosine B, while nonviable or dead cells uptake these dyes [11, 12]. Upon exposure to a dye, cell viability can be determined by cell count on a hemocytometer followed by total live cells to total cells ratio calculation. The staining procedure is quite simple, rapid, and requires a small number of cells; however, experimental procedures on a large number of samples can be difficult and time-consuming [10].

Fluorometric assays are based on a nonfluorescent dye that, in living cells, will be reduced to a strong fluorescent product, facilitating the measure of cell viability using a fluorescence microscope, fluorescence microplate reader, or flow cytometer. An example of this type of assay is the Alamar Blue assay [13, 14]. Resazurin, a component of Alamar Blue reagent, is a nonfluorescent dye permeable in viable cells. Inside the cells, resazurin is reduced to pink fluorescent resorufin by the activity of cellular enzymes (mitochondrial, for example) [13, 14]. The amount of resorufin produced can be measured and related to the number of viable cells.

There are many fluorescent DNA-binding dyes that also penetrate dead cells, which can be used to measure cell viability. Propidium iodide (PI) is an example of a small fluorescent molecule that only binds to the DNA of cells that have lost membrane integrity. Nonviable or dead cells will fluoresce red upon uptake of PI and can be monitored by flow cytometry or fluorescence microscopy [15].

7.4 Cell Metabolic Assays

Cell metabolic assays are intermediate duration experiments that measure a cell metabolic response at the time of, or shortly after, exposure to a compound [5]. In this case, cell viability will be related to the measurement of a metabolic product that only viable cells will produce. These assays can be classified according to the endpoint of measurement into colorimetric or luminescent assays [10].

Colorimetric assays require incubation of a reagent that will suffer an enzymatic reduction only in viable cells. The reagent reduced form results in a colored product, which can be measured to quantitate viable cells using a spectrophotometer [16]. There are multiple colorimetric methods based on the reduction of colorless tetrazolium salts into colored formazans. The MTT (3-(4,5-dimethylthiazol-2-yl)-2,5-diphenyltetrazolium bromide) assay is a fast and commonly used colorimetric assay that measures cell viability and compound cytotoxicity [17]. In viable cells, mitochondrial or cell plasma enzymes (dehydrogenases, oxidases, and peroxidases) are able to reduce MTT salt into purple-colored formazan crystals [18, 19]. The amount of formazan directly correlates to the number of metabolically active cells in culture. However, formazan crystals are insoluble in water and require an organic solvent for solubilization, such as dimethyl sulfoxide or isopropanol, before absorbance measurement [20, 21]. This step not only limits the assay but also is more time-consuming. In order to produce a water-soluble formazan product, other colorimetric assays have been developed, such as MTS (3-(4,5-dimethylthiazol-2-yl)-5-(3-carboxymethoxyphenyl)-2-(4-sulfophenyl)-2H-tetrazolium, inner salt) assay [22], XTT (2,3-bis(2-methoxy-4-nitro-5-sulphophenyl)-5-carboxanilide-2H-tetrazolium, monosodium salt) assay [23], and WST-1 (2-(4-iodophenyl)-3-(4-nitrophenyl)-5-(2,4-disulfophenyl)-2H-tetrazolium, monosodium salt) assay [24]. Regarding the MTT assay, only viable cells are able to reduce the salts into colored products; however, these products are formazans that solubilize in the culture medium itself, without the need of a solubilization step. Despite this advantage, these colorimetric assays have not replaced the well-established MTT assay [25].

Cell metabolic activity can also be measured in luminescent assays, based on the persistent and stable glow-type signal produced after cell exposure to a reagent. Since ATP serves as the principal donor of energy and is present in similar concentrations within all metabolically active cells, the number of living cells can be assessed based on the amount of ATP available [26]. The quantification of ATP in a cell culture sample requires the addition of a suitable firefly luciferase reagent. This enzyme catalyzes the conversion of luciferin to oxyluciferin in the presence of Mg^{2+} ions and ATP, yielding a luminescent signal [27]; the intensity of this signal will correlate to the ATP concentration [26]. ATP assay is advantageous when many growth conditions should be evaluated because it can be automated [27]. However, the assay endpoint destroys the cells and is incompatible with real-time analysis of the compound toxicity.

The real-time cell viability assay is a bioluminescent assay developed to measure cytotoxicity in real-time and uses an engineered luciferase enzyme derived from a marine shrimp and a cell viability probe that is a luciferase prosubstrate [28]. When both reagents are added to the cell culture medium, viable cells will reduce the prosubstrate into a form that is used by the luciferase in the medium, generating a luminescent signal [28]. If the cells divide and expand, the luminescent signal increases, whereas if the cells die, the signal decreases. This assay can perform a real-time measurement of cytotoxicity as the reagents are nontoxic to cells, the turnover of the prosubstrate is slow, and a continuous read can be done for at least 72 hours without replenishing the prosubstrate [8, 28].

7.5 Cell Survival Assays

Cell survival assays evaluate the long-term self-renewal capacity of cells upon exposure to a compound. Contrary to the tests that reveal dead cells (only) at the time of the assay, survival assays can measure the effects of compounds in cell proliferation capacity several hours or days later [5]. The clonogenic assay, or colony formation assay, is an example of a widely used *in vitro* cell survival assay. Although initially described for studying the effects of radiation on cells, this assay can also be used to determine the effectiveness of other cytotoxic agents [29, 30].

Loss of reproductive integrity and the inability to proliferate indefinitely are common features of cell death. A cell that is unable to divide and produce a large number of progeny can be considered dead, while a cell that has retained the capacity to divide and proliferate indefinitely, producing cell colonies, is referred to as "clonogenic" [31]. The clonogenic assay detects all cells that have retained the capacity for producing progeny after treatment with compounds that can cause cell reproductive death [29]. Thereby, in this assay, compound cytotoxicity increases with the decrease of cell ability to produce new colonies.

7.6 Cell Transformation Assays

For safety reasons, before the release of a compound for human use, it is necessary to test its potential to induce phenotypic alterations, characteristics of tumorigenic cells [5, 32]. Cell transformation assays are used to measure cell survival in an altered condition, which expresses a genetic mutation responsible for controlling cell growth, or for malignant transformation, for example [5]. Transformation assays make it possible to detect potential genotoxic or mutagenic compounds [33]. An example is the soft agar colony formation assay that can identify cells with anchorage-independent growth, that is cells with the ability to grow in a semisolid medium, which is a characteristic of malignantly transformed cells [34].

7.7 Cell Irritation Assays

Before releasing a pharmaceutical or cosmetic to the public, it is necessary to test whether it induces inflammatory responses, being (consequently) potentially responsible for many forms of allergy or irritation [5]. *In vitro* assays used to assess skin irritancy are being developed as alternatives to *in vivo* Draize tests [35, 36]. Although this step is considerably difficult to perform *in vitro*, the use of coculture of different cell types and the measurement of cytokine release enabled these tests to be carried out *in vitro*.

The coculture of keratinocytes and fibroblasts, representing the epidermal and dermal skin structure, respectively, is an example of a culture that simulates the human skin [37, 38]. When compounds reach the keratinocytes, these cells initiate a paracrine response that triggers the release of cytokines from the fibroblast cells, and an inflammatory response is propagated [37, 39]. Thus, monitoring the release of cytokines in a cell coculture by ELISA is a method that can determine irritation *in vitro*. These *in vitro* assays are, however, in early stages of development and need to be further improved to replace the current consolidated *in vivo* assays.

There is a variety of *in vitro* assays available to measure compound cytotoxicity in cell culture. Each one has its own set of advantages and disadvantages. The selection of the ideal method should be guided by a rapid, safe, reliable, efficient, and cost-effective assay. Moreover, it should not interfere with the test compounds. Regardless of the model system, establishing a consistent and reproducible procedure is important. Factors for reproducibility and success of *in vitro* assays include establishing an *in vitro* cell culture model, parameterization of reagents and compound concentrations, and determining the incubation time for each experimental model system.

In vitro assays are cheaper, require less time, and provide valuable information about the cytotoxicity of compounds with therapeutic properties. However, they are a control and an initial step toward the next stage of the search for bioactive substances. Because it is challenging to monitor systemic and physiological effects of compounds *in vitro*, the results of *in vitro* assays still need to be validated *in vivo*.

7.8 Heterologous Expression of Recombinant Proteins of Biomedical Relevance

Neglected diseases continue to take their toll in low-income countries. Tuberculosis (TB) and malaria are among the most lethal infectious diseases around the world, especially in developing nations. *Mycobacterium tuberculosis* is a weak, gram-positive bacterium that affects mainly the pulmonary system, being responsible for TB in humans. The main symptoms include prolonged cough, chest pain, weakness or fatigue, weight loss, fever, and night sweats. Every year, over 10 million people fall ill from TB, and about 1.5 million people die worldwide, which makes this disease one of the most fatal infections according to reports of the WHO [40]. TB treatment lasts a minimum of 6 months and uses four different antibiotics, including rifampicin and isoniazid, two extremely effective first-line drugs [41]. Although chemotherapy is available for TB patients, the generation of multidrug-resistant strains is imposing great challenges for treatment success.

Malaria is another well-established neglected disease and is currently known to be the deadliest human disease caused by a protozoan. The infection takes place as *Plasmodium falciparum* is injected through the human skin by a female *Anopheles* mosquito. This tropical disease has most cases distributed in Africa. In 2018, more than 228 million people were infected, resulting in 405 000 deaths from malaria. Likewise, *P. falciparum* has developed multidrug resistance to current chemotherapy, becoming a worldwide health concern [42].

As a consequence of drug resistance, TB and malaria treatments are becoming more extensive, expensive, and complex. Hence, the search for new strategies and alternative treatments is crucial. Targeting essential enzymes from these pathogens might be an ideal step toward the development of (new and) more efficient drugs with selective and alternative mechanisms of action.

The remarkable development in the biotechnology field has guided efforts from the pharmaceutical industry to improve the health of numerous patients [43]. The discovery of the structure and mechanism of DNA replication ultimately allowed the development of techniques which make it possible to synthesize proteins not naturally found in other organisms [44]. Recombinant proteins can be used as therapeutic agents in different diseases, in vaccines progress, and in different pathology diagnosis. Genetic engineering techniques have made manipulation of recombinant DNA possible, and one can now study the viability of pathogens from the heterologous production of their enzymes. Expressing the desired enzyme will depend on the chosen microorganism and chemical properties of the protein to be produced. The most common hosts for heterologous expression of recombinant enzymes are bacteria, especially *Escherichia coli*. Comprehending the host's system, choosing the ideal expression vector, producing the protein, and choosing the purification strategy are the basic steps to make a recombinant enzyme available. An important factor to keep in mind is that one should rely on a system capable of producing the chosen protein on a commercial scale [44]. Thus, developing protein drugs and validating new drug targets is extremely important for human health. This approach has the potential to improve the treatment of multidrug-resistant infected patients.

Expanding experimental strategies to treat TB are emerging, and component enzymes of essential pathways constitute interesting targets for the development of a new class of drugs [45]. The shikimate pathway is responsible for the biosynthesis of aromatic rings from carbohydrate precursors in algae, fungi, and bacteria; this pathway is, however, absent from mammals [46]. Thus, the inhibition of enzymes in this pathway could represent an efficient mechanism for starving bacteria and clearing infection. A better understanding of these enzymes and their kinetic/chemical mechanisms should provide essential knowledge on which to base selective enzyme inhibitor design.

Enzymes from metabolic pathways involved in persistence can be targeted in *M. tuberculosis* and *P. falciparum* [46–49]. One of the most studied proteins in *M. tuberculosis* is purine nucleoside phosphorylase (PNP), a component of the purine salvage pathway [48]. PNP catalyzes the reversible phosphorolysis of N-glycosidic bond of purine (deoxy)-ribonucleosides to generate (deoxy)ribose 1-phosphate and the corresponding purine bases. Experimental studies have shown that PNP inhibition generates dGTP accumulation, leading to ribonucleotide reductase inhibition and reducing DNA and RNA synthesis [47]. The accumulation of dGTP is known to be toxic to T-cells; T-cell deficiency induces apoptosis, making human PNP a vital target for T-cell cancer and autoimmune drug development. Due to the accumulation of guanine nucleotide and hyperphosphorylated guanosine moieties, *M. tuberculosis* PNP could also stand out as an important target, since purine analogues were already found to be toxic to cell growth, making bacteria suffer from nutritional stress [47, 50].

The purine salvage also plays a fundamental role in *P. falciparum*, as extensive DNA replication occurs during the blood and liver stages of the malaria parasite. *P. falciparum* has a high demand for purine nucleotides owing to the fact that most parasitic protozoa lack a *de novo* purine nucleotide biosynthetic pathway. Once more, the PNP enzyme plays an essential role in purine salvage, being essential for the formation of hypoxanthine in protozoa [49]. As this enzyme is highly expressed in the host's erythrocytes and in *P. falciparum*, there is a clear requirement for compounds to have selective activity against the parasite's homolog.

Owing to the fact that hypoxanthine is the key precursor for purine nucleotide synthesis, the most important enzyme in the purine salvage pathway in malaria is hypoxanthine-guanine-xanthine phosphoribosyltransferase [49]. This enzyme catalyzes the reversible Mg^{2+}-dependent conversion of 6-oxopurine bases and phosphoribosyl pyrophosphate to their respective nucleotides and inorganic pyrophosphate. Again, due to the lack of *de novo* purine synthesis in *P. falciparum*, this enzyme is being intensively studied, aiming for the development of new drugs against malaria.

Several excellent published reviews have demonstrated the importance of recombinant therapeutic proteins. Understanding the function of enzymes should guide the development of selective and efficient inhibitors [51]. An excess or deficiency of a specific metabolite can generate a diverse spectrum of diseases, and many of these problems can be solved by selective enzyme inhibition. An inhibitor is a compound that blocks or modulates a reaction, where the metabolic products can no longer be produced in sufficient amounts. If the product of an enzyme is important to carry out a physiological function, blocking this enzyme with a specific compound will potentially cause metabolic limitations, including inhibition of cell growth or replication to a microorganism or aberrant cell. Enzyme inhibition is an extremely promising approach for the discovery of new and more efficient drugs to treat human diseases, including TB and malaria.

7.9 Lactic Acid Bacteria and the Production of Metabolites with Therapeutic Roles

Lactic acid bacteria (LAB) belong to a heterologous group of non-sporulating gram-positive bacteria that are used in the production and preservation of food, as well as probiotic agents that provide benefits to human health, having a considerable and promising biotechnological potential due to their recognized safety and health-promoting properties. From an economic standpoint, the most important LAB applications lie in the production of fermented foods, such as dairy, vegetable, and meat products. Aside from their use in the dietary area, LAB are also capable of producing high added-value metabolites with potential for therapeutic application, such as postbiotics, which include microbial cells, cell constituents, and bioactive metabolites [52], such as short-chain fatty acids, bacteriocins, exopolysaccharides (EPSs), vitamins, polycationic molecules and neurotransmitters produced by probiotics [53–55]. Moreover, LAB-secreted metabolites have shown to promote protective properties against many cancer types via modulation of the host's natural defense mechanisms, inhibition of carcinogenesis onset and progression, improvement of apoptosis, and downregulation of genes involved in cancer cell proliferation [56–58]. Aside from the natural production of bioactive metabolites, genetic engineering systems allow LAB to be a source for the production of recombinant proteins [59].

The benefits provided by probiotics to human health are usually associated with the viability of microorganisms undergoing the gastrointestinal tract. However, growing evidence shows new perspectives on the health-promoting effects of LAB-derived metabolites. The potential synergic relationship between probiotics and prebiotics on maximizing the biosynthesis of postbiotics [52] is gaining popularity due to their beneficial properties in the gastrointestinal tract, playing a functional role in gut microbiota, and to their activities in different organs and tissues [55].

Several LAB-derived metabolites have been studied in order to evaluate their effects in human health, e.g. gamma-aminobutyric acid (GABA), an amino acid with a non-protein structure, which has been gaining significant attention in the biotechnological field. GABA has been largely used as both a pharmaceutical and a nutraceutical product, as well as a precursor in the synthesis of materials for industrial use. The bioproduction of GABA using LAB, among other microorganisms, is replacing its chemical synthesis, mostly due to the higher yield, lower cost, and lower environmental impact. Biotechnological studies have been focusing on the development of GABA-rich food supplements and fermented food as this amino acid provides benefits to human health, including gut modulation, neurostimulation, and cardioprotection [60–62].

LAB synthesize antimicrobial peptides (12–100 amino acids) released extracellularly in order to inhibit the growth of other microorganisms. Bacteriocins stand out (among these peptides) due to their spectrum of antimicrobial activities and may be exploited in medicine, e.g. replacing antibiotics in the treatment of patients infected with resistant bacterial strains or promoting the activity of these antibiotics, thus reducing the rise of resistant strains [63, 64]. LAB-derived bacteriocins are assumed to have either low or no cytotoxicity, based on tests performed with eukaryotic cell lines [65, 66]. The purification of bacteriocins, along with large-scale production, is a concern that directly affects the applications and price of the end product. Several stages are required to purify bacteriocins, e.g. the production of cell-free supernatant, ammonium sulfate precipitation, and higher concentration via preparative isoelectric focusing/chromatography methods [67–69], which renders their cost-effective purification a major bottleneck in large-scale applications. The small size of these proteins and the presence of low-molecular-weight protein impurities require the development of more advanced tools, such as more selective and specific membrane separation systems for purification of smaller-sized proteins and peptides, as suggested by Juturu and Wu [70].

A high selection of LAB has been studied due to their ability to produce EPSs, which are repeating mono/oligosaccharide units with several chemical compositions and properties. There is a great diversity of LAB-produced EPSs, which can be distinguished by their chemical structure, molecular size, charge, presence of side chains, and rigidity [57]. The nature of bacterial EPSs depends primarily on the strain, culture composition, and cultivation condition, and their biosynthesis involves several enzymes and proteins, which might be controlled through the regulation of gene expression [71]. EPSs have potential applications in the pharmaceutical and food industries and can have health-promoting properties, e.g. antioxidant, antiviral, anticancer, and immunomodulatory [59, 72]. The anticancer mechanism of action of these polysaccharides might be related to the stimulation of certain components in the immune system, particularly T- and B-lymphocytes, macrophages, of natural killer (NK)-induced interleukin-6 [73, 74], and to autophagic cell death, promoted (either directly or indirectly) by the expression of Beclin-1 and GRP78, for example [75].

The use of LAB as microbial cell factories for recombinant proteins, as well as a safety delivery platform for the treatment of a wide range of diseases, has been studied over the last decades due to its promising approach [76–79] The development of important tools for the production of recombinant proteins, such as suitable expression vectors, modified strains, and enhanced induction and secretion systems, has made LAB an attractive alternative for the production of biologically useful proteins. These LAB-produced recombinant proteins could be used for biomedical purposes, e.g. inflammatory bowel disease, diabetes, *Rotavirus* infection, and cancer, among others [79]. Although LAB include microorganisms of different genera, such as *Lactococcus*, *Lactobacillus*, *Enterococcus, Leuconostoc, Pediococcus,* and *Streptococcus*, *Lactococcus lactis* has been the most frequently used LAB in studies involving cloning and production of recombinant proteins, as it has several lineages largely characterized and its genome has been completely sequenced [80–84]. Aside from *L. lactis*, *Lactobacillus acidophilus*, *Lactobacillus casei*, *Lactobacillus paracasei,* and *Lactobacillus plantarum* are examples of LAB that have been genetically modified to produce recombinant proteins for biomedical purposes [79].

A well-established LAB expression system is the nisin-controlled expression (NICE) system, based on the control of a strong nisin-inducible promoter (P_{nisA}) [78]. It is usually employed as an expression host, as its chromosome contains the signal transduction genes *nisR* and *nisK*, involved in the nisin-induced transcriptional control of the P_{nisA} promoter. Any genes cloned downstream of P_{nisA} can be expressed in a controlled manner by adding nisin to the bacterial culture [85]. Aside from the genetic engineering expression systems already established, new approaches have been used to allow the use of LAB as vectors for the delivery of therapeutic molecules, such as Clustered Regularly Interspaced Short Palindromic Repeats (CRISPR) associated with the Cas enzyme (CRISPR-Cas). Regarding the possibilities of editing LAB genome, a key challenge will be to use these probiotic species for the delivery of mucosal vaccines to improve human health. Mucosal vaccine delivery has been proposed as an alternative to traditional vaccines due to the reduction of secondary effects, ease of administration, and the modulation of mucosal immune responses and systemic responses. Utilizing CRISPR-Cas systems in LAB for the rational design of vaccines and therapeutic strains will provide opportunities for precise genome engineering [86].

These topics showed the potential of LAB for the production of several high added-value metabolites and its important role for therapeutic applications. These applications explicit the potential use of this microbial group in the pharmaceutical and food industries. Moreover, advanced molecular biology techniques have improved therapeutic roles for LAB. In this context, although the safety of using wild strains from this group of microorganisms has been confirmed, using recombinant LAB (genetically modified organisms) is being addressed with caution by regulating agencies. Genetically modified LAB represent a potential vector for *in situ* delivery of therapeutic molecules and a potential option for clinical treatment [54].

7.10 Preclinical Studies

The process of drug development includes, aside from planning, the identification and synthesis of the target molecule, preclinical and clinical research, and review of the drug by the appropriate agency, which may vary according to the nation. The most important agencies responsible for protecting and promoting public health are the Food and Drug Administration (FDA), the European Medicines Agency (EMA), the Pharmaceuticals and Medical Devices Agency (PMDA), and the International Conference on Harmonization of Drugs for Human Use (ICH) [87, 88]. However, only 5–7% of potential drugs reach the final stage of approval by the agency supervisory, and a relatively low number of (new) drugs are approved annually by the FDA [87, 89–91]. A review published in 2014 reported that only 10 drugs were approved for lung cancer treatment since 1998, while 167 other therapies failed clinical trials [89]. Therefore, preclinical phase or nonclinical trials is a crucial step in drug development. During this stage, target molecules are investigated in animal models (*in vivo*), allowing assessment of the drug's safety, efficacy, and adverse effects. This is achieved through toxicological tests, reducing the probability of failure of the new drug in subsequent steps, without putting human health at risk [91–96].

Such an important step requires methodological quality. Hence, the procedure needs to be rigorous and appropriate, otherwise, the results may be compromised, as in the case of a study conducted by Sibbald (2000), where it was sought to investigate the effects of an experimental therapy on sepsis using a monoclonal antibody (mAb) against tumor necrosis factor or endotoxin. While evaluating these effects, the researchers used a model of baboon bacteremia based mainly on intoxication, as the relevant pathological process, rather than infection (which is more characteristic of clinical sepsis), making the results limiting [97]. Thus, a study should have well-designed protocols and clear objectives and be conducted according to the requirements of the aforementioned supervisory bodies and, additionally, Handbook: Good Laboratory Practice (Quality practices for regulated non-clinical research and development) by the WHO. The methods used in preclinical toxicological tests should be based on the toxicological information and the relevance of the data obtained for the knowledge of the toxicological properties of the test compound [88, 98].

It is important to emphasize that the extrapolation of data obtained from animal models to humans is not linear. Hence, two distinct models are used, usually a rodent species (rats or mice) and a nonrodent species (rabbits, pigs, beagle dogs, baboons, etc.), thus obtaining information on any overlapping similarities and differences [87, 88, 98]. Additionally, it is worth noting that although the use of animals in drug development is already well established, they should only be used when there are direct benefits to human healthcare, no other methodological alternatives, and when there is no unnecessary suffering of the animals [92, 93, 97].

Preclinical trials are grouped into three categories: pharmacological, pharmacokinetic, and toxicological [98], where pharmacodynamic studies, i.e. effects of drugs on human physiology, involve mainly analyzing the effects of drugs on the organs of cardiovascular, respiratory, and nervous systems. In addition, pharmacokinetic effects, such as absorption, distribution, metabolism, and elimination of the drug, are investigated in toxicology trials in most cases. These trials aim to interpret the pharmacological and toxicological effects of the drug and compare the possible metabolic pathways (in animals/humans) and their clinical relevance [87, 88, 98].

Preclinical toxicological tests include mainly studies to evaluate the general toxicity (single dose and repeated dose), genotoxicity, carcinogenicity, and reproductive toxicity. There is a high number of analysis, which can be performed and will vary according to the expected result, but will test behavioral assessment, body weight, hematological, biochemical, histological, and mortality [98]. Complementary toxicological tests are less common but may be additionally performed depending on the type of drug molecule, route of administration, or effect to be investigated; these tests include immunotoxicity, toxicity to the pediatric population, and local tolerance. The data collected during these trials are used to establish the initial dose to be administered, the therapeutic interval to be used during clinical trials, and the potential adverse effects that may occur [98, 99].

7.10.1 Acute Toxicity

Acute toxicity trials (single dose) correspond to tests where a dose of the potential drug is administered within a period of up to 24 hours, along which the results and data are obtained regarding the drug's intrinsic (acute) toxicity, dose-effect relationship, appropriate dose, signs and symptoms of overdose, and appropriate species for the trials. These results are observed for 14 days after the administration, and the data obtained include lethal dose 50 (LD_{50}) or lethal concentration 50 (LC_{50}) where applicable, the signs of toxicity, and cause of death [87, 88].

7.10.2 Repeated Dose Toxicity

Toxicological trials with repeated administration (chronic toxicity) allow researchers to analyze the relevance of this toxicity in humans to determine the safety margins by identifying the toxic effects, target organs, effects on animal physiology and hematological, biochemical, anatomical, and histopathological effects. In addition, these trials also determine the no-observed-adverse-effect level and lowest-observed-adverse effect level. The duration of a repeated-dose toxicity trial is directly related to the duration of treatment, the therapeutic indication, and the duration of the clinical trial. It is subdivided into three types of trials: subacute dose toxicity trial (14–28 days), subchronic dose toxicity trial (90 days), or chronic dose toxicity trial (6–12 months) [87, 88, 99].

7.10.3 Genotoxicity

The assessment of genotoxicity or mutagenic potential includes *in vitro* trials for studying chromosomal changes or mutations and additional trials in case of positive or inconclusive results. These additional trials are divided into two categories. The first category comprises tests for a genetic mutation in bacteria, including a cytogenetic test to assess chromosomal damage (*in vitro* micronucleus assay) or an *in vitro* genetic mutation test in mouse lymphoma cells, an *in vivo* test for genotoxicity, and a chromosomal damage assay in rodent hematopoietic cells. The second category includes a bacterial gene mutation test and an *in vivo* test for genotoxicity in two tissues, usually a mouse hematopoietic cell micronucleus assay and a second *in vivo* assay. Although the results of all genotoxicity tests should be available before initiating clinical trials, the possible genotoxic effects of the potential drug, such as tumor development, are not always evident as such effects may only be expressed after multiple years. Therefore, in addition to these tests, carcinogenicity tests are required, which demand a longer duration [87, 88].

7.10.4 Carcinogenicity

Carcinogenicity tests are intended to evaluate the tumorigenic potential of a drug over a period of 24 months, or more, if required, until drug approval. However, these tests are expensive and time-consuming, and they are only performed when the genotoxicity tests indicate that the new drug has a mutagenic potential, the new compound (potential drug) falls within a class of drugs with known carcinogenicity, the new drug is intended for chronic treatments for longer than six months, the structure–activity relationship shows carcinogenic potential or the new drug (or its metabolites, or both) are retained in tissues and organs that have undergone physiopathological changes. In these tests, the mortality and morbidity rates are observed, along with the adverse effects that occur after 1 h of dose administration: changes in body weight, food consumption pattern and blood parameters, the appearance of diseases, and modifications of metabolization and excretion processes [87, 88, 99].

7.10.5 Reproductive Toxicity

The need for specific reproductive toxicity studies depends on the population that will be exposed to the drug. These trials can reveal any effect on mammalian reproduction, and they investigate the effects of drug exposure from conception to sexual maturity. Such trials mainly investigate fertility, early embryonic development, and pre- and postnatal development, including maternal function and development of the embryo and fetus. Infertile individuals, such as postmenopausal women, may still be included in phase I and II trials, provided that a relevant assessment of the reproductive organs has been made during the repeated-dose toxicity trials [87, 88].

Although preclinical trials are considered to be costly, they are extremely important for the development of a therapeutic target to ensure the safety and efficacy of the drug and to minimize adverse effects and the risk that the exposure of the new drug has for human beings. In addition, the tests chosen must be in accordance with the expected result in order to reduce costs but guaranteeing safety, effectiveness, and quality. Once data relating to a new drug becomes sufficiently studied, i.e. when there is enough supporting information in relation to obtaining the active ingredient, its chemical structure, stability, galenic formulation, as well as its pharmacological, toxicological, and pharmacokinetic properties, the last stage of the drug research and development process begins, which is clinical trials [87, 88, 100].

7.11 Computer-aided Drug Design

Computational methods emerged as essential tools for the process of drug design and development, consisting of an array of different techniques and methods. The field of bioinformatics provides an *in silico* approach for the identification, modeling, and prediction of the behavior of macromolecules and/or small molecules not only to the pharmaceutical industry purposes but also to academic research. While these techniques do not replace *in vitro* or *in vivo* testing, computational approaches are aimed at reducing the cost and number of potential molecules to be tested experimentally and speed up the process of discovery, identification, and/or development.

When looking for hypothetically bioactive or effective molecules, virtual screening methods can be used to search libraries for potential candidates according to a set of desired properties. These candidates can advance to subsequent testing, either by following computational or experimental methods. The screening of larger macromolecules, such as proteins and nucleic acids, is usually based on crystallographic information. In these cases, the three-dimensional conformation of the molecule is directly tied to its purpose and functionality. Such structures are usually determined through X-ray crystallography or nuclear magnetic resonance, which can then be built and calculated by different softwares [101–103]. A more advanced clarification of target-centered virtual screening (VS) techniques can be primarily divided into two approaches: ligand-based and structure-based methods. In ligand-based VS, physicochemical properties of two- or three-dimensional chemical structures (of which activities are known) can be utilized to retrieve other compounds from libraries and databases through a measure of similarity or a common substructure. On the other hand, in structure-based VS, the compounds of interest are docked into the active site of the protein of interest via computer algorithms, being assigned a score according to their affinity. As such, the choice of which approach to take is dependent on what level of information is available at the start of the VS tests. Ideally, the application of both types of approaches, if employed in parallel, should yield a most diverse set of reasonable starting points for chemical optimization [104].

Molecular docking is likely the most common structure-based method. In other words, it is a technique through which the preferred binding orientations between a small molecule and a macromolecule can be predicted. The workflow usually starts with the target selection, the evaluation of its structure, and the identification of the binding site – a "hollow" part (or "pocket") of the target where the small molecule (or ligand) binds. The most important features to look for in a binding site are the potential hydrogen-bond donors and acceptors, hydrophobic surfaces and size. The molecular docking methodologies usually consist of searching algorithms and scoring functions that are responsible for investigating likely conformations of the small molecules while they interact with the binding site, which is ranked according to the scoring functions' classification/category.

Among the classical scoring functions are force field-based, empirical-based and knowledge-based [103]. Force-field-based functions compute the binding affinity according to the physical atomic contacts in the protein–ligand complex. In general, force-field-based functions can be described as the sum of several nonbonded interaction terms computed by a given function, such as van der Waals forces, and electrostatic and hydrogen bonding. Empirical-based scoring functions usually adopt similar functional forms to the force-field-based ones, but with some additional terms, such as the number of rotatable bonds of the ligand and solvent accessible surface area, in order to better characterize the protein–ligand interactions. Knowledge-based functions, on the other hand, can be considered a group of inherently different methods, offering a good balance between accuracy and speed in comparison to the previous classifications. The main difference between knowledge-based functions and the others is in their principle; specifically, the three-dimensional coordinates of a set of protein–ligand structures and complexes are regarded as a knowledge base, and the pairwise potentials can be acquired from the occurrence frequency of atom–atom pairs in databases [105, 106].

The VS approach is, however, only possible if the three-dimensional structure of the target macromolecule has been previously elucidated. When it is not the case, such structure must be solved before the application of VS techniques for the discovery of potential ligands. Homology modeling (also called comparative or template-based modelling) is a computational method for structure prediction that is used to determine a protein's three-dimensional structure from its amino acid sequence, considered to be the most accurate method among its kind. It is based on the observation that most (if not all) members of a protein family continuously exhibit the same folding patterns, characterized by a core structure, which is robust against sequence mutations or modifications. Thus, it relies on experimentally determined structures of homologous proteins – the templates – which enable the generation of models starting from these proteins' sequences – the targets.

Despite the high accuracy displayed by models obtained through homologous design, even sequences with low similarity can generate suitable models [107, 108].

Ligand-based VS utilizes the similarity between ligands to retrieve potential compounds of interest from small-molecule databases, while structure-based methods measure the fitness of a compound against the target structure. Ligand-based VS looks for "new" compounds that are similar to the ones activity is known, based on the assumption that molecules with similar structure will also share similar properties, functions, and/or activity – which is not always true due to what is called an activity cliff. The main differences between the multiple ligand-based methods are the various representations of a molecule and the measurement, or criteria, of similarity between them [109]. Ligand-based models can even be simple matched-molecular pair models, where molecular transformations are associated with changes in molecular properties. On the level of whole molecules, entire chemical structures can be represented via sets of descriptors or features in two or three dimensions (such as 2D "fingerprints" or 3D shape comparison). Some of these models include complex quantitative activity-structural relationship (QSAR) models, which mathematically express the links between the potential biological activity of a given small molecule and its respective structure and descriptors. There's also pharmacophore mapping, in which certain molecular features (such as steric and electronic features), necessary for a molecule to modulate a particular biological response or target, are identified. These large sets of computed molecular properties are used to train machine-learning algorithms that correlate descriptors with experimental data and measurements. In contrast to three-dimensional protein–ligand models, these statistical models require a certain amount of caution because the compounds of interest and their properties might not be well represented in training sets. Even so, they are usually a faster calculation system and can be improved through the addition of more experimental data [110, 111]. Quite a few research groups have established QSAR models that show the potential inhibiting effects and the relationships between chalcone, quinoline, 7-methylquinoline, and pyrrole molecules, and their respective bioactive properties against *M. tuberculosis*. For example, Adeniji and coworkers [112] used a combination of QSAR models and molecular docking to find a quinoline derivative with an even higher binding affinity (lower binding energy) to the DNA gyrase receptor than commercial isoniazid (drug).

After the initial screening and subsequent discovery of potentially active compounds, molecular dynamics simulations are often applied to guide further chemical optimization, particularly in the absence of a crystal structure of the protein–ligand complex – even in cases where said structure is known, molecular dynamics is capable of providing valuable insight as to which interactions are the most stable over time and contribute the most to binding, through the characterization of flexible binding sites, pathways, kinetics, and thermodynamics [113]. In a typical molecular dynamics simulation, the movements of the atoms of the ligand and its target in the physiological solvent are simulated over time in a series of short steps (lasting for hundreds of nanoseconds or more). At each step, the forces acting on each atom are computed according to the atom's position and velocity, and their resulting trajectories are used to calculate, study, and investigate the behaviors of the system, as well as predicting the overall binding affinity of the ligand–target complex, its potential influence on mutation effects, and computing the free energy of binding [114].

Therefore, it gradually becomes clear that the use of computational methods has become a paramount feature of drug design and development through the selection, design, identification, and optimization of (novel) bioactive compounds. This development is closely related to both technological advancements and growing empirical knowledge, as processing power and chemical-structural information of micro and macromolecules walk hand-to-hand. It is thus expected that computational-aided drug design will continue to grow in the next few years and continue to improve research quality and quantity, by accelerating and facilitating the discovery of new, and perhaps more potent, drugs.

7.12 Conclusions and Perspectives

Biological and computational methods are crucial for the production of bioactive compounds with the potential to take a role in human health. However, it is extremely important to establish or develop more databases for research on new target molecules for therapeutic and nutritional applications. Such an approach is expected to bring significant improvement to the treatment of human diseases and give research global importance is expected to boost health and food industries.

References

1 Guido, R., Andricopulo, A.D., and Oliva, G. (2010). Drug design, biotechnology and medicinal chemistry: applications to infectious diseases. *Advanced Studies 24* (70): 81–98.

2 Piccirillo, E. and Amaral, A.T.D. (2018). Virtual screening of bioactive compounds: concepts and applications. *Química Nova 41* (6): 662–677.

3 Di Nunzio, M., Valli, V., Tomás-Cobos, L. et al. (2017). Is cytotoxicity a determinant of the different *in vitro* and *in vivo* effects of bioactives? *BMC Complementary and Alternative Medicine 17* (1): 453.

4 Ediriweera, M.K., Tennekoon, K.H., and Samarakoon, S.R. (2019). *in vitro* assays and techniques utilized in anticancer drug discovery. *Journal of Applied Toxicology 39* (1): 38–71.

5 Freshney, R.I. (2010). Cytotoxicity. In: *Culture of Animal Cells: A Manual of Basic Technique and Specialized Applications* (ed. I. Freshney), 365–381. New Jersey: Wiley.

6 Rodrigues, F., Gaspar, C., Palmeira-de-Oliveira, A. et al. (2016). Application of Coffee Silverskin in cosmetic formulations: Physical/antioxidant stability studies and cytotoxicity effects. *Drug Development and Industrial Pharmacy 42* (1): 99–106.

7 Zhao, H., Zhao, X., Lei, S. et al. (2019). Effect of cell culture models on the evaluation of anticancer activity and mechanism analysis of the potential bioactive compound, iturin A, produced by: *Bacillus subtilis*. *Food & Function* 10 (3): 1478–1489.

8 Riss, T., Niles, A., Moravec, R. et al. (2016). Cytotoxicity assays: in vitro methods to measure dead cells. In: *Assay Guidance Manual* (eds. S. Markossian, G.S. Sittampalam, A. Grossman, et al.), 295–319. Bethesda: Eli Lilly & Company and the National Center for Advancing Translational Sciences.

9 Riss, T., Niles, A., Moravec, R. et al. (2019). Cytotoxicity assays: in vitro methods to measure dead cells. In: *Assay Guidance Manual* (eds. S. Markossian, G.S. Sittampalam, A. Grossman, et al.), 759–773. Bethesda: Eli Lilly & Company and the National Center for Advancing Translational Sciences.

10 Aslantürk, Ö.S. (2018). in vitro Cytotoxicity and cell viability assays: principles, advantages, and disadvantages. In: *Genotoxicity – A Predictable Risk to Our Actual World* (eds. M.L. Larramendy and S. Soloneski). London: IntechOpen.

11 Strober, W. (2015). Trypan blue exclusion test of cell viability. *Current Protocols in Immunology* 111: A3.B.1–A3.B.3.

12 Yip, D.K. and Auersperg, N. (1972). The dye-exclusion test for cell viability: Persistence of differential staining following fixation. *in vitro: Journal of the Tissue Culture Association* 7 (6): 323–329.

13 O'brien, J., Wilson, I., Orton, T. et al. (2000). Investigation of the Alamar Blue (resazurin) fluorescent dye for the assessment of mammalian cell cytotoxicity. *European Journal of Biochemistry* 267 (17): 5421–5426.

14 Page, B., Page, M., and Noel, C. (1993). A new fluorometric assay for cytotoxicity measurements *in vitro*. *International Journal of Oncology* 3 (3): 473–476.

15 Dengler, W.A., Schulte, J., Berger, D.P. et al. (1995). Development of a propidium iodide fluorescence assay for proliferation and cytotoxicity assays. *Anti-Cancer Drugs* 6 (4): 522–532.

16 Präbst, K., Engelhardt, H., Ringgeler, S. et al. (2017). Basic colorimetric proliferation assays: MTT, WST, and Resazurin. In: *Cell Viability Assays. Methods in Molecular Biology* (eds. D.F. Gilbert and O. Friedrich), 1–17. New Jersey: Humana Press Inc.

17 Mosmann, T. (1983). Rapid colorimetric assay for cellular growth and survival: application to proliferation and cytotoxicity assays. *Journal of Immunological Methods* 65 (1-2): 55–63.

18 Berridge, M.V. and Tan, A.S. (1993). Characterization of the cellular reduction of 3-(4,5-dimethylthiazol-2-yl)-2,5-diphenyltetrazolium Bromide (MTT): subcellular localization, substrate dependence, and involvement of mitochondrial electron transport in MTT reduction. *Archives of Biochemistry and Biophysics* 303 (2): 474–482.

19 Liu, Y., Peterson, D.A., Kimura, H. et al. (1997). Mechanism of cellular 3-(4,5-Dimethylthiazol-2-yl)-2,5-Diphenyltetrazolium Bromide (MTT) reduction. *Journal of Neurochemistry* 69 (2): 581–593.

20 Berridge, M.V., Herst, P.M., and Tan, A.S. (2005). Tetrazolium dyes as tools in cell biology: new insights into their cellular reduction. *Biotechnology Annual Review* 11: 127–152.

21 Stockert, J.C., Horobin, R.W., Colombo, L.L. et al. (2018). Tetrazolium salts and formazan products in cell biology: viability assessment, fluorescence imaging, and labeling perspectives. *Acta Histochemica* 120 (3): 159–167.

22 Barltrop, J.A., Owen, T.C., Cory, A.H. et al. (1991). 5-(3-carboxymethoxyphenyl)-2-(4,5-dimethylthiazolyl)-3-(4-sulfophenyl)te trazolium, inner salt (MTS) and related analogs of 3-(4,5-dimethylthiazolyl)-2,5-diphenyltetrazolium bromide (MTT) reducing to purple water-soluble formazans as cell-viability indicators. *Bioorganic and Medicinal Chemistry Letters* 1 (11): 611–614.

23 Paull, K.D., Shoemaker, R.H., Boyd, M.R. et al. (1988). The synthesis of XTT: a new tetrazolium reagent that is bioreducible to a water-soluble formazan. *Journal of Heterocyclic Chemistry* 25 (3): 911–914.

24 Tan, A.S. and Berridge, M.V. (2000). Superoxide produced by activated neutrophils efficiently reduces the tetrazolium salt, WST-1 to produce a soluble formazan: a simple colorimetric assay for measuring respiratory burst activation and for screening anti-inflammatory agents. *Journal of Immunological Methods* 238 (1-2): 59–68.

25 Van Tonder, A., Joubert, A.M., and Cromarty, A.D. (2015). Limitations of the 3-(4,5-dimethylthiazol-2-yl)-2,5-diphenyl-2H-tetrazolium bromide (MTT) assay when compared to three commonly used cell enumeration assays. *BMC Research Notes* 8 (1): 1–10.

26 Crouch, S.P.M., Kozlowski, R., Slater, K.J. et al. (1993). The use of ATP bioluminescence as a measure of cell proliferation and cytotoxicity. *Journal of Immunological Methods* 160 (1): 81–88.

27 Lundin, A., Hasenson, M., Persson, J. et al. (1986). Estimation of biomass in growing cell lines by adenosine triphosphate assay. *Methods in Enzymology* 133: 27–42.

28 Duellman, S.J., Zhou, W., Meisenheimer, P. et al. (2015). Bioluminescent, nonlytic, real-time cell viability assay and use in inhibitor screening. *Assay and Drug Development Technologies* 13 (8): 456–465.

29 Franken, N.A., Rodermond, H.M., Stap, J. et al. (2006). Clonogenic assay of cells *in vitro*. *Nature Protocols* 1 (5): 2315–2319.

30 Puck, T.T., Morkovin, D., Marcus, P.I. et al. (1957). Action of x-rays on mammalian cells. II. Survival curves of cells from normal human tissues. *Journal of Experimental Medicine* 106 (4): 485–500.

31 Munshi, A., Hobbs, M., and Meyn, R.E. (2005). Clonogenic cell survival assay. In: *Chemosensitivity. Methods in Molecular Medicine* (ed. R.D. Blumenthal), 20–28. New Jersey: Humana Press Inc.

32 Creton, S., Aardema, M.J., Carmichael, P.L. et al. (2012). Cell transformation assays for prediction of carcinogenic potential: state of the science and future research needs. *Mutagenesis* 27 (1): 93–101.

33 Kirkland, D., Aardema, M., Müller, L. et al. (2006). Evaluation of the ability of a battery of three *in vitro* genotoxicity tests to discriminate rodent carcinogens and non-carcinogens. II. Further analysis of mammalian cell results, relative predictivity and tumour profiles. *Mutation Research, Genetic Toxicology and Environmental Mutagenesis* 608 (1): 29–42.

34 Macpherson, I. and Montagnier, L. (1964). Agar suspension culture for the selective assay of cells transformed by polyoma virus. *Virology* 23 (2): 291–294.

35 Kim, C.W., Kim, C.D., and Choi, K.C. (2017). Establishment and evaluation of immortalized human epidermal keratinocytes for an alternative skin irritation test. *Journal of Pharmacological and Toxicological Methods* 88: 130–139.

36 Miles, A., Berthet, A., Hopf, N.B. et al. (2014). A new alternative method for testing skin irritation using a human skin model: a pilot study. *Toxicology in vitro* 28 (2): 240–247.

37 Canton, I., Cole, D.M., Kemp, E.H. et al. (2010). Development of a 3D human *in vitro* skin co-culture model for detecting irritants in real-time. *Biotechnology and Bioengineering* 106 (5): 794–803.

38 Ikuta, S., Sekino, N., Hara, T. et al. (2006). Mouse epidermal keratinocytes in three-dimensional organotypic coculture with dermal fibroblasts form a stratified sheet resembling skin. *Bioscience, Biotechnology, and Biochemistry* 70 (11): 2669–2675.

39 Coquette, A., Berna, N., Vandenbosch, A. et al. (2003). Analysis of interleukin-1α (IL-1α) and interleukin-8 (IL-8) expression and release in *in vitro* reconstructed human epidermis for the prediction of *in vivo* skin irritation and/or sensitization. *Toxicology in vitro* 17 (3): 311–321.

40 World Health Organization, Geneva (2019). Global Tuberculosis Report 2019.

41 Ducati, R.G., Ruffino-Netto, A., Basso, L.A. et al. (2006). The resumption of consumption – a review on tuberculosis. *Memórias do Instituto Oswaldo Cruz* 101 (7): 697–714.

42 World Health Organization, Geneva (2019). World Malaria Report 2019.

43 Kumar, P., Shaunak, I., and Verma, M.L. (2020). Biotechnological application of health promising bioactive molecules. In: *Biotechnological Production of Bioactive Compounds* (eds. M.L. Verma and A.K. Chandel), 165–189. Amsterdam: Elsevier.

44 Overton, T.W. (2014). Recombinant protein production in bacterial hosts. *Drug Discovery Today* 19 (5): 590–601.

45 Villela, A.D., Sánchez-Quitian, Z.A., Ducati, R.G. et al. (2011). Pyrimidine salvage pathway in *Mycobacterium tuberculosis. Current Medicinal Chemistry* 18 (9): 1286–1298.

46 Nunes, J.E., Duque, M.A., de Freitas, T.F. et al. (2020). *Mycobacterium tuberculosis* Shikimate pathway Enzymes as targets for the rational design of anti-tuberculosis drugs. *Molecules* 25 (6): 1259.

47 Ducati, R.G., Breda, A., Basso, L.A. et al. (2011). Purine salvage pathway in *Mycobacterium tuberculosis. Current Medicinal Chemistry* 18 (9): 1258–1275.

48 Ducati, R.G., Namanja-Magliano, H.A., and Schramm, V.L. (2013). Transition-state inhibitors of purine salvage and other prospective enzyme targets in malaria. *Future Medicinal Chemistry* 5 (11): 1341–1360.

49 Ducati, R.G., Firestone, R.S., and Schramm, V.L. (2017). Kinetic isotope effects and transition state structure for Hypoxanthine-Guanine-Xanthine Phosphoribosyltransferase from *Plasmodium falciparum. Biochemistry* 56 (48): 6368–6376.

50 Caceres, R.A., Timmers, L.F., Ducati, R.G. et al. (2012). Crystal structure and molecular dynamics studies of purine nucleoside phosphorylase from *Mycobacterium tuberculosis* associated with acyclovir. *Biochimie* 94 (1): 155–165.

51 Ducati, R.G., Namanja-Magliano, H.A., Harijan, R.K. et al. (2018). Genetic resistance to purine nucleoside phosphorylase inhibition in *Plasmodium falciparum. Proceedings of the National Academy of Sciences of the United States of America* 115 (9): 2114–2119.

52 Wegh, C.A.M., Geerlings, S.Y., Knol, J. et al. (2019). Postbiotics and their potential applications in early life nutrition and beyond. *International Journal of Molecular Sciences* 20 (19): 4673.

53 Alvarez-Sieiro, P., Montalbán-López, M., Mu, D. et al. (2016). Bacteriocins of lactic acid bacteria: extending the Family. *Applied Microbiology and Biotechnology* 100 (7): 2939–2951.

54 Plavec, T.V. and Berlec, A. (2019). Engineering of lactic acid bacteria for delivery of therapeutic proteins and peptides. *Applied Microbiology and Biotechnology* 103 (5): 2053–2066.

55 Diez-Gutiérrez, L., Vicente, L.S., Barrón, L.J.R. et al. (2020). Gamma-aminobutyric acid and probiotics: multiple health benefits and their future in the global functional food and nutraceuticals Market. *Journal of Functional Foods* 64: 103669.

56 Deepak, V., Ram Kumar Pandian, S., Sivasubramaniam, S.D. et al. (2016). Optimization of anticancer exopolysaccharide production from probiotic *Lactobacillus acidophilus* by response surface methodology. *Preparative Biochemistry and Biotechnology* 46 (3): 288–297.

57 Zannini, E., Waters, D.M., Coffey, A. et al. (2016). Production, properties, and industrial food application of lactic acid bacteria-derived exopolysaccharides. *Applied Microbiology and Biotechnology* 100 (3): 1121–1135.

58 Saber, A., Alipour, B., Faghfoori, Z. et al. (2017). Cellular and molecular effects of yeast probiotics on cancer. *Critical Reviews in Microbiology* 43 (1): 96–115.

59 Saadat, Y.R., Khosroushahi, A.Y., and Gargari, B.P. (2019). A comprehensive review of anticancer, immunomodulatory and health beneficial effects of the lactic acid bacteria exopolysaccharides. *Carbohydrate Polymers* 217: 79–89.

60 Auteri, M., Zizzo, M.G., and Serio, R. (2015). GABA and GABA receptors in the gastrointestinal tract: from motility to inflammation. *Pharmacological Research* 93: 11–21.

61 Lim, H.S., Seo, D.H., Cha, I.T. et al. (2018). Expression and characterization of glutamate decarboxylase from *Lactobacillus brevis* HYE1 isolated from *kimchi*. *World Journal of Microbiology and Biotechnology* 34 (3): 1–10.

62 Kim, J.A., Park, M.S., Kang, S.A. et al. (2014). Production of γ-aminobutyric acid during fermentation of *Gastrodia elata* Bl. by co-culture of *Lactobacillus brevis* GABA 100 with *Bifidobacterium bifidum* BGN4. *Food Science and Biotechnology* 23 (2): 459–466.

63 Hols, P., Ledesma-Garcia, L., Gabant, P. et al. (2018). Mobilization of microbiota commensals and their bacteriocins for therapeutics. *Cell Press* 27 (8): 690–702.

64 Cotter, P.D., Hill, C., and Ross, R.P. (2005). Bacteriocins: developing innate immunity for food. *Nature Reviews Microbiology* 3 (10): 777–788.

65 Dicks, L.M., Dreyer, L., Smith, C. et al. (2018). Review: the fate of bacteriocins in the human gastro intestinal tract: do they cross the gut-blood barrier? *Frontiers in Microbiology* 9: 2297.

66 Maher, S. and McClean, S. (2006). Investigation of the cytotoxicity of eukaryotic and prokaryotic antimicrobial peptides in intestinal epitelial cells *in vitro*. *Biochemical Pharmacology* 71 (9): 1289–1298.

67 Sabo, S.S., Lopes, A.M., Santos-Ebinuma, V.C. et al. (2018). Bacteriocin partitioning from a clarified fermentation broth of *Lactobacillus plantarum* ST16Pa in aqueous two-phase systems with sodium sulfate and choline-based salts as additives. *Process Biochemistry* 66: 212–221.

68 Kumar, V. and Tiwari, S.K. (2017). Halocin HA1: an archaeocin produced by the haloarchaeon *Haloferax larsenii* HA1. *Process Biochemistry* 61: 202–208.

69 Lee, S.G. and Chang, H.C. (2018). Purification and characterization of mejucin, a new bacteriocin produced by *Bacillus subtilis* SN7. *LWT – Food Science and Technology* 87: 8–15.

70 Juturu, V. and Wu, J. (2018). Microbial production of bacteriocinas: latest research development and applications. *Biotechnology Advances* 36 (8): 2187–2200.

71 Ciszek-Lenda, M. (2011). Biological functions of exopolysaccharides from probiotic bacteria. *Central European Journal of Immunology* 36 (1): 51–55.

72 Abid, Y., Casillo, A., Gharsallah, H. et al. (2018). Production and structural characterization of exopolysaccharides from newly isolated probiotic lactic acid bacteria. *International Journal of Biological Macromolecules* 108: 719–728.

73 Wasser, S.P. (2002). Medical mushrooms as a source of antitumor and immunomodulating polysaccharides. *Applied Microbiology and Biotechnology* 60 (3): 258–274.

74 Lin, Z. and Zhang, H. (2004). Anti-tumor and immunoregulatory activities of *Ganoderma lucidum* and its possible mechanisms. *Acta Pharmacologica Sinica* 11 (11): 1387–1395.

75 Kim, Y., Oh, S., Yun, H.S. et al. (2010). Cell-bound exopolysaccharide from probiotic bacteria induces autophagic cell death of tumour cells. *Letters in Applied Microbiology* 51 (2): 123–130.

76 García-Fruitós, E. (2012). Lactic acid bacteria: a promising alternative for recombinant protein production. *Microbial Cell Factories* 11: 157.

77 Konings, W.N., Kok, J., Kuipers, O.P. et al. (2000). Lactic acid bacteria: the bugs of the new millennium. *Current Opinion in Microbiology* 3 (3): 276–282.

78 Bermúdez-Humarán, L.G., Aubry, C., Motta, J.-P. et al. (2013). Engineering lactococci and lactobacilli for human health. *Current Opinion in Microbiology* 16 (3): 278–283.

79 Cano-Garrido, O., Seras-Franzoso, J., and Garcia-Fruitós, E. (2015). Lactic acid bacteria: reviewing the potential of a promising delivery live vector for biomedical purposes. *Microbial Cell Factories* 14 (1): 1–12.

80 Pontes, D.S., de Azevedo, M.S., Chatel, J.M. et al. (2011). *Lactococcus lactis* as a live vector: heterologous protein production and DNA delivery systems. *Protein Expression and Purification* 79 (2): 165–175.

81 Douillard, F.P., O'Connell-Motherway, M., Cambillau, C. et al. (2011). Expanding the molecular toolbox for *Lactococcus lactis*: construction of an inducible thioredoxin gene fusion expression system. *Microbial Cell Factories* 10 (1): 66.

82 Morello, E., Bermudez-Humaran, L.G., Llull, D. et al. (2008). *Lactococcus lactis*, an efficient cell factory for recombinant protein production and secretion. *Journal of Molecular Microbiology and Biotechnology* 14 (1-3): 48–58.

83 Nouaille, S., Ribeiro, L.A., Miyoshi, A. et al. (2003). Heterologous protein production and delivery systems for *Lactococcus lactis*. *Genetics and Molecular Research* 2 (1): 102–111.

84 Tremillon, N., Issaly, N., Mozo, J. et al. (2010). Production and purification of staphylococcal nuclease in *Lactococcus lactis* using a new expression-secretion system and a pH-regulated mini-reactor. *Microbial Cell Factories* 9 (37): 1–12.

85 Kuipers, O.P., de Ruyter, P.G.G.A., Kleerebezem, M. et al. (1998). Quorum sensing-controlled gene expression in lactic acid bacteria. *Journal of Biotechnology* 64 (1): 15–21.

86 Hidalgo-Cantabrana, C., O'Flaherty, S., and Barrangou, R. (2017). CRISPR-based engineering of next-generation lactic acid bacteria. *Current Opinion in Microbiology* 37: 79–87.

87 U.S. Food and Drug Administration. https://www.fda.gov/.

88 European Medicines Agency. https://www.ema.europa.eu/en.

89 Pan, E., Bogumil, D., Cortessis, V. et al. (2020). A systematic review of the efficacy of preclinical models of lung cancer drugs. *Frontiers in Oncology* 10: 591.

90 Day, C.P., Merlino, G., and Van Dyke, T. (2015). Preclinical mouse cancer models: a maze of opportunities and challenges. *Cell* 163 (1): 39–53.

91 Hay, M., Thomas, D.W., Craighead, J.L. et al. (2014). Clinical development success rates for investigational drugs. *Nature Biotechnology* 32 (1): 40–51.

92 Zeng, X., Zhang, Y., Kwong, J.S. et al. (2015). The methodological quality assessment tools for preclinical and clinical studies, systematic review and meta-analysis, and clinical practice guideline: a systematic review. *Journal of Evidence-Based Medicine* 8 (1): 2–10.

93 Perel, P., Roberts, I., Sena, E. et al. (2007). Comparison of treatment effects between animal experiments and clinical trials: systematic review. *BMJ* 334 (7586): 197.

94 Dorato, M.A. and Buckley, L.A. (2006). Toxicology in the drug discovery and development process. *Current Protocols in Pharmacology* 32 (1): 10.3.

95 ICH Official web site : ICH. https://www.ich.org/page/ich-guidelines.

96 Hingorani, A.D., Kuan, V., Finan, C. et al. (2019). Improving the odds of drug development success through human genomics: modelling study. *Scientific Reports* 9 (1): 1–25.

97 Sibbald, W.J. (2000). An alternative pathway for preclinical research in fluid management. *Critical Care* 4 (2): 1–8.

98 Denny, K.H. and Stewart, C.W. (2017). Acute, subacute, subchronic, and chronic general toxicity testing for preclinical drug development. In: *A Comprehensive Guide to Toxicology in Nonclinical Drug Development* (ed. Ali S. Faqi), 109–127. Academic Press.

99 Issa, N.T., Wathieu, H., Ojo, A. et al. (2017). Drug metabolism in preclinical drug development: a survey of the discovery process, toxicology, and computational tools. *Current Drug Metabolism* 18 (6): 556–565.

100 Health, Wellness, and Nutrition – IControlMyHealth, Inc. https://icontrolmyhealth.org/.

101 de Almeida, J.S., Samir, F.D.A., Dolezal, R. et al. (2019). Surface screening, molecular modeling and *in vitro* studies on the interactions of aflatoxin M1 and human enzymes acetyl- and butyrylcholinesterase. *Chemico-Biological Interactions* 308: 113–119.

102 González-Méndez, I., Aguayo-Ortíz, R., Sorroza-Martínez, K. et al. (2020). Conformational analysis by NMR and molecular dynamics of Adamantane-Doxorubicin prodrugs and their assemblies with beta-cyclodextrin: a focus on the design of platforms for controlled drug delivery. *Bioorganic & Medicinal Chemistry* 28 (13): 115510.

103 Hardianto, A., Yusuf, M., Liu, F. et al. (2018). Structure-based drug design workflow. In: *Encyclopedia of Bioinformatics and Computational Biology*, vol.3, (eds. Shoba Ranganathan, Kenta Nakai and Christian Schonbach), 273–282. Elsevier.

104 Clark, D.E. (2008). What has virtual screening ever done for drug discovery? *Expert Opinion on Drug Discovery* 3 (8): 841–851.

105 Shen, C., Ding, J., Wang, Z. et al. (2020). From machine learning to deep learning: advances in scoring functions for protein-ligand docking. *Computational Molecular Science* 10 (1): e1429.

106 Huang, S.Y., Grinter, S.Z., and Zou, X. (2010). Scoring functions and their evaluation methods for protein-ligand docking: recent advances and future directions. *Physical Chemistry Chemical Physics* 12 (40): 12899–12908.

107 Muhammed, M.T. and Aki-Yalcin, E. (2019). Homology modeling in drug discovery: overview, current applications, and future perspectives. *Chemical Biology & Drug Design* 93 (1): 12–20.

108 Schimidt, T., Bergner, A., and Schwede, T. (2014). Modelling three-dimensional protein structures for applications in drug design. *Drug Discovery Today* 19 (7): 890–897.

109 Du, J., Guo, J., Kang, D. et al. (2020). New techniques and strategies in drug discovery. *Chinese Chemical Letters* 31 (7): 1695–1708.

110 Neves, B.J., Braga, R.C., Melo-Filho, C.C. et al. (2018). QSAR-based virtual screening: advances and applications in drug discovery. *Frontiers in Pharmacology* 9: 1275.

111 Zhang, J.D., Sach-Peltason, L., Kramer, C. et al. (2020). Multiscale modelling of drug mechanism and safety. *Drug Discovery Today* 25 (3): 519–534.

112 Adeniji, S.E., Shallangwa, G.A., Arthur, D.E. et al. (2020). Quantum modelling and molecular docking evaluation of some selected quinoline derivatives as anti-tubercular agents. *Heliyon* 6 (3): e03639.

113 Sledz, P. and Caflisch, A. (2018). Protein structure-based drug design: from docking to molecular dynamics. *Current Opinion in Structural Biology* 48: 93–102.

114 Choi, S.B., Yap, B.K., Choong, Y.S. et al. (2019). Molecular dynamics simulations in drug discovery. In: *Encyclopedia of Bioinformatics and Computational Biology*, vol. 3 (eds. S. Ranganathan, M. Gribskov, K. Nakai and C. Schonbach), 652–665. Elsevier.

8

Microbial Production of Antimicrobial and Anticancerous Biomolecules

M. Indira, T. C. Venkateswarulu, S. Krupanidhi, and K. Abraham Peele

Department of Biotechnology, Vignan's Foundation for Science, Technology & Research, Vadlamudi, Andhra Pradesh, India

8.1 Introduction

Microorganisms are present everywhere on earth, and they have the ability to survive in extreme environments also [1]. Microorganisms served as a valuable source of bioactive compounds that are developed into therapeutic drug molecules [2]. The novel bioactive compounds produced from microorganisms are structurally diverse, and the newly discovered drugs are used to treat various diseases [3]. Nearly 50% of the pharmaceutical drugs available in the market are derived from natural sources, of which microorganisms are the prolific producer compared to plants and animals [4]. Nevertheless, in the microorganism category, only 1% of the microbial biodiversity was explored for a beneficial purpose, but 99% of the microorganisms are yet to be explored [5]. The microorganisms provide a new platform for research and development of new molecules due to their diverse role in the ecological system [6]. Moreover, the microorganisms act as reservoirs for the synthesis of novel compounds and maintain sustainable development by removing toxic pollutants from the environment [7]. Traditionally, the bioactive compounds produced by microorganisms are discovered via conventional bioprospecting based on isolation of potential microbe, and screening of their extracts for potential bioactive compounds through bioassays [2].

Bioprospecting is defined as the exploration of biological diversity for useful products derived from microorganisms, plants, and animals [8]. It is the process of searching for novel species in nature, and these species produce valuable products for human purposes. The products can include genes, chemical compounds, enzymes, metabolites, novel strains, and more [2]. Bioprospecting is diverse, but the focus is required to explore commercial compounds with therapeutic applications. The exploration is driven by evolution/ecology where the adaptations have evolved but not the number of species located in a habitat [9]. Currently, novel bioactive compounds are in high demand due to emerging infections,0 age-related diseases, and the development of antibiotic resistance [10]. The novel bio active compounds can act as antimicrobial, anti-inflammatory, anticancer, antiviral,

Bioprospecting of Microorganism-Based Industrial Molecules, First Edition.
Edited by Sudhir P. Singh and Santosh Kumar Upadhyay.

immunomodulatory, and role in several metabolic diseases [11, 12]. The bioprospecting of novel molecules from microorganisms are enzymes and secondary metabolites, which are having industrial, pharmaceutical, biotechnological, and therapeutic applications [13]. Bioprospecting mainly focuses on: (i) development of novel lead molecules [4], (ii) biodiversity conservation [14], and (iii) sustainable development of the biodiversity [15]. In the last decade, the microbial species are used in industrial processes, and the number of microbial products kept growing as high-value molecules with special characteristics [16]. Furthermore, microbial compounds have great potential in various industries and commercial value. The compounds serve as potential ingredients for different formulations currently being developed on an industrial scale [17]. The live microbial strains provide numerous characteristic features for applications in agriculture, food processing, environment, bioremediation, catalytic reactions, public health, and pharmaceutical industries. In the pharmaceutical sector, these microbial resources have a prominent role in the problems of antibiotic resistance of bacteria to antibiotics and lack of novel molecules to treat the diseases [18]. In this chapter, we have given an overview of bioprospecting process methods with a special focus on bioactive compounds and their industrial importance.

8.2 Microbial Sources

Since ancient times, microbes are used for the production of valuable products and as probiotic microorganisms [19]. Penicillin is an example of the development of therapeutic drug molecules and thus paves the way for exploring the antibacterial compounds during World War II [20]. Several naturally occurring molecules were developed using this bioprospecting technology. This continued for several years to explore the antibiotic compounds from different habitats and niches. The microorganisms have been exploited for natural molecules with various applications [21]. From the last decade, a few microorganisms have been used as expression systems for the production of recombinant therapeutic drug molecules by using *Saccharomyces cerevisiae* and *Escherichia coli* [22, 23]. Microorganisms have high biotechnological relevance for the synthesis of new molecules with anticancer, antibacterial, antifungal, antioxidant, bioenergy compounds, and bioplastics activities to be applied in industrial, agricultural, bioremediation, and biofuels applications [24]. Microorganisms also have the ability to genetically modify themselves to produce novel molecules of interest, and also, the increased production contributes to large-scale industrial applications [25].

8.2.1 Bacteria

For more than 100 years, bacteria have contributed to various industries not only for the treatment of diseases and also for other applications [26]. The previous studies reported that bacterial secondary metabolites are diverse with different biological effects [27]. Several different gram-positive and gram-negative bacteria were screened for the production of bioactive compounds. The various species are *Bacillus subtilis*, *E. coli*, *Staphylococcus aureus*, *Pseudomonas aeruginosa*, and *Micrococcus luteus* [28]. Bacteria play a vital role in various industries. The bacteria are used for the manufacture of foods, probiotics, enzymes, organic solvents, biofuels, insecticides, antibiotics, starter cultures, and medicinal compounds [29, 30].

The unusual environments are poorly explored, and they are valuable resources for novel products. For the past several decades, bacteria are widely used in large-scale cultivation in industries to manufacture bioactive compounds, the transformation of chemicals, and also recombinant products [16].

Bacteria have more advantages compared to plants and animals. They are as follows:

1) Adaptation to a wide and different environment
2) Nutrients uptake for high metabolism and biosynthesis of metabolites
3) Variety of metabolic reactions
4) Simple screening procedures
5) Different enzymes are produced
6) More biomass production and high yield
7) Ease of genetic manipulation to increase the yield of valuable compounds

The genetically modified organisms are produced for enhancement of yield. The bacteria used for recombinant DNA technology are *E. coli*, *B. subtilis*, and *Lactococcus lactis* [31]. The bacteria producing secondary metabolites are important for our nutrition and health. The secondary metabolites are produced after the growth of the bacteria, and they will not play any role in that organism [32]. The secondary metabolites produced by one organism act as lead molecules for treating various diseases in other organisms [16]. The secondary metabolites act as anticancer agents, antimalarial, antibacterial, cholesterol-lowering activity, insecticidal, and pest control [33]. Apart from these, they play a role in the food industry as functional food ingredients, vitamins, amino acids, and growth factors. Some of the beneficial bacteria act as probiotic, and they are used in the food industry for the development of probiotic products [34, 35].

8.2.2 Fungi

Fungi are one of the unexplored resources for novel compounds. Most of the fungi grow in extreme environments are diverse in nature [36]. Due to their adaptability to extreme environments, they act as reservoirs for valuable bioactive compounds [37]. The bioactive compounds identified from this group are used as antibacterial, antiviral, anti-inflammatory, anticancer, and anti-plasmodial agents [38]. The fungal biodiversity has several species consists of both culturable and nonculturable species [39]. Previous studies reported that the culturable fungi produce natural molecules from the marine environment used as antimicrobial, antioxidant, antitumor potential, antihyperlipidemic, anti-hypercholesterolemic, hepatoprotective, antidiabetic, and immunomodulatory [40]. Apart from these, the fungi are the natural sources for the production of enzymes such as proteases, lipases, cellulases, hydrolases, xylanases, and so on [33]. The first-generation antibiotic penicillin was produced by fungi *Penicillium notatum*. The beta-lactam antibiotics dominate the pharmaceutical market for several decades, and they are the footstep for the development of next-generation antibiotics [20]. Furthermore, the fungal compound cyclosporine, which is an antirejection substance, developed as pharmaceutical drug in organ transplant surgery [41].

The cholesterol-lowering drugs lovastatin and compactin are in clinical use. It is evident that the novel compounds are still to be explored from the fungal diversity of extreme environments for various industrial applications [42]. Another application of fungi in the pharmaceutical industry is used as the expression system for the development of therapeutic proteins. The fungal species used as expression systems for the production of therapeutic molecules are *Saccharomyces cerevisiae*, *Pichia pastoris*, *Pichia methanolica*, *Hansenula polymorpha*, and *Kluyveromyces lactis* [43].

8.2.3 Actinomycetes

The phylum actinomycetes are the gram-positive microorganisms. They produce secondary metabolites and exhibit a wide variety of activities such as anticancer, antibiotic, antimalarial, antifungal, and anti-inflammatory [44]. For several decades, these microorganisms were explored for the screening of active metabolites (antibiotics) from the soil sample. The genome sequences data revealed that the actinomycetes species have the ability to produce an average of around 10–15 antimicrobial agents. However, when the species are grown under laboratory conditions, 1 or 2 bioactive compounds were detected [45]. This is because of antibiotic gene clusters are not expressed under laboratory conditions. The dormant gene clusters are the resource materials for the development of novel molecules as antimicrobial compounds [45]. To overcome this, scientists identified a key approach wherein the understanding of gene regulation and expression of these compounds provides manipulation in the laboratory. In this, the manipulation of biosynthetic pathways using a rational approach leads to an increase in the yield of the compound [46]. In this rational approach, the overexpression of positive regulators and the deletion of negative regulators are maintained. The development of actinomycetes genetic tools provides information to enhance the overexpression or reduction of expression of genes [47]. In the genome of the actinomycetes, I-SecI mega nuclease system, Cre-*lox*P system, and CRISPR/Cas are the genome editing systems [48]. Another approach for overexpression of compounds is the induction of biosynthetic gene clusters in heterologous host's results in the high yields of the compound of interest [49]. The various genera include in this group are *Streptomyces*, *Actinoplanes*, *Saccharopolyspora*, *Amycolatopsis*, and *Micromonospora* [50]. Among all, the streptomyces is the initial antibiotic maker produced several secondary metabolites used in medicine and veterinary fields [51]. The *Streptomyces* sp. produces clinically useful compounds such as streptomycin, streptothricin, actinomycin, tetracycline, erythromycin, vancomycin, leucomycin, and chloramphenicol. The compounds isolated from other genera are gentamicin, tricoplamin, rosamicin, nocardicin, and fortimicin. The anticancer drug salinosporamide A is produced by the strain *Salinospora tropica* strain CNB-440 [52].

8.2.4 Extremophiles

Extremophiles are the microorganisms that can resist adverse conditions in their surrounding environment [53]. The microorganisms are evolving continuously and adapting to extreme conditions such as high salt concentration, pressure, temperature (hot, and cold), pH (acidic, and alkaline), oxidative stress, radiation, water availability, altitude, ocean environment, polar region, and glacier environment. The microorganisms survive in extreme high or low conditions and are called as extremophiles [54]. The organisms have opened a new way for bioprospecting of novel bioactive molecules with applications and commercialization aspects. One such example is the Taq DNA polymerase from thermophilic organism *Thermus aquaticus* [55]. The extreme cold environments are highly attractive for novel compounds, and they find applications in food industries to store the fermented and frozen products [56]. In addition to this, the microorganisms find applications in various industries that include bioremediation, biofuels, textile, biopolymers, agriculture, medicine, and value-added products [57].

The extremophiles are categorized into the following:

i) Based on temperature – Psychrophiles, thermophiles, and hyperthermophiles
ii) Based on pH – Acidophiles and alkaliphiles
iii) Based on pressure – Barophiles
iv) Based on salt concentration – Halophiles

The bacteria are isolated from different extreme environments – hot springs, geothermal sites, ocean depth, volcanic sites, deep biospheres, polar regions, and mining areas [58]. There is an increase in demand for industrial applications of extremophiles due to the enzymes active in harsh conditions and metabolic activities carried out by them. With the advancement of molecular biology and genetic engineering tools, extensive scale screening for potential genes can be done using metagenomic approach [59]. The bioactive compounds produced by extremophiles act as antibiotics, anticancer compounds, enzymes, pigments, xenobiotic degrading enzymes, plant growth promoters, and recycling of agriculture residue [60].

8.3 Microbial Bioprospecting Methods

Microorganisms cannot be grown easily in a laboratory (<1%). The biotechnological potential of microorganisms is clearly underused. The traditional and advanced bioprospecting technologies have been developed to explore novel microbial ecosystems and also for the identification and use of new biotechnological products [6]. Bioprospecting consists of systematic identification, evaluation, and exploitation of diversity of life on earth for commercial purposes [61]. Bioprospecting of microorganisms consists of three strategies to explore new molecules:

1) Cultural bioprospecting of microorganisms
2) Nonculturable bioprospecting of microorganisms
3) In silico bioprospection of genome sequences of microorganisms

All these methods are used to identify microorganisms, genes, enzymes, and metabolic pathways for biotechnological applications in industry.

8.3.1 Cultural Bioprospecting

This is the traditional method of approach used for culturing the microorganisms. This method involves the collection of samples, cultivation, and screening of microorganisms for selected metabolites [62]. The selection of microorganisms depends on nutrient selection for specific growth, as well as specific metabolites of interest. In the case of culturable bioprospecting, maintenance of in situ conditions in the laboratory is difficult to maintain. The growth of the organism and the medium composition can be a problem in this method in order to get a good result [63].

The outline for culturable bioprospecting of microorganisms is shown in Figure 8.1.

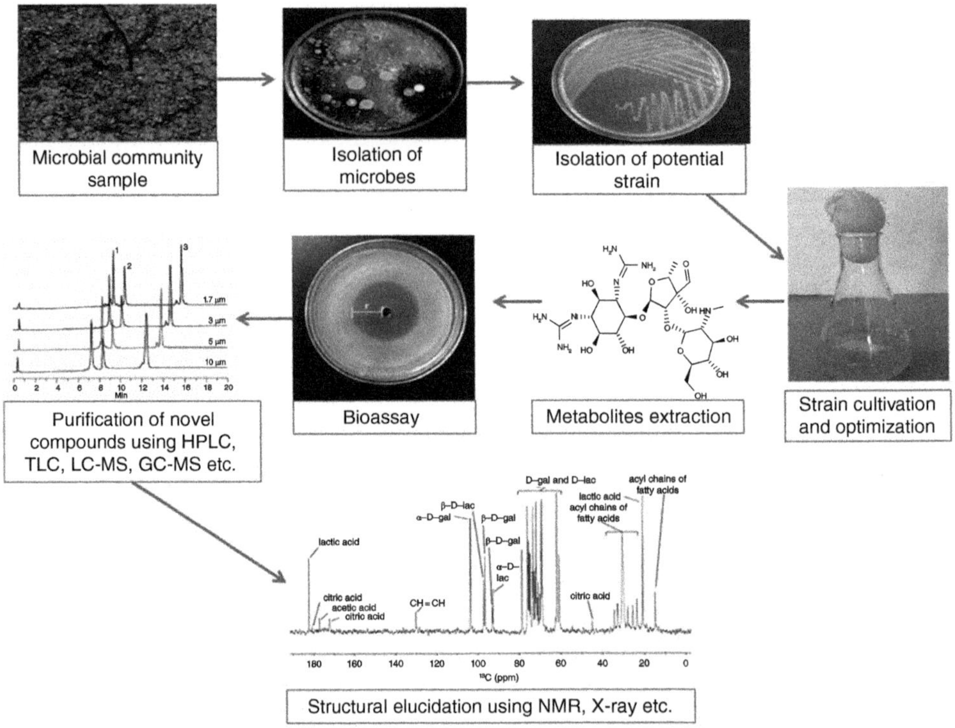

Figure 8.1 Culturable bioprospecting of novel bioactive compounds from microorganisms.

8.3.2 Nonculturable Microorganism's Bioprospecting

The microorganisms that cannot be cultivated with the current microbiological techniques are screened for specific bioactive compounds using metagenomics approach [64]. In this technique, the genetic material from the microbial community sample from certain environments is collected and isolated using specific protocols. The isolated DNA is amplified, and cDNA libraries are developed. The DNA libraries have prospected for specific activities of metabolites. The examples for metagenomic libraries from distinct environments are marine environment, gut microbiomes, insect intestines, sewage samples, etc. The metagenomic libraries are screened for bioactive compounds used for industrial applications [65]. The steps involved in nonculturable bioprospecting are outlined in the Figure 8.2.

8.3.3 In Silico Bioprospecting of Microorganisms

In silico bioprospecting is based on the discovery of new genes and metabolic pathways through genome sequence databases. Nowadays, the decrease in sequencing costs and also advanced techniques for sequencing have enabled the generation of a large quantity of information related to bioactive compounds of microbial genomes [66]. The whole-genome sequencing studies paved the way for the generation of a large quantity of data related to bacteria, fungi, and archaea. The bioprospecting of novel genes from these genomes through comparative genomics and evolutionary relationship analysis with the help of

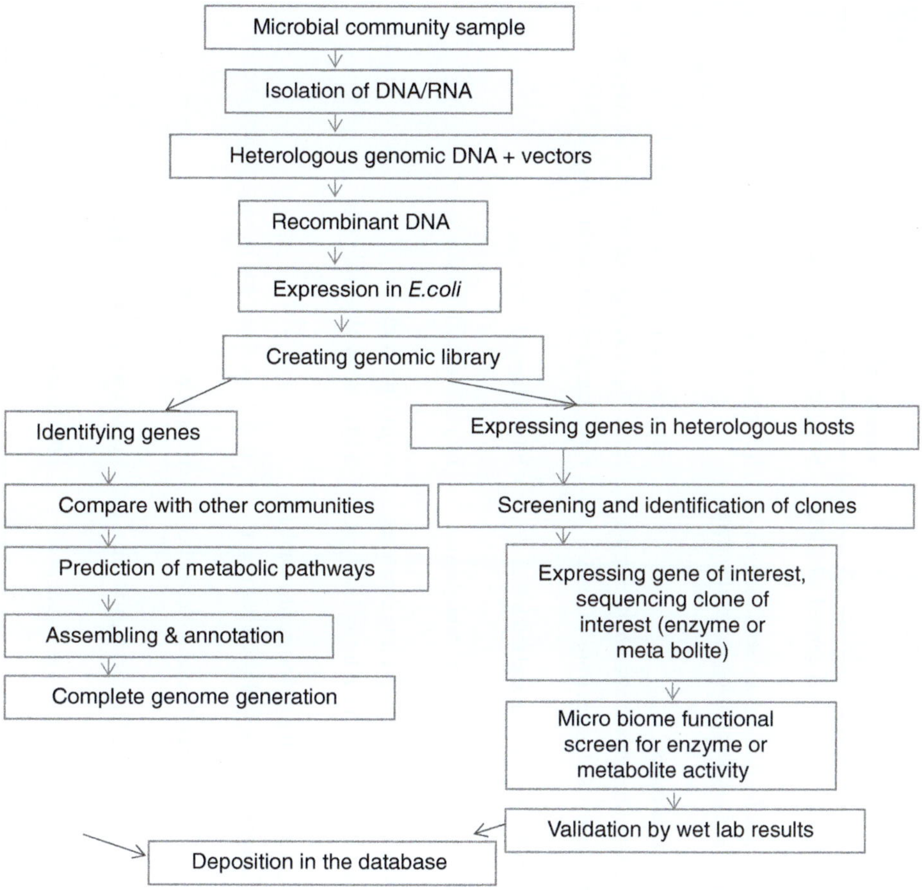

Figure 8.2 Steps involved in nonculturable bioprospecting using metagenomics approach.

bioinformatics tools allows quick identification of novel genes and metabolites with a wide range of functions and applications [67]. The genome sequences in the databases are searched for novel compounds to develop industrially essential molecules. The databases used for in silico approaches are represented in Table 8.1. The method is speedy, cost-effective, and good results are produced. Bioinformatics tools are very useful in performing analysis of a large number of data sets to get valuable lead molecules [68].

The combination of different approaches for bioprospecting of microorganisms helps in the identification of novel bioactive compounds. The in silico bioprospecting is used for designing the molecular probes that are employed in culturable and nonculturable strategies for the development of new compounds.

8.4 Bioactive Compounds

Bioactive compounds are substances that promote health benefits in the body. They are naturally occurring compounds available in low quantities. The compounds have biological activity and contributed to beneficial effects in various industries [69]. The bioactive

Table 8.1 The databases and software tools used for in silico approaches for bioprospecting of bioactive compounds.

S. No	Database/ Software tool	Description of database/software tool	Website link
1	MBGD	A database for comparative analysis of completely sequenced microbial genomes	http://mbgd.genome.ad.jp/
2	ExPASy	A database resource portal provides scientific databases and software tools related to proteomics, genomics, phylogeny, systems biology, transcriptomics, and population genetics	https://www.expasy.org/
3	Microbes online	Online tool for annotation of microbial genes	http://www.microbesonline.org/
4	Microbes GPS	Used for assigning relatedness among microbes in microbiota sample	https://sourceforge.net/projects/microbegps/
5	Meta microbes online	Used for performing phylogenetic analysis of genes from microbial genomes and metagenomes	http://meta.microbesonline.org/
6	HAMAP	A database of completely sequenced microbial sets and microbial protein families	https://hamap.expasy.org/
7	BLAST	Assigning relationship to a given sequence by searching homologous sequencing	https://blast.ncbi.nlm.nih.gov/Blast.cgi
8	Pfam	Gives information about protein families and relatedness in database	https://pfam.xfam.org/
9	PDB	A database of the tertiary (or) quaternary structure of biological molecules	https://www.rcsb.org/
10	MOST	Prediction of targets of ligand compounds by using fingerprint similarity	
11	KEGG	A database deals with genomes, diseases, drugs, biological pathways, and chemical substances	https://www.genome.jp/kegg/pathway.html
12	MetaCyc	A metabolic database contains enzymes, metabolites, and metabolic pathways of all domains of life	https://metacyc.org/
13	Proteome Tools	The tool is used for drug discovery, personalized medicine, and life science research	http://www.proteometools.org/
14	Genome tools	It is a collection of free bioinformatics tools related to genomes of all domains of life	http://genometools.org/

compounds are metabolites, which include antibiotics, bacteriocins, enzymes, biofuels, exopolysaccharides, vitamins, and biopolymers [35, 70] The compounds are exploited for various industrial applications, including food and nutraceuticals, beverages, agriculture as plant growth promoters, biofertilizers and controlling pests/insects, dairy industry, textile, pharmaceutical, and clinical fields [71].

In the medical and health field, the bioactive compounds contribute to control the physiological and pathological state of the body. The compounds control the physiological processes and act as cholesterol-lowering agents, obesity control, hypertension and hypotension regulators, hyperglycemia, neurological issues like anxiety, and depression [72, 73].

8.4.1 Antibiotics

Today, the antibiotic-resistant bacteria are evolving continuously and become a severe threat to infections. The discovery of novel antibiotics to kill the antibiotic-resistant microorganisms is essential to save the modern medicine [74]. The strains of *Pseudomonas aeruginosa*, *Acinetobacter baumanii*, *Enterobacter aerogenes*, *S. aureus*, and *Klebsiella pneumoniae* have been found resistant to carbapenem, cephalosporins, methicillin, and vancomycin antibiotics [75]. The organisms cause life-threatening infections due to the presence of several resistant genes and becoming superbugs. Initially, the actinomycetes genus was explored for the production of antibiotics [76]. The actinomycetes group have been shown to produce several molecules that are clinically used for treating several infections. The antibiotics find applications in veterinary, agriculture, medicine, pharmaceutical fields, food, and dairy industries [33]. Different approaches are required to explore novel antibiotic compounds. The metagenomics approach is a culture-independent approach where the metagenomic libraries are used for finding the novel antibiotic compounds [77]. The heterologously expressed antibiotic gene clusters are used for identifying novel compounds. The unexplored valuable repository is used to find the antibiotic applications in various industries [78].

8.4.2 Bacteriocins

Bacteriocins are the industrially important potential molecules used as antimicrobials to treat a broad range of infections [79]. The bacteriocins are the ribosomally synthesized short-length peptides produced naturally by microorganisms [80]. The bacteriocins molecules have no health risks and considered to be safe to use. The bacteriocins are produced by genera lactic acid bacteria, Bifidobacteria, Lactococcus, Streptococcus, Bacillus, and Enterococcus [81, 82]. These molecules are used as an alternative to chemical preservatives in the food and dairy industries. For preservation, they are used either as pure form or as crude form of the substance. They act as antibacterial and control the pathogens in food and dairy products such as cheese, milk products, and yogurt [83]. They have a narrow spectrum of activity by inhibiting the closely related bacteria and a broad spectrum of activity by inhibiting the wide variety of bacteria. Till to date, nisin and pediocin PA1 have been commercialized as food preservatives and approved by food and drug administration (FDA) [84]. The bioprospecting of bacteriocin molecules is through the traditional approach agar diffusion assay. A large number of bacteriocin molecules screened for antimicrobial activity through this assay [85]. The bioinformatics approach helps in the identification of novel bacteriocin molecules and the bacteriocin coding sequences [86]. The online genome databases such as BAGEL, BACTIBASE, bacteriocin operon and gene bock associator help in the identification of bacteriocin coding sequences. Further, the molecular-based evidences are required to confirm the antimicrobial activity of identified novel compounds [87]. For their use as therapeutics, the bacteriocins must be stable, soluble in nature, and proteolytic resistance. The bacteriocin amino acid prediction tools help the researchers to identify the protease-resistant peptides from genome sequences. This makes bacteriocins an attractive therapeutic alternative molecule to conventional antibiotics, particularly to antibiotic resistance crisis [88, 89].

8.4.3 Biosurfactants

Biosurfactants are the amphiphilic molecules that consist of both hydrophilic and hydrophobic parts. The molecules produced by microorganisms are biodegradable, nontoxic, and biocompatible. The biosurfactants are used as food additives, antimicrobial agents, anti-adhesive molecules, and stabilizers [90]. In the bakery industry, the biosurfactant molecules are used to adjust the consistency of creams, cakes, and ice creams. The biosurfactant obtained from *Bacillus subtilis* has been evaluated for food stabilizers, and it showed significant stabilizing activity. It improves the consistency, changes in the rheological properties, and maintains the shelf life of the product [91]. The biosurfactants are used as emulsifiers produced by microorganisms generally recognized as safe (GRAS). The biosurfactants are rhamnolipids, polysaccharide–protein complexes, lipopeptides, and emulsan. The emulsifier improves the texture of creams and reduces the viscosity [92]. The rhamnolipids are added as food additives to improve the stability of the food products. In addition to these, the biosurfactants play a role in bioremediation [93]. They help in the degradation of hydrocarbons and remove the contaminants in the polluted environment. The biosurfactants also play a role in nanotechnology [94]. This is the emerging field where the nano emulsions find applications in paintings, lubricants, agrochemicals, foods, cosmetics, personal care products, bioremediation, and pharmaceutical industries [95].

8.4.4 Exopolysaccharides

Exopolysaccharides are the naturally occurring molecules produced by microorganisms released into the surrounding environment in stressful conditions [96]. They constitute part of sugar units, proteins, lipids, and DNA. Some exopolysaccharides consist of nonsugar units such as pyruvate, acetate, succinate, and phosphate [97]. The exopolysaccharides find applications in food and pharmaceutical industries. In the medical field, the clinical trials were done on dextran as a plasma expander [98]. The exopolysaccharides produced from microorganisms are:

i) Xanthan gum – food additive
ii) Dextran – plasma expander
iii) Alginates – matrix for tablets, dental impression
iv) Bacterial cellulose – wound dressings
v) Hyaluronic acid – arthritis treatment, wound healing
vi) Pullulan – oral care products
vii) Gellan – food stabilizer

The EPS molecules are nontoxic, biodegradable, and biocompatible materials used as drug carriers and vaccine adjuvants. Microbial exopolysaccharides offer a large field of medical applications due to their unique properties [99]. The EPS is essential for biofilm formation and pathogenicity nature of gram-negative bacteria [100, 101]. The biosynthetic pathways for EPS production are

i) Wzx/Wzy-dependent pathway
ii) ATP-binding cassette ABC transporter-dependent pathway
iii) Extracellular synthesis pathway
iv) Synthase pathway

With the advancement of high-throughput genome sequencing, comparative genomics, functional genomics, protein structure analysis database tools aid in the identification of new pathways for EPS formation [102]. By regulating the overexpression of genes involved in EPS formation, the EPS production can be enhanced. Another strategy is the engineering of EPS material, in which altering of molecular structure by addition/deletion of monomeric sugar units to the side chain group results in a change in properties of EPS materials [103].

8.4.5 Enzymes

Enzymes are the biocatalysts used to convert the substrates into products. The microorganisms secrete the enzyme molecules into their surrounding medium [104]. Currently, there is a considerable demand for industrial enzymes that are alternative to chemical synthesis. In these aspects, the extreme environments are the valuable resources that provide bioactive compounds with industrial applications [105]. The industry-relevant enzymes are proteases, hydrolases, lipases, phosphatases, acylases, xylanases, and oxidoreductases [106]. The microbes capable of growth under extreme conditions are extremophiles, and the enzymes are extremozymes. The enzymes are resistant to extreme conditions such as high temperature, pH, pressure, and radiation [57]. The unique features of these enzymes act as catalysts in industrial bioprocesses. After the discovery of *Taq* polymerase from *Thermus aquaticus*, thermophilic organisms, scientists started exploring the new enzymes from extreme environments [107]. The microorganisms have the ability to produce potential biocatalysts suggest their role in the biotechnological field. The enzymes produced by psychrophilic bacteria have considerable importance in the food and dairy industries due to the storage of frozen products. The cold-adapted enzymes play a role in the food sector [108]. The metagenomic profiling of microbes from different habitats gave new insights to technological advancements in the biotechnological field. Recent reports suggested that the metagenomic libraries are screened for thousands of clones to identify the genes encoding enzymes of interest [109]. The metagenomic approaches used for the study of microbial communities without culturing and isolation of strains revealed many unknown enzymes of interest in industries. One of the approaches in metagenomics is the functional genomics in which the activity of the enzyme or metabolites is screened for their activity using functional genomics and confirmation with wet-lab studies [110]. Sequence-based approaches also help in the identification of genes encoding enzymes. The sequence comparison of homologous proteins provides information about phylogenic relatedness and the activity of the enzymes [109]. The enzymes isolated from extreme environments have high catalytic rates. The enzymes have unique features and overcome the limitations of several industrial processes such as stability of the enzyme, cost of the pure enzyme, narrow reaction conditions, and half-life of the enzyme. Hence, the extremophiles will serve as the largest resource for the development of many industrially important enzymes [111].

8.4.6 Biopolymers

Biopolymers are the natural polymer materials produced by living organisms. The biopolymers are biodegradable, nontoxic, biocompatible materials [112]. The biopolymers have found wide applications in various industries that include agriculture, medicine, automotive, food packaging, biomedical, and pharmaceutical field [113]. The market for biopolymers is increasing globally and expected that the biopolymers replace synthetic polymers [114]. The biopolymers are produced by microorganisms require specific nutrients and controlled environmental conditions [115]. Based on the polymer backbone, the biopolymers are classified into polyesters, polysaccharides, polyamides, polycarbonates, and vinyl polymers. The biopolymers are produced by chemical polymerization or by fermentation process [112]. The biopolymers are starch, cellulose, chitin, chitosan, DNA, RNA, polylactic acid, polyhydroxy butyrate, etc. Metagenomics enable us to screen the clones to synthesize the biopolymer materials with industrial importance [116].

8.4.7 Bioenergy Compounds

With an increasing demand for energy consumption globally, there is an urgent need to develop alternative energy resources. Biofuels provide a means to reduce greenhouse gases into the environment [117]. Biofuel is derived from natural renewable resources like plants, microbes, and algae. Biofuels reduce the risk of climate change. Cellulose is the most abundant molecule in the plant cell walls. Cellulose-based ethanol can be used as an alternate fuel in industries [118]. Cellulose consists of a long chain of sugars called glucose molecules. The glucose molecule is the starting material for fermentation. The cellulose breakdown is by the action of the cellulase enzyme produced by bacteria and fungi. The finding of cellulose-producing novel strains from nature is a key element in bioprospecting of bioenergy compounds [119]. Currently, efforts are going on to develop cellulose-degrading novel enzymes from microorganisms. Various bacteria, fungi, and archaeal organisms have received great attention as potential resources for thermostable cellulase enzymes [120]. Further, this is a rapidly evolving area, where biofuels become an emerging field for future needs. Alternate fuel development reduces climate change and greenhouse gases released into the environment. Metagenomic approaches help in search of novel enzymes for industrial production of biofuels from plant feed stocks. Biofuel DB is a database used for searching for novel enzymes involved in biofuel production [121].

8.4.8 Anticancer Compounds

Microorganisms have the ability to kill the cancerous cells by producing proteins, peptides, and active metabolites. These biomolecules act as antiproliferative agents and contribute to the regulation of cancer. The actinomycetes group and marine bacteria are the leading producers for bioactive metabolites and several compounds undergone for clinical trials [122]. The toxin molecules produced by bacterial cells destruct the tumor cells, as well as the immunomodulating effect. The genetically modified bacteria for selective tumor destruction are also an emerging field where the engineered bacteria have novel characteristics for controlling the cancer cells [123]. *Salmonella* is the well-studied organism for anticancer activity and expression systems to target the tumor. The metagenomic approach helps in the screening of unculturable microorganisms for novel metabolites for the development of therapeutic drug molecules [124]. The compounds identified as anticancer compounds from marine biodiversity are eleutherobin, sarcodictyin, discodermolide, and bryostatins [125].

The bioactive compounds from different categories and their applications in various industries are listed in Table 8.2.

Table 8.2 Bioactive compounds from the microbial origin and their applications in various industries.

S. No	Name of the compound	Description of the compound	Industrial application	References
Antibiotics				
1	Teixobactin	Antibacterial	Pharmaceutical drug	[126]
2	Tigecycline	Antibacterial	Pharmaceutical drug	[127]
3	Alrubicin	Antitumor activity	Pharmaceutical drug	[4]
4	Telithromycin	Upper respiratory tract infections	Pharmaceutical drug	[128]
Bacteriocins				
5	Nisin	Antibacterial peptide	Food preservative	[129]
6	Pediocin SA1	Bacteriocin against *L. monocytogenes*	Food industry	[130]
7	ABC transporter ATP-binding protein	*Pseudomonas aeruginosa*	Food industry	[131]
8	Enterocins S37	Active against gram positive bacteria	Pharmaceutical	[132]
9	Aureocin A 70	Heat stable bacteriocin	Food bio preservation	[133]
Biosurfactants				
10	Rhamnolipids	Antibacterial activity	Biosurfactant	[134]
11	Emulsan	Emulsification of crude oil in water	Biosurfactant	[135]
12	Phosphatidyl ethanolamine	Bioremediation of hydrocarbons	Biosurfactant	[136]
Exopolysaccharides				
13	Dextran	Blood plasma expander	Medical field	[137]
14	Xanthan gum	Food additive	Food	[138]
15	Alginate	Stabilizer in creams, matrix for tablets	Pharmaceutical, cosmetics	[139]
16	Hyaluronic acid	Wound dressings, osteoarthritis	Pharmaceutical	[140]
17	Gellan	Food stabilizer and thickener	Food industry	[141]

(Continued)

Table 8.2 (Continued)

S. No	Name of the compound	Description of the compound	Industrial application	References
Enzymes				
18	Hydrolases	Hydrolytic activity	Biofuels and bio chemicals	[142]
19	Proteases	Proteolysis activity	Pharmaceuticals, food industry	[143]
Biopolymers				
20	Polyhydroxyl alkanoate	Implant material	Medical devices	[144]
21	Polyhydroxyl butyrate	Implant material	Medical devices	[145]
22	Polylactic acid	Packing of food materials	Food packaging	[146]
23	Poly e-caprolactone	Tissue scaffolds	Tissue engineering	[147]
Bio energy compounds				
24	Thermostable cellulases	Lignocellulosic biomass conversion	Biofuels	[148]
25	Thermostable hemicellulases	Lignocellulosic biomass conversion	Biofuels	[149]
Anticancer compounds				
26	Bryostatin	Oxygenated macrolide for treating cancer	Pharmaceutical	[150]
27	Eleutherobin	Cytotoxic agent	pharmaceutical	[151]

8.5 Future Prospects

Bioprospecting of microorganisms provides a valuable resource for the development of novel compounds from untapped and extreme environments. In the future, the bioprospecting will maintain the socioeconomic development of the nation, commercialization of the bioresources, and sustainable development of the biological resources. The bioprospecting industry is a continuously growing field, which includes biotechnology industries, pharmaceutical, agriculture, bioremediation, bioenergy, food industries, dairy industries, and cosmetics. However, the conservation of biodiversity is the most important concern due to threats to biodiversity, species loss, and loss of potential strains may affect the development of bioactive compounds as lead molecules. The market demand for bioprospecting of potential molecules from microorganisms is expected to increase in the coming years and offers new possibilities for the commercialization of novel products.

8.6 Conclusions and Perspectives

With the advancement of molecular biology, genetic engineering, and bioinformatics tools, the bioprospecting of bioactive compounds gained importance in various industries. The majority of the drugs in the market are derived from natural sources through culturable bioprospecting. Currently, various industries involved are biologics, pharmaceuticals, food, dairy, cosmetics, bioremediation, agriculture, medicine, and clinical fields. The industries are looking for more useful and unique features of bioactive compounds from unexplored environments. Moreover, there is an increasing trend toward the development of pharmaceuticals for the treatment of various diseases. Hence, unexplored extreme environments may fulfill the requirements of various industries in the near future.

Acknowledgments

The authors acknowledge the management of VFSTR (Deemed to be University) for providing a platform to carry the research work. Also, thanks to DST FIST Project LSI-576/2013 for assistance during the tenure of this work.

References

1 Singh, R.P. (2020). *Microbial Versatility in Varied Environments: Microbes in Sensitive Environments*, 3–12. Springer Nature.

2 Sekurova, O.N., Schneider, O., and Zotchev, S.B. (2019). Novel bioactive natural products from bacteria via bioprospecting, genome mining and metabolic engineering. *Microbial Biotechnology* 12 (5): 828–844.

3 Abdel-Razek, A.S., El-Naggar, M.E., Allam, A. et al. (2020). Microbial natural products in drug discovery. *Processes* 8 (4): 470.

4 Cragg, G.M. and Newman, D.J. (2013). Natural products: a continuing source of novel drug leads. *Biochimica et Biophysica Acta (BBA) – General Subjects* 1830 (6): 3670–3695.

5 Gibbons, S.M. and Gilbert, J.A. (2015). Microbial diversity – exploration of natural ecosystems and microbiomes. *Current Opinion in Genetics & Development* 35: 66–72.

6 Vitorino, L.C. and Bessa, L.A. (2017). Technological microbiology: development and applications. *Frontiers in Microbiology* 8: 827.

7 Ahmad, M., Pataczek, L., Hilger, T.H. et al. (2018). Perspectives of microbial inoculation for sustainable development and environmental management. *Frontiers in Microbiology* 9: 2992.

8 Oyemitan, I.A. (2017). African medicinal spices of genus piper. In: *Medicinal Spices and Vegetables from Africa* (ed. Victor Kuete), 581–597. Academic Press.

9 Beattie, A.J., Hay, M., Magnusson, B. et al. (2011). Ecology and bioprospecting. *Austral Ecology* 36 (3): 341–356.

10 Cheesman, M.J., Ilanko, A., Blonk, B., and Cock, I.E. (2017). Developing new antimicrobial therapies: are synergistic combinations of plant extracts/compounds with conventional antibiotics the solution? *Pharmacognosy Reviews* 11 (22): 57.

11 Mushtaq, S., Abbasi, B.H., Uzair, B., and Abbasi, R. (2018). Natural products as reservoirs of novel therapeutic agents. *EXCLI Journal* 17: 420.

12 Kang, H.K., Lee, H.H., Seo, C.H., and Park, Y. (2019). Antimicrobial and immunomodulatory properties and applications of marine-derived proteins and peptides. *Marine Drugs* 17 (6): 350.

13 Giubergia, S., Schleissner, C., de la Calle, F. et al. (2016). Screening microorganisms for bioactive compounds. In: *The Marine Microbiome*, 345–376. Cham: Springer.

14 Chiarolla, C., Lapeyre, R., and Pirard, R. (2013). Biodiversity conservation: how can the regulation of bioprospecting under the Nagoya Protocol make a difference? *Studies* 06: 13.

15 Harvey, A.L. and Gericke, N. (2011). Bioprospecting: creating a value for biodiversity. *Research in Biodiversity–Models and Applications*: 323–338.

16 Park, S.R., Yoon, Y.J., Pham, J.V. et al. (2019). A review of the microbial production of bioactive natural products and biologics. *Frontiers in Microbiology* 10: 1404.

17 Verma, M.L. and Chandel, A. (Eds.). (2019). *Biotechnological production of bioactive compounds*. Elsevier.

18 Kraemer, S.A., Ramachandran, A., and Perron, G.G. (2019). Antibiotic pollution in the environment: from microbial ecology to public policy. *Microorganisms* 7 (6): 180.

19 Song, D., Ibrahim, S., and Hayek, S. (2012). Recent application of probiotics in food and agricultural science. *Probiotics* 10: 1–34.

20 Lobanovska, M. and Pilla, G. (2017). Focus: drug development: penicillin's discovery and antibiotic resistance: lessons for the future? *The Yale Journal of Biology and Medicine* 90 (1): 135.

21 Mantravadi, P.K., Kalesh, K.A., Dobson, R.C. et al. (2019). The quest for novel antimicrobial compounds: emerging trends in research, development, and technologies. *Antibiotics* 8 (1): 8.

22 Jozala, A.F., Geraldes, D.C., Tundisi, L.L. et al. (2016). Biopharmaceuticals from microorganisms: from production to purification. *Brazilian Journal of Microbiology* 47: 51–63.

23 Vieira Gomes, A.M., Souza Carmo, T., Silva Carvalho, L. et al. (2018). Comparison of yeasts as hosts for recombinant protein production. *Microorganisms* 6 (2): 38.

24 Du, J., Shao, Z., and Zhao, H. (2011). Engineering microbial factories for synthesis of value-added products. *Journal of Industrial Microbiology & Biotechnology* 38 (8): 873–890.

25 Deckers, M., Deforce, D., Fraiture, M.A., and Roosens, N.H. (2020). Genetically modified micro-organisms for industrial food enzyme production: an overview. *Foods* 9 (3): 326.

26 Acharya, T., Kennedy, R., Daar, A.S., and Singer, P.A. (2004). Biotechnology to improve health in developing countries: a review. *Memórias do Instituto Oswaldo Cruz* 99 (4): 341–350.

27 Singh, B.P., Rateb, M.E., Rodriguez-Couto, S. et al. (2019). Microbial secondary metabolites: recent developments and technological challenges. *Frontiers in Microbiology* 10: 914.

28 Gupta, V.K., Tuohy, M.G., O'Donovan, A., and Lohani, M. (Eds.). (2015). *Biotechnology of Bioactive Compounds: Sources and Applications*. Wiley.

29 Vieco-Saiz, N., Belguesmia, Y., Raspoet, R. et al. (2019). Benefits and inputs from lactic acid bacteria and their bacteriocins as alternatives to antibiotic growth promoters during food-animal production. *Frontiers in Microbiology* 10: 57.

30 Hug, J.J., Krug, D., and Müller, R. (2020). Bacteria as genetically programmable producers of bioactive natural products. *Nature Reviews Chemistry*: 1–22.

31 Adrio, J.L. and Demain, A.L. (2010). Recombinant organisms for production of industrial products. *Bioengineered Bugs* 1 (2): 116–131.

32 Thirumurugan, D., Cholarajan, A., Raja, S.S., and Vijayakumar, R. (2018). An introductory chapter: secondary metabolites. *Secondary Metabolites: Sources and Applications*: 1–21.

33 Singh, R., Kumar, M., Mittal, A., and Mehta, P.K. (2017). Microbial metabolites in nutrition, healthcare and agriculture. *3 Biotech* 7 (1): 15.

34 Syngai, G.G., Gopi, R., Bharali, R. et al. (2016). Probiotics-the versatile functional food ingredients. *Journal of Food Science and Technology* 53 (2): 921–933.

35 Indira, M., Venkateswarulu, T.C., Peele, K.A. et al. (2019). Bioactive molecules of probiotic bacteria and their mechanism of action: a review. *3 Biotech* 9 (8): 306.

36 Ibrar, M., Ullah, M.W., Manan, S. et al. (2020). Fungi from the extremes of life: an untapped treasure for bioactive compounds. *Applied Microbiology and Biotechnology*: 104 (7): 1–25.

37 Neifar, M., Maktouf, S., Ghorbel, R.E. et al. (2015). Extremophiles as source of novel bioactive compounds with industrial potential. In: *Biotechnology of Bioactive Compounds: Sources and Applications* (eds. Vijai Kumar Gupta and Maria G. Tuohy), 245–268. Hoboken: Wiley.

38 Stierle, A.A. and Stierle, D.B. (2015). Bioactive secondary metabolites produced by the fungal endophytes of conifers. *Natural Product Communications* 10 (10): 1671–1682.

39 Grządziel, J. and Gałązka, A. (2019). Fungal biodiversity of the most common types of polish soil in a long-term microplot experiment. *Frontiers in Microbiology* 10: 6.

40 Shazhni, J.A., Renu, A., and Vijayaraghavan, P. (2018). Insights of antidiabetic, anti-inflammatory and hepatoprotective properties of antimicrobial secondary metabolites of corm extract from Caladium x hortulanum. *Saudi Journal of Biological Sciences* 25 (8): 1755–1761.

41 Anjum, T., Azam, A., and Irum, W. (2012). Production of cyclosporine A by submerged fermentation from a local isolate of Penicillium fellutanum. *Indian Journal of Pharmaceutical Sciences* 74 (4): 372.

42 Barrios-González, J. and Miranda, R.U. (2010). Biotechnological production and applications of statins. *Applied Microbiology and Biotechnology* 85 (4): 869–883.

43 Baghban, R., Farajnia, S., Rajabibazl, M. et al. (2019). Yeast expression systems: overview and recent advances. *Molecular Biotechnology* 61 (5): 365–384.

44 Subramani, R. and Sipkema, D. (2019). Marine rare actinomycetes: a promising source of structurally diverse and unique novel natural products. *Marine Drugs* 17 (5): 249.

45 Hug, J.J., Bader, C.D., Remskar, M. et al. (2018). Concepts and methods to access novel antibiotics from actinomycetes. *Antibiotics* 7 (2): 44.

46 Wei, J., He, L., and Niu, G. (2018). Regulation of antibiotic biosynthesis in actinomycetes: perspectives and challenges. *Synthetic and Systems Biotechnology* 3 (4): 229–235.

47 Grasso, L.L., Martino, D.C., and Alduina, R. (2016). Production of antibacterial compounds from Actinomycetes. *Actinobacteria-Basics and Biotechnological Applications*: 177–198.

48 Zhao, Y., Li, G., Chen, Y., and Lu, Y. (2020). Challenges and advances in genome editing technologies in streptomyces. *Biomolecules* 10 (5): 734.

49 Baral, B., Akhgari, A., and Metsa-Ketela, M. (2018). Activation of microbial secondary metabolic pathways: avenues and challenges. *Synthetic and Systems Biotechnology* 3 (3): 163–178.

50 Barka, E.A., Vatsa, P., Sanchez, L. et al. (2016). Taxonomy, physiology, and natural products of Actinobacteria. *Microbiology and Molecular Biology Reviews* 80 (1): 1–43.

51 Harir, M., Bendif, H., Bellahcene, M. et al. (2018). Streptomyces secondary metabolites. *Basic Biology and Applications of Actinobacteria* 6: 99–122.

52 Takahashi, Y. and Nakashima, T. (2018). Actinomycetes, an inexhaustible source of naturally occurring antibiotics. *Antibiotics* 7 (2): 45.

53 Merino, N., Aronson, H.S., Bojanova, D.P. et al. (2019). Living at the extremes: extremophiles and the limits of life in a planetary context. *Frontiers in Microbiology* 10: 780.

54 Orellana, R., Macaya, C., Bravo, G. et al. (2018). Living at the frontiers of life: extremophiles in Chile and their potential for bioremediation. *Frontiers in Microbiology* 9: 2309.

55 Zhang, R., Bellenberg, S., Neu, T. R., Sand, W. and Vera, M. (2016) *Biotechnology of Extremophiles: Advances and Challenges*. Springer.

56 Jorquera, M.A., Graether, S.P., and Maruyama, F. (2019). Bioprospecting and biotechnology of extremophiles. *Frontiers in Bioengineering and Biotechnology* 7: 204.

57 Dumorné, K., Córdova, D.C., Astorga-Eló, M., and Renganathan, P. (2017). Extremozymes: a potential source for industrial applications. *Journal of Microbiology and Biotechnology* 27 (4): 649–659.

58 Rampelotto, P. H. (2013) Extremophiles and extreme environments.

59 Sarmiento, F., Peralta, R., and Blamey, J.M. (2015). Cold and hot extremozymes: industrial relevance and current trends. *Frontiers in Bioengineering and Biotechnology* 3: 148.

60 Shivlata, L. and Tulasi, S. (2015). Thermophilic and alkaliphilic Actinobacteria: biology and potential applications. *Frontiers in Microbiology* 6: 1014.

61 Cox, P. A. and King, S. (2013) *Bioprospecting*. Elsevier.

62 Vester, J.K., Glaring, M.A., and Stougaard, P. (2015). Improved cultivation and metagenomics as new tools for bioprospecting in cold environments. *Extremophiles* 19 (1): 17–29.

63 Abida, H., Ruchaud, S., Rios, L. et al. (2013). Bioprospecting marine plankton. *Marine Drugs* 11 (11): 4594–4611.

64 Stewart, E.J. (2012). Growing unculturable bacteria. *Journal of Bacteriology* 194 (16): 4151–4160.

65 Mahapatra, G.P., Raman, S., Nayak, S. et al. (2019). Metagenomics approaches in discovery and development of new bioactive compounds from marine actinomycetes. *Current Microbiology*: 1–12.

66 Kamble, A., Srinivasan, S., and Singh, H. (2019). In-silico bioprospecting: finding better enzymes. *Molecular Biotechnology* 61 (1): 53–59.

67 Rutledge, P.J. and Challis, G.L. (2015). Discovery of microbial natural products by activation of silent biosynthetic gene clusters. *Nature Reviews Microbiology* 13 (8): 509–523.

68 Milshteyn, A., Schneider, J.S., and Brady, S.F. (2014). Mining the metabiome: identifying novel natural products from microbial communities. *Chemistry & Biology* 21 (9): 1211–1223.

69 Ortega, A.M.M. and Campos, M.R.S. (2019). Bioactive compounds as therapeutic alternatives. In: *Bioactive Compounds* (ed. Maira Rubi Segura Campos), 247–264. Woodhead Publishing.

70 da Silva Sabo, S., Vitolo, M., González, J.M.D., and de Souza Oliveira, R.P. (2014). Overview of *Lactobacillus plantarum* as a promising bacteriocin producer among lactic acid bacteria. *Food Research International* 64: 527–536.

71 Babbar, N., Oberoi, H.S., and Sandhu, S.K. (2015). Therapeutic and nutraceutical potential of bioactive compounds extracted from fruit residues. *Critical Reviews in Food Science and Nutrition* 55 (3): 319–337.

72 Kyrou, I., Randeva, H. S., Tsigos, C., et al. (2018). Clinical problems caused by obesity. In *Endotext*. MDText. com, Inc.

73 Choudhury, H., Pandey, M., Hua, C.K. et al. (2018). An update on natural compounds in the remedy of diabetes mellitus: a systematic review. *Journal of Traditional and Complementary Medicine* 8 (3): 361–376.

74 Ventola, C.L. (2015). The antibiotic resistance crisis: part 1: causes and threats. *Pharmacy and Therapeutics* 40 (4): 277.

75 Santajit, S. and Indrawattana, N. (2016). Mechanisms of antimicrobial resistance in ESKAPE pathogens. *BioMed Research International* 2016: 1–8.

76 Chevrette, M.G. and Currie, C.R. (2019). Emerging evolutionary paradigms in antibiotic discovery. *Journal of Industrial Microbiology & Biotechnology* 46 (3-4): 257–271.

77 de Castro, A.P., Fernandes, G.D.R., and Franco, O.L. (2014). Insights into novel antimicrobial compounds and antibiotic resistance genes from soil metagenomes. *Frontiers in Microbiology* 5: 489.

78 Chen, R., Wong, H.L., and Burns, B.P. (2019). New approaches to detect biosynthetic gene clusters in the environment. *Medicines* 6 (1): 32.

79 Simons, A., Alhanout, K., and Duval, R.E. (2020). Bacteriocins, antimicrobial peptides from bacterial origin: overview of their biology and their impact against multidrug-resistant bacteria. *Microorganisms* 8 (5): 639.

80 Sankar, N.R., Priyanka, V.D., Reddy, P.S. et al. (2012). Purification and characterization of bacteriocin produced by Lactobacillus plantarum isolated from cow milk. *International Journal of Microbiology* 3 (2): 133–137.

81 Bali, V., Panesar, P.S., Bera, M.B., and Kennedy, J.F. (2016). Bacteriocins: recent trends and potential applications. *Critical Reviews in Food Science and Nutrition* 56 (5): 817–834.

82 Mikkili, I., John, B.D., Chaitanya, M., and Kodali, V.P. (2015). Isolation and characterization of bacteriocin producing lactic acid bacteria from curd. *International Journal of PharmTech Research* 8 (1): 388–396.

83 Silva, C.C., Silva, S.P., and Ribeiro, S.C. (2018). Application of bacteriocins and protective cultures in dairy food preservation. *Frontiers in Microbiology* 9: 594.

84 Yang, S.C., Lin, C.H., Sung, C.T., and Fang, J.Y. (2014). Antibacterial activities of bacteriocins: application in foods and pharmaceuticals. *Frontiers in Microbiology* 5: 241.

85 Indira, M., Venkateswarulu, T.C., Chakravarthy, K. et al. (2015). Isolation and characterization of bacteriocin producing lactic acid bacteria from diary effluent. *Research Journal of Pharmacy and Technology* 8 (11): 1560–1565.

86 Arakawa, K. (2019). Basic antibacterial assay to screen for bacteriocinogenic lactic acid bacteria and to elementarily characterize their bacteriocins. In: *Lactic Acid Bacteria* (ed. Makoto Kanauchi), 15–22. New York, NY: Humana Press.

87 Tietz, I.J. and Mitchell, A. (2016). Using genomics for natural product structure elucidation. *Current Topics in Medicinal Chemistry* 16 (15): 1645–1694.

88 Indira, M., Venkateswarulu, T.C., Peele, K.A. et al. (2019). Characterization of bacteriocin producing probiotic properties of Enterococcus casseliflavus MI001 isolated from curd sample. *Current Trends in Biotechnology and Pharmacy* 13 (1): 64–71.

89 Meade, E., Slattery, M.A., and Garvey, M. (2020). Bacteriocins, potent antimicrobial peptides and the fight against multi drug resistant species: resistance Is Futile? *Antibiotics* 9 (1): 32.

90 Santos, D.K.F., Rufino, R.D., Luna, J.M. et al. (2016). Biosurfactants: multifunctional biomolecules of the 21st century. *International Journal of Molecular Sciences* 17 (3): 401.

91 Nitschke, M. and Silva, S.S.E. (2018). Recent food applications of microbial surfactants. *Critical Reviews in Food Science and Nutrition* 58 (4): 631–638.

92 Alizadeh-Sani, M., Hamishehkar, H., Khezerlou, A. et al. (2018). Bioemulsifiers derived from microorganisms: Applications in the drug and food industry. *Advanced Pharmaceutical Bulletin* 8 (2): 191.

93 Karlapudi, A.P., Venkateswarulu, T.C., Tammineedi, J. et al. (2018). Role of biosurfactants in bioremediation of oil pollution-a review. *Petroleum* 4 (3): 241–249.

94 Olasanmi, I.O. and Thring, R.W. (2018). The role of biosurfactants in the continued drive for environmental sustainability. *Sustainability* 10 (12): 4817.

95 Le Guenic, S., Chaveriat, L., Lequart, V. et al. (2019). Renewable surfactants for biochemical applications and nanotechnology. *Journal of Surfactants and Detergents* 22 (1): 5–21.

96 Kambourova, M., Oner, E.T., and Poli, A. (2015). Exopolysaccharides from prokaryotic microorganisms – promising sources for white biotechnology processes. In: *Industrial Biorefineries & White Biotechnology* (eds. Ashok Pandey, Rainer Höfer, Mohammad Taherzadeh et al., 554), 523. Elsevier.

97 Schmid, J., Sieber, V., and Rehm, B. (2015). Bacterial exopolysaccharides: biosynthesis pathways and engineering strategies. *Frontiers in Microbiology* 6: 496.

98 Moscovici, M. (2015). Present and future medical applications of microbial exopolysaccharides. *Frontiers in Microbiology* 6: 1012.

99 Yildiz, H. and Karatas, N. (2018). Microbial exopolysaccharides: resources and bioactive properties. *Process Biochemistry* 72: 41–46.

100 Sujana, K., Abraham, K.P., Mikkili, I., and Prabhakar, K.V. (2013). Biochemical and molecular characterization of biofilm producing bacteria. *International Journal of Pharma and Bio Sciences* 4: 702–712.

101 Indira, M., Venkateswarulu, T.C., Chakravarthy, K. et al. (2016). Morphological and biochemical characterization of exopolysaccharide producing bacteria isolated from dairy effluent. *Journal of Pharmaceutical Sciences and Research* 8 (2): 88.

102 Nouha, K., Kumar, R.S., Balasubramanian, S., and Tyagi, R.D. (2018). Critical review of EPS production, synthesis and composition for sludge flocculation. *Journal of Environmental Sciences* 66: 225–245.

103 Zafar, S.B., Siddiqui, N.N., Shahid, F. et al. (2018). Bioprospecting of indigenous resources for the exploration of exopolysaccharide producing lactic acid bacteria. *Journal, Genetic Engineering & Biotechnology* 16 (1): 17–22.

104 Robinson, P.K. (2015). Enzymes: principles and biotechnological applications. *Essays in Biochemistry* 59: 1–41.

105 Adrio, J.L. and Demain, A.L. (2014). Microbial enzymes: tools for biotechnological processes. *Biomolecules* 4 (1): 117–139.

106 Singh, R., Kumar, M., Mittal, A., and Mehta, P.K. (2016). Microbial enzymes: industrial progress in 21st century. *3 Biotech* 6 (2): 174.

107 Straub, C.T., Counts, J.A., Nguyen, D.M. et al. (2018). Biotechnology of extremely thermophilic archaea. *FEMS Microbiology Reviews* 42 (5): 543–578.

108 Hamdan, A. (2018). Psychrophiles: ecological significance and potential industrial application. *South African Journal of Science* 114 (5–6): 1–6.

109 Alves, L.D.F., Westmann, C.A., Lovate, G.L. et al. (2018). Metagenomic approaches for understanding new concepts in microbial science. *International Journal of Genomics* 2018: 1–15.

110 Ufarté, L., Potocki-Veronese, G., and Laville, E. (2015). Discovery of new protein families and functions: new challenges in functional metagenomics for biotechnologies and microbial ecology. *Frontiers in Microbiology* 6: 563.

111 Littlechild, J.A. (2015). Enzymes from extreme environments and their industrial applications. *Frontiers in Bioengineering and Biotechnology* 3: 161.

112 Mohan, S., Oluwafemi, O.S., Kalarikkal, N. et al. (2016). Biopolymers–application in nanoscience and nanotechnology. *Recent Advances in Biopolymers* 1 (1): 47–66.

113 Wroblewska-Krepsztul, J., Rydzkowski, T., Michalska-Pożoga, I., and Thakur, V.K. (2019). Biopolymers for biomedical and pharmaceutical applications: recent advances and overview of alginate electrospinning. *Nanomaterials* 9 (3): 404.

114 Andreeßen, C. and Steinbüchel, A. (2019). Recent developments in non-biodegradable biopolymers: precursors, production processes, and future perspectives. *Applied Microbiology and Biotechnology* 103 (1): 143–157.

115 Ivanov, V. and Stabnikov, V. (2016). Basic concepts on biopolymers and biotechnological admixtures for eco-efficient construction materials. In: *Biopolymers and Biotech Admixtures for Eco-Efficient Construction Materials* (eds. Fernando Pacheco-Torgal, Volodymyr Ivanov, Niranjan Karak, et al.), 13–35. Wood head Publishing.

116 Angelini, S., Cerruti, P., Immirzi, B. et al. (2015). From microbial biopolymers to bioplastics: sustainable additives for PHB processing and stabilization. In: *Microbial factories* (ed. Vipin Chandra Kalia), 139–160. New Delhi: Springer.

117 Owusu, P.A. and Asumadu-Sarkodie, S. (2016). A review of renewable energy sources, sustainability issues and climate change mitigation. *Cogent Engineering* 3 (1): 1167990.

118 Robak, K. and Balcerek, M. (2018). Review of second generation bioethanol production from residual biomass. *Food Technology and Biotechnology* 56 (2): 174–187.

119 Sethi, S., Datta, A., Gupta, B.L., and Gupta, S. (2013). Optimization of cellulase production from bacteria isolated from soil. *International Scholarly Research Notices* 2013: 1–7.

120 Acharya, S. and Chaudhary, A. (2012). Bioprospecting thermophiles for cellulase production: a review. *Brazilian Journal of Microbiology* 43 (3): 844–856.

121 Chaudhary, N., Gupta, A., Gupta, S., and Sharma, V.K. (2017). BioFuelDB: a database and prediction server of enzymes involved in biofuels production. *PeerJ* 5: e3497.

122 Karpinski, T.M. and Adamczak, A. (2018). Anticancer activity of bacterial proteins and peptides. *Pharmaceutics* 10 (2): 54.

123 Sedighi, M., Zahedi Bialvaei, A., Hamblin, M.R. et al. (2019). Therapeutic bacteria to combat cancer; current advances, challenges, and opportunities. *Cancer Medicine* 8 (6): 3167–3181.

124 Song, S., Vuai, M.S., and Zhong, M. (2018). The role of bacteria in cancer therapy–enemies in the past, but allies at present. *Infectious Agents and Cancer* 13 (1): 1–7.

125 Khalifa, S.A., Elias, N., Farag, M.A. et al. (2019). Marine natural products: a source of novel anticancer drugs. *Marine Drugs* 17 (9): 491.

126 Piddock, L.J. (2015). Teixobactin, the first of a new class of antibiotics discovered by iChip technology? *Journal of Antimicrobial Chemotherapy* 70 (10): 2679–2680.

127 Greer, N.D. (2006). Tigecycline (Tygacil): the first in the glycylcycline class of antibiotics. *Baylor University Medical Center Proceedings* 19 (2): 155–161. Taylor & Francis.

128 Fogarty, C.M., Buchanan, P., Aubier, M. et al. (2006). Telithromycin in the treatment of pneumococcal community-acquired respiratory tract infections: a review. *International Journal of Infectious Diseases* 10 (2): 136–147.

129 Kitagawa, N., Otani, T., and Inai, T. (2019). Nisin, a food preservative produced by Lactococcus lactis, affects the localization pattern of intermediate filament protein in HaCaT cells. *Anatomical Science International* 94 (2): 163–171.

130 Amado, I.R., Fuciños, C., Fajardo, P., and Pastrana, L. (2016). Pediocin SA-1: a selective bacteriocin for controlling Listeria monocytogenes in maize silages. *Journal of Dairy Science* 99 (10): 8070–8080.

131 Mikkili, I., Venkateswarulu, T.C., Karlapudi, A.P. et al. (2019). Characterization of bacteriocin ABC transporter ATP-binding protein produced by a newly isolated Enterococcus casseliflavus MI001 strain. *Beni-Suef University Journal of Basic and Applied Sciences* 8 (1): 5.

132 Belguesmia, Y., Choiset, Y., Prévost, H. et al. (2010). Partial purification and characterization of the mode of action of enterocin S37: a bacteriocin produced by Enterococcus faecalis S37 isolated from poultry feces. *Journal of Environmental and Public Health*: 1–8.

133 Fagundes, P.C., de Farias, F.M., da Silva Santos, O.C. et al. (2016). The four-component aureocin A70 as a promising agent for food biopreservation. *International Journal of Food Microbiology* 237: 39–46.

134 Magalhaes, L. and Nitschke, M. (2013). Antimicrobial activity of rhamnolipids against Listeria monocytogenes and their synergistic interaction with nisin. *Food Control* 29 (1): 138–142.

135 Amani, H. and Kariminezhad, H. (2016). Study on emulsification of crude oil in water using emulsan biosurfactant for pipeline transportation. *Petroleum Science and Technology* 34 (3): 216–222.

136 Janek, T., Krasowska, A., Czyżnikowska, Ż., and Lukaszewicz, M. (2018). Trehalose lipid biosurfactant reduces adhesion of microbial pathogens to polystyrene and silicone surfaces: an experimental and computational approach. *Frontiers in Microbiology* 9: 2441.

137 Leung, A.D., Wong, K.H., and Tien, J. (2012). Plasma expanders stabilize human microvessels in microfluidic scaffolds. *Journal of Biomedical Materials Research Part A* 100 (7): 1815–1822.

138 Mortensen, A., Aguilar, F., Crebelli, R. et al. (2017). Re-evaluation of xanthan gum (E 415) as a food additive. *EFSA Journal* 15 (7): e04909.

139 Spadari, C.D.C., Lopes, L.B., and Ishida, K. (2017). Potential use of alginate-based carriers as antifungal delivery system. *Frontiers in Microbiology* 8: 97.

140 Litwiniuk, M., Krejner, A., Speyrer, M.S. et al. (2016). Hyaluronic acid in inflammation and tissue regeneration. *Wounds* 28 (3): 78–88.

141 Iurciuc, C.E., Savin, A., Lungu, C. et al. (2016). Gellan. Food applications. *Cellulose Chemistry and Technology*: 1–13.

142 Peng, X., Su, H., Mi, S., and Han, Y. (2016). A multifunctional thermophilic glycoside hydrolase from Caldicellulosiruptor owensensis with potential applications in production of biofuels and biochemicals. *Biotechnology for Biofuels* 9 (1): 98.

143 Razzaq, A., Shamsi, S., Ali, A. et al. (2019). Microbial proteases applications. *Frontiers in Bioengineering and Biotechnology* 7: 110.

144 Chen, G.Q. and Zhang, J. (2018). Microbial polyhydroxyalkanoates as medical implant biomaterials. *Artificial Cells, Nanomedicine, and Biotechnology* 46 (1): 1–18.

145 Ali, I. and Jamil, N. (2016). Polyhydroxyalkanoates: current applications in the medical field. *Frontiers in Biology* 11 (1): 19–27.

146 Muller, J., González-Martínez, C., and Chiralt, A. (2017). Combination of poly (lactic) acid and starch for biodegradable food packaging. *Materials* 10 (8): 952.

147 Huang, A., Jiang, Y., Napiwocki, B. et al. (2017). Fabrication of poly (ε-caprolactone) tissue engineering scaffolds with fibrillated and interconnected pores utilizing microcellular injection molding and polymer leaching. *RSC Advances* 7 (69): 43432–43444.

148 Arora, R., Behera, S., Sharma, N.K., and Kumar, S. (2015). Bioprospecting thermostable cellulosomes for efficient biofuel production from lignocellulosic biomass. *Bioresources and Bioprocessing* 2 (1): 1–12.

149 Sweeney, M.D. and Xu, F. (2012). Biomass converting enzymes as industrial biocatalysts for fuels and chemicals: recent developments. *Catalysts* 2 (2): 244–263.

150 Kollar, P., Rajchard, J., Balounová, Z., and Pazourek, J. (2014). Marine natural products: bryostatins in preclinical and clinical studies. *Pharmaceutical Biology* 52 (2): 237–242.

151 Alarif, W.M., Abdel-Lateff, A., Alorfi, H.S., and Alburae, N.A. (2019). Alcyonacea: a potential source for production of nitrogen-containing metabolites. *Molecules* 24 (2): 286.

9

Microbial Fuel Cells and Plant Microbial Fuel Cells to Degradation of Polluted Contaminants in Soil and Water

Chung-Yu Guan[1] *and Chang-Ping Yu*[2]

[1] *Department of Environmental Engineering, National Ilan University, Yilan City, Taiwan*
[2] *Graduate Institute of Environmental Engineering, National Taiwan University, Taipei, Taiwan*

9.1 Introduction

Bioelectrochemical technology integrates microorganisms and electrochemical systems to create green technology that produces power from wastewater, polluted soils, and sediment. A conventional microbial fuel cell (MFC) contains the anode chamber, cathode chamber, and membrane. The microbes are located at the anode to decompose organic matters to produce CO_2. The metabolic electrons are transferred to the anode and delivered to the cathode through the loop. Hydrogen ions are produced in the anode chamber through the proton exchange membrane and react with oxygen to generate H_2O in the cathode chamber [1, 2] (Figure 9.1). Plant microbial fuel cells (PMFCs) are a novel technology developed based on the above concept. This technology combines plants and bioelectrochemical systems to convert solar energy into electricity [3]. In PMFCs, plants absorb CO_2 via photosynthesis under sunlight and secrete root exudates. Microbes in soils decompose organic matter in the rhizosphere to generate electrons [4]. The electrons in the soils are transferred to the anode and then to the cathode through the circuit. The protons moving from the anode to the cathode react with oxygen to produce H_2O [5] (Figure 9.2).

Although created over 100 years ago, bioelectrochemical technology had not developed rapidly for a long time [6]. However, in these recent years, microbial electrochemical systems have rapidly changed and been utilized in the environmental engineering field. This treatment has some advantages of microbial and electrochemical technology. Compared with other methods, bioelectrochemical technology for environmental engineering areas can degrade pollutions under mild reactions and generate electricity [7]. These years, research about MFCs and PMFCs for treating environmental problems has increased. The mechanism of electrochemical reactions of electricigens through interaction with electrodes in bioelectrochemical systems has been determined. The materials of MFCs and PMFCs in anodes, cathodes, and membranes affect the current density and pollution removal rates. Furthermore, the configuration of MFCs and PMFCs affect power density and pollution removal rates.

Bioprospecting of Microorganism-Based Industrial Molecules, First Edition.
Edited by Sudhir P. Singh and Santosh Kumar Upadhyay.

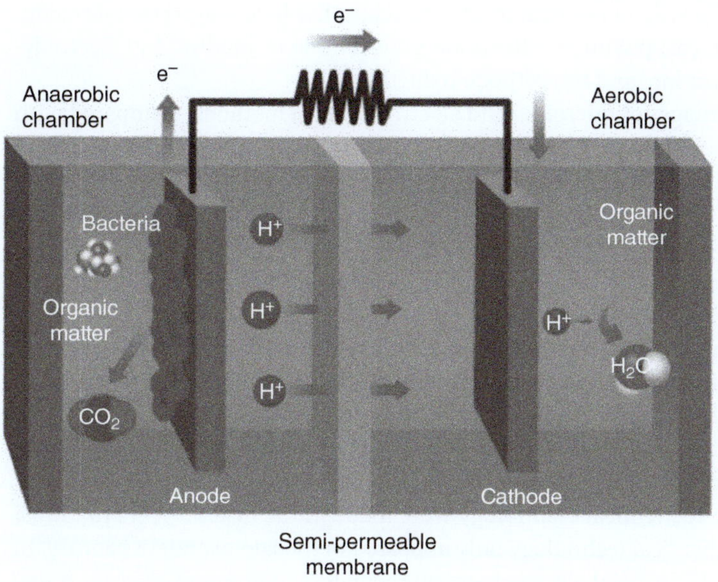

Figure 9.1 Schematic diagrams of MFCs. *Sources:* Based on [1, 2].

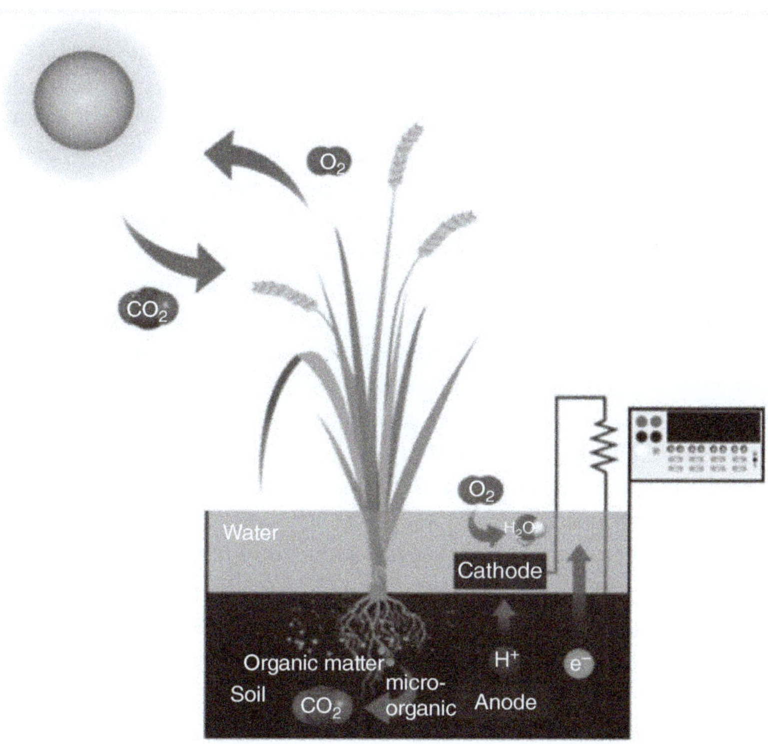

Figure 9.2 Schematic diagrams of PMFCs. *Source:* Based on [5].

The main mechanisms of pollutant removal by MFCs include the following: (i) anode oxidation, (ii) anode reduction, (iii) pollution adsorption on electrode or biofilm, (iv) electrodynamics, (v) cathode alkalization, and (vi) cathode reduction [8].

MFCs possess the advantages of microbial and electrochemical methods. Compared with other methods and technologies, MFCs consume less energy and generate more electricity by degrading contaminants.

9.2 History

MFCs and PMFCs are the new technologies that generate electricity and energy through the microbial activities of biodegradable organic compounds. In 1911, the British botanist, Potter, experimented with yeast and *Escherichia coli*. He used a platinum electrode placed in the bacterial culture fluid and discovered that microbes could generate electricity. This experiment is considered to be the first MFC in human history [9]. However, no major discoveries about bioelectrochemical technology were made in the next 55 years [6]. Most researches on bioelectrochemical technology only appeared in the late twentieth century. In the 1980s, researchers used artificial electrochemical mediators to promote electron transfer between microbes and electrodes. Some research teams added such electron mediators as thionine, phenothiazines, and neutral red to the anode to increase the power because the electron mediators were capable of receiving the electrons in bacteria cells and transferring the electrons to the electrodes. In 1985, Thruston's team used thionine as a redox mediator in dual-chamber MFCs and *Proteus vulgaris* as the catalyst to evaluate the coulombic yield produced by glucose oxidation [10]. In 1999, some researchers found that MFCs without mediators could also produce electricity, which caused a major breakthrough in the bioelectrochemical area. Kim and his team demonstrated that electrons could be transferred from bacterial cells to electrodes without adding exogenous mediators. In this first mediator-less MFCs using Fe(III), *Shewanella putrefaciens* IR-1's cell suspension generated electricity without redox mediators in the case of the lactate as the main carbon sources [11]. In 2003, another important of bioelectrochemical technology breakthrough was that *Geobacter sulferreducens*, *Geobacter meetallireducens*, and *Rhodoferax ferrireducens* were able to directly transfer electrons within the cell to the outside of the cell under the condition that no high current was produced. These microbes not only directly transferred the electrons inside the cell to the outside but were also grown on anodes to improve the stability of operation and coulombic efficiency [12–14]. Furthermore, microbes in the anode and cathode, such as *Klebsiella pneumoniae* and *Leptothrix discophora*, also impacted the electricity production of bioelectrochemical systems [15].

After media-free MFCs were discovered, researchers became more interested in bioelectrochemical areas, especially in wastewater treatment, because media-free MFCs provided a more practical and promising method for removing organic pollutions and generating electricity from wastewater systems by using different configurations and materials of MFCs [16–22]. In the future, MFC technology may be the solution to energy shortages and wastewater treatment. Therefore, the field of MFCs has attracted attention as a new green technology for alternative energy production and wastewater treatment.

Sediment MFCs are designed such that the cathode and anode are placed at the air–liquid phase junction and the sediment, respectively. The electricity generation occurs when the electron is generated from the decomposition of organic matters by sediment microbes and passed to the cathode through the external circuit. In 2006, some researchers applied sediment MFCs to marine meteorological monitoring stations to provide the trace power required for data transmission [23]. Reimers' team and Dumas' team set up sediment MFCs in the ocean and continuously operated for 125 and 18 days [24, 25]. Rezaei et al. set up single-chamber sediment MFCs with marine sediment [26]. In 2010, a team used lake sediment as the material to operate sediment MFCs under freshwater conditions [27].

Soils contain various microbes (electrogenic bacteria), sugars, and nutrients accumulated by the decay of animals and plants [28]. Soil is a huge power generation resource that is widely distributed and easily obtained. Therefore, the electricity generation is from the result of a combination of soils and MFCs. Since soil MFCs process redox reactions at the anode and cathode parts, people can apply soil MFCs to soil remediation. In 2011, Huang et al. used tubular air-diffusion anode MFCs for the remediation of phenol contaminated soils [29]. In 2014, Deng's team constructed a single-chamber soil MFC without a membrane for microbe activity in soils [30]. In 2015, Jian's team used soil MFCs to treat soil contaminated with Cd^{2+} [31].

Plants and bacteria play vital roles in converting solar energy into electricity. People can combine plants with bioelectrochemical systems to create PMFCs. The concept of PMFCs was originally proposed by De Schamphelaire and Strik's groups in 2008 [3, 32]. Their work showed the proof that plant rhizodeposits were used as substrates by microbes to produce electricity in PMFCs. In 2008, some researchers used rice in PMFCs for sustainable power production [32]. In 2010, Helder et al. evaluated bioelectricity and biomass production in several PMFCs using different plant species [4]. In 2013, Helder et al. applied PMFCs on the roof during winter [33]. In 2016, Habibul et al. removed Cr(VI) in the liquid phase using PMFCs [34]. In 2019, Guan's group used PMFCs with wetland plants in the remediation of soils contaminated with heavy metals [35]. Some researchers investigated polycyclic aromatic hydrocarbons (PAH) and oil degradation in soils using PMFCs [36].

9.3 Electricigens

As a biocatalyst of bioelectrochemical systems, electricigens play a key role in MFCs and PMFCs. At present, isolated electricigens mainly consist of Proteobacteria and Firmicutes. More and more studies have shown that the genus of electricigens in bioelectrochemical systems demonstrates diversified trends. Many electricigens occur in the natural environment, and many microorganisms with electricity generation functions are waiting to be discovered [37]. To better understand the differences and similarities of electricigens, it is necessary to summarize the genera electricigens information.

9.3.1 Electricigens of Bacteria

Geobacter is strictly anaerobic bacteria and only survive under anaerobic conditions. This kind of bacteria can be highly enriched on the anode of bioelectrochemical systems. The electrode can be used both as the acceptor and the electron donor in the redox system in the absence of an external electron acceptor. So far, the entire genome of the strain *Geobacter sulfurreducens* has been obtained, and then researchers have the work sequence of

Geobacter metallireducens genome, which will help obtain microorganisms with higher electricity production characteristics through the molecular method in the future [38, 39].

The *Shewanella* genus is a well-studied electricigens. They belong to the Progeobacteria, gram-negative bacteria, and have iron-reducing functions. Under aerobic conditions, these microbes can produce CO_2 in respiratory metabolism processes. Under anaerobic conditions, they can obtain the nutrients necessary for survival.

In addition to the two major genera of *Shewanella* and *Geobacter*, electricigens are also present in *Citrobacter*, *Clostridium*, *Enterobacter*, *Geothrix*, *Klebsiella*, *Ochrobactrum*, *Pseudomonas*, *Rhodoferax*, etc. It summarizes the research about electrocigens of bacteria (Table 9.1).

Table 9.1 Electrocigens and functions.

Kingdoms	Genus	Microorganisms	Functions
Bacteria	*Geobacter*	*Geobacter metallireducens*	Metal reduction, reduce pollutants, directly transfer electrons between electrodes [40]
		Geobacter sulfurreducens	Good electricity production [41]
	Shewanella	*Shewanella decolorationis* S12	Using Fe(III) in anaerobic conditions [42]
		Shewanella japonica	The biosynthesized extracellular mediators for electron transfer [43]
		Shewanella oneidensis DSP10	Production in an aerobic environment, utilize organic matters, expanding the substrate utilization range through domestication and screening [44]
	Citrobacter	*Citrobacter* sp. LAR-1	High iron-reducing strain [45]
	Clostridium	*Clostridium butyricum* EG3	Decomposing and oxidizing organic matters under anaerobic conditions [46]
	Geothrix	*Geothrix fermentans*	Decomposing substrates for electricity generation [47]
	Klebisella	*K.* sp. IR21	The dissimilatory iron-reducing bacteria to produce electricity without artificial medium [48]
	Ochrobactrum	*Ochrobactrum anthropic* YZ-1	A pathogenic bacterium with good electricity generation efficiency [49]
	Pseudomonas	*Pseudomonas aeruginosa*	The earliest discovered microorganism that can produce an electron shuttle mediator [50]
	Rhodoferax	*Rhodoferax ferrireducens*	Completely oxidize and decomposed glucose to produce electricity [13]
	Tolumonas	*Tolumonas osonensis* P2-A-1	The maximum power density is 424 mV/m^2 [51]
Fungi	*Arxula*	*Arxula adeninivorans*	Secreting extracellular redox molecular and an ideal biocatalyst for MFCs [52]
	Pichia	*Pichia anomala*	Directly transfer electrons to the electrode through electrochemically enzymes [53]
	Saccharomyces	*Saccharomyces cerevisiae*	High vitality, wide distribution, easy access, and environmental friendly [54]

9.3.2 Electrocigens of Fungi

In the twenty-first century, some researchers have begun to study eukaryotic organisms with electricity-generating properties in bioelectrochemical systems. Because the structure of eukaryotic cells is complicated, the fungal electricigens isolated from MFC systems are far fewer than prokaryotes [55]. However, yeasts have certain advantages for developing bioelectrochemical technology: first, yeast cells are easy to control and can decompose complex organic substances, such as starch and cellulose, and, second, yeast can survive under anaerobic conditions because anode chambers are mostly anaerobic or a facultative anaerobic environment. Therefore, the isolation and purification of yeast fungi with electrochemical activity have a far-reaching significance for the development of bioelectrochemical technology [56, 57]. The literature Table 9.1 includes some reports of fungi electricigens.

9.4 Electron Generation and Transfer Mechanisms of Electricigens

In the catalytic oxidation processes of microbes, electricigens decomposed organics in an anode chamber to generate electrons and protons. The rate at which electrons are transferred from microbial cells to the anode surface is closely related to the electricity generation capacity of the bioelectrochemical systems. Research on the mechanisms of electron generation and electron transfer by electrocigens is of great significance in improving the electricity production of MFCs and PMFCs. The following summarizes the main mechanisms of current research, which facilitates the understanding of the operating principles of bioelectrochemical technology, as well as lays the theoretical foundation for related research.

9.4.1 Electron Generation Mechanism

In MFCs and PMFCs, the metabolic reaction of electricigens is closely related to the generation and transfer of electrons. Electricigens in anode parts oxidize and decompose organic matter to generate electrons. The bio-oxidation process of electricigens consists of three main stages. First, organic matter is decomposed under the action of microbial dehydrogenase to generate electrons. The electrons are transferred to the cathode through the electron transfer chain by the action of multiple electron carriers. Then, the electrons interact with the final electron acceptor to form a complete route for producing electricity. At the same time, electricigens obtain energy to maintain their own growth [58].

9.4.2 Electron Transfer Mechanism

Research on the electron transfer mechanisms of electricigens mainly relies on such electrochemical techniques as cyclic voltammetry and chronoamperometry. Furthermore, research on the electron transfer mechanism is evaluated by the molecular biology method. For the entire genome of genus *Shewanella* and *Geobacter*, the mechanism of electron transfer of genus electricigens has been primarily studied. The biofilm mechanism and the electron shuttle mechanism are currently the most studied electron transfer mechanisms [59]. With a greater understanding of the whole genome of microbes, more microbial electron transfer mechanisms will be discovered in the future.

9.4.3 Biofilm Mechanism

Microorganisms can concentrate on the anode surface of bioelectrochemical systems to form a thin biofilm, which can directly transfer electrodes through the nanowire or cell surface without an electron intermediary. The biofilm mechanism has two main transmission methods: (i) Direct transmission means that microorganisms are in direct contact with the anode surface. The strain S. putrefactions IR-1 is the first reported electricigen capable of direct electron transfer. Electrochemically active biofilms on the anode surface can promote direct electron transfer [60]. Some researchers have found that the internal resistance of electroactive biofilms is also significantly different at growth stages. The appropriate methods at different stages of the biofilm reduce its internal resistance to improve the electricity generation of bioelectrochemical systems [61]. (ii) Nanowire transmission means that microbes can synthesize fimbriae or flagella with conductivity themselves. Electrons can indirectly contact the anode through nanowire to carry them over long distances. Microorganisms require more energy to synthesize nanowires, so adding too much substrate may affect the performance of the bioelectrochemical technology. Reguera et al. found that strain *G. sulfurreducens* produced nanowires similar to fimbriae during the reduction of trivalent iron and has very good electrical conductivity [62]. The bacteria can secrete a large amount of organic and inorganic matters under the anoxic environment. The bacteria's nanowire was initially considered to be based on the fimbriae structure, but further research proved that *Shewanella oneidensis* did not produce fimbriae. Its "nanowire" was an extension of the cell's outer membrane and periplasm [63].

9.4.4 Electron Shuttle Mechanism

The electron shuttle mechanism refers to microorganisms using some catalytic, electronic intermediaries, which are exogenous intermediaries, some small molecules, or substances for transferring electrons. In comparison with the biofilm mechanism, this mechanism is unlimited by distance and does not need more energy to synthesize the nanowire. The electronic shuttle mechanism has mediator electron transfer of exogenous mediators, microbial secondary metabolites for electron transfer, and oxidative transfer of reduced primary metabolites.

9.4.5 Electron Transfer by Exogenous Mediators

Exogenous electron shuttles in bioelectrochemical technology are quinones and humus. Humus is a natural organic mixture widely distributed in soils. Because humus enriches quinone groups and electrochemically active substances, it has been a popular research topic in recent years. Although humus has significant advantages over other external electron shuttles, the structure of humus and external environmental factors affect the capacity of electron transfer [64].

9.4.6 Microbial Secondary Metabolites for Electron Transfer

Electrocigens can secrete such secondary metabolites such as pyocyanin, bacterial phenazines, and riboflavin during electricity production. These substances have redox properties for promoting the electricity generation of bioelectrochemical technology [65]. This study found that microorganisms can secrete multiple secondary metabolites. These different secondary metabolites play a synergistic role to enhance the electron transfer efficiency and improve the power generation of bioelectrochemical systems [66].

9.4.7 Oxidation of Reduced Primary Metabolites

One type of microorganism uses H_2, H_2S, and other gases released from metabolism as redox mediators for the electron transfer process between electrodes [67]. For example, Schröder et al. used *E. coli* K12 to generate H_2, and then H_2 was oxidized on the platinum-coated catalytic electrode to obtain a maximum current of 150 mA. Though early in the research, it is preliminarily speculated that H_2 plays a vital role in an electronic shuttle in the subsequent power generation process [68].

9.5 Materials

Bioelectrochemical systems are generally composed of an anode, cathode, or PEM. These materials vary in different constructions and play vital roles in the cost and performance of MFCs and PMFCs.

9.5.1 Anode Materials

Electricity generation, electron migration rate, and internal resistance all affect the performance of biotechnology systems. The anode of MFCs and PMFCs is the accumulation sites of electricigens and the intermediary of electron transfer. It is closely related to the electron transfer, polarization internal resistance, and electrode surface reaction. An anode considered to have higher performance anode has high conductivity, large specific surface area, biological affinity, low cost, hydrophilicity, and corrosion resistance. Great anode materials can reduce internal polarization resistance and provide active sites for electrochemical reactions. So far, improving anode materials is based on traditional base materials with surface decoration and modification.

9.5.2 Base Materials of the Anode

Carbon-based materials with different configurations were the earliest and the most widely used in bioelectrochemical systems. Their advantages include high conductivity, bio-affinity, and low cost. Planar carbon-based anodes are sheet materials, carbon fiber yarn, and carbon cloth. Three-dimensional materials include carbon brushes, graphite rods, granular activated carbon (GACR), and foamed carbon [69–71]. Although carbon-based materials contribute to the accumulation of electricigens, their high energy status on the anode surface loses electrons and becomes reductive, which hinders electron migration. However, metal-based electrodes have higher catalytic oxidation and conductivity to improve the performance of bioelectrochemical technology [41].

9.5.3 The Modification of Anode Materials

Every base material of anodes has limitations. Although the development of new materials can fundamentally change anodes' performance, such work takes considerable time and effort. Surface decoration of the anode in MFCs can change the performance of those material surfaces, while the surface decoration of MFC materials can change the performance of those materials. While improving the adhesion of electricigens, it also reduces the overpotential of anode chamber reaction and promotes electron migration to improve MFC performance. The nitrogen treatment changes the surface morphology of the anode material to increase hydrophilic functional groups, such as hydroxyl groups, carboxyl groups, or nitrogen-containing functional groups. This method improves the biocompatibility of the anode surface and reduces the impedance to improve biotechnology system performance [72]. Ammonia and amine groups are the most common compounds for nitrogen treatment [73, 74]. The chemical oxidation method increases the specific surface area of the base material and attaches hydrophilic functional groups to the anode surface to improve the anode's electrochemical performance [75]. The chemical oxidation method includes inorganic acid, an inorganic base, and inorganic salt treatment. Traditional chemical oxidation combined with high-temperature heat treatment can significantly improve the anode efficiency [76, 77]. This method is currently a hotspot for anode surface decoration. In the electrochemical oxidation treatment, three types of electrolytes, acid, alkali, and salt, are often used to add functional groups on the surface of the anode materials to promote the formation of peptide bonds for accelerating the electronic coupling between electricigens and the anode. Some research has demonstrated that strong oxidizing substances for modifying the anode base material can effectively improve the power generation of biotechnology systems. Compared with the nitrogen electrochemical oxidation method, it improves the effective area and biocompatibility of the anode surface to promote e-donor electron coupling and electron migration [78]. Meanwhile, using high-temperature heat treatment can effectively improve the anode surface decoration.

To change the physical and chemical properties, the anode surface was modified to improve electricigens accumulation and enhance the electron transfer rate. More and more research groups have been using high-performance materials to modify the anode surface for combining the base material and modified materials, such as nanocarbon materials, conductive polymers, and metal oxides. Nanocarbon materials not only inherit the advantages of carbon materials, such as electrical conductivity, biocompatibility, and chemical stability [79], but also have a large surface area, effective space, and empty structure for more contact sites with extracellular electron transfer [80]. Electron transfer of anode microorganisms to electrodes needs to cross the energy barrier. Due to the relatively stable electron energy state of the dominant bacterial in the anode, the anode activation overpotential is mainly affected by the anode surface energy state [81]. The surface energy states of traditional carbon-based materials severely limit the performance of bioelectrochemical technology. Therefore, more and more researchers try to modify metals or metal ions to the surface of carbon-based materials to improve bioelectrochemical system performance [82]. Conductive polymers have the biological affinity, the binding properties of metal and organic polymers, and electrochemical redox properties compared with nanocarbon materials and metal oxides. Therefore, using the single high-performance materials to modify anode materials can improve the biocompatibility, electron transfer rate, and redox activity [83, 84]. However, all materials have their advantages and disadvantages. To maximize the advantages of modification methods, more and more researchers are using two or more modified materials to improve the performance of bioelectrochemical technology.

9.5.4 Cathode Materials

The cathode is a part of bioelectrochemical systems, and its cost accounts for most of the entire system. The cathode catalyst performance directly determines the efficiency of electricity generation. Since they are irreversible reacting processes, the cathode material has a significant impact on the performance of bioelectrochemical technology [85]. Therefore, the low cost and high-performance materials are critical to developing bioelectrochemical technology. This study introduces the cathode substrate materials and catalysts. Cathode substrate materials contain carbon-based materials and non-carbon-based materials. Cathode catalysts include precious metals, non-precious metals, transition metals, and conductive polymer catalysts. The physical properties, catalytic activity, and redox reactions all affect the power generation of bioelectrochemical systems.

9.5.5 Carbon-Based Materials of Cathodes

In recent years, many studies have found that the performance of MFCs is influenced by the cathode's material properties, such as conductivity, oxygen reduction rate, catalytic performance, and stability [86]. Most of the cathode materials are carbon materials, including carbon fiber brushes, carbon felts, carbon paper, and graphite rods. The advantages of carbon-based materials include good electrical conductivity, low cost, and industrialization. Comparing graphite, carbon cloth, and stainless steel mesh as cathode materials, researchers found that MFCs using graphite as the cathode material have the highest current density, power density, and Coulomb efficiency [87]. In the aeration of algae, the power density of MFCs using a carbon fiber brush as the cathode was 30 mW/m^2, higher than a pure carbon cloth [88]. Although carbon-based materials have many advantages, they are limited in applications due to their respective disadvantages, such as high cost for carbon cloth, low durability for carbon paper, high resistance for carbon felts, and clogged properties for carbon fiber brushes.

9.5.6 Non-Carbon-Based Materials

Some metal materials have a low cost, stable properties, and low toxicity for use as cathode materials of bioelectrochemical systems, such as nickel foam, stainless steel, and other metals [88]. The high porosity, three-dimensional structure, and high electrical conductivity of nickel foam reduce the internal resistance of MFCs and improve the utilization rate of the catalyst. Due to the high corrosiveness of nickel foam, some researchers have used polytetrafluoroethylene (PTFE) and activated carbon-modified nickel foam as the cathode materials of MFCs. It was found that the power density of MFCs using nickel foam as the cathode reached $1190 \pm 50\ mW/m^2$, which is equivalent to the power density of MFCs with carbon cloth as the cathode materials [89]. Stainless steel mesh has been used in MFCs as a gas diffusion electrode with good conductivity, chemical stability, and low cost [90]. Some researchers found that Co_3O_4 particle-modified stainless steel mesh cathode has high electrocatalysis, selectivity, and stability. Using this material as the cathode improves the redox reaction of MFCs [91]. The maximum power density of MFCs reached 2676 mW/m^2 utilizing carbon nanotubes (CNT), MnO_2, polymethylphenylsiloxane-modified stainless steel mesh as the cathode of MFCs [92].

9.5.7 Cathode Catalyst

Platinum with an empty d orbital can adsorb various charged substances to form active substances and promote the reaction. It can reduce reaction activation energy and increase the reaction rate in the cathode. Although platinum is often used as the cathode catalyst of bioelectrochemical systems, its cost is an obstacle to development and application [93]. Therefore, researching a catalyst with high catalytic activity, low cost, and stable chemical properties is the hotspot [94]. Quan et al. used Pt/Pd alloy as a catalyst and the maximum power density of 1274 mV/m^2, which was equivalent to the power density produced by using Pt/C as a catalyst under the same conditions. Therefore, Pd is feasible as a replacement for Pt as a redox catalyst in cheaper MFCs due to its high catalytic activity and lower cost than Pt [95].

Due to the wide source and low cost of transitional metal oxides, they have been used as the cathode catalyst of bioelectrochemical systems. MnO_2 and TiO_2 are two transition metal oxides that can be used as the cathode catalyst in MFCs. Compared to Pt, MnO_2 with a manganese–potassium octahedral molecular sieve structure had high conductivity, low cost, and a large specific surface area [96]. MFCs that use a stainless steel mesh as the cathode loaded with MnO_2/CNT composite had a higher output voltage and power density of MFCs with no catalyst cathode [97]. TiO_2 has a low cost, stable properties, and strong redox reactions. Its photocatalytic performance in MFCs increases the migration of cathode electrons [98]. Hosseini et al. used Pd-TiO_2 as the cathode catalyst to test MFCs through cyclic voltammetry and electrochemical impedance spectroscopy. They found that the Pd-TiO_2 nanostructured electrode with a high specific surface area is a cathode catalyst material that can reduce dissolved oxygen in water for long-term operation [99].

Conductive polymers like polyaniline (PANI) and polypypyrrole (PPy) are widely used as MFC cathode catalysts due to their easy synthesis, good conductivity, and stability. Ghoreishi et al. used V205-PANI composite as the cathode catalyst to obtain a maximum power density of 79 mW/m^2, higher than that of pure V205 and pure PANI as the cathode catalyst [100]. Esmaeili et al. used PPy-carrageenan nano (KC) composite as the cathode catalyst to obtain a maximum power density of 72.1 mV/m^2, higher than the maximum power density of 28.8 mW/m^2 using pure PPy as the cathode catalyst [101]. CNT has a large specific surface area, high catalytic activity, good conductivity, good corrosion resistance, high mechanical strength, and ductility. It is commonly used as a cathode catalyst in MFCs [102]. Because the hydroxyl groups on the surface of MWNTs can increase the chemical reactivity of metal nanoparticles, the electricity generation of a single-chamber MFC composed of electrodes modified with MWNTs is much better than one with unmodified electrodes [103]. Since the Mn-PPy-CNT composite prepared by solvothermal methods has redox catalytic performance in the neutral electrolyte as a cathode catalyst, the active site of Mn-N of a cathode demonstrates long-term stability [104]. Zhang et al. used CNT coated with MnO_2 as the cathode catalyst. MnO_2 on the surface of CNT increased the catalytic activity of the catalyst and the electron transfer rate of the redox reaction [105]. However, large-scale application is still limited due to the high cost and complicated preparation processes.

9.5.8 Biocathode

Biocathode is a new method that utilizes microbes as the catalyst to accept electrons for cathode reduction without the use of precious metals or non-precious metals as the catalyst. Compared with chemical cathodes, biological cathodes have the following advantages: improving the stability of MFCs, reducing the operating cost of MFCs, avoiding catalyst poisoning problems, no chemical catalyst deactivation, microbes grow in cathode chamber for wastewater treatment, and mild processes in the reaction systems. Electron transfer mechanisms in the MFC biocathode have two main types: direct and indirect electron transfer mechanisms [106]. Some researchers modified the biocathode with PANI and its copolymer for biocathode application. After coating the conductive polymer, the cathode biofilm became thicker. The leading microorganism on the cathode biofilm changed from β-proteobacteria before modification to α, γ-proteobacteria afterward [107].

9.5.9 Separator Materials

Partition materials isolate the reduction and oxidation reactions in the cathode and anode chamber, respectively. The ions in bioelectrochemical systems can pass this material, which prevents oxygen going from the cathode to the anode chamber. The design of separator materials will affect the performance of MFCs or PMFCs.

9.5.9.1 Conventional Separator Materials

The ideal separator materials in bioelectrochemical technology can improve the power density, coulombic efficiency, and pollution removal rate. Furthermore, the research and development of separator materials is the core technology for the design of the performance of bioelectrochemical systems. Currently, the most used separator materials in MFCs include proton exchange membranes [108], cation exchange membranes [109], anion exchange membranes [110], bipolar membranes [111], and ultrafiltration membranes [112].

9.5.9.2 New Separator Materials

In recent years, new separation materials with high proton conductivity, bio-fouling, and high water absorption have been developed to reduce internal resistance and costs and improve output voltage, output power, and treatment efficiency. Therefore, the research and development of new separation materials is key to improving the efficiency of bioelectrochemical systems. Sulfonated polyether ketone (SPEEK) is a polyether ether ketone (PEEK) membrane regenerated after high concentration H_2SO_4 sulfonation [113]. SPEEK membranes are becoming increasingly popular and are replacing Nafion membranes in MFCs because of its high proton conductivity, simple production, high chemical stability, and low price. The polyvinylidene fluoride (PVDF) membrane production method is simple and easy to merge with other materials for modification. The high-porosity PVDF membrane with Nafion added under electrospinning improves the dispersion of Nafion to increase proton conductivity. After PVDF is combined with other polymers, the ion exchange capacity, water absorption, and internal resistance are optimized to increase the power density of MFCs [114]. Polyvinyl alcohol (PVA) can be dissolved in hot water, so the formation method is simple and has antibacterial properties [115]. Compared with other membranes, the PVA membrane has a low cost, a simple film formation process, and easy merging properties with other materials [116]. Agar and chitosan are natural polymer substances extracted from the shells of plants and animals, so this resource is abundant in the natural environment. The membrane properties are easy to process and synthesize [117, 118]. Ionic liquids are salts composed of organic cations or inorganic anions in the liquid phase at 100 °C. The ionic liquid membrane has high ion conductivity and has been widely used in the field of electrochemistry. The ionic liquid is composed of anions and cations, so the conduction efficiency rate increases as the ionic concentration increases [119]. Clay is widely distributed, is rich in reserves, has a low cost, and is easy to obtain, with high mechanical strength, acid and alkali resistance, and high chemical and thermal stability [120]. The ion selectivity and proton conductivity of clay membranes are low and cannot inhibit the deposition of microorganisms. This material is brittle and not easily deformed, resulting in the low efficiency of MFCs [121]. In the future, natural clay materials can be mixed with other substrates to change the hydrophilicity and hydrophobicity and increase the proton selectivity to achieve higher output voltage.

9.6 Design and Operation of Bioelectrochemical Systems

9.6.1 MFC Configuration

9.6.1.1 Two-Compartment MFCs

Two-chamber MFCs are composed of a cathode chamber, anode chamber, and ion exchange membrane and the most popular devices for detecting optimizing conditions and microbial activity. H-type MFCs and U-shaped MFCs are widely used and low cost [122, 123] (Figure 9.3a,b). For high internal resistance and electrode losses, H-type MFCs have lower power density. Some researchers developed the miniature configuration of MFCs (Mini-MFCs). The mini-MFCs generate high power density because of graphite felt electrodes' high ratios of the surface area and volume [124] (Figure 9.3c). Furthermore, some researchers have adopted flat-plate MFCs (FPMFCs) to treat wastewater (Figure 9.3d). This configuration possessed higher power density because wastewater and dry air continuously flow through the anode and cathode chamber, respectively [18]. Another reactor design is up-flow MFCs (UMFC), which use the interior cathode to achieve continuous treatment and improve performance [21] (Figure 9.3e). Based on the above configuration, another UMFCs

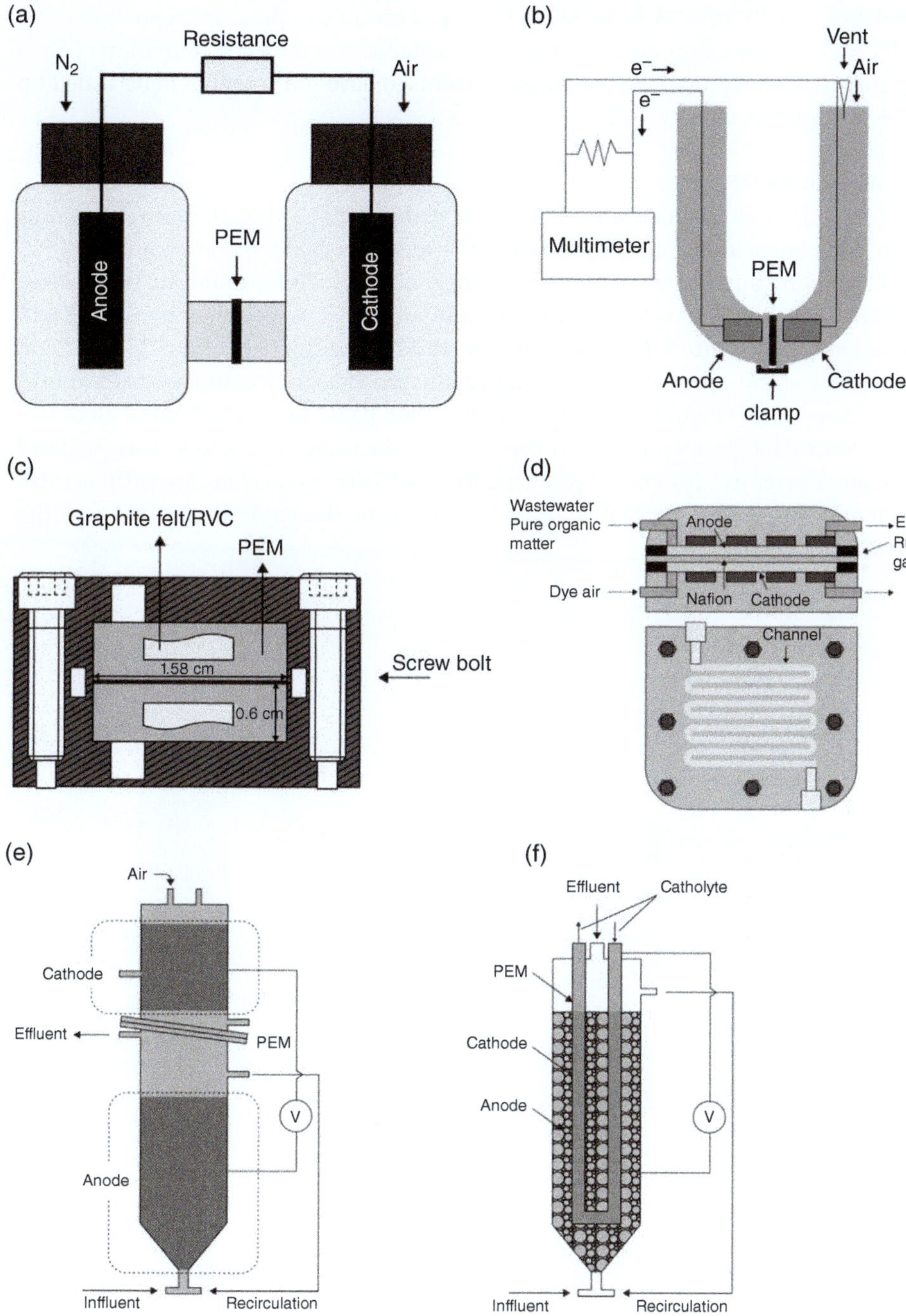

Figure 9.3 Schematic of typical dual-chamber MFCs: (a) H-type MFCs. *Source:* Based on [122]; (b) U-type MFCs. *Source:* Milliken and May [123] / with permission of Springer Nature; (c) mini-MFCs. *Source:* Modified from [124]; (d) flat-plate MFCs; (e) upflow MFCs. *Source:* He et al. [21] / with permission of American Chemical Society; (f) upflow MFCs with U-shaped cathode. *Source:* He et al. [125] / with permission of American Chemical Society.

was developed, which features an installed U-shaped cathode in the anode chamber [125] (Fig. 9.3f). This system mentioned that internal resistance was the main limitation of electricity generation. However, this configuration seems to have the potential to be scaled up for practical implementation.

9.6.1.2 Air Cathode MFC

The structure of an air cathode MFC is a single-compartment MFC that only possesses an anode chamber without an air chamber. Liu et al. provided single-chamber MFCs with eight graphite anode electrodes with a single air cathode for domestic wastewater [16] (Figure 9.4a). Some researchers designed a tubular MFC with an outer cathode and an inner anode [19] (Figure 9.4b). Removing the PEM of the single-chamber MFCs could increase the power output. Liu et al. developed air cathode MFCs equipped with a cathode and an anode [23] (Figure 9.4c), using rigid carbon paper or a carbon-cloth electrode with a proton exchange membrane as the cathode electrode. The anode was made of carbon paper. Therefore, despite the oxygen diffusion to the anode chamber with no PEM system, it increases the proton flux from the anode to the cathode and reduces the cost of PEM.

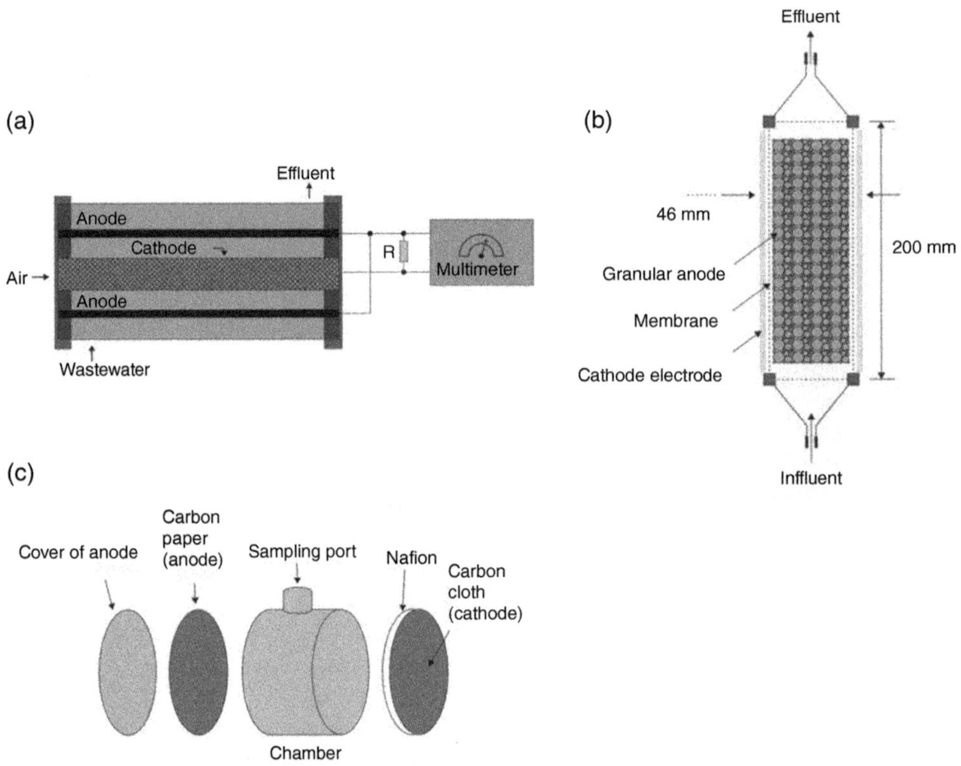

Figure 9.4 Schematics of type single-chamber MFCs: (a) eight graphite anode electrodes with air cathode MFCs. *Source:* Liu et al. [16] / with permission of American Chemical Society; (b) tubular MFCs. *Source:* Based on [19]; (c) lab-scale single-chamber MFCs. *Source:* Based on [23].

9.6.1.3 Other Configurations

These basic designs are modified to achieve different purposes, such as power density, continuous flow, and nutrient removal. Stacked MFCs in series can improve the output voltage [20]. Perforated baffles instead of PEM in MFCs achieve a continuous flow mode for anoxic/oxic wastewater treatment [126]. By installing MFCs inside an algae bioreactor, this system realizes the removal of organic matter and nutrients from the synthetic solution [127].

9.6.2 Soil MFC and PMFC Configurations

The configuration of soil MFCs includes a dual chamber, a single chamber, air-diffusion cathode systems, and other configurations.

9.6.2.1 Dual-Chamber of Soil MFCs and PMFCs

Dual-chambers have three parts: one cathode chamber, one anode chamber, and the proton exchange membrane, which functions as separation (Figure 9.5a). Soils are installed in the anode chamber and maintain a high moisture content. Soil microbes decompose organic matters to produce protons and electrons. Afterward, protons and electrons reach the cathode through the wire and ion exchange membrane for the reduction reaction. Through these processes, soil MFCs generate electricity. Because the ion exchange membrane prevents the electron acceptor of the cathode chamber from diffusing into the anode chamber, dual-chamber systems are beneficial for maintaining the anaerobic environment of the anode chamber.

This PMFC structure originally adopted by Strik et al. was the systems of dual-chamber MFCs with a cation exchange membrane that separated the anode and cathode compartments [3] (Figure 9.5b). Furthermore, researchers have used multiple materials in PMFCs, such as carbon felt for electrode materials, graphite particles for supporting

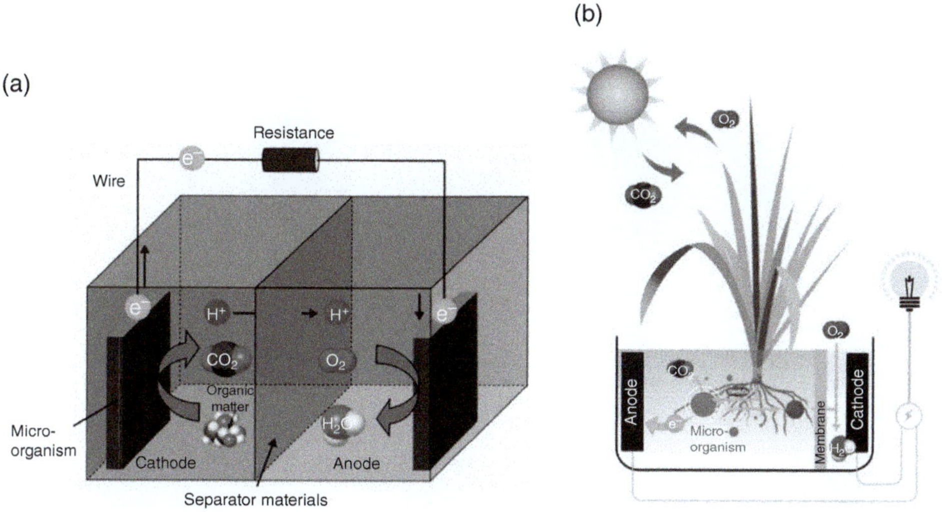

Figure 9.5 Schematics of dual-chamber of soil MFCs and PMFCs: (a) soil MFCs; (b) PMFCs. *Source:* Modified from [3].

materials for plant growth and current collection, and iron hydride or oxygen for electron acceptors. This configuration has been adopted in my studies to evaluate PMFC mechanisms.

9.6.2.2 Single-Chamber MFCs

The anode and cathode are buried in the flooded soils and immersed in the water layer [128] (Figure 9.6a). After the electron and hydrogen ion both reach the cathode with the dissolved O_2 in water to produce H_2O, soil MFCs generate electricity through the reduction reaction. Single-chamber systems have no ion exchange membrane, so the cost is lower than the dual-chamber system. Furthermore, this system uses O_2 as the electron acceptor, thus being a green and sustainable electricity production system. However, compared to the dual-chamber system, this configuration has a higher internal resistance and lower anaerobic effect because it relies on soils to block O_2 in water to the anode parts.

Since many PMFC studies have been performed based on sediment MFCs, this PMFC configuration is the same as the sediment MFCs, which have no separation between the cathode and anode compartments (Figure 9.6b). The anode is generally located at the plant roots to favor collecting electrons, while the cathode is located at the water vapor interface or soil surface to favor the use of oxygen as an electron acceptor [33, 128].

9.6.2.3 Air-Diffusion Cathode System

One side of an anode chamber is a cation exchange membrane, and the other side is an air-diffusion cathode, using O_2 in the air as the electron acceptor (Figure 9.7). Since the cathode is attached to the ion exchange membrane, the distance between the electrodes is shorter, and the internal resistance is low. The ion exchange membrane is suitable for maintaining the anaerobic condition of the anode chamber. In some applications and research of the air-diffusion cathode system, the tube of this system improves the utilization efficiency of the membrane [129].

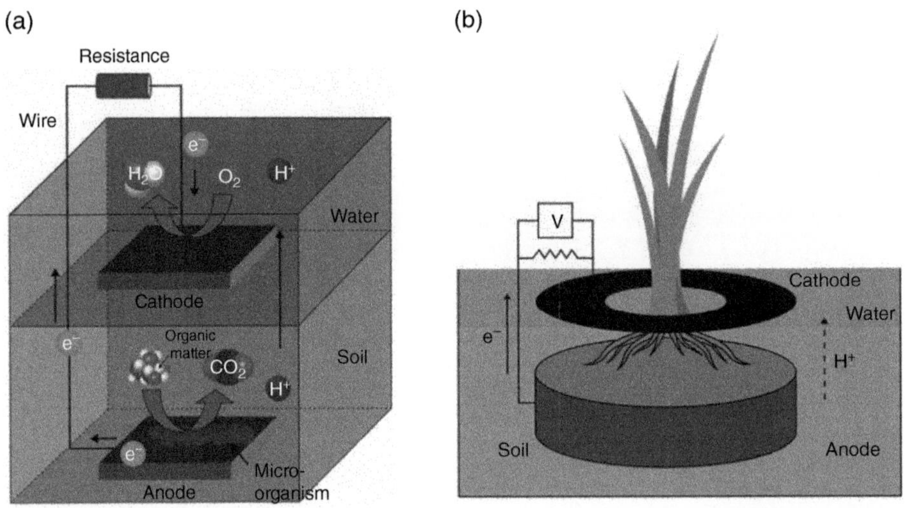

Figure 9.6 Schematics of single chamber of soil MFCs and PMFCs: (a) soil MFCs. *Source:* Based on [128]; (b) PMFCs.

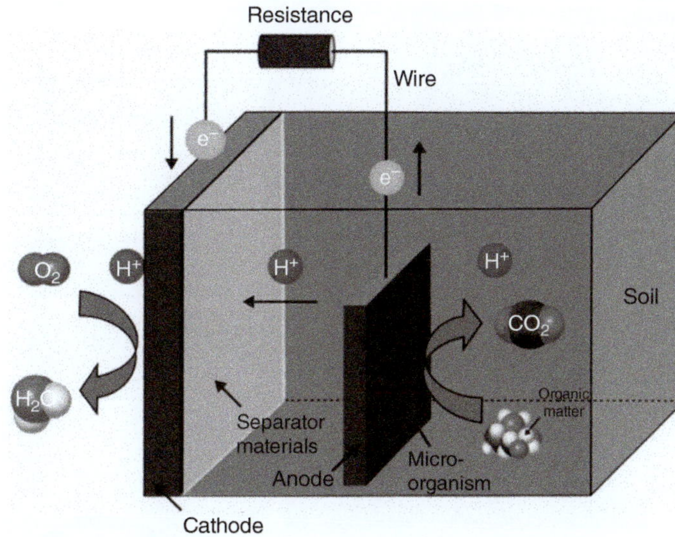

Figure 9.7 Schematics of air-diffusion cathode systems.

9.6.2.4 Other Configuration of PMFCs

For the short distance between the anode and cathode of PMFCs, some researchers have designed a flat-plant design of PMFCs to reduce the internal resistance of this system and improve the efficiency of electricity production (Figure 9.8a). In this design, the membrane is placed vertically to increase the membrane area of the cultivation space [130]. Compared to other designs in the literature, the efficiency of electricity generation in the flat-plant design of PMFCs is increased.

Timmers' group designed a new configuration of PMFCs. This new tube-type PMFC adopts carbon felt as the electrode material and can thus able to significantly reduce the usage and cost of the electrode material [131] (Figure 9.8b). Due to the smaller size electrode, the tube-type PMFCs can reduce disturbances to soils in the natural environment, such as paddy fields and wetlands. In some areas, this configuration of PMFCs is considered more practical.

9.7 Performances of the MFCs in Actual Wastewater Treatment

9.7.1 Industrial Wastewater

MFCs can applied to wastewater treatment for several reasons. Wastewater in sugar factory contains a high organic matter concentration, which can be used as a substance for microbes in MFC systems. Starch processing wastewater consists of high sugar and protein concentrations to serve as the energy sources of MFCs [132]. For example, many researchers have utilized MFCs to treat beer brewing wastewater [25].

Many industrial wastewaters contain a large number of synthetic chemical dyes, such as dye and textile wastewater. Many studies have shown that MFCs can decolorize certain dyes. Kalathil et al. developed MFCs based on GACR without any expensive materials for synchronous actual dye decolorization and bioelectricity generation [133]. Their results

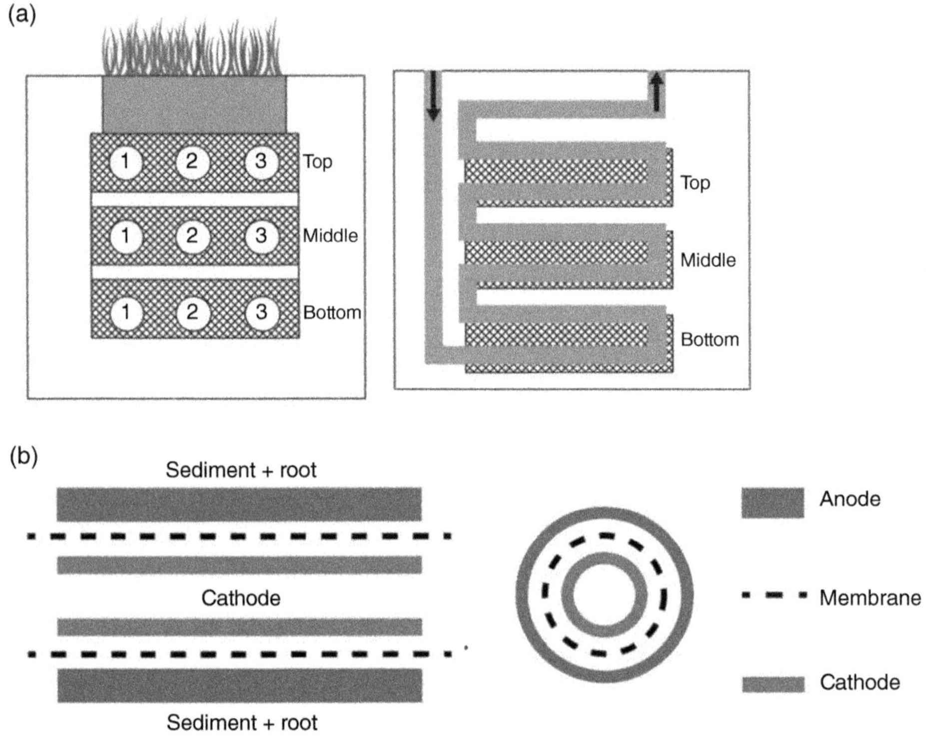

Figure 9.8 Other configuration of PMFCs: (a) flat-plant design of PMFCs. *Source:* Helder et al. [130] / with permission of Elsevier; (b) tube-type PMFCs. *Source:* Timmers et al. [131] / with permission of Elsevier.

demonstrated that the anode and cathode parts removed nearly 73.0 and 77.0% of the color. At the same time, 71.0 and 76.0% of the COD removal rates were obtained at the anode and cathode parts, respectively. Toxicity measurements indicate that the wastewater at the cathode and anode parts were nontoxic and lower than original wastewater after 48 hours of operation.

9.7.2 Domestic and Livestock Wastewater

Domestic wastewater not only includes a huge amount of raw materials but also produces a variety of pollutants. Due to the reduced toxicity, it can be used for MFCs to generate electricity [18]. Many researchers apply MFC systems to remove different contaminants from wastewater. Furthermore, the activated sludge from conventional aerobic treatment is considerable, and thus the costs of wastewater treatment are very high [134]. Therefore, using bioelectrochemical technology in domestic wastewater reduced the cost of sludge treatment. Large volumes of animal manure wastewater are part of the livestock industry. The polluted water contains high concentrations of organic matter, nitrate, and phosphate [135]. This wastewater must be treated to remove contamination and odor problems. Some MFCs applications in agriculture wastewater can be observed during recent years [14, 136].

9.8 Soil MFCs for Soil Remediation

9.8.1 Remediation of Organic Contaminated Soils

In bioelectrochemical systems, organic pollutants can be oxidized and eliminated around the anode parts. Microbes in the anode can directly oxidize organic matter to generate electrons and hydrogen ions and then react with the electron accepters of cathode parts, such as O_2, to form H_2O. SMFCs are used to remediate dibenzothiophene (DBT) contaminated soils. After the operation period, the degradation efficiency of DBT by SMFC was three times higher than natural degradation, and soil toxicity was reduced by 50% [28]. Previous researchers found that 90.1% of phenol in soils was removed after 10 days of operation. In contrast, the removal of phenol in soils under open circuit and without bioelectrochemical conditions were 27.6 and 12.3%, respectively [29]. Some researchers utilized SMFCs to remove hard-to degrade hexachlorobenzene (HCB) in soils and found the removal rate of HCB was 71.15% with a maximum power density of 77.5 mW/m^2 under 1000 ohm external resistance [137].

Because complex organic pollutants are hard to reduce, petroleum-contaminated soils pose environmental problems. The configuration of SMFCs provides more promising on-site treatment technology for remediation of petroleum-contaminated soils. SMFCs with U-type tubular columns can be used to remediate actually contaminated soils. The total petroleum degraded rate around anode parts increased from 6.9 to 15.2% during a 25-day operation [138]. Furthermore, some studies have shown that the composition of organic matter in contaminated soils affects the bioelectrochemical processes in SMFCs. Li et al. used spiking glucose on SMFC systems, which significantly improved the removal rate of total petroleum hydrocarbon degradation in salt land to 200% [139]. Furthermore, changing the physical and chemical properties of soils can improve the soil remediation efficiency of SMFCs. Li et al. flushed soil contaminated saline soils and added carbon fiber for the soil salinity and conductivity effect in SMFCs. They found that the degradation rate of soil hydrocarbons after this treatment was increased to 153–484%, especially for compounds with large molecular weights [140].

9.8.2 Remediation of Heavy Metal Contaminated Soils

The working principle of bioelectrochemical systems is to dissolve low-valent metal ions in the anode into insoluble high-valent metal oxides. The highly oxidizing metal ions in the cathode are reduced into low valence and low-toxic substances. In theory, metals with a higher standard potential greater than zero can be reduced. However, metals with a higher standard electrode potential are reduced and precipitated on the cathode to facilitate recovery [141–144].

Wang et al. used MFCs with vertical glass cylinders to remediate Cu-contaminated soils. After 58 days of operation, the concentration of soluble Cu and total Cu gradually increased from the anode to the cathode parts due to the competitive migration of Cu^{2+} and H^+. The main purpose of this SMFC application was to improve the Cr^{2+} migration more than Cu^{2+} reduction [145]. Some researchers used microbial desalination cells derived from SMFCs for the remediation of Zn^{2+}-contaminated soils.

Dual-chamber MFCs with electrokinetics have been used to remediate soils contaminated with Cd and Pb. After microbes carry out oxidation to acetate in the anode parts, the metal ions in the soils migrated from the anode to the cathode under electric field processes. Part of the metal ions was reduced in the cathode in SMFC systems. The removal rate of Cd and Pb around the anode parts was 31.0 and 44.1%, respectively [146].

9.9 PMFCs for Environmental Remediation

PMFCs have attracted attention in renewable energy fields in recent years. This system contains many mechanisms, such as bioelectrochemical processes, redox reactions, microbial actions, plant uptake, and electrode adsorption [147]. Currently, more and more researchers are adopting this system to remediate pollutions. Due to the elements of PMFC systems, this technology possesses the advantages of conventional phytoremediation and bioelectrochemical technology.

9.9.1 PMFCs for Wastewater Treatment

PMFCs have been used to generate electricity from a various wastewater forms. Compared with other treatment methods, PMFCs have been proven to be environmentally friendly with a high organic matter removal rate and stable power generation. Mohan et al. used floating macrophyte ecosystems to remove 86.67% COD and 72.32% volatile fatty acids, as well as to generate 224.93 mA/m^2 of power density [148]. In upflow CW-MFC with a cattail for wastewater, the pollution removal was as follows: 100% of COD, 40% of NO^{3-}, and 91% of NH^{4+}, with a coulombic efficiency of 8.6% and the maximum power density of 6.12 $mWm/^2$ [149]. The efficiency of removing Cr(VI) in PMFCs is 98% [35]. Compared with sediment MFCs without plants and single phytoremediation without bioelectrochemical systems, PMFCs increased the degradation of benzo[a]pyrene to 70% [150]. Although the power of PMFCs is very low, PMFCs have the promising ability to provide stable energy for low energy consumption equipment.

9.9.2 PMFCs for Soil Remediation

PMFCs have garnered more attention to soil remediation through the degradation of organic matter of root rhizosphere and solids. In one experiment, PMFCs with submerged aquatic plants removed 25.3% nitrogen and 4.42 mW/m^2 power density. [151]. In PMFCs systems, Cr(VI)-polluted soils were degraded by bioelectrochemical reduction, plant uptake, and electrode adsorption. Bioelectrochemical reduction and plant uptake make up a large proportion of the Cr(VI) removal, which are 57.1 and 29.4%, respectively [35]. On the other hand, plants, carbon sources, and surfactants affect PMFCs to degrade petroleum and PAHs in soils. With the increase of surfactant concentration and the addition of plants, the degradation rate of PAHs (phenanthrene and pyrene) increased significantly [36]. The above research shows that soils and wetland systems can be continuously used to generate electricity without harvesting plants. This system can be integrated into the aquatic ecosystem to promote ecological restoration and aesthetic value.

9.10 Prospectives

Today, applying MFCs in a real environment still has many challenges. Most of the MFC research is done at the laboratory scale, while the actual application of MFCs is still scarce. MFC size can be enlarged, but this method increases the internal resistance of the reactor. Stacking systems increase the system volume but will cause voltage reversal to additional losses. MFCs have been extensively studied and are regarded as an innovative technology with unique advantages for wastewater treatment. Some areas are recommended to conduct research in the future. Since MFC configuration affects its performance, the MFC configuration can be optimized, and the membrane modified. The material properties of the anode are critical for the widespread application of MFCs; such properties include low resistance, corrosion resistance, high porosity, and specific surface area.

Soil MFCs have a great potential in electricity generation and soil remediation. Furthermore, this technology can be applied to greenhouse gas emission reduction and pollution testing. However, due to the lower mobility of ions in soils and the high internal resistance of soils caused by a lack of water or salt to inhibit the transport of protons to the cathode, both the performance and efficiency of soil MFCs are limited. The operation of soil MFCs may cause uneven soil remediation in different soil properties in controlled sites. Therefore, the application of soil MFCs needs further exploration. The research topics worth exploring in the future include developing suitable configurations of soil MFCs or applications, reducing the impact of high resistance on soil MFCs, and applying soil MFCs in soil monitoring.

PMFC technology is an emerging technology that uses solar energy to generate electricity. From the current research level, the advantages of PMFCs include multiple functions. PMFCs can be combined with green roofs for urban greening and farmlands for electricity, as well as reduce greenhouse gas emissions and be used in wetlands for wastewater treatment. The operation of PMFCs does not require additional energy supply, nor does it affect the ecological environment, and it can even improve the ecological environment. Despite the gap between the expected results and the actual situation of electricity generation efficiency, PMFCs can be applied in small spaces and low consumption.

9.11 Conclusions

This chapter discussed the application of MFCs on wastewater treatment, soil MFCs for soil remediation, and PMFCs for wastewater and soil remediation. The research of bioelectrochemical technology in the areas of environmental remediation is quite popular. However, certain challenges remain to be conquered, such as cost and small scale. Despite some problems in application, MFCs and PMFCs are promising methods that can be developed and applied in many areas.

References

1 Allen, R.M. and Bennetto, H.P. (1993). Microbial fuel-cells: electricity production from carbohydrates. *Applied Biochemistry and Biotechnology* 39: 27–40.
2 Logan, B.E., Hamelers, B., Rozendal, R. et al. (2006). Microbial fuel cells: methodology and technology. *Environmental Science & Technology* 40: 5181–5192.
3 Strik, D.P.B.T.B., Hamelers, H.V.M., Snel, J.F.H., and Buisman, C.J.N. (2008). Green electricity production with living plants and bacteria in a fuel cell. *International Journal of Energy Research* 32: 870–876.
4 Helder, M., Strik, D.P.B.T.B., Hamelers, H.V.M. et al. (2010). Concurrent bio-electricity and biomass production in three plant-microbial fuel cells using *Spartina anglica*, *Arundinella anomala* and *Arundo donax*. *Bioresource Technology* 101: 3541–3547.
5 Timmers, R.A., Strik, D.P.B.T.B., Hamelers, H.V.M., and Buisman, C.J.N. (2010). Long-term performance of a plant microbial fuel cell with *Spartina anglica*. *Applied Microbiology and Biotechnology* 86: 973–981.
6 Lewis, K. (1966). Symposium on bioelectrochemistry of microorganisms: IV. Biochemical fuel cells. *Bacteriological Reviews* 30: 101–113.
7 Zhang, Q., Hu, J., and Lee, D.J. (2016). Microbial fuel cells as pollutant treatment units: research updates. *Bioresource Technology* 217: 121–128.
8 Wu, Y., Jing, X., Gao, C. et al. (2018). Recent advances in microbial electrochemical system for soil bioremediation. *Chemosphere* 211: 156–163.
9 Potter, M.C. (1911). Electrical effects accompanying the decomposition of organic compounds. *Proceedings of the Royal Society B* 84: 260–276.
10 Thurston, C.F., Bennetto, H.P., Delaney, G.M. et al. (1985). Glucose metabolism in a microbial fuel cell Stoichiometry of product formation in a thionine-mediated *Proteus vulgaris* fuel cell and its relation to coulombic yields. *Journal of General Microbiology* 131: 1393–1401.
11 Kim, H., Hyun, M., Chang, I., and Kim, B.H. (1999). A microbial fuel cell type lactate biosensor using a metal-reducing bacterium, Shewanella putrefaciens. *Journal of Microbiology* 9: 365–367.
12 Bond, D.R. and Lovley, D.R. (2003). Electricity production by *Geobacter sulfurreducens* attached to electrodes. *Applied and Environmental Microbiology* 69: 1548–1555.
13 Chaudhuri, S.K. and Lovley, D.R. (2003). Electricity generation by direct oxidation of glucose in mediatorless microbial fuel cells. *Nature Biotechnology* 21: 1229–1232.
14 Min, B., Kim, J., Oh, S. et al. (2005). Electricity generation from swine wastewater using microbial fuel cells. *Water Research* 39: 4961–4968.
15 Rhoads, A., Beyenal, H., and Lewandowski, Z. (2005). Microbial fuel cell using anaerobic respiration as an anodic reaction and biomineralized manganese as a cathodic reactant. *Environmental Science & Technology* 39: 4666–4671.
16 Liu, H., Ramnarayanan, R., and Logan, B.E. (2004). Production of electricity during wastewater treatment using a single chamber microbial fuel cell. *Environmental Science & Technology* 38: 2281–2285.
17 Min, B. and Logan, B.E. (2004). Continuous electricity generation from domestic wastewater and organic substrates in flat plate microbial fuel cells. *Environmental Science & Technology* 38: 5809–5814.
18 Rabaey, K., Clauwaert, P., Aelterman, P., and Verstraete, W. (2005). Tubular microbial fuel cells for efficient electricity generation. *Environmental Science & Technology* 39: 8077–8082.

19 Aelterman, P., Rabaey, K., Pham, H.T. et al. (2006). Continuous electricity generation at high voltages and currents using stacked microbial fuel cells. *Environmental Science & Technology* 40: 3388–3394.

20 Liu, H. and Logan, B.E. (2004). Electricity generation using an air-cathode single chamber microbial fuel cell in the presence and absence of a proton exchange membrane. *Environmental Science & Technology* 38: 4040–4046.

21 He, Z., Minteer, S.D., and Angenent, L.T. (2005). Electricity generation from artificial wastewater using an upflow microbial fuel cell. *Environmental Science & Technology* 39: 5262–5267.

22 Logan, B., Cheng, S., Watson, V., and Estadt, G. (2007). Graphite fiber brush anodes for increased power production in air-cathode microbial fuel cells. *Environmental Science & Technology* 41: 3341–3346.

23 Lovley, D.R. (2006). Bug juice: harvesting electricity with microorganisms. *Nature Reviews: Microbiology* 4: 497–508.

24 Reimers, C.E., Girguis, P., Stecher, H.A. III et al. (2006). Microbial fuel cell energy from an ocean cold seep. *Geobiology* 4: 123–136.

25 Dumas, C., Mollica, A., Féron, D. et al. (2007). Marine microbial fuel cell: use of stainless steel electrodes as anode and cathode materials. *Electrochimica Acta* 53: 468–473.

26 Rezaei, F., Richard, T.L., Brennan, R.A., and Logan, B.E. (2007). Substrate-enhanced microbial fuel cells for improved remote power generation from sediment-based systems. *Environmental Science & Technology* 41: 4053–4058.

27 Song, T.S., Yan, Z.S., Zhao, Z.W., and Jiang, H.L. (2011). Construction and operation of freshwater sediment microbial fuel cell for electricity generation. *Bioprocess and Biosystems Engineering* 34: 621–627.

28 Ringelberg, D.B., Foley, K.L., and Reynolds, C.M. (2011). Electrogenic capacity and community composition of anodic biofilms in soil-based bioelectrochemical systems. *Applied Microbiology and Biotechnology* 90: 1805–1815.

29 Huang, D.Y., Zhou, S.G., Chen, Q. et al. (2011). Enhanced anaerobic degradation of organic pollutants in a soil microbial fuel cell. *Chemical Engineering Journal* 172: 647–653.

30 Deng, H., Wu, Y.C., Zhen, F. et al. (2014). Factors affecting the performance of single-chamber soil microbial fuel cells for power generation. *Pedosphere* 24: 330–338.

31 Jiang, Y.B., Deng, H., Sun, D.M., and Zhong, W.H. (2015). Electrical signals generated by soil microorganisms in microbial fuel cells respond linearly to soil Cd^{2+} pollution. *Geoderma* 255-256: 35–41.

32 De Schamphelaire, L., Van den Bossche, L., Dang, H.S. et al. (2008). Microbial fuel cells generating electricity from rhizodeposits of rice plants. *Environmental Science & Technology* 42: 3053–3058.

33 Helder, M., Strik, D.P.B.T.B., Timmers, R.A. et al. (2013). Resilience of roof-top plant-microbial fuel cells during Dutch winter. *Biomass and Bioenergy* 51: 1–7.

34 Habibul, N., Hu, Y., Wang, Y.K. et al. (2016). Bioelectrochemical chromium(VI) removal in plant-microbial fuel cells. *Environmental Science & Technology* 50: 3882–3889.

35 Guan, C.Y., Tseng, Y.H., Tsang, D.C.W. et al. (2019). Wetland plant microbial fuel cells for remediation of hexavalent chromium contaminated soils and electricity production. *Journal of Hazardous Materials* 365: 137–145.

36 Zhao, L., Deng, J., Hou, H. et al. (2019). Investigation of PAH and oil degradation along with electricity generation in soil using an enhanced plant-microbial fuel cell. *Journal of Cleaner Production* 221: 678–683.

37 Koch, C. and Harnisch, F. (2016). Is there a specific ecological niche for electroactive microorganisms? *ChemElectroChem* 3: 1282–1295.

38 Coppi, M.V., Leang, C., Sandler, S.J., and Lovley, D.R. (2001). Development of a genetic system for *Geobacter sulfurreducens*. *Applied and Environmental Microbiology* 67: 3180–3187.

39 Aklujkar, M., Krushkal, J., DiBartolo, G. et al. (2009). The genome sequence of *Geobacter metallireducens*: features of metabolism, physiology and regulation common and dissimilar to *Geobacter sulfurreducens*. *BMC Microbiology* 9: 1–22.

40 Douglas, F. and Logan, B.E. (2011). A method for high throughput bioelectrochemical research based on small scale microbial electrolysis cells. *Biosensors and Bioelectronics* 26: 4526–4531.

41 Yi, H., Nevin, K.P., Kim, B.C. et al. (2009). Selection of a variant of Geobacter Sulfurreducens with enhanced capacity for current production in microbial fuel cells. *Biosensors & Bioelectronics* 24: 3498–3503.

42 Xu, M., Guo, J., Cen, Y. et al. (2005). *Shewanella decolorationis* sp. nov., a dye-decolorizing bacterium isolated from activated sludge of a waste-water treatment plant. *International Journal of Systematic and Evolutionary Microbiology* 55: 363–368.

43 Biffinger, J.C., Fitzgerald, L.A., Ray, R. et al. (2011). The utility of *Shewanella japonica* for microbial fuel cells. *Bioresource Technology* 102: 290–297.

44 Sekar, R., Shin, H.D., and DiChristina, T.J. (2016). Activation of an otherwise silent xylose metabolic pathway in *Shewanella oneidensis*. *Applied and Environmental Microbiology* 82: 3996–4005.

45 Liu, L., Lee, D.J., Wang, A. et al. (2016). Isolation of Fe(III)-reducing bacterium, *Citrobacter* sp. LAR-1, for startup of microbial fuel cell. *International Journal of Hydrogen Energy* 41: 4498–4503.

46 Pham, A.P., Jung, S.J., Phung, N.T. et al. (2003). A novel electrochemically active and Fe(III)-reducing bacterium phylogenetically related to *Aeromonas hydrophila*, isolated from a microbial fuel cell. *FEMS Microbiology Letters* 223: 129–134.

47 Bond, D.R. and Lovley, D.R. (2005). Evidence for involvement of an electron shuttle in electricity generation by *Geothrix fermentans*. *Applied and Environmental Microbiology* 71: 2186–2189.

48 Lee, Y.Y., Kim, T.G., and Cho, K.S. (2016). Enhancement of electricity production in a mediatorless air-cathode microbial fuel cell using *Klebsiella* sp. IR21. *Bioprocess and Biosystems Engineering* 39: 1005–1014.

49 Zuo, Y., Xing, D., Regan, J.M., and Logan, B.E. (2008). Isolation of exoelectrogenic bacterium *Ochrobactrum anthropic* YZ-1 by using a U-Tube microbial fuel cell. *Applied and Environmental Microbiology* 74: 3130–3137.

50 Rabaey, K., Boon, N., Siciliano, S.D. et al. (2004). Biofuel cells select for microbial consortia that self-mediate electron transfer. *Applied and Environmental Microbiology* 70: 5373–5382.

51 Luo, J., Yang, J., He, H. et al. (2013). A new electrochemically active bacterium phylogenetically related to *Tolumonas osonensis* and power performance in MFCs. *Bioresource Technology* 139: 141–148.

52 Williams, J., Trautwein-Schult, A., Jankowska, D. et al. (2014). Identification of uric acid as the redox molecule secreted by the yeast *Arxula adeninivorans*. *Applied Microbiology and Biotechnology* 98: 2223–2229.

53 Prasad, D., Arun, S., Murugesan, M. et al. (2007). Direct electron transfer with yeast cells and construction of a mediatorless microbial fuel cell. *Biosensors and Bioelectronics* 22: 2604–2610.

54 Raghavulu, S.V., Goud, R.K., Sarma, P.N., and Mohan, S.V. (2011). *Saccharomyces cerevisiae* as anodic biocatalyst for power generation in biofuel cell: Influence of redox condition and substrate load. *Bioresource Technology* 102: 2751–2757.

55 Hubenova, Y. and Mitov, M. (2010). Potential application of *Candida melibiosica* in biofuel cells. *Bioelectrochemistry* 78: 57–61.

56 Mao, L. and Verwoerd, W.S. (2013). Selection of organisms for systems biology study of microbial electricity generation: a review. *International Journal of Energy and Environmental Engineering* 4: 1–18.

57 Schaetzle, O., Barrière, F., and Baronian, K. (2008). Bacteria and yeasts as catalysts in microbial fuel cells: electron transfer from microorganisms to electrodes for green electricity. *Energy & Environmental Science* 1: 607–620.

58 Kim, B.H., Park, H.S., Kim, H.J. et al. (2004). Enrichement of microbial community generating electricity using a fuel-cell-type electrochemical cell. *Applied Microbiology and Biotechnology* 63: 672–681.

59 Logan, B.E. (2009). Exoelectrogenic bacteria that power microbial fuel cells. *Nature Reviews Microbiology* 7: 375–381.

60 Kim, H.J., Park, H.S., Hyun, M.S. et al. (2002). A mediator-less microbial fuel cell using a meal reducing bacterium, *Shewanella putrefaciens*. *Enzyme and Microbial Technology* 30: 145–152.

61 Peng, L., Zhang, X.T., Yin, J. et al. (2016). *Geogbacter sulfurreducens* adapts to low electrode potential for extracellular electron transfer. *Electrochimica Acta* 191: 743–749.

62 Reguera, G., McCarthy, K.D., Mehta, T. et al. (2005). Extracellular electron transfer via microbial nanowires. *Nature* 435: 1098–1101.

63 Kumar, R., Singh, L., and Zularisam, A.W. (2016). Exoelectrogens: recent advances in molecular drivers involved in extracellular electron transfer and strategies used in improve it for microbial fuel cell applications. *Renewable & Sustainable Energy Reviews* 56: 1322–1336.

64 Hubenova, Y., Bakalska, R., Hubenova, E., and Mitov, M. (2016). Mechanisms of electron transfer between a styrylquinolinium dye and yeast in biofuel cell. *Bioelectrochemistry* 112: 158–165.

65 Xia, X., Cao, X.X., Liang, P. et al. (2010). Electricity generation from glucose by a *Klebsiella* sp. in microbial fuel cells. *Bioenergy and Biofuels* 87: 383–390.

66 Okamoto, A., Hashimoto, K., Nealson, K.H., and Nakamura, R. (2013). Rate enhancement of bacterial extracellular electron transport involves bound flavin semiquinones. *Proceedings of the National Academy of Sciences of the United States of America* 110: 7856–7861.

67 Rabaey, K. and Rozendal, R.A. (2010). Microbial electrosynthesis-revisiting the electrical route for microbial production. *Nature Reviews Microbiology* 8: 706–716.

68 Schröder, U., Niessen, J., and Scholz, F. (2003). A generation of microbial fuel cells with current outputs boosted by more than one order of magnitude. *Angewandte Chemie International Edition* 115: 2986–2989.

69 Liu, Y., Harnisch, F., Fricke, K. et al. (2011). The study of electrochemically active microbial biofilms on different carbon-based anode materials in microbial fuel cells. *Biosensors and Bioelectronics* 25: 2167–2171.

70 Chen, S., Hou, H., Harnisch, F. et al. (2011). Electrospun and solution blown three-dimensional carbon fiber nonwovens for application as electrodes in microbial fuel cells. *Energy & Environmental Science* 4: 1417–1421.

71 Li, J., Zhang, J., Ye, D. et al. (2014). Optimization of inner diameter of tubular bamboo charcoal anode for a microbial fuel cell. *International Journal of Hydrogen Energy* 39: 19242–19248.

72 Cheng, S. and Logan, B.E. (2007). Ammonia treatment of carbon cloth anodes to enhance power generation of microbial fuel cells. *Electrochemistry Communications* 9: 492–496.

73 Watson, V.J., Delgado, C.N., and Logan, B.E. (2013). Improvement of activated carbons as oxygen reduction catalysts in neutral solutions by ammonia gas treatment and their performance in microbial fuel cells. *Journal of Power Sources* 242: 756–761.

74 Song, Y.C., Kim, D.S., Woo, J.H. et al. (2015). Effect of surface modification of anode with surfactant on the performance of microbial fuel cell. *International Journal of Energy Research* 39: 860–868.

75 Peng, X., Chu, X., Liu, W. et al. (2015). Bioelectricity-generating behavior of a chemically modified carbon black anode in microbial fuel cells. *Chinese Journal of Catalysis* 36: 1326–1332.

76 Peng, X., Chen, S., Liu, L. et al. (2016). Modified stainless steel for high performance and stable anode in microbial fuel cells. *Electrochimica Acta* 194: 246–252.

77 Feng, Y., Yang, Q., Wang, X., and Logan, B.E. (2010). Treatment of carbon fiber brush anodes for improving power generation in air-cathode microbial fuel cells. *Journal of Power Sources* 195: 1841–1844.

78 Tang, J., Chen, S., Yuan, Y. et al. (2015). *In situ* formation of graphene layers on graphite surfaces for efficient anodes of microbial fuel cells. *Biosensors and Bioelectornics* 71: 387–395.

79 Hou, J., Liu, Z., Yang, S., and Zhou, Y. (2014). Three-dimensional microporous anodes based on stainless steel fiber felt for high-performance microbial fuel cells. *Journal of Power Sources* 258: 204–209.

80 Erbay, C., Pu, X., Choi, W. et al. (2015). Control of geometrical properties of carbon nanotube electrodes towards high-performance microbial fuel cells. *Journal of Power Sources* 280: 347–354.

81 Guo, W., Pi, Y., Song, H. et al. (2012). Layer-by layer assembled gold nanoparticles modified anode and its application in microbial fuel cells. *Colloids and Surfaces A: Physicochemical and Engineering Aspects* 415: 105–111.

82 Lv, Z., Xie, D., Yue, X. et al. (2012). Ruthernium oxide-coated carbon felt electrode: a highly active anode for microbial fuel cell applications. *Journal of Power Sources* 210: 26–31.

83 Niessen, J., Schröder, U., Rosenbaum, M., and Scholz, F. (2004). Fluorinated polyanilines as superior materials for electrocatalytic anodes in bacterial fuel cells. *Electrochemistry Communications* 6: 571–575.

84 Ma, X., Feng, C., Zhou, W., and Yu, H. (2016). Municipal sludge-derived carbon anode with nitrogen- and oxygen-containing functional groups for high-performance microbial fuel cells. *Journal of Power Sources* 307: 105–111.

85 Lu, M. and Li, S.F.Y. (2012). Cathode reactions and applications in microbial fuel cells: a review. *Critical Reviews in Environmental Science and Technology* 42: 2504–2525.

86 Zhou, M., Chi, M., Luo, J. et al. (2011). An overview of electrode materials in microbial fuel cells. *Journal of Power Sources* 196: 4427–4435.

87 Zhang, Y., Sun, J., Hu, Y. et al. (2012). Bio-cathode materials evaluation in microbial fuel cells: a comparison of graphite felt, carbon paper and stainless steel mesh materials. *International Journal of Hydrogen Energy* 37: 16935–16942.

88 Kakarla, R. and Min, B. (2014). Evaluation of microbial fuel cell operation using algae as an oxygen supplier: carbon paper cathode vs. carbon brush cathode. *Bioprocess and Biosystems Engineering* 37: 2453–2461.

89 Cheng, S. and Wu, J. (2013). Air-cathode preparation with activated carbon as catalyst, PTFE as binder and nickel foam as current collector for microbial fuel cells. *Bioelectrochemistry* 92: 22–26.

90 Gong, X.B., Yu, S.J., Wang, X.H. et al. (2013). Silver-tungsten carbide nanohybrid for efficient electrocatalysis of oxygen reduction reaction in microbial fuel cell. *Journal of Power Sources* 255: 330–337.

91 Gong, X.B., You, S.J., Wang, X.H. et al. (2014). A novel stainless steel mesh/cobalt oxide hybrid electrode for efficient catalysis of oxygen reduction in a microbial fuel cell. *Biosensors & Bioelectronics* 55: 237–241.

92 Chen, Y., Lv, Z., Xu, J. et al. (2012). Stainless steel mesh coated with MnO_2/carbon nanotube and polymethylphenyl siloxane as low-cost and high-performance microbial fuel cell cathode materials. *Journal of Power Sources* 201: 136–141.

93 Rabaey, K. and Verstraete, W. (2005). Microbial fuel cells: novel biotechnology for energy generation. *Trends in Biotechnology* 23: 291–298.

94 Ahmed, J., Yuan, Y., Zhou, L., and Kim, S. (2012). Carbon supported cobalt oxide nanoparticles-iron phthalocyanine as alternative cathode catalyst for oxygen reduction in microbial fuel cells. *Journal of Power Sources* 208: 170–175.

95 Quan, X., Mei, Y., Xu, H. et al. (2015). Optimization of Pt-Pd alloy catalyst and supporting materials for oxygen reduction in air-cathode microbial fuel cells. *Electrochimica Acta* 165: 72–77.

96 Li, X., Hu, B., Suib, S. et al. (2010). Manganese dioxide as a new cathode catalyst in microbial fuel cells. *Journal of Power Sources* 195: 2586–2561.

97 Amade, R., Vila-Costa, M., Hussain, S. et al. (2015). Vertically aligned carbon nanotubes coated with manganese dioxide as cathode material for microbial fuel cells. *Journal of Materials Science* 50: 1214–1220.

98 Ding, H., Li, Y., Lu, A. et al. (2010). Photocatalytically improved azo dye reduction in a microbial fuel cell with rutile-cathode. *Bioresource Technology* 101: 3500–3505.

99 Hosseini, M.G. and Ahadzadeh, I. (2012). A dual-chambered microbial fuel cell with Ti/nano-TiO_2/Pd nano-structure cathode. *Journal of Power Sources* 220: 292–297.

100 Ghoreishi, K.B., Ghasemi, M., Rahimnejad, M. et al. (2014). Development and application of vanadium oxide/polyaniline composite as a novel cathode catalyst in microbial fuel cell. *International Journal of Energy Research* 38: 70–77.

101 Esmaeili, C., Ghasemi, M., Heng, L.Y. et al. (2014). Synthesis and application of polypyrrole/carrageenan nano-bio composite as a cathode catalyst in microbial fuel cells. *Carbohydrate Polymers* 114: 253–259.

102 Ghasemi, M., Ismail, M., Kamarudin, S.K. et al. (2012). Carbon nanotube as an alternative cathode support and catalyst for microbial fuel cells. *Applied Energy* 102: 1050–1056.

103 Tsai, H.Y., Wu, C.C., Lee, C.Y., and Shih, E.P. (2009). Microbial fuel cell performance of multiwall carbon nanotubes on carbon cloth as electrodes. *Journal of Power Sources* 194: 199–205.

104 Lu, M., Guo, L., Kharkwal, S. et al. (2013). Manganese-polypyrrole-carbon nanotube, a new oxygen reduction catalyst for air-cathode microbial fuel cells. *Journal of Power Sources* 221: 381–386.

105 Zhang, Y., Hu, Y., Li, S. et al. (2011). Manganese dioxide-coated carbon nanotubes as an improved cathodic catalyst for oxygen reduction in a microbial fuel cell. *Journal of Power Sources* 196: 9284–9289.

106 Huang, L.P., Regan, J.M., and Quan, X. (2011). Electron transfer mechanisms, new applications, and performance of biocathode microbial fuel cells. *Bioresource Technology* 102: 316–323.

107 Li, C., Ding, L., Cui, H. et al. (2012). Application of conductive polymers in biocathode of microbial fuel cells and microbial community. *Bioresource Technology* 116: 459–465.

108 Jacobson, K.S., Drew, D.M., and He, Z. (2011). Efficient salt removal in a continuously operated upflow microbial desalination cell with an air cathode. *Bioresource Technology* 102: 376–380.

109 ElMekawy, A., Hegab, H.M., Domingues-Benetton, X., and Pant, D. (2013). Internal resistance of microfluidic microbial fuel cell: Challenges and potential opportunities. *Bioresource Technology* 142: 672–682.

110 Varcoe, J.R., Atanassov, P., Dekel, D.R. et al. (2014). Anion-exchange membranes in electrochemical energy systems. *Energy & Environmental Science* 7: 3135–3191.

111 Harnisch, F., Schröder, U., and Scholz, F. (2008). The suitability of monopolar and bipolar ion exchange membranes as separators for biological fuel cells. *Environmental Science & Technology* 42: 1740–1746.

112 Zuo, Y., Cheng, S., and Logan, B.E. (2008). Ion exchange membrane cathodes for scalable microbial fuel cells. *Environmental Science & Technology* 42: 6967–6972.

113 Ghasemi, M., Daud, W.R.W., Alam, J. et al. (2016). Treatment of two different water resources in desalination and microbial fuel cell processes by poly sulfone/Sulfonated poly ether ether ketone hybrid membrane. *Energy* 96: 303–313.

114 Suzuki, K., Owen, R., Mok, J. et al. (2016). Comparison of electrochemical and microbiological characterization of microbial fuel cells equipped with SPEEK and Nafion membrane electrode assemblies. *Journal of Bioscience and Bioengineering* 122: 322–328.

115 Pandit, S., Khilari, S., Bera, K. et al. (2014). Application of PVA-PDDA polymer electrolyte composite anion exchange membrane separator for improved bioelectricity production in a single chambered microbial fuel cell. *Chemical Engineering Journal* 257: 138–147.

116 Rudra, R., Kumar, V., Pramanik, N., and Kundu, P.P. (2017). Graphite oxide incorporated crosslinked polyvinyl alcohol and sulfonated styrene nanocomposite membrane as separating barrier in single chambered microbial fuel cell. *Journal of Power Sources* 341: 285–293.

117 Hernández-Florez, G., Poggi-Varaldo, H.M., Solorza-Feria, O. et al. (2015). Characteristics of a single chamber microbial fuel cell equipped with a low cost membrane. *International Journal of Hydrogen Energy* 40: 17380–17387.

118 Srinophakun, P., Thanapimmetha, A., Plangsri, S. et al. (2017). Application of modified chitosan membrane for microbial fuel cell: roles of proton carrier site and positive charge. *Journal of Cleaner Production* 142: 1274–1282.

119 Breitsprecher, K., Szuttor, K., and Holm, C. (2015). Electrode models for ionic liquid-based capacitors. *The Journal of Physical Chemistry C* 119: 22445–22451.

120 Behera, M., Jana, P.S., and Ghangrekar, M.M. (2010). Performance evaluation of low cost microbial fuel cell fabricated using earthen pot with biotic and abiotic cathode. *Bioresource Technology* 101: 1183–1189.

121 Pasternak, G., Greenman, J., and Ieropoulos, I. (2016). Comprehensive study on ceramic membranes for low-cost microbial fuel cells. *ChemSusChem* 9: 88–96.

122 Min, B., Cheng, S., and Logan, B.E. (2005). Electricity generation using membrane and salt bridge microbial fuel cells. *Water Research* 39: 1675–1686.

123 Milliken, C. and May, H. (2007). Sustained generation of electricity by the spore-forming, gram-positive, *Desulfitobacterium hafniense* strain DCB2. *Applied Microbiology and Biotechnology* 73: 1180–1189.

124 Ringeisen, B.R., Henderson, E., Wu, P.K. et al. (2006). High power density from a miniature microbial fuel cell using *Shewanella oneidensis* DSP 10. *Environmental Science & Technology* 40: 2629–2634.

125 He, Z., Wagner, N., Minteer, S.D., and Angenent, L.T. (2006). An upflow microbial fuel cell with an interior cathode: assessment of the internal resistance by impedance spectroscopy. *Environmental Science & Technology* 40: 5212–5217.

126 Feng, C., Hu, A., Chen, S., and Yu, C.P. (2013). A decentralized wastewater treatment system using microbial fuel cell techniques and its response to a copper shock load. *Bioresource Technology* 143: 76–82.

127 Xiao, L., Young, E.B., Berges, J.A., and He, Z. (2012). Integrated photo-bioelectrochemical system for contaminants removal and bioenergy production. *Environmental Science & Technology* 46: 11459–11466.

128 Goto, Y., Yoshida, N., Umeyama, Y. et al. (2015). Enhancement of electricity production by graphene oxide in soil microbial fuel cells and plant microbial fuel cells. *Frontiers in Bioengineering and Biotechnology* 3: 1–8.

129 Wang, X., Cai, Z., Zhou, Q. et al. (2011). Bioelectrochemical stimulation of petroleum hydrocarbon degradation in saline soil using u-Tube microbial fuel cells. *Biotechnology and Bioengineering* 109: 426–433.

130 Helder, M., Strik, D.P.B.T.B., and Hamelers, H.V.M. (2012). New plant-growth medium for increased power output of the plant-microbial fuel cell. *Bioresource Technology* 104: 417–423.

131 Timmers, R.A., Strik, D.P.B.T.B., Hamelers, H.V.M., and Buisman, C.J.N. (2013). Electricity generation by a novel design tubular plant microbial fuel cells. *Biomass and Bioenergy* 51: 60–67.

132 Gil, G.C., Chang, I.S., Kim, B.H. et al. (2003). Operational parameters affecting the performance of a mediator-less microbial fuel cell. *Biosensors and Bioelectronics* 18: 327–334.

133 Kalathil, S., Lee, J., and Cho, M.H. (2011). Granular activated carbon based microbial fuel cell for stimultaneous decolorization of real dye wastewater and electricity generation. *New Biotechnology* 29: 32–37.

134 Xiao, B., Yang, F., and Liu, J. (2011). Enhancing simultaneous electricity production and reduction of sewage sludge in two-chamber MFC by aerobic sludge digestion and sludge pretreatments. *Journal of Hazardous Materials* 189: 444–449.

135 He, L., Huang, G.H., and Lu, H.W. (2010). A stochastic optimization model under modeling uncertainty and parameter certainty for groundwater remediation design- Part I. Model development. *Journal of Hazardous Materials* 176: 521–526.

136 Kim, J.R., Dec, J., Bruns, M.A., and Logan, B.E. (2008). Removal of odors from swine wastewater by using microbial fuel cells. *Applied and Environmental Microbiology* 74: 2540–2543.

137 Cao, X., Song, H.L., Yu, C.Y., and Li, X.N. (2015). Simultaneous degradation of toxic refractory organic pesticide and bioelectricity generation using a soil microbial fuel cell. *Bioresource Technology* 189: 87–93.

138 Li, X.J., Wang, X., Wang, L.L. et al. (2016). Enhanced biodegradation of aged petroleum hydrocarbons in soils by glucose addition in microbial fuel cells. *Journal of Chemical Technology and Biotechnology* 91: 267–275.
139 Li, X.J., Wang, X., Zhang, Y.Y. et al. (2014). Extended petroleum hydrocarbon bioremediation in saline soil using Pt-free multianodes microbial fuel cells. *RSC Advances* 4: 59803–59808.
140 Li, X., Wang, X., Zhang, Y. et al. (2016). Salinity and conductivity amendment of soil enhanced the bioelectrochemical degradation of petroleum hydrocarbons. *Scientific Reports* 6: 32861.
141 Wang, G., Huang, L.P., and Zhang, Y.F. (2008). Cathodic reduction of hexavalent chromium [Cr(VI)] coupled with electricity generation in microbial fuel cells. *Biotechnology Letters* 30: 1959–1966.
142 Tao, H.C., Liang, M., Li, W. et al. (2011). Removal of copper from aqueous solution by electrodeposition in cathode chamber of microbial fuel cell. *Journal of Hazardous Materials* 189: 186–192.
143 Tao, H.C., Gao, Z.Y., Ding, H. et al. (2012). Recovery of silver from silver(I)-containing solutions in bioelectrochemical reactors. *Bioresource Technology* 111: 92–97.
144 Wang, Z., Lim, B., and Choi, C. (2011). Removal of Hg^{2+} as an electron acceptor coupled with power generation using a microbial fuel cell. *Bioresource Technology* 102: 6304–6307.
145 Wang, H., Song, H., Yu, R. et al. (2016). New process for copper migration by bioelectricity generation in soil microbial fuel cells. *Environmental Science and Pollution Research International* 23: 13147–13154.
146 Cao, X.X., Huang, X., Liang, P. et al. (2009). A new method for water desalination using microbial desalination cells. *Environmental Science & Technology* 43: 7148–7142.
147 Guan, C.Y., Hu, A., and Yu, C.P. (2019). Stratified chemical and microbial characteristics between anode and cathode after long-term operation of plant microbial fuel cells for remediation of metal contaminated soils. *Science of the Total Environment* 670: 585–594.
148 Mohan, S.V., Mohanakrishna, G., and Chiranjeevi, P. (2011). Sustainable power generation from floating macrophytes based ecological microenvironment through embedded fuel cells along with simultaneous wastewater treatment. *Bioresource Technology* 102: 7036–7042.
149 Oon, Y.L., Ong, S.A., Ho, L.N. et al. (2015). Hybrid system up-flow constructed wetland integrated with microbial fuel cell for simultaneous wastewater treatment and electricity generation. *Bioresource Technology* 186: 270–275.
150 Yan, Z., Jiang, H., Cai, H. et al. (2015). Complex interactions between the Macrophyte *Acorus Calamus* and microbial fuel cells during pyrene and benzo[a] pyrene degradation in sediments. *Science Reports* 5: 1–11.
151 Xu, P., Xiao, E.R., Xu, D. et al. (2017). Internal nitrogen removal from sediments by the hybrid system of microbial fuel cells and submerged aquatic plants. *PLoS One* e0172757: 1–20.

10

Microalgae-Based UV Protection Compounds

Jorge Alberto Vieira Costa[1], *Juliana Botelho Moreira*[2], *Gabrielle Guimarães Izaguirres*[1], *Liliane Martins Teixeira*[2], *and Michele Greque de Morais*[2]

[1] *Laboratory of Biochemical Engineering, College of Chemistry and Food Engineering, Federal University of Rio Grande (FURG), Rio Grande, RS, Brazil*
[2] *Laboratory of Microbiology and Biochemistry, College of Chemistry and Food Engineering, Federal University of Rio Grande (FURG), Rio Grande, RS, Brazil*

10.1 Introduction

Exposure to ultraviolet (UV) radiation from sunlight is the cause of several harmful effects, such as burns, pigmentation, immune response, and skin cancer [1, 2]. UV-B rays directly induce the formation of dimers and photoproducts of cyclobutene pyrimidine. These damages cause the stimulation of reactive oxygen species (ROS), which leads to photoaging and alteration of transcription control or genetic mutation [2–5].

Sunscreen formulations have been developed to protect the skin from damage caused by UV-B [6]. Many synthetic chemical compounds have been used in sunscreen to obtain a higher sun protection factor (SPF). However, these compounds can cause side effects, such as skin irritation and carcinogenesis. In recent years, natural active compounds have been applied in formulations to reduce the possible side effects of synthetic compounds. Besides, consumers have shown greater acceptance for natural compounds [7].

Microalgae have a high capacity to fix atmospheric carbon dioxide (CO_2) through photosynthesis, with cell growth that can be 100 times greater than that of terrestrial plants [8]. These microorganisms are considered promising and sustainable sources of biomass for the extraction of high-value compounds. The photosynthetic efficiency of microalgae is extremely sensitive to several environmental stresses. The demand for light for life-sustaining biological phenomena exposed microalgae to solar radiation, including harmful doses of UV-A (315–400 nm) and UV-B (280–315 nm) radiations [9].

UV radiation, particularly UV-B (280–315 nm), can affect physiology and biochemistry of the cell either directly or by the photosensitized reactions mediated by the UV-induced ROS [10–13]. However, with the increasing evolution of studies concerning microalgae, defense mechanisms, or tolerance to UV stress have been noticed. Some examples of this mechanisms include DNA repair, antioxidants synthesis, such as phenolic compounds and

Bioprospecting of Microorganism-Based Industrial Molecules, First Edition.
Edited by Sudhir P. Singh and Santosh Kumar Upadhyay.

phycocyanins, in addition to enzymatic/nonenzymatic compounds, such as mycosporine-like amino acids (MAAs), carotenoids, scytonemin, and polyamines [9, 10, 12]. There is a dynamic balance between damage adaptation and stress in microalgae that allows adaptation to the harmful effects of UV radiation. In addition to physiological, biochemical, and molecular resilience mechanisms, microalgae develop self-regulatory processes to combat the negative impacts of UV radiations [9].

In this context, the chapter addresses the potential of microalgae biomass for the extraction of compounds with potential UV protection capacity. The extraction techniques, cultivation strategies, the effects of UV stress, as well as the prospects for application for microalgae biomass and its metabolites, are also presented.

10.2 UV Radiation

The radiation emitted by the sun differs by the electromagnetic wavelength. Among UV rays are included UV-C (200–280 nm), UV-B (280–320 nm), and UV-A (320–400 nm). The Earth's atmosphere fully absorbs UV-C radiation. Even with the protection provided by the ozone layer, UV-A and UV-B rays reach the biosphere [14]. UV-A rays are responsible for the occurrence of skin cancer, managing to penetrate the epidermis deeply. UV-B rays are necessary for the synthesis of vitamin D. However, UV-B radiation in excess can be highly harmful [15].

UV-B radiation can reach depths of the ocean, where species of low trophic level survive. High levels of UV-B exposure can cause damage to proteins, lipids, and DNA [13], forming thymine dimers, consequently affecting metabolism and cell division [16]. With ROS production, photosynthetic organisms develop protective compounds to deal with these adversities [11].

UV-A and UV-B radiation represent a photobiological stressor for microalgae, with adverse effects on metabolic processes. Proteins and nucleic acids, due to the preferential absorption of UV-B, change the three-dimensional molecular structure and disturbances of essential biochemical functions, such as transcription, replication, and translation [17]. Thus, UV radiation plays a vital role in the photosynthetic efficiency of microalgae. In certain geographic regions (different latitudes), due to increased levels of UV-B radiation, microalgae may reduce the ability to perform photosynthesis of the chloroplast to protect cells [18].

10.3 Protection Compounds Induced by UV Radiation

UV irradiation not only stimulates the production of ROS but also stimulates the secretion of pro-inflammatory cytokines, which can increase the skin's sensitivity to some allergens [5, 19]. Therefore, natural active compounds that demonstrate photoprotective activities and biological prevention capacity for UV-induced damage should be further explored for pharmaceutical or cosmetic applications [2]. In this context, the tolerance of microalgae to environmental adversities, mainly UV exposure, along with other parameters such as temperature and nutrients, increases the natural defenses of these microorganisms, producing metabolites that can absorb UV radiation act as antioxidants.

10.3.1 Mycosporine-Like Amino Acids

Mycosporine-like amino acids (MAAs) are hydrophilic compounds found in several taxonomic groups, with high molar extinction coefficients ($\mathcal{E}$ = 28 000–50 000 M/cm) and UV absorption (307–362 nm). These compounds have the potential to be used as efficient natural sunscreens [20]. Structurally, MAAs are characterized by a cyclohexenone or cyclohexenimine chromophore conjugated to the nitrogenous substituent of amino acid [21].

MAAs occur more frequently in marine organisms, such as cyanobacteria, algae, invertebrates, and terrestrial organisms [22]. In microalgae, MAAs are essential for cell survival, with the potential to prevent 3 out of 10 photons from reaching cytoplasmic targets [23]. There are approximately 20 types of MAAs, including mycosporin-glycine, palytine, palytinol, asterina-330, porphyra-334, and shinorine (most commonly found in microalgae) [21]. Due to their high absorption in the UV region, MAAs are used as activators of cell proliferation, inhibiting lipid peroxidation, in addition to having anti-inflammatory activity [24].

MAAs have great potential for the biotechnological industries, since these compounds are biodegradable, nontoxic, have low molecular weight, absorption in the UV region, high melting point, solubility in organic solvents and water, and stability over a wide range of pH and temperature. Moreover, MAAs transform the UV rays absorbed into thermal energy without producing ROS and without losing protection efficiency [25].

10.3.2 Phenolic Compounds

The molecular structure of phenolic compounds has an aromatic ring containing hydroxyl groups, which may be a simple or polymerized molecule [26]. Phenolic compounds are divided into several groups, although the main ones are phenolic acids, flavonoids, and tannins [27, 28]. The bioactivities of these compounds are related to their chemical characteristics, particularly the presence of an aromatic structure and several hydroxyl groups capable of donating electrons or hydrogen atoms, neutralizing free radicals and other ROS [26]. The presence of a hydroxyl group connected directly to the aromatic ring allows these compounds to interact with other biomolecules. The presence of a hydrophobic aromatic ring enables the absorption of UV radiation [29].

Phenolic compounds of microalgal origin are promising and innovative substances to replace compounds of synthetic origin in the production of sunscreens and anti-aging products for the skin. High-molecular-weight phenolic compounds present photoprotective property and the ability to prevent the adverse effects of UV radiation [2]. In addition to their photoprotective function, phenolic compounds have been reported to have depigmenting actions [30]. In microalgae, the accumulation of phenolic compounds in the cell is induced due to the stress caused by UV-B radiation [2].

10.3.3 Carotenoids

Carotenoids are synthesized by some bacteria and fungi and photosynthetic organisms. These compounds are divided into two groups based on direct or indirect involvement in photosynthesis: primary, such as light-capturing molecules (α-carotene, β-carotene, and lutein), and secondary, acting against ROS (astaxanthin and canthaxanthin) [31].

Carotenoids are considered the most found pigments in nature, with colors ranging from yellow to red [9].

Carotenoid staining is determined from the chromophore length and configuration of the conjugated double-bond. Structurally, carotenoids are lipophilic tetraterpenoids composed of eight isoprenoid units on a 40-carbon aliphatic polyene chain with conjugated double-bond [32]. Carotenoids have light-capturing functions and act as photoprotectors [9]. These compounds function as protective agents, extinguishing singlet oxygen that is highly reactive and dissipating excess excitation energy under high irradiance [33]. When the content of intracellular chlorophylls in microalgae is low, carotenoids transfer energy, playing an essential role in photosynthesis [34].

Carotenoids are effective compounds in protecting the photosynthetic apparatus against UV-B [35]. Carotenoids accumulate outside the cell system's thylakoid membrane after exposure to a stress condition, such as UV radiation, to protect cells from oxidative damage [9]. Moreover, these compounds have beneficial effects in preventing degenerative disease [36] and cancer [37], and act to protect the skin against sunlight [38]. Due to their antioxidant property, carotenoids protect cells from reactive radicals and prevent lipid peroxidation [39].

10.3.4 Phycocyanin

Proteins are targets of UV radiation due to their absorption maximums in the UV region. Among proteins, most phycobiliproteins are affected due to dependence on solar radiation [10]. Phycocyanin is a phycobiliprotein, as well as allophycocyanin and phycoerythrin [40]. Phycocyanin is a photosynthetic pigment found in microalgae [41], mainly in *Spirulina*.

Phycocyanin is a fluorescent protein due to linear prosthetic groups (bilins) that are linked to specific cysteine residues [40]. As a light-capturing complex, phycocyanin, a hydrophilic protein pigment (λ max 610–620 nm) [42], acts by absorbing light in regions of the visible spectrum that are poorly absorbed by chlorophyll [41].

Phycocyanin has promising antioxidant, anti-inflammatory, immunomodulatory, and photoprotective properties [40]. In vitro studies have shown that phycocyanin hijacks peroxyl, hydroxyl, and alkoxy radicals. Phycocyanin also inhibits lipid peroxidation [43, 44], decreases nitric oxide, and suppresses inducible nitric oxide synthase [45]. The photoprotection responses observed in phycocyanin can be attributed, in part, to its antioxidant capacity.

10.3.5 Polyamines

Polyamines are low-molecular-weight aliphatic amines [46] present in all living cells, in a protonated form, in concentrations in the magnitude of mM [47, 48]. Polyamines are polycationic compounds, whose functions are achieved according to bonds with nucleic acids, phospholipids, and some proteins [49]. Therefore, numerous roles have been reported for these compounds, including protection from nucleic acids, regulation of gene expression and protein translation, modulation of signal transduction, and stabilization of the cell membrane [9].

Polyamines play fundamental roles in maintaining physiological functions or against stress conditions. Microalgae are capable of producing a variety of polyamines [50]. Putrescine (Put), spermidine (Spd), and spermine (Spm) are normally found in most cells as common polyamines. In addition to Put, Spd, and Spm, microalgae also contain polyamines, such as diaminopropane (Dap), cadaverine (Cad), nor-spermidine (NSpd), and thermospermine (TSpm). NSpd is the most abundant polyamine in many microalgae [9, 50]. Some polyamines produced by microalgae have medical or nanotechnological applications [50].

Studies report the role of different polyamines in the UV stress tolerance of microalgae [9, 51, 52]. The high accumulation of polyamines in microalgae seems to be found in the early and late stages of cell growth [53]. Moreover, cell levels of polyamines can be regulated by light [54]. Its elevation during microalgal cultivation confers tolerance to UV-B radiation [55].

The positively charged polyamines under physiological ionic conditions occur as free molecular bases or conjugated or bound with negatively charged molecules inside the cell [56]. Polyamines can act as a free radical scavenging agent and protect against ROS-induced DNA damage [57]. The polyamine-DNA interaction has a direct influence on DNA stabilization and replication, which further regulates cell division, differentiation, and proliferation. In addition, polyamine's biosynthesis, conjugation, transport, and catabolism are highly influenced by environmental factors [56, 58]. Thus, the synthesis of polyamines induced by UV-B and the tolerance to UV-B stress suggests the potential of using these compounds of microalgal origin to mitigate oxidative stress induced by UV.

10.3.6 Scytonemin

Scytonemin is a fat-soluble yellowish-brown compound synthesized by microalgae [9, 59]. Chemically, scytonemin is a dimeric compound composed of indolic and phenolic subunits [60]. Scytonemin plays an essential role in photoprotection, minimizing damage to cellular machinery caused by UV radiation, as a result of its high ability to absorb a broad spectrum of UV radiations at 384 ± 2, 300, 278, and 252 nm [9]. Therefore, the production of scytonemin occurs due to high exposure to UV-A radiation, and its synthesis can be improved by exposure to high temperatures and oxidative stress of the cell [61].

Scytonemin can protect target cells by dissipating the absorbed UV radiation in the form of thermal energy. Scytonemin blocks the formation of the thymine dimer and acts as a potent antioxidant, preventing cell damage resulting from UV-induced ROS [9, 62]. This photoprotective compound can maintain normal physiology and biochemistry of the cell, preventing the incidence of up to 90% solar UVA radiation [63].

In addition to its UV absorption capacity, scytonemin has anti-inflammatory and antiproliferative properties through the inhibition of human polo-like kinase [64–66]. Thus, the high stability against stress factors such as temperature and UV-A radiation and its ability to perform UV screening, even after prolonged physiological inactivity, such as desiccation, make scytonemin a strong candidate for use as a natural UV protection compound for humans [66]. According to Fuentes-Tristan et al. [67], scytonemin can be applied in the pharmaceutical industry, in cosmetics, and in biomedical research.

10.4 Microalgal Biotechnology for the Production of Photoprotective Compounds

Microalgae are photosynthetic organisms found abundantly in freshwater, seawater, and soils. Many microalgae species can tolerate extreme conditions due to their unique metabolic pathways [68]. These microorganisms have strategies to avoid excessive excitation of the photosynthetic mechanism, such as the balance between light energy input and the activity of the energy sink [69]. Exposures to abiotic stresses, including solar UV radiation, compel microalgae to synthesize UV-absorbing compounds with photoprotective activity to maintain their cell viability [9].

Nitrogen is an essential nutrient since its availability significantly affects the growth and metabolism of microalgae [70]. The production of MAAs, for example, is induced by exposure to UV light and depends on the availability of nitrogen [71]. Chandra et al. [72] found that the abundant availability of nitrogen, oxidative stress, and UV-B showed an increase in the production of MAAs in *Lyngbya purpurea*. In a study by Rath et al. [73], the authors observed that the MAAs content in *Lyngbya aestuarii* increased with the increase in salinity. According to Klisch and Häder [74], organisms that grow in saline environments tend to have the highest concentrations of MAAs, which indicates that these compounds may also be part of osmotic regulation.

As for the production of MAAs, the supply of nitrogen combined with exposure to UV radiation has been shown to influence the content of phenolic compounds in microalgae. Al-Rashed et al. [12] cultivated *Spirulina platensis* and *Dunaliella salina* in BG-11 medium exposed to UV-B lamp (290–320 nm) at 0.50W/m^2 irradiance with and without nitrogen deprivation. The authors noted that the total phenolic content in *S. platensis* and *D. salina* increased significantly by 81.9% (5.33 mg/g) and 82.3% (5.87 mg/g), respectively, under the combination of UV-B exposure and nitrogen deprivation.

Another defense mechanism of microalgae, when exposed to UV radiation, is to change the composition of photosynthetic pigments [75]. Therefore, carotenoid levels may be one of the reasons for the diverse responses of *D. salina* to different doses of UV-B radiation. UV-B-induced damage may be associated with a dynamic balance between photodamage and adaptation [76].

Microcystis aeruginosa was grown in a limited culture medium (0.05 mg/l) or not (2.53 mg/l) by phosphorus and exposed to an intensity of 1.33 W/m^2 UV-B radiation. The carotenoid content increased significantly after nine days of exposure to UV-B for both treatments (with and without phosphorus limitation). However, the photosynthetic apparatus of the *M. aeruginosa* cell with phosphorus limitation did not function with prolonged exposure to UV-B radiation. After 15 days, exposure to UV-B rays and phosphorus limitation led to a significant decline in pigment content (phycocyanin and carotene) [77]. Kavitha et al. [78] cultivated *S. platensis* in Zarrouk medium, the light intensity of 30/μm^2/s, 12 hours light/dark photoperiod at 24 ± 2 °C and exposure to UV-B radiation (3 W/m^2). The microalgae produced UV protectors with maximum concentrations of carotenoid in 30 min (140 mg/l) and phycocyanin in 10 min (22 mg/l). Kokabi et al. [75] cultivated *Leptolyngbya cf. fragilis* in BG-11 medium, in these conditions: 12 hours light/dark photoperiod under 18 ± 2 W/m^2 illumination, UV-B radiation (1 W/m^2) and photosynthetically active radiation (PAR), for 12, 24, 48, and

72 hours. The carotenoid content increased in response to UV stress. On the other hand, phycocyanin content reduced. The microalgae produced maximum concentrations of carotenoid in 24 and 72 hours (0.29 mg/g) and phycocyanin in 12 hours (0.025 mg/ml).

Singh et al. [23] evaluated the cultivations of *Anabaena doliolum* and *Anabaena* strain L31 after 3, 6, 9, and 12 hours of exposure to UV-B radiation. The authors observed a reduction in the phycocyanin concentration up to three times for both microalgae after 12 hours of UV-B stress. Kannaujiya and Sinha [79] demonstrated a reduction of approximately 61 and 89% of fluorescence, after 48 hours, of exposure to UV-B in phycocyanin from *Nostoc* sp. HKAR-2 and *Nostoc* sp. HKAR-11, respectively. The reduction in fluorescence intensity of phycocyanin may be related to photodestruction of cysteine-associated chromophore in phycobiliproteins complex.

The effects of UV radiation on *Lyngbya* sp. A09DM were observed. The absorption spectra showed a reduction in the phycocyanin content after 5 hours of UV-A (6.88% $\pm$ 3.62) and UV-B (51.65% $\pm$ 2.86). This decrease was associated with light capture complexes absorb more than 99% of solar UV-B in photosynthetic organisms [10]. Kavitha et al. [78] reported that there was the dissociation of the protein aggregate and a 37.5% reduction of c-phycocyanin in *S. platensis* (9 mg/l) compared to the control (24 mg/l) after 60 minutes of exposure to UV-B radiation. This statement can be associated with the direct absorption of UV-B radiation on the outer surface of the thylakoid membrane, characteristic of the protein.

Polyamine content in microalgae is also influenced by adapting to environmental stress [50]. *Scenedesmus obliquus*, for example, when is grown, under UV radiation, resulted in a decrease in Put levels associated with thylakoids, while Spm levels increased [55]. UV-induced cell damage in *Synechocystis* sp. PCC 6803 resulted in decreased Spd content [80]. Moreover, the stress caused by high temperatures and salinity caused an increase in the polyamine content [81].

Regardless of the synthetic pathway used by microalgae, polyamine synthesis requires energy, in addition to other essential nutrients, such as amino acids. Thus, the nutritional limitation is one of the factors that affect the content of this photoprotective compound in microalgae. Conditions of nitrogen limitation result in a reduction of the polyamine content [82]. Moreover, some microalgae, such as *Thalassiosira pseudonana*, excrete specific polyamines in the surrounding environment, both under standard growing conditions and cultivations with reduction nitrogen [81].

In addition to UV radiation, production of scytonemin is affected by factors such as desiccation [61, 67], temperature, light intensity, salinity [67, 83, 84], light-dark cycle, and photooxidative and osmotic stresses [67]. In a study developed by Rastogi et al. [62], UV radiation showed a stimulative effect on scytonemin production by *Scytonema* sp. R77DM. Besides, a significant synergistic improvement in the synthesis of scytonemin was verified under the combination of heat and UV radiation.

In this way, considering that the production of microalgae metabolites using standardized culture conditions can be expensive and time-consuming, novel cultivation strategies must be explored. Thus, aiming at economic viability and increasing the production of photoprotective compounds, exposure to UV radiation can be an adequate alternative for converting biomass into a high-value metabolite [85], as shown in Table 10.1.

Table 10.1 Production of UV protector compounds by microalgae.

Microalgae	UV protector compound	Stress factors	Maximum concentration of biocompound	References
Polarella glacialis	Carotenoid: Xanthophyll	UV-B (0–0.20 W/m^2)	0.0010 ng/cell	[86]
Chaetoceros dichaeta	Carotenoid: Xanthophyll	UV-B (0–0.20 W/m^2)	0.0017 ng/cell	[86]
Dunaliella bardawil	Carotenoid: β-carotene	UV-A (8.7 W/m^2) PAR[a] (250 W/m^2), nitrogen starvation	51.5 pg/cell	[87]
Dunaliella bardawil	Carotenoid: violaxanthin	UV-A (8.7 W/m^2) PAR[a] (250 W/m^2), nitrogen starvation	0.04 pg/cell	[87]
Dunaliella bardawil	Carotenoid: zeaxanthin	UV-A (8.7 W/m^2) PAR[a] (250 W/m^2), nitrogen starvation	0.25 pg/cell	[87]
Haematococcus pluvialis	Carotenoid: astaxanthin	UV-B (0.73 W/m^2)	23 μg/cell	[88]
Coccomyxa onubensis	Carotenoid: lutein	PAR*a* + UV-A (8.7 W/m^2)	7.07 mg/g DW[b]	[34]
Coccomyxa onubensis	Carotenoid: lutein	PAR*a* + UV-B (0.16 W/m^2)	6.10 mg/g DW[b]	[34]
Coccomyxa onubensis	Carotenoid: β-carotene	PAR*a* + UV-A (8.7 W/m^2)	1.32 mg/g DW[b]	[34]
Coccomyxa onubensis	Carotenoid: β-carotene	PAR*a* + UV-B (0.16 W/m^2)	0.95 mg/g DW[b]	[34]
Anabaena doliolum	MAAs: shinorine, Porphyra-334 and mycosporine-glycine	UV-B (1 W/m^2), 9 h	1600 nmol/g DW[b]	[23]
Anabaena strain L31	MAA: shinorine	UV-B (1 W/m^2), 9 h	1400 nmol/g DW[b]	[23]
Chlamydomonas hedleyi	MAA: porphyra-334	UV-A (23 mW/cm^2)	0.12 μmol/g	[21]
Chlamydomonas hedleyi	MAA: shinorine	UV-A (23 mW/cm^2)	0.17 μmol/g	[21]
Chlamydomonas hedleyi	MAA: mycosporine-glycine	UV-A (23 mW/cm^2)	0.32 μmol/g	[21]
Lyngbya purpurem	MAAs: porphyra-334 and shinorine	UV-B (5–6.3 W/m^2)	17.40 mg/ml	[72]
Lyngbya purpurem UTEX LB 2716	MAAs: porphyra-334 and Shinorine	UV-B (6.8 W/m^2), 12 h	17.5 mg/g DW*b*	[85]

Table 10.1 (Continued)

Microalgae	UV protector compound	Stress factors	Maximum concentration of biocompound	References
Lyngbya purpurem UTEX LB 2716	MAAs: porphyra-334 and shinorine	UV-A (7 W/m^2), 12 h	10.5 mg/g DW[b]	[85]
Leptolyngbya cf. *fragilis*	Scytonemin	PAR*a* + UV-B (1 W/m^2), 48 h	0.42 mg/g DW[b]	[75]
Fischerella sp. strain HKAR-13	Scytonemin	PAR*a* + UV-A + UV-B (12, 3.5 and 0.5 W/m^2, respectively), 72 h	1.1 mg/g DW[b]	[24]
Chlorella sp.	Phenolic compounds	UV-B (470 $\mu W/cm^2$), for three days (1 h/day)	0.39 μg/mg	[89]
Spirulina platensis	Phenolic compounds: polyphenols	Artificial UV radiation (20 W fluorescent lamp), 3 h	33.2 mg gallic acid/g	[90]
Spirulina platensis	Phenolic compounds: phenolics	UV-B (0.50 W/m^2), for three days (2 h/day), nitrogen starvation	5.33 mg/g	[12]
Dunaliella salina	Phenolic compounds: phenolics	UV-B (0.50 W/m^2), for three days (2 h/day), nitrogen starvation	5.87 mg/g	[12]
Spirulina platensis	Phenolic compounds: flavonoids	UV-B (0.50 W/m^2), for three days (2 h/day), nitrogen starvation	1.95 mg/g	[12]
Dunaliella salina	Phenolic compounds: flavonoids	UV-B (0.50 W/m^2), for three days (2 h/day), nitrogen starvation	2.29 mg/g	[12]
Spirulina sp.	Phenolic compounds: flavonoids	UV-C (8000 μW/cm)	0.0022 catechin/μg	[91]
Spirulina sp.	Phenolic compounds	UV-C (8000 μW/cm)	0.0025 gallic acid/μg	[91]

[a] PAR = Photosynthetically active radiation.
[b] DW = Dry weight.
Source: Based on [85].

10.5 Effects of UV Radiation on the Growth, Morphology, and Production of Lipids, Proteins, and Carbohydrates

In addition to harmful effects on nucleic acids and proteins, several biological functions including motility and photo-orientation, cell differentiation, membrane permeability, productivity, pigmentation, photosynthesis, CO_2 assimilation, enzyme metabolism, and N_2 fixation are affected by the incidence of UV radiation in microalgae [9, 92, 93]. Microalgae develop mechanisms of molecular resilience, physiological and biochemical mechanisms, and self-regulatory processes to maintain normal biological functioning. Therefore, these microorganisms overcome the stress and adversities of UV light [9].

Studies mention that low doses of UV radiation stimulate the growth of microalgae, while high doses cause an inhibitory effect. High doses of UV radiation usually cause stress, leading to DNA damage, impaired photosynthesis, lipid peroxidation, and the formation of cyclobutane pyrimidine dimer [76, 94–96]. The high UV radiation (280–400 nm) tends to damage the cellular structure of microalgae. With this, the genomic stability is altered, and the photochemical and photobiological processes are inhibited [97]. Nevertheless, microalgae showed a tendency to acquire a high degree of adaptation against UV radiation [98].

Kumar et al. [99] evaluated the growth, biomass, and accumulation of lipids, carbohydrates, and proteins of the microalgae *Chlorella sorokiniana* UUIND6 grown under LED light (control) and UV light (280–320 nm). The cultivation with UV treatment showed an adaptation phase of four days, and after that period, there was an increase in the growth rate, probably due to DNA repair. According to Hull et al. [100], DNA repair of microalgae cells occurs after 24 hours of UV light treatment. Besides, low productivity in the UV condition (43 mg/l/d) was observed compared to the control (55 mg/l/d). The accumulation of lipids and carbohydrates by microalga was reduced by approximately 20% with UV radiation. On the other hand, UV treatment showed an increase in protein content (23%).

Kurinjimalar et al. [101] evaluated the impact of UV-B stress on microalga *Botryococcus braunii* treated at 1 and 5 W/m^2, also analyzing the time of radiation exposure. The biomass yield gradually decreased with an increase in the intensity of UV-B treatment. Moreover, the longer the exposure period to UV treatment, the higher the effect of the decrease in biomass concentration. The levels of primary metabolites, such as total carbohydrates, proteins, and lipids, have been drastically reduced under high doses of UV-B radiation. However, with UV-B treatment for 15 min at 1 W/m^2, it was observed that the carbohydrate content (295 ± 3.54 mg/l) was equal to the control (285 ± 4.28 mg/l) on the 15th day. Regarding proteins, a reduction of 57% (55.02 ± 2.11 mg/l) and 79% (28.76 ± 2.88 mg/l) was observed when the microalga was treated with 1 and 5 W/m^2 UV-B for 60 minutes, on the 15th day, respectively, compared to the control (132.36 ± 4.53 mg/l).

In another study, UV radiation was applied to four microalgae genera, *Scenedesmus* sp., *Pseudokrichneriella* sp., *Nannochloropsis* sp., and *Chlorella* sp., for lipid induction. The highest lipid content (49.4 μg/ml) was observed in *Chlorella* with 60 minutes of exposure to UV light. Besides, the authors found that all microalgae showed rapid growth during the exponential phase when compared to cultures without UV treatment. This study highlighted the UV mutation as a viable approach to improve the productivity of biomass and lipids in microalgae. According to the authors, the stress caused by UV radiation would have disrupted the cells that released the lipids in the medium, maximizing the content of this macromolecule [102].

Some studies have addressed the photosynthetic performance of microalgae when exposed to UV light. Zhang et al. [76] cultivated *D. salina* exposed to intensity of the UV-B irradiance of 0.16 W/m^2. Four daily irradiation treatments were performed (0.25–1.00 kJ/m^2d) to elucidate what occurs during photosynthesis when the microalga is subjected to UV-B radiation. The results indicated that the UV-B treatment interferes in the light absorption of *D. salina*. UV-B stress can decrease the light energy utilization of photosystem II. Moreover, it may cause damage to the photosynthetic capacity in light reaction.

Compared to the control condition, the authors noted an increase in Rubisco (ribulose-1,5-bisphosphate carboxylase/oxygenase) activity in treated groups. This discovery indicates that the microalga has an adaptive mechanism to moderate UV-B radiation.

Singh et al. [103] verified that UV-B radiation reduced the photosynthetic performance of photosystem II and increased cell size by ~15–65%, being more pronounced in the case of *Chlorococcum humicola* at 4 hours of UV-B. Furthermore, there was an increase in the content of microelements (C, O, Na, and K) in *Chlorella vulgaris*. UV-B treatment reduced the protein content in microalgae. Related to carbohydrates, and total lipids, both microalgae showed an increasing trend up to UV-B exposure for 1 hour. Moreover, after 1h of UV-B radiation, there was a reduction of these macromolecules.

The high adaptability of microalgae against UV radiation is relevant for the production of lipids, accumulation of nutrients, carbon sequestration, environmental remediation, and oxygen production in the environment [104]. Therefore, with the incidence of UV radiation increasing continuously, it is necessary to explore and study the various responses obtained by microalgae under these stress conditions. Thus, it will be possible to identify the best strain to meet the demand for high biomass productivity and the synthesis of photoprotective metabolites in the future.

10.6 Extraction Methods of Photoprotective Compounds

The process of extracting biomolecules from microalgae biomass begins with the selection of the appropriate microalgae strain, as well as the cultivation conditions, followed by downstream processing (pretreatment, harvesting, extraction, and purification). For efficient extraction and purification of high-value compounds from microalgae biomass, the harvesting must be carried out with the appropriate technique to separate the biomass from the culture medium [105].

The choice of method has a significant influence on the yield and activity of the extracted compounds. Ideal conditions of the extraction process, such as solvent, pH, temperature, and so on, must be based on the nature of the extracted compound. Conventional extraction techniques include extraction in Soxhlet apparatus, solid–liquid extraction, and liquid–liquid extraction [106, 107]. However, traditional methods have some disadvantages, such as the use of a high volume of toxic solvents, arduous separation of the compounds from the medium, and long extraction times. Thus, the search for solvent-free extraction methods, such as extraction with a supercritical fluid, microwave-assisted and ultrasound-assisted extractions, enzyme-assisted extraction, and pressurized liquid extraction, have been growing in recent years [29, 107, 108]. These techniques are more ecological and characterized by higher extraction yield, shorter process time, where there is no loss of activity of the compounds [109].

Most MAAs extraction processes, as well as other compounds, primarily use centrifugation of microalgae biomass. From the sediment, the MAAs are extracted with 100% methanol by incubation overnight at 4 °C. After extraction, the aliquots are subjected to centrifugation, followed by evaporation of the methanolic extract. The dry residue is redissolved in water, followed by the addition of chloroform. The samples are centrifuged, and after the upper aqueous phase is filtered through 0.2 μm microcentrifuge filters to obtain the partially purified MAAs. The aqueous phase is analyzed by UV spectroscopy between 200 and 400 nm, and Fourier Transform Infrared pectroscopy is used to confirm the presence of UV protector compounds [72, 110, 111]. Lalegerie et al. [22] proposed to extract MAAs with 70% ethanol, together with the analysis by high-performance liquid chromatography (HPLC), and this method provided reliable data. The use of less toxic solvent, such as ethanol, allowed the valorization of the extracted MAAs since these compounds have significant potential for applications in human health and cosmetics.

There are many methods for isolating phenolic compounds from raw materials of natural origin. Organic solvents, such as alcohols (methanol, ethanol), ethyl acetate, acetone, acetonitrile, or combination between them in an aqueous solution, are used mainly for the isolation of phenolic compounds. Novel methods are studied to reduce costs and energy, as well as to minimize the volume of solvents used. In this context, Korzeniowska et al. [29] evaluated different types of extraction (solvent extraction by shaking, ultrasonic assisted-extraction using an ultrasonic bath, and microwave-assisted extraction) to isolate phenolic compounds. Besides, for each extraction method, five different polar extractors were used: water, methanol, ethanol, and aqueous solutions of these alcohols [112]. Phenolic compounds can easily undergo hydrolysis or oxidation at high temperatures. Therefore, a range between 20 and 50 °C and short extraction time are recommended [113]. In this way, Korzeniowska et al. [29] performed the extraction for 2 hours at a maximum temperature of 40 °C and concluded that ethanol, being safer, can be an alternative to methanol. Moreover, the microwave-assisted extraction with water was the most effective in isolating phenolic compounds.

Studies related to the solubility of carotenoids in solvents such as acetone, petroleum ether, hexane, diethyl ether, ethanol, methanol, and dichloromethane were carried out to evaluate the ability to extract carotenoids from microalgae. However, the use of dichloromethane is restricted because it is corrosive and carcinogenic [114, 115]. Hexane is used to meet commercial specifications for large-scale production [116, 117]. Meanwhile, for the use of hexane, more than eight extraction steps are required, and its use wastes high volumes of solvent. Furthermore, hexane is not able to recover all types of polar carotenoids, making it necessary to use a solvents mixture of different polarities [116]. Thus, the objective of Cerón-García et al. [116] was to optimize the extraction of microalgal carotenoids to establish a relatively easy and reliable method for analyzing them by HPLC. For this, saponification was performed after the rupture of the cells, varying the temperature and concentrations of sodium hydroxide (KOH). According to the authors, saponification conditions must be selected for each strain, and the recovery solvent must be defined based on the carotenoid to be purified.

Conventional carotenoid extraction techniques are time-consuming and generally require the use of dry algae biomass with a moisture content of less than 10% [118]. In a study by Damergi et al. [119], the extraction of carotenoids was carried out with wet biomass and more ecological extraction techniques, using green solvents, such as 2-methyl tetrahydrofuran (MTHF), combined or not with ethanol. The combined use of MTHF and ethanol provided an increase of 70% in the extraction yield.

Temperature, pH, light, and isoelectric point are parameters that must be adjusted during the extraction process of active compounds to maintain their stability. In this case, it can be considered antioxidants compounds as phycocyanin [40]. The appropriate method of cell disruption will depend on the characteristics of the cell wall of each species of microalga [120]. Cell disruption can be performed by physical methods (freezing and thawing cycles, precipitation, sonication, osmotic shocks, nitrogen lysis, high-pressure homogenization, cavitation, and supercritical fluids), chemical methods (acids, alkalis, detergents, and enzymes), or by a combination of both [40]. As phycocyanin is soluble in water, a solid–liquid extraction is performed by adding water and buffers [121]. Methods for purifying phycocyanin include precipitation with ammonium sulfate, biphasic aqueous system, membrane separation, and chromatographic methods in a fixed-bed or expanded bed column [40, 122-123].

Polyamines exist in different forms in the cells in which may be in free form, and insoluble form bound to structural molecules, or covalently conjugated to proteins and peptides. Therefore, the extraction method can influence the detection of cellular polyamines. Free-form polyamines transported between cellular organelles are the most easily extracted and analyzed [50]. In general, the analysis of polyamines obtained from biological samples is carried out by chromatographic separation before detection with selective detectors or mass spectrometers. Hyphenated techniques, such as liquid chromatography/mass spectroscopy (LC/MS) [124–127], gas chromatography/mass spectroscopy (GC/MS) [128] or capillary electrophoresis coupled to mass spectrometry (MS) [129], are among the methods for analyzing the metabolites. Hamana et al. [130] used HPLC and high-performance gas chromatography to determine the polyamines of various microalgae extracted with 6M hydrochloric acid.

On the other hand, many of these methods have disadvantages related to long chromatographic execution times, used for the determination of polyamines only in a specific matrix, higher limits of detection, or requirement for sample pretreatment. In this context, Magnes et al. [127] developed a robust, high-throughput solid-phase extraction online coupled to a liquid chromatography–tandem mass spectrometry (SPE–LC/MS/MS) approach for the simultaneous determination of various polyamines in biological samples. The main advantages are the short process time (4 minutes), and the fact that the use of two SPE cartridges online-coupled to LC/MS/MS minimizes sample pretreatment to a single derivatization step.

According to Dionisio [66], scytonemin can be extracted with nonpolar solvents, such as acetone or ethyl acetate, and authenticated by spectrophotometry or chromatography. Generally, scytonemin is extracted with methanol and ethyl acetate (1:1, v/v) by overnight incubation at 4 °C followed by homogenization and centrifugation. The residues are redissolved in methanol or a mixture of solvents, such as methanol: ethyl acetate [62, 110, 111]. Before the centrifugation step, the sample can also be subjected to sonication. Samples are evaluated by spectroscopic analyzes between 200 and 700 nm, assuming that the ones with absorption between 325 and 425 nm and maximum absorption at 384 nm contain scytonemin [111], or analyzed by HPLC. For this analysis, the residues redissolved in methanol are centrifuged to remove impurities insoluble in methanol. The supernatants are evaporated and redissolved in methanol:ethyl acetate (1:1, v/v). Next, they are filtered through a disposable microcentrifuge syringe filter with a pore size of 0.2 μm before HPLC analysis [62].

10.7 Prospects for Commercial Applications

Microalgae present rapid growth and require relatively simple requirements such as carbon dioxide, light energy, and nutritional elements. In this way, microalgal biomass is an attractive source for the production of metabolites with bioactive properties. With this, the synthesis of high-value bioproducts by microalgae for commercial applications has been explored [131]. Several species have been grown commercially, and the biomass produced is made available as a source of bioproducts for consumption, such as food and cosmetics [132].

Products containing mycosporine-like amino acids (MAAs), such as Helioguard® and Helionori®, extracted from *Porphyra umbilicalis*, have been studied and are commercially available. Helioguard® 365 contains the MAAs shinorin and porphyrin-334 and is a natural UV screening compound and ideal for daily use, ensuring skin protection against photoaging [133]. Helionori® contains shinorin, palitine, and porphyrin-334 MAAs and provides natural protection of the skin and its components against UVA-induced damage, such as membrane lipids and DNA, avoiding early photoaging [134]. Both products are of algal origin. Therefore, the production and commercialization of these products from microalgae will be more explored.

Microalgae are also promising and renewable sources of active compounds [132, 135], such as phycocyanin and carotenoids [136]. Phycocyanin is a natural, nontoxic, biodegradable, and noncancerous pigment [40], which yields around US $10–50 million per year to the world market [137]. In 2019, the global market was valued at US $60 million with manufacturers and suppliers of phycobiliproteins, such as Algapharma BiotechCorp, Columbia Bioscience, and QuantaPhy Inc. [138]. In the food industry, phycocyanin was the first natural dye authorized by the Food and Drug Administration in the United States. Companies like Norland Biotech and Ozone Naturals™ market phycocyanin for applications such as food coloring and cosmetics. The cost of phycocyanin can reach from US $160 to 180/kg [40].

Microalgae carotenoids represent a successful model in terms of economic viability [81]. β-Carotene is produced by microalgae such as *D. salina, Dunaliella bardawil, Botryococcus braunii, Chlamydomonas nivalis, Chlamydomonas acidophila, Chlorococcum* sp., *Chlamydocapsa* sp., and *Coelastrella striolata* [71], providing health benefits such as corneal protection, antiaging, and prevention of cardiovascular diseases [139]. Lutein is another carotenoid produced by microalgae, mainly *Chlorella* sp., *Coelastrum proboscideum*, *C. acidophila, Neospongiococcum, Chlamydocapsa* sp., *Muriellopsis* sp. [71].

The cost of β-carotene produced from *D. salina* ranges from US $300 to 500/kg [140], while the cost of microalgae lutein ranges from US $910 to 15,000/kg [138]. Industries have been producing carotenoids from microalgae biomass [141]. Shaivaa Algaetech is a company that cultivates microalgae to produce natural antioxidants for use as a food supplement and cosmetics. Shaivaa Algaetech cultivates microalgae *D. salina* and *Haematococcus*, which are rich sources of carotenoids such as β-carotene and astaxanthin, respectively.

Mapoung et al. [2] evaluated the UV-B absorption capacity of *S. platensis*, determining its SPF. The authors revealed that the extract of *S. platensis* contained high concentrations of phenolic compounds ($23{,}465.71 \pm 167.72$ mg of gallic acid kg extract), with SPF of 30.39 ± 3.37 (3 mg/ml). In addition to *Spirulina* sp. extract, *Chlorella* sp. was also applied in the formulation of sunscreen [142]. The extract of *Chlorogloeopsis* sp. protects the keratinous tissue, preventing damage caused by UV-A and UV-B and, consequently, photoaging, the formation of wrinkles, and sagging skin. *Isochrysis* sp. is also applied in sunscreen formulations that contain organic and inorganic filters with SPF 15 to prevent the transmission of UV rays. Furthermore, *Nannochloropsis* sp. showed effective measures against UV-A and UV-B radiations [143]. Therefore, microalgal extracts are considered prospective sources of UV-A and UV-B absorbers with an antioxidant capacity to serve as an additive in photoprotective cosmetic formulations to prevent the harmful effects of UV radiation.

10.8 Conclusion and Perspectives

UV radiation has been considered one of the environmental stressors that alter the physiology and metabolism of photosynthetic microorganisms such as microalgae. However, microalgae are capable of developing resilience mechanisms and rapid regulatory processes to adapt to UV stress conditions. In response to adaptation, these microorganisms synthesize UV-absorbing compounds with photoprotective activity, such as MAAs, carotenoids, phenolic compounds, phycocyanin, scytonemin, and polyamines. In addition to growth and morphology, UV radiation has a significant effect on the accumulation of macromolecules, such as proteins, carbohydrates, and lipids by microalgae. Concerning the extraction of photoprotective compounds from microalgae biomass, more ecological methods have been applied, as they provide reliable data, do not use toxic materials, and are characterized by higher extraction yield and shorter process time. Microalgal biomass, as well as its metabolites and photoprotective compounds, have a great potential to be used in the future to allow an industrial production to be commercialized or used for the development of new products.

References

1 Simo, A., Kawal, N., Paliyath, G., and Bakovic, M. (2014). Botanical antioxidants for skin health in the world of cosmeceuticals. *International Journal of Nutrition and Health Sciences* 2: 67–88.

2 Mapoung, S., Arjsri, P., Thippraphan, P. et al. (2020). Photochemoprotective effects of *Spirulina platensis* extract against UVB irradiated human skin fibroblastos. *South African Journal of Botany* 130: 198–207.

3 Farage, M.A., Miller, K.W., Elsner, P., and Maibach, H.I. (2008). Functional and physiological characteristics of the aging skin. *Aging Clinical and Experimental Research* 20: 195–200.

4 Poljšak, B., Dahmane, R.G., and Godić, A. (2012). Intrinsic skin aging: the role of oxidative stress. *Acta Dermatovenerol Alp Pannonica Adriat* 21: 33–36.

5 Mesa-Arango, A.C., Flórez-Muñoz, S.V., and Sanclemente, G. (2017). Mechanisms of skin aging. *Iatreia* 30: 160–170.

6 Dutra, E.A., Oliveira, D.A.G.V., Kedor-Hackmann, E.R.M., and Santoro, M.I.R.M. (2004). Determination of sun protection factor (SPF) of sunscreens by ultraviolet spectrophotometry. *Brazilian Journal of Pharmaceutical Sciences* 40: 381–385.

7 Rodrigues, M.J., Matkowski, A., Slusarczyk, S. et al. (2019). Sea knotgrass (*Polygonum maritimum* L.) as a potential source of innovative industrial products for skincare applications. *Industrial Crops and Products* 128: 391–398.

8 Costa, J.A.V., Moraes, L., Moreira, J.B. et al. (2017). Microalgae-based biorefineries as a promising approach to biofuel production. In: *Prospects and Challenges in Algal Biotechnology* (eds. B.N. Tripathi and D. Kumar), 113–140. Singapore: Springer.

9 Rastogi, R.P., Madamwar, D., Nakamoto, H., and Incharoensakdi, A. (2020). Resilience and self-regulation processes of microalgae under UV radiation stress. *Journal of Photochemistry and Photobiology, C: Photochemistry Reviews* 43: 100322.

10 Rastogi, R.P., Sonani, R.R., and Madamwar, D. (2015). Effects of PAR and UV radiation on the structural and functional integrity of phycocyanin, phycoerythrin and allophycocyanin isolated from the marine cyanobacterium Lyngbya sp. A09DM. *Photochemistry and Photobiology* 91: 837–844.

11 Zhang, X., Tang, X., Wang, M. et al. (2017). ROS and calcium signaling mediated pathways involved in stress responses of the marine microalgae *Dunaliella salina* to enhanced UV-B radiation. *Journal of Photochemistry and Photobiology B: Biology* 173: 360–367.

12 Al-Rashed, S.A., Ibrahim, M.M., El-Gaaly, G.A. et al. (2016). Evaluation of radical scavenging system in two microalgae in response to interactive stresses of UV-B radiation and nitrogen starvation. *Saudi Journal of Biological Sciences* 23: 706–712.

13 Ramos, B., Costa, G.B., Ramlov, F. et al. (2019). Ecophysiological implications of UV radiation in the interspecific interaction of *Pyropia acanthophora* and *Grateloupia turuturu* (Rhodophyta). *Marine Environmental Research* 144: 36–45.

14 Bonaventura, R. and Matranga, V. (2017). Overview of the molecular defense systems used by sea urchin embryos to cope with UV radiation. *Marine Environmental Research* 128: 25–35.

15 Shanuja, S.K., Iswarya, S., and Gnanamani, A. (2018). Marine fungal DHICA as a UVB protectant: assessment under in vitro and in vivo conditions. *Journal of Photochemistry and Photobiology B: Biology* 179: 139–148.

16 Yarosh, D.B. (2002). Enhanced DNA repair of cyclobutane pyrimidine dimers changes the biological response to UV-B radiation. *Mutation Research, Fundamental and Molecular Mechanisms of Mutagenesis* 509: 221–226.

17 Hartmann, A., Glaser, K., Holzinger, A. et al. (2020). Klebsormidin A and B, Two New UV-Sunscreen Compounds in Green Microalgal *Interfilum* and *Klebsormidium* Species (Streptophyta) From Terrestrial Habitats. *Frontiers in Microbiology* 11: 1–12.

18 Lai, J.W.S., Lim, P.E., Wong, C.Y. et al. (2019). Photosynthetic response and DNA mutation of tropical, temperate and polar *Chlorella* under short-term UVR stress. *Polar Science* 20: 35–44.

19 Limtrakul, P., Yodkeeree, S., Punfa, W., and Srisomboon, J. (2016). Inhibition of the MAPK signaling pathway by red rice extract in UVB-irradiated human skin fibroblasts. *Natural Product Communications* 11: 1877–1882.

20 Rastogi, R.P., Sonani, R.R., Madamwar, D., and Incharoensakdi, A. (2016). Characterization and antioxidant functions of mycosporine-like amino acids in the cyanobacterium *Nostoc* sp. R76DM. *Algal Research* 16: 110–118.

21 Suh, S.S., Hwang, J., Park, M. et al. (2014). Anti-inflammation activities of mycosporine-like amino acids (MAAs) in response to UV radiation suggest potential anti-skin aging activity. *Marine Drugs* 12: 5174–5187.

22 Lalegerie, F., Lajili, S., Bedoux, G. et al. (2019). Photo-protective compounds in red macroalgae from Brittany: considerable diversity in mycosporine-like amino acids (MAAs). *Marine Environmental Research* 147: 37–48.

23 Singh, G., Babele, P.K., Sinha, R.P. et al. (2013). Enzymatic and non-enzymatic defense mechanisms against ultraviolet-B radiation in two *Anabaena* species. *Process Biochemistry* 48: 796–802.

24 Rajneesh, Rajeshwar, J., Richa et al. (2019). Impacts of ultraviolet radiation on certain physiological and biochemical processes in cyanobacteria inhabiting diverse habitats. *Environmental and Experimental Botany* 161: 375–387.

25 Jofre, J., Celis-Plá, P.S.M., Figueroa, F.L., and Navarro, N.P. (2020). Seasonal variation of mycosporine-like amino acids in three subantarctic red seaweeds. *Marine Drugs* 18: 1–17.

26 Lima, M.C., Sousa, C.P., Fernandez-Prada, C. et al. (2019). A review of the current evidence of fruit phenolic compounds as potential antimicrobials against pathogenic bacteria. *Microbial Pathogenesis* 130: 259–270.

27 Andrade, M.A., Lima, V., Silva, A.S. et al. (2019). Pomegranate and grape by-products and their active compounds: are they a valuable source for food applications? *Trends in Food Science & Technology* 86: 68–84.

28 Singh, B., Singh, J.P., Kaur, A., and Singh, N. (2018). Phenolic compounds as beneficial phytochemicals in pomegranate (*Punica granatum* L.) peel: a review. *Food Chemistry* 261: 75–86.

29 Korzeniowska, K., Łęska, B., and Wieczorek, P.P. (2020). Isolation and determination of phenolic compounds from freshwater *Cladophora glomerata*. *Algal Research* 48: 101912.

30 Działo, M., Mierziak, J., Korzun, J. et al. (2016). The Potential of plant phenolics in prevention and therapy of skin disorders. *International Journal of Molecular Sciences* 160 (17): 1–41.

31 Hamidi, M., Safarzadeh Kozani, P., Safarzadeh Kozani, P. et al. (2020). Marine bacteria versus microalgae: who is the best for biotechnological production of bioactive compounds with antioxidant properties and other biological applications? *Marine Drugs* 18: 1–38.

32 Wong, H.J., Mohamad-Fauzi, N., Rizman-Idid, M. et al. (2019). Protective mechanisms and responses of micro-fungi towards ultraviolet-induced cellular damage. *Polar Science* 20: 19–34.

33 Shourie, A., Tomar, P., Srivastava, D., and Chauhan, R. (2014). Enhanced biosynthesis of quercetin occurs as a photoprotective measure in *Lycopersicon esculentum* Mill. under acute UV-B exposure. *Brazilian Archives of Biology and Technology* 57: 317–325.

34 Bermejo, E., Ruiz-Domínguez, M.C., Cuaresma, M. et al. (2018). Production of lutein, and polyunsaturated fatty acids by the acidophilic eukaryotic microalga *Coccomyxa onubensis* under abiotic stress by salt or ultraviolet light. *Journal of Bioscience and Bioengineering* 125: 669–675.

35 Rakhimberdieva, M.G., Stadnichuk, I.N., Elanskaya, I.V., and Karapetyan, N.V. (2004). Carotenoid-induced quenching of the phycobilisome fluorescence in photosystem II-deficient mutant of *Synechocystis* sp. *FEBS Letters* 574: 85–88.

36 Rodriguez-Amaya, D.B. (2015). *Food Carotenoids: Chemistry, Biology and Technology*. Oxford: IFT Press-Wiley.

37 Jain, A., Sharma, G., Kushwah, V. et al. (2017). Fabrication and functional attributes of lipidic nanoconstructs of lycopene: an innovative endeavour for enhanced cytotoxicity in MCF-7 breast cancer cells. *Colloids and Surfaces. B, Biointerfaces* 152: 482–491.

38 Stahl, W. and Sies, H. (2012). β-Carotene and other carotenoids in protection from sunlight. *The American Journal of Clinical Nutrition* 96: 1179S–1184S.

39 Grossman, A.R., Lohr, M., and Im, C.S. (2004). *Chlamydomonas reinhardtii* in the landscape of pigments. *Annual Review of Genetics* 38: 119–173.

40 Morais, M.G., Prates, D.F., Moreira, J.B. et al. (2018). Phycocyanin from microalgae: properties, extraction and purification, with some recent applications. *Industrial Biotechnology* 14: 30–37.

41 Schipper, K., Fortunati, F., Oostlander, P.C. et al. (2020). Production of phycocyanin by *Leptolyngbya* sp. in desert environments. *Algal Research* 47: 101875.

42 Kumar, D., Kannaujiya, V.K., Richa et al. (2018). Composition and functional property of photosynthetic pigments under circadian rhythm in the cyanobacterium *Spirulina platensis*. *Protoplasma* 255: 885–898.

43 Romay, C.H., Gonzaléz, R., Ledón, N. et al. (2003). C-Phycocyanin: a biliprotein with antioxidant, anti-inflammatory and neuroprotective effects. *Current Protein & Peptide Science* 4: 207–216.

44 Shalaby, E.A. and Shanab, S.M.M. (2013). Antiradical and antioxidant activities of different *Spirulina platensis* extracts against DPPH and ABTS radical assays. *Journal of Marine Biology & Oceanography* 2: 1–8.

45 Cherng, S.C., Cheng, S.N., Tarn, A., and Chou, T.C. (2007). Anti-inflammatory activity of c-phycocyanin in lipopolysaccharide-stimulated RAW 264.7 macrophages. *Life Sciences* 81: 1431–1435.

46 Burczyk, J., Zych, M., Ioannidis, N.E., and Kotzabasis, K. (2014). Polyamines in Cell Walls of Chlorococcalean Microalgae. *Zeitschrift fur Naturforschung C: Journal of Biosciences* 69: 75–80.

47 Bregier-Jarzebowska, R., Hoffmann, S.K., Łomozik, L. et al. (2019). Studies of ternary complexes formed in the biocoordination systems including copper (II) ions, polyamines and L-lysine. *Polyhedron* 173: 114137.

48 Esmer, I., Tüney, I., Özakça, D.Ü., and Sukatar, A. (2017). Protective effects of polyamines against UV-A andUV-B illumination in *Physcia semipinnata* thalli. *Botanica Serbica* 41: 17–24.

49 Kabir, A., Jash, C., Paygahan, P.V. et al. (2020). Polyamines and its analogue modulates amyloid fibrillation in lysozyme: a comparative investigation. *BBA – General Subjects* 1864: 129557.

50 Lin, H.-Y. and Lin, H.-J. (2019). Polyamines in microalgae: something borrowed, something new. *Marine Drugs* 17: 1–23.

51 Pothipongsa, A., Jantaro, S., and Incharoensakdi, A. (2012). Polyamines induced by osmotic stress protect *Synechocystis* sp PCC 6803 cells and arginine decarboxylase transcripts against UV-B radiation. *Biochemistry and Biotechnology* 168: 1476–1488.

52 Jantaro, S., Baebprasert, W., Piyamawadee, C. et al. (2014). Exogenous spermidine alleviated UV-induced growth inhibition of *Synechocystis* sp. PCC 6803 via reduction of hydrogen peroxide and malonaldehyde levels. *Applied Biochemistry and Biotechnology* 173: 1145–1156.

53 Jantaro, S. and Kanwal, S. (2017). Low-molecular-weight nitrogenous compounds (GABA and Polyamines) in blue green algae. In: *Algal Green Chemistry: Recent Progress in Biotechnology* (eds. R.P. Rastogi, D. Madamwar and A. Pandey), 149–169. Elsevier.

54 Kotzabasis, K., Strasser, B., Navakoudis, E. et al. (1999). The regulatory role of polyamines in structure and functioning of the photosynthetic apparatus during photoadaptation. *Journal of Photochemistry and Photobiology B: Biology* 50: 45–52.

55 Sfichi-Duke, L., Ioannidis, N.E., and Kotzabasis, K. (2008). Fast and reversible response of thylakoid-associated polyamines during and after UV-B stress: a comparative study of the wild type and a mutant lacking chlorophyll b of unicellular green alga *Scenedesmus obliquus*. *Planta* 228: 341–353.

56 Bouchereau, A., Aziz, F., Larher, J., and Martin-Tanguy, J. (1999). Polyamines and environmental challenges: recent development. *Plant Science* 140: 103–125.

57 Ha, H.C., Sirisoma, N.S., Kuppusamy, P. et al. (1998). The natural polyamine spermine functions directly as a free radical scavenger. *Proceedings of the National Academy of Sciences of the United States of America* 95: 11140–11145.

58 Groppa, M.D. and Benavides, M.P. (2008). Polyamines and abiotic stress: recent advances. *Amino Acids* 34: 35–45.

59 Rastogi, R.P., Richa, Sinha, R.P. et al. (2010). Photoprotective compounds from marine organisms. *Journal of Industrial Microbiology & Biotechnology* 37: 537–558.

60 Proteau, P.J., Gerwick, W.H., Garcia-Pinchel, F., and Castenholz, R.W. (1993). The structure of scytonemin, an ultraviolet sunscreen pigment from the sheaths of cyanobacteria. *Experientia* 49: 825–829.

61 Fleming, E.D. and Castenholz, R.W. (2007). Effects of periodic desiccation on the synthesis of the UV-screening compound, scytonemin, in cyanobacteria. *Environmental Microbiology* 9: 1448–1455.

62 Rastogi, R.P., Sonani, R.R., and Madamwar, D. (2014). The high-energy radiation protectant extracellular sheath pigment scytonemin and its reduced counterpart in the cyanobacterium *Scytonema* sp. R77DM. *Bioresource Technology* 171: 396–400.

63 Garcia-Pichel, F. and Castenholz, R.W. (1991). Characterization and biologicalimplications of scytonemin, a cyanobacterial sheath pigment. *Journal of Phycology* 27: 395–409.

64 Stevenson, C.S., Capper, E.A., Roshak, A.K. et al. (2002a). The identification and characterization of the marine natural product scytonemin as a novel antiproliferative pharmacophore. *The Journal of Pharmacology and Experimental Therapeutics* 303: 858–866.

65 Stevenson, C.S., Capper, E.A., Roshak, A.K. et al. (2002b). Scytonemin – a marine natural product inhibitor of kinases key in hyperproliferative inflammatory diseases. *Inflammation Research* 51: 112–114.

66 Dionisio-Sese, M.L. (2010). aquatic microalgae as potential sources of UV-screening compounds. *The Philippine Journal of Science* 139: 5–16.

67 Fuentes-Tristan, S., Parra-Saldivar, R., Igbal, H.M.N., and Carrillo-Nieves, D. (2019). Bioinspired biomolecules: Mycosporine-like amino acids and scytonemin from *Lyngbya* sp. with UV-protection potentialities. *Journal of Photochemistry and Photobiology, B: Biology* 201: 111684.

68 Kim, I., Choi, G.G., Nam, S.W. et al. (2020). Enhanced removal of cesium by potassium-starved microalga, *Desmodesmus armatus* SCK, under photoheterotrophic condition with magnetic separation. *Chemosphere* 252: 126482.

69 Grund, M., Jakob, T., Wilhelm, C. et al. (2019). Electron balancing under different sink conditions reveals positive effects on photon efficiency and metabolic activity of *Synechocystis* sp. PCC 6803. *Biotechnology for Biofuels* 12 (1): 1–14.

70 Hu, Q. (2004). Environmental effects on cell composition. In: *Handbook of Microalgal Culture: Biotechnology and Applied Phycology* (ed. A. Richmond), 83–93. Oxford: Blackwell.

71 Sathasivam, R., Radhakrishnan, R., Hashem, A., and Abd_Allah, E.F. (2019). Microalgae metabolites: a rich source for food and medicine. *Saudi Journal of Biological Sciences* 26: 709–722.

72 Chandra, R., Das, P., Vishal, G., and Nagra, S. (2019). Factors affecting the induction of UV protectant and lipid productivity in *Lyngbya* for sequential biorefinery product recovery. *Bioresource Technology* 278: 303–310.

73 Rath, J., Mandal, S., and Adhikary, S.P. (2014). Ecophysiology of the estuarine cyanobacterium *Lyngbya aestuarii* to varying salinity in vitro. *Acta Physiologiae Plantarum* 36: 409–419.

74 Klisch, M. and Häder, D.P. (2008). Mycosporine-like amino acids and marine toxins – The common and the diferent. *Marine Drugs* 6: 147–163.

75 Kokabi, M., Yousefzadi, M., Soltani, M., and Arman, M. (2019). Effects of different UV radiation on photoprotective pigments and antioxidant activity of the hot-spring cyanobacterium *Leptolyngbya cf. fragilis*. *Phycological Research* 67: 215–220.

76 Zhang, X., Tang, X., Zhou, B. et al. (2015). Effect of enhanced UV-B radiation on photosynthetic characteristics of marine microalgae *Dunaliella salina* (Chlorophyta, Chlorophyceae). *Journal of Experimental Marine Biology and Ecology* 469: 27–35.

77 Yang, Z., Kong, F., Shi, X. et al. (2014). UV-B radiation and phosphorus limitation interact to affect the growth, pigment content, and photosynthesis of the toxic cyanobacterium Microcystis aeruginosa. *Journal of Applied Phycology* 26: 1669–1674.

78 Kavitha, G., Kurinjimalar, C., Thevanathan, R., and Rengasamy, R. (2015). Evaluation of Ultraviolet -B radiation induced changes in biochemical response of *Arthrospira platensis* (Gomont). *International Research Journal of Biological Sciences* 4: 52–59.

79 Kannaujiya, V.K. and Sinha, R.P. (2015). Impacts of varying light regimes on phycobiliproteins of *Nostoc* sp. HKAR-2 and *Nostoc* sp. HKAR-11 isolated from diverse habitats. *Protoplasma* 252: 1551–1561.

80 Jantaro, S., Pothipongsa, A., Khanthasuwan, S., and Incharoensakdi, A. (2011). Short-term UV-B and UV-C radiations preferentially decrease spermidine contents and arginine decarboxylase transcript levels of *Synechocystis* sp. PCC 6803. *Current Microbiology* 62: 420–426.

81 Liu, Q., Nishibori, N., Imai, I., and Hollibaugh, J.T. (2016). Response of polyamine pools in marine phytoplankton to nutrient limitation and variation in temperature and salinity. *Marine Ecology Progress Series* 544: 93–105.

82 Park, J.J., Wang, H., Gargouri, M. et al. (2015). The response of *Chlamydomonas reinhardtii* to nitrogen deprivation: a systems biology analysis. *The Plant Journal* 81: 611–624.

83 Mushir, S. and Fatma, T. (2012). Monitoring stress responses in cyanobacterial scytonemin – screening and characterization. *Environmental Technology* 33: 153–157.

84 Rath, J., Mandal, S., and Adhikary, S.P. (2012). Salinity induced synthesis of UV-screening compound scytonemin in the cyanobacterium *Lyngbya aestuarii*. *Journal of Photochemistry and Photobiology B* 115: 5–8.

85 Chandra, R., Pons-Faudoa, F.P., Parra Saldívar, R., and Rittmann, B.E. (2020). Effect of ultra-violet exposure on production of mycosporine-like amino acids and lipids by *Lyngbya purpurem*. *Biomass and Bioenergy* 134: 105475.

86 Buma, A.G.J., Wright, S.W., Van Den Enden, R. et al. (2006). PAR acclimation and UVBR-induced DNA damage in Antarctic marine microalgae. *Marine Ecology Progress Series* 315: 33–42.

87 Mogedas, B., Casal, C., Forján, E., and Vílchez, C. (2009). β-Carotene production enhancement by UV-A radiation in *Dunaliella bardawil* cultivated in laboratory reactors. *Journal of Bioscience and Bioengineering* 108: 47–51.

88 Suyono, E.A., Aminin, P., Mu'avatun, L. et al. (2015). Combination of blue, red, white, and ultraviolet lights for increasing carotenoids and biomass of microalga *Haematococcus pluvialis*. *Procedia Environmental Sciences* 28: 399–405.

89 Copia, J., Gaete, H., Zúñiga, G. et al. (2012). Efecto de la radiación ultravioleta B en la producción de polifenoles en la microalga marina *Chlorella* sp. *Latin American Journal of Aquatic Research* 40: 113–123.

90 Casazza, A.A., Ferrari, P.F., Aliakbarian, B. et al. (2015). Effect of UV radiation or titanium dioxide on polyphenol and lipid contents of *Arthrospira* (*Spirulina*) *platensis*. *Algal Research* 12: 308–315.

91 Oguzkan, S.B., Guroy, B.K., Tonus, S.S. et al. (2018). The bioactive component and DNA protective capability of cultured *Spirulina* in Turkey (Marmara Region). *Genetics of Aquatic Organisms* 2: 7–12.

92 Häder, D.-P., Williamson, C.E., Wängberg, S.A. et al. (2015). Effects of UV radiation on aquaticecosystems and interactions with other environmental factors. *Photochemical & Photobiological Sciences* 14: 108–126.

93 Shi, K., Yu, J., Liu, C. et al. (2017). Impacts of enhanced UVB radiation on photosynthetic characteristics of the marine diatom *Phaeodactylum tricornutum* (Bacillariophyceae, Heterokontophyta). *Journal of Applied Phycology* 29: 1287–1296.

94 Ma, Z.L., Helbling, E.W., Li, W. et al. (2012). Motility and photosynthetic responses of the green microalga Tetraselmis subcordiformis to visible and UV light levels. *Journal of Applied Phycology* 24: 1613–1621.

95 Gao, K.S., Guan, W.C., and Helbling, E.W. (2007). Effects of solar ultraviolet radiation on photosynthesis of the marine red tide alga *Heterosigma akashiwo* (Raphidophyceae). *Journal of Photochemistry and Photobiology B: Biology* 86: 140–148.

96 Vimalabai, C.P.M. and Kulandaivelu, G. (2002). Effect of prolonged UV-B enhanced fluorescente radiation on some marine microalgae. *Biologia Plantarum* 45: 389–394.

97 Wong, C.-Y., Teoh, M.-L., Phang, S.-M. et al. (2015). Interactive Effects of temperature and UV radiation on photosynthesis of *Chlorella* strains from polar, temperate and tropical environments: differential impacts on damage and repair. *PLoS One* 10: e0139469.

98 Holzinger, A., Albert, A., Aigner, S. et al. (2018). Arctic, Antarctic, and temperate green algae Zygnema spp. Under UV-B stress: vegetative cells perform better than pre-akinetes. *Protoplasma* 255: 1239–1252.

99 Kumar, V., Nanda, M., Kumar, S., and Chauhan, P.K. (2018). The effects of ultraviolet radiation on growth, biomass, lipid accumulation and biodiesel properties of microalgae. *Energy Sources, Part A: Recovery, Utilization, and Environmental Effects* 40: 787–793.

100 Hull, N.M., Isola, M.R., Petri, B. et al. (2017). Algal DNA repair kinetics support culture-based enumeration for validation of ultraviolet disinfection ballast water treatment systems. *Environmental Science & Technology Letters* 4: 192–196.

101 Kurinjimalar, C., Kavitha, G., Thevanathan, R. et al. (2019). Impact of ultraviolet-B radiation on growth and biochemical composition of *Botryococcus braunii* Kutz. *Current Science* 116: 89–95.

102 Saranya, A., Sudha, M., Selvakumar, G., and Sivakumar, N. (2016). Stress response analysis on lipid production by microalgae isolated from estuaries by UV radiation. *International Journal of Advanced Research* 4: 2203–2207.

103 Singh, R., Upadhyay, A.K., Singh, D.V. et al. (2019). Photosynthetic performance, nutrient status and lipid yield of *microalgae Chlorella vulgaris* and *Chlorococcum humicola* under UV-B exposure. *Current Research in Biotechnology* 1: 65–77.

104 Brennan, L. and Owende, P. (2010). Biofuels from microalgae a review of technologies for production, processing and extractions of biofuels and co-products. *Renewable and Sustainable Energy Reviews* 14: 557–577.

105 Tang, D.Y.Y., Khoo, K.S., Chew, K.W. et al. (2020). Potential utilization of bioproducts from microalgae for the quality enhancement of natural products. *Bioresource Technology* 304: 122997.

106 Ibañez, E., Herrero, M., Mendiola, J.A., and Castro-Puyana, M. (2011). Extraction and characterization of bioactive compounds with health benefits from marine resources: macro and micro algae, cyanobacteria, and invertebrates. In: *Marine Bioactive Compounds* (ed. M. Hayes), 55–98. Boston: Springer.

107 Singh, S.K., Kaur, R., Bansal, A. et al. (2020). Biotechnological exploitation of cyanobacteria and microalgae for bioactive compounds. In: *Biotechnological Production of Bioactive Compounds* (eds. M.L. Verma and A.K. Chandel), 221–259. Elsevier.

108 Michalak, I., Dmytryk, A., Wieczorek, P.P. et al. (2015). Supercritical algal extracts: a source of biologically active compounds from nature. *Journal of Chemistry* 2015: 1–14.

109 Herrero, M., Mendiola, J.A., Cifuentes, A., and Iba, E. (2010). Supercritical fluid extraction recente advances and applications. *Journal of Chromatography A* 1217: 2495–2511.

110 Rastogi, R.P. and Incharoensakdi, A. (2014). Characterization of UV-screening compounds, mycosporine-like amino acids, and scytonemin in thecyanobacterium *Lyngbya* sp. CU2555. *FEMS Microbiology Ecology* 87: 244–256.

111 Singh, A., Tyagi, M.B., and Kumar, A. (2017). Cyanobacteria growing on tree barks possess high amount of sunscreen compound mycosporine-like amino acids (MAAs). *Plant Physiology and Biochemistry* 119: 110–120.

112 Reichardt, C. and Welton, T. (2011). *Solvents and Solvent Effects in Organic Chemistry*. Wiley.

113 Korzeniowska, K., Zmudzki, S., Ambroziak, K., and Wieczorek, P.P. (2017). Use of plant extracts containing phenolic compounds in organic agriculture. *Przemysl Chemiczny* 96: 1518–1522.

114 Ryckebosch, E., Muylaert, K., and Foubert, I. (2012). Optimization of an analytical procedure for extraction of lipids from microalgae. *Journal of the American Oil Chemists Society* 89: 189–198.

115 Ryckebosch, E., Cuéllar Bermúdez, S.P., Termote-Verhalle, R. et al. (2013). Influence of extraction solvent system on the extractability of lipid compounds from the biomass of *Nannochloropsis gaditana*. *Journal of Applied Phycology* 26: 1501–1510.

116 Cerón-García, M.C., González-López, C.V., Camacho-Rodríguez, J. et al. (2018). Maximizing carotenoid extraction from microalgae used as food additives and determined by liquid chromatography (HPLC). *Food Chemistry* 257: 316–324.

117 Fernández-Sevilla, J.M., Acién, F.G., and Molina-Grima, E. (2010). Biotechnological production of lutein and its applications. *Applied Microbiology and Biotechnology* 86: 27–40.

118 Elsayed, S., Boukis, N., Patzelt, D. et al. (2016). Gasification of microalgae using supercritical water and the potential of effluent recycling. *Chemical Engineering & Technology* 39: 335–342.

119 Damergi, E., Schwitzguébel, J.-P., Refardt, D. et al. (2017). Extraction of carotenoids from *Chlorella vulgaris* using green solvents and syngas production from residual biomass. *Algal Research* 25: 488–495.

120 Siegelman, H.W. and Kycia, H.J. (1978). Algal biliproteins. In: *Handbook of Phycological Methods* (eds. J.A. Hellebust and J.S. Craigie), 71–79. Cambridge: Cambridge University Press.

121 Silveira, S.T., Burkert, J.F.M., Costa, J.A.V. et al. (2007). Optimization of phycocyanin extraction from *Spirulina platensis* using factorial design. *Bioresource Technology* 98: 1629–1634.

122 Chaiklahan, N. and Bunnag, B. (2012). Stability of phycocyanin extracted from *Spirulina* sp.: influence of temperature, pH and preservatives. *Process Biochemistry* 47: 659–664.

123 Sonani, R.R., Patel, S., Bhastana, B. et al. (2017). Purification and antioxidant activity of phycocyanin from *Synechococcus* sp. R42DM isolated from industrially polluted site. *Bioresource Technology* 245: 325–331.

124 Gosetti, F., Mazzucco, E., Gianotti, V. et al. (2007). High performance liquid chromatography/tandem mass spectrometry determination of biogenic amines in typical Piedmont cheeses. *Journal of Chromatography A* 1149: 151–157.

125 Byun, J.A., Lee, S.H., Jung, B.H. et al. (2008). Analysis of polyamines as carbamoyl derivatives in urine and serum by liquid chromatography-tandem mass spectrometry. *Biomedical Chromatography* 22: 73–90.

126 Gosetti, F., Mazzucco, E., Gennaro, M.C., and Marengo, E. (2013). Simultaneous determination of sixteen underivatized biogenic amines in human urine by HPLC-MS/MS. *Analytical and Bioanalytical Chemistry* 405: 907–916.

127 Magnes, C., Fauland, A., Gander, E. et al. (2014). Polyamines in biological samples: rapid and robust quantification by solid-phase extraction online-coupled to liquid chromatography–tandem mass spectrometry. *Journal of Chromatography A* 1331: 44–51.

128 Morgan, M.A., Griffith, C.M., Dinges, M.M. et al. (2019). Evaluating sub-lethal stress from Roundup® exposure in Artemia franciscana using 1H NMR and GC–MS. *Aquatic Toxicology* 212: 77–87.

129 Simó, C., Moreno-Arribas, M.V., and Cifuentes, A. (2008). Ion-trap versus time-of-flight mass spectrometry coupled to capillary electrophoresis to analyze biogenic amines in wine. *Journal of Chromatography A* 1195: 150–156.

130 Hamana, K., Kobayashi, M., Furuchi, T. et al. (2017). Polyamine analysis of brown-algal seaweeds (class Phaeophyceae) from food markets – distribution of diaminohexane, penta-amines, and hexa-amine. *Microbial Resources and Systematics* 33: 3–8.

131 Ermis, H., Guven-Gulhan, U., Cakir, T., and Altinbas, M. (2020). Effect of iron and magnesium addition on population dynamics and high value product of microalgae grown in anaerobic liquid digestate. *Scientific Reports* 10: 1–12.

132 Vaz, B.S., Moreira, J.B., Morais, M.G., and Costa, J.A.V. (2016). Microalgae as a new source of bioactive compounds in food supplements. *Current Opinion in Food Science* 7: 73–77.

133 Chrapusta, E., Kaminski, A., Duchnik, K. et al. (2017). Mycosporine-like amino acids: potential health and beauty ingredients. *Marine Drugs* 15: 1–29.

134 Orfanoudaki, M., Hartmann, A., Alilou, M. et al. (2020). Absolute configuration of mycosporine-like amino acids, their wound healing properties and in vitro anti-aging effects. *Marine Drugs* 18: 1–16.

135 Yusuf, M., Shabbir, M., and Mohammad, F. (2017). Natural colorants: historical, processing and sustainable prospects. *Natural Products and Bioprospecting* 7: 123–145.

136 Saini, D.K., Chakdar, H., Pabbi, S., and Shukla, P. (2020). Enhancing production of microalgal biopigments through metabolic and genetic engineering. *Critical Reviews in Food Science and Nutrition* 60: 391–405.

137 Kannaujiya, V.K. and Sinha, R.P. (2016). Thermokinetic stability of phycocyanin and phycoerythrin in food-grade preservatives. *Journal of Applied Phycology* 28: 1063–1070.

138 Hu, I.-C. (2019). Production of potential coproducts from microalgae. In: *Biomass, Biofuels and Biochemicals: Biofuels from Algae* (eds. A. Pandey, J.-S. Chang, C.R. Soccol, et al.), 345–358. Elsevier.

139 Barkia, I., Saari, N., and Manning, S.R. (2019). Microalgae for high-value products towards human health and nutrition. *Marine Drugs* 17: 304.

140 Velea, S., Oancea, F., and Fischer, F. (2017). Heterotrophic and mixotrophic microalgae cultivation. In: *Microalgae-Based Biofuels and Bioproducts: From Feedstock Cultivation to End Products* (eds. C. Gonzalez-Fernandez and R. Muñoz), 45–65. Woodhead Publishing.

141 Panis, G. and Rosales Carreon, J. (2016). Commercial astaxanthin production derived by green alga *Haematococcus pluvialis*: a microalgae process model and a techno-economic assessment all through production line. *Algal Research* 18: 175–190.

142 Jha, D., Jain, V., Sharma, B. et al. (2017). Microalgae-based pharmaceuticals and nutraceuticals: an emerging field with immense market potential. *ChemBioEng Reviews* 4: 1–16.

143 Ariede, M.B., Candido, T.M., Jacome, A.L.M. et al. (2017). Cosmetic attributes of algae-A review. *Algal Research* 25: 483–487.

11

Microorganisms as a Potential Source of Antioxidants

Ayerim Hernández-Almanza[1], Nathiely Ramírez-Guzman[2], Gloria A. Martínez-Medina[3], Araceli Loredo-Treviño[3], Deepak Kumar Verma[4], and Cristobal N. Aguilar[3]

[1] *School of Biological Science, Autonomous University of Coahuila, Torreón, Coahuila, México*
[2] *Center for Interdisciplinary Studies and Research, Universidad Autónoma de Coahuila, Saltillo, México*
[3] *Bioprocesses and Bioproducts Research Group, Food Research Department, School of Chemistry, Autonomous University of Coahuila, Saltillo, Coahuila, México*
[4] *Agricultural and Food Engineering Department, Indian Institute of Technology, Kharagpur, West Bengal, India*

11.1 Introduction

Microorganisms are a group of living beings that, in addition to their key characteristics, often have complex metabolic processes from which they receive the energy and nutrients they need to their survival and reproduction [1]. They also synthesize certain organic compounds that have complementary functions, such as aesthetics, defense against predator attacks, defense against radiation or freezing, and so on [2]. These biochemical reactions, which occur in parallel with primary metabolism, are known as secondary metabolism. Some of these secondary metabolites can be antibiotics, pigments, aromatic compounds, phenols, and other compounds [3, 4].

The production of secondary metabolites is of particular industrial interest in a variety of areas. These metabolites can delay or prevent the oxidation of other molecules that can produce free radicals that cause cell-damage reactions. Over the years, they have been synthetically produced, including BHT, BHA, TBHQ, etc. [5]. However, the shift that current society has taken to become aware of the need for natural and organic products will help to visualize the use of natural alternatives, such as microorganisms.

11.2 Antioxidant-Producing Microorganisms

Today, microorganisms with the capacity to produce secondary metabolites, such as antioxidant compounds, are of interest to different industries due to the advantages they have over conventional production methods. For example, the use of *Saccharomyces*

Bioprospecting of Microorganism-Based Industrial Molecules, First Edition.
Edited by Sudhir P. Singh and Santosh Kumar Upadhyay.

cerevisiae is well known in the bakery and brewing industries. However, it also produces phenolic compounds (a type of flavonoid, such as naringenin and its derivatives) [6], which have antimicrobial and antioxidant potential against food pathogens such as *Escherichia coli* and *Staphylococcus aureus*. Cipolatti et al. [6] reported antioxidants to be used in the food industry from wild strains of *Rhodotorula mucilaginosa, Sporidiobolus pararoseus*, and *Pichia fermentans* isolated from Escudo Sul-Rio Grandense, Brazil, by liquid-state fermentation. In addition, the use of *Rhodotorula glutinis* and *Blakeselea trispora* strains has been reported for the production of lycopene, a carotenoid of high added value since it has biological activities such as anti-inflammatory and antioxidant properties [7]. On the other hand, the production of phenolic compounds using mango seed as a substrate of *Aspergillus niger* has been reported [8]. Some important lactic acid bacteria are also examples (such as *Lactobacillus casei, Lactobacillus acidophilus*, and *Lactococcus lactis*) that are capable of producing antioxidant compounds [9]. This group of lactic acid bacteria is widely used to enhance the antioxidant properties of different products such as wines, coffee pulp, cheeses, and cereals. Quinoa flour is fortified with *Lactobacillus plantarum*, which releases antioxidant peptides by proteolysis of native quinoa proteins [10, 11].

11.3 Production of Some Microbial Antioxidants and Their Action Mechanisms

The availability of information has recently made it clear to consumers how various ingredients affect their health in food and other products. This raises interest in the use of natural ingredients and has beneficial effects on human health. They also focus on the antioxidant activity of peptides, pigments, and polyphenols that are beneficial to prevent diseases like cancer, atherosclerosis, etc.

11.3.1 Peptides

Bioactive peptides are fragments of amino acids that are generally 2–20 units long. These can be produced by hydrolysis of proteins with microbial enzymes or by proteolytic microorganisms [12, 13]. They show different biological activities, such as antitumor, antioxidant, anticancer, and so on, in which peptides depend on their structural properties such as chain length and physicochemical characteristics of the residual amino acids (such as their hydrophobicity, molecular charge, and side-chain bulkiness) [13, 14]. Free radicals that target lipids in membranes, proteins, and DNA can be considered a choice for minimizing risks to health for health disorders such as diabetes, cancer, neurodegenerative, and inflammatory diseases [14]. Peptides that may have antioxidant activity must have at least one antioxidant amino acids, such as tryptophan, tyrosine, methionine, and cysteine. The presence of antioxidant amino acid (Tyr or Tpr) in the C-terminus is generally mandatory for scavenging activity. A study by Babini et al. [13] reported that phenylalanine seemed important to inhibit lipid peroxidation. On the other hand, the most reactive amino acids were the nucleophilic sulfur-containing amino acids Cys and Met, and the aromatic amino acids Trp, Tyr, and Phe (Figure 11.1) [14].

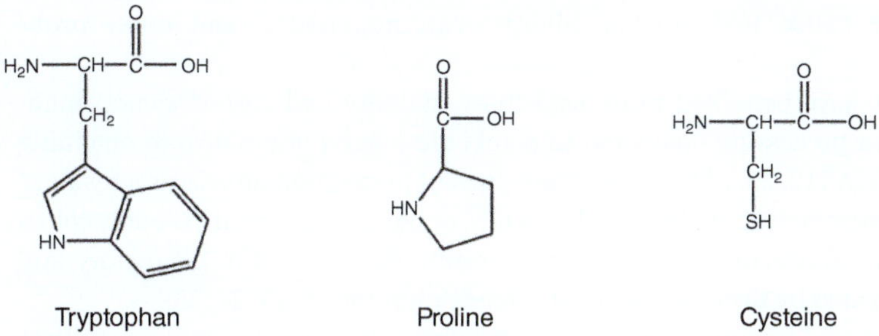

Figure 11.1 Chemical structure of amino acids associated with antioxidant activity. *Source:* Based on Sila and Bougatef [14].

These peptides can be obtained either by the digestive procedure or by the enzymatic or microbial hydrolysis. This can be induced or occurs "naturally" in processes such as cheese maturation [14]. Rice is also used with microbial proteases (because their proteins made up of alkali-soluble glutelins and water-soluble albumins) [12]. Rapeseed has been fermented in SSF with *Bacillus subtilis* and with the help of enzymes to obtain bioactive peptides [12]. Animal sources are also used to produce these peptides. Milk, whey, eggs, chicken, bones, blood, and fish waste are widely used sources of animal origin for bioactive peptides [12, 14, 15].

Bioprocesses must be eco-friendly, cost-effective, and high quality in order to meet consumer needs. Microbial production has a range of problems. Scaling and purification techniques are the most challenging among them [12, 16]. The production strategies include enzymatic hydrolysis, fermentation using proteolytic microorganisms, and the combination of both techniques [12, 14, 17]. The factors influencing peptides are the nature of the protein substrate, the specificity of the used enzyme, the hydrolytic temperature and time conditions, and the enzyme-substrate relationship, affecting the molecular weight and amino acid composition of the peptide obtained, and consequently its biological activity [14]. For the pharmaceutical and food industry, the enzymatic hydrolysis method is preferred more than chemical methods because there are no residual substances or toxic substances [14]. The specificity of the enzyme is, therefore, of the utmost importance. Fermentation methods are cheaper, but peptides with homogeneous characteristics are more difficult to obtain. In both cases, methodologies such as the surface response methodology are used to optimize conditions [18].

11.3.2 Pigments

Colors are one of the first parameters to be noticed and one of the most pleasing for humans [19]. Artificial or synthetic colors are commonly used in the food and cosmetics industry. They are reliable and economical compared to natural dyes, which are more expensive, less stable, and less intense, precisely because of their natural origin, usually from plants, making them more sensitive to extraction and storage conditions [19, 20]. However, due to their negative side effects, some artificial dyes have been banned because they cause toxicological allergy, carcinogenicity, and other problems [19, 21, 22].

Microorganisms that have benefited from availability, stability and cost-efficiency, manpower, and downstream processing compared to plants are a good alternative to obtaining pigments of natural origin [12, 20, 21]. Pigments are present in microorganisms as a result of circumstances where stress factors, including pH, metals, or light, are covered. Not only colors but also potential biological activities such as anticancer, antioxidants, anti-inflammatory, and antimicrobials are suggested by these pigments, as already mentioned [19, 21, 23].

Pigments can be produced by fungi, bacteria, yeasts, algae, and protozoa, and the cultivation techniques used are in solid-state (SSF) or liquid (SmF). The use of different agro-industrial wastes such as rice liqueurs, mung beans, corn, buttermilk, soybeans, molasses, sugar cane bagasse, corn bran, rice bran, grapes, and so on, must be explored in order to avoid the use of expensive synthetic culture media [12, 19, 21]. The microorganisms used must be nonpathogenic and nontoxic to produce pigments for food or pharmacy use and must have the ability to use a wide range of carbon and nitrogen sources. They must also have a tolerance to salt, temperature, and pH, and purification processes should preferably be simple [21]. Some pigments, such as *Monascus* pigments; astaxanthin from *Xanthophyllomyces dendrorhous*; Arpink Red from *Penicillium oxalicum*; riboflavin from *Ashbya gossypii*; and *β*-carotene from *Blakeslea trispora*, are currently used in food [19, 22, 24, 25].

On the other hand, the factors affecting pigment production in cultivation are temperature, pH, carbon source, fermentation type, minerals, nitrogen source, moisture content, and aeration rate [21]. For example, in the production of actinomycete melanin [26], Plackett-Burman experiments were designed. It was found that the incubation period, protease-peptone, and ferric ammonium citrate were factors, which had a positive effect on melanin production. This is because the Fe^{+2} ion is a cofactor required for the activity of various aromatic hydrolases. Melanin is a polymer capable of giving or accepting an electron.

Once the pigments are produced, their stability depends on UV light as it degrades them. In the same way, high temperatures, oxygen presence, and long processes with solvents lead to degradation and unwanted formation of isomers [12, 21]. Because pigments have a high conjugation of dual bonds in their molecules, they break down in light, temperature, and oxygen. Light and heat favor photochemical reactions and alter the carotenoid's original structure, while oxygen promotes double-bond oxygenation for the functionality of hydroxyl and peroxide [12]. In this circumstance, we have focused on the antioxidant activity as it is closely related to all other activities. Toxic oxygen molecules, such as reactive oxygen and nitrogen species, and other nonbiological radicals formed in cells, can be effectively reduced by pigments. Antioxidant activity is related to chronic diseases such as cancer, diabetes, cardiovascular disease, and autoimmune disorders, which, in turn, are related to free radical action. Microbial pigments have shown antioxidant potential by inhibiting lipid peroxidation in liposomes and providing protection against UV light, reducing the effects of oxidative stress and free radicals [21, 27].

Similar to the action of peptides, areas with double bonds in pigments may be responsible for the ability of free radicals to be absorbed. Red pigments (carotenes, and riboflavin) are positively charged and can prevent oxidative stress in humans. Fucoxanthin is capable of protecting against oxidative stress caused by UV light [12]. Melanins are negatively charged with black or brown pigments and hydrophobic pigments. These are molecules formed by the polymerization of phenolic and indole compounds. These pigments of microbial origin can chelate metals (which are catalysts for fatty acid oxidation), provide structural rigidity, help to store water, absorb UV light, are cation exchangers, and absorb X-rays and γ-rays [26, 28, 29]. Melanin (Figure 11.2) is a polymer capable of giving or accepting an electron. It interacts quickly with free radicals and other reactive species due to the presence of unpaired electrons in their molecules and acts as antioxidants. It also interacts with free radicals through simple one-electron transfer processes [26]. The antioxidants potential of polyphenols, anthraquinones, and by-products is large. This is because of its characteristics that allow them to work in metal chelators and absorb and neutralize free radicals, combined with the oxidation of agents with different potential [20].

11.3.3 Polyphenols

Polyphenols molecules (like pedunculagin, depicted in Figure 11.3) are present in a wide variety of plants, both inedible and inedible. Grapes, walnuts, chestnuts, strawberries, cranberries, Brazilian nuts, pecan nuts, pomegranates, and olives are among the most popular [30, 31]. They contain aromatic units in their structure and may or may not be linked to glucose molecules. They serve as protection and also have biological activity, such as pigments and bioactive peptides, and this activity is related to antioxidant activity. The mode of action of polyphenols is similar to that of pigments and peptides. They are all

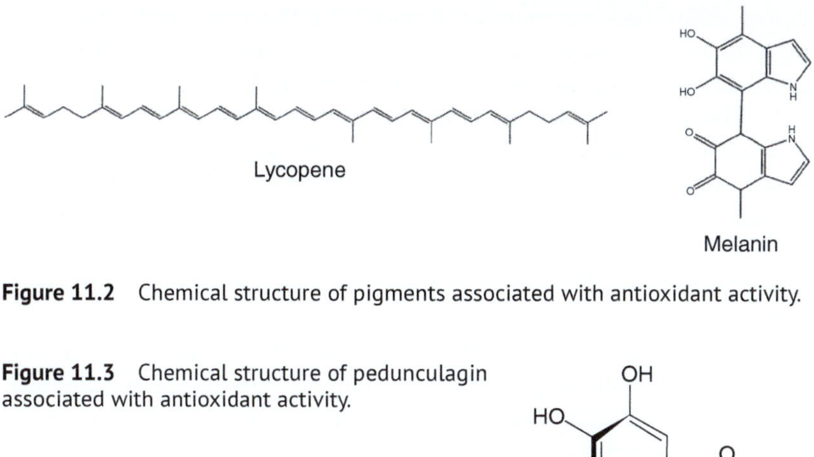

Figure 11.2 Chemical structure of pigments associated with antioxidant activity.

Figure 11.3 Chemical structure of pedunculagin associated with antioxidant activity.

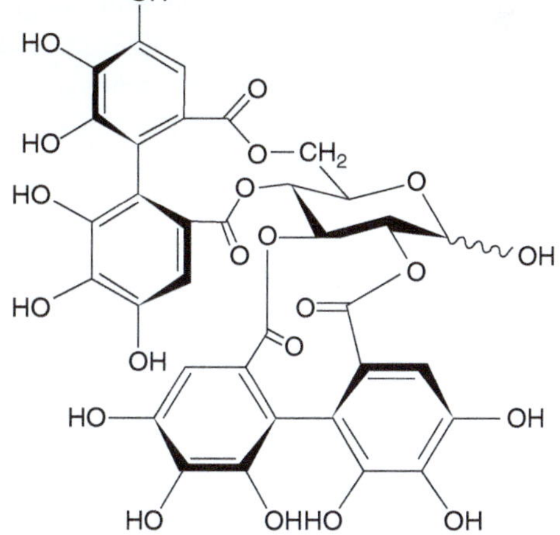

based on trapping free radicals through zones in their molecules with a high electron charge.

The extraction of these compounds can be carried out by physical, chemical, and biological means, similar to the extraction of pigments. In the case of extraction by microorganisms, these break the bonds that unite the aromatic units with the glucose molecules. This allows the accumulation of phenolic compounds that can be recovered in a similar way to pigments [30–33]. Similarly, for pigments and peptides, polyphenols have many double bonds that can have a scavenging effect.

11.4 Extraction and Purification of Microbial Antioxidants

11.4.1 Extraction of Microbial Antioxidants

Once the different microorganisms produce antioxidant activity compounds, they must be recovered. In order to choose the most appropriate extraction method, it is necessary to consider whether the compound is produced intra-or extracellularly, as well as the chemical structure, pH stability, temperature, and light of the bio-compound, cost, and complexity of the production processes. Isolation methods must be selected based on the molecules' physicochemical properties to be separated [34].

If the production of antioxidant compounds is intracellular, the recovery of antioxidant compounds involves various downstream processes, the most critical stage of cell disruption [35]. Several cell disruption techniques have been described; this method is classified as mechanical and nonmechanical (such as chemical, physical, and enzymatic) [36]. Liu et al. [37] indicated that mechanical methods favor large-scale cell disruption, and nonmechanical methods are more favorable and common on a laboratory scale due to economic and operational limitations. On the other hand, if the production of antioxidant compounds is extracellular, extraction is more comfortable, and the use of organic solvents or equipment is cheaper. Some methods of cell disruption and extraction have been discussed in the following paragraph.

The use of hydrolytic enzymes to break down cell walls has drawn attention to the fact that they break down the cell wall integrity, generate cell lyses, and release intracellular components [12]. Sonication performs well due to cavitation caused by ultrasonic waves. The effect is mechanical and breaks down the cells. This method is cheaper and simpler than the extraction of supercritical fluid and microwave-assisted extraction (MAE). Other mechanical methods are the use of blades and pearls. This method must be applied in a cold environment that allows the pigments to remain stable. This method is nontoxic and does not use solvents; furthermore, it is more suitable for industrial quantities [12]. There has also been evidence of such methodologies as high hydrostatic pressure, pulse electric field, sonication-assisted extraction, gamma irradiation, enzymatic extraction, and membrane technology [21]. Subsequently, a partition is made of a nonpolar solvent; acetone, hexane, petroleum ether, methanol, and ethanol, as well as mixtures thereof, are normally used. Pressurized liquids that have the advantage of increasing solubility and temperature have been used to better diffusion of the analyte from the matrix to the solvent and a lowering of the solvent's viscosity. Supercritical fluids have been used in extraction, purification, crystallization, and fractionation operations in different industries without the use of solvents. These advantages are due to the volatility of supercritical fluids and improved transport characteristics [38–40].

11.4.2 Purification of Microbial Antioxidants

In general, when we produce a compound via microbial sources, we obtain a mixture of components that have a similar characteristic; therefore, we must perform a separation process that allows us to obtain a compound of interest in a pure form. Recovery and purification account for 20–60% of the overall cost of the process. Purification processes may be based on molecular weight, the affinity of any chromatographic filler, and the bioactive compound's polarity. Among the most widely used methods for the purification of antioxidant compounds are HPLC, GC-MS, thin-layer chromatography, dialysis, crystallization, electrophoresis, etc. High-performance liquid chromatography is a widely used method of separation; the components of the mixture must be divided into two phases, one stationary and another mobile phase. These are based on differences in each compound's partition coefficient, which causes differential retention in the stationary phase [34].

Papadaki et al. [41] reported fumaric acid production by *Rhizopus arrhizus* NRRL-2582 under submerged fermentation, which determines the individual phenolic compounds by HPLC coupled with a diode array detector (HPLC-DAD). In addition, the HPLC method was used to identify ferulic acid produced by the biotransformation of rice brains with lactic acid bacteria [42]. Moreover, Li et al. [20] isolated an endophytic fungus that produced red pigments; after production, the red pigment components were analyzed using ultraviolet (UV), Raman, FTIR, and GC-MS, and using these methods, phenols and anthraquinones were identified as pigment mixtures. On the other hand, methods such as the ion-exchange layer system and gel filtration have been used to purify microbial enzymes and proteins [43]. Some of the methods reported for insolation and purification of microbial antioxidants are summarized in Table 11.1.

11.5 Evaluation of Antioxidant Activity

The natural respiratory chain produces several redox reactions, as shown in Figure 11.4; this process develops a set of intermediate molecules, called the reactive oxygen species (ROS), including unpaired electron molecules (radicals) and other non-radical substances. These chemical and natural oxidizing species are highly reactive [55] and usually contribute to the maintenance of homeostasis organisms, but this could also be caused by exogenous sources, causing a disturbance, reacting with fundamental molecules, and generating a cascade of side effects [56].

The antioxidant capacity is a chemical concept that is associated with the fields of nutrition, medicine, and epidemics [57]. This has also described important information and discoveries in recent years, including antioxidant mechanisms, health benefits, and novel sources with appropriate demand, bioassay methods, etc. Many research teams have described and reported bioassay techniques for antioxidant studies [57–59], also summarized in Table 11.2.

Table 11.1 Summary of some critical studies on isolation, purification, and identification of microbial antioxidant compounds.

Microbial antioxidant compounds	Employed microorganism	Extraction approaches	Purification/ identification methods	References
Pigments	*Penicillium purpurogenum*	Filtration/ crystallization	GC-Mass spectrometry	[44]
Pigments	Himalayan soils isolated	Acetone:methanol mixture	Chromatography	[45]
Monascus Pigments	*Monascus*	Filtration	Scanning spectrophotometry	[46]
Orange-Red pigments	*Penicillium mallochii*	Ethanol/filtration	Spectrophotometry	[47]
Carotenoids	*Gordonia terrae* TWRH01	Sonication	UV-visible and GC-MS	[48]
Carotenoids	*Rhodotorula gracilis* ATCC 10788	DMSO and zirconium beads	HPLC	[49]
Phenolic compounds	*Inonotus obliquus*	Ultrasonication	HPLC-DAD-ESI-MS/ MS	[50]
Phenolic compounds	*Ganoderma lipsiense*		HPLC-ESI-MS/MS	[51]
Exopolysaccharide	*Bacillus sp.* S-1	DEAE anion-exchange column	Sephadex G-74 chromatography	[52]
Biosurfactants	*Bacillus* strains	Centrifugation	Dialysis	[53]
Pyocyanin	*Pseudomonas aeruginosa*	Organic solvents		[54]

$$H_2O_2 + Fe^{2+} \longrightarrow Fe^{3+} + OH^- + OH^\cdot$$

Figure 11.4 The chemical reaction for the development of different radical developed during the respiratory chain.

11.5.1 Classical Methods

First, it must be noted that the antioxidant methods usually have two main mechanics, which may sometimes be combined with the hydrogen atom transference (HAT) or the single electron transference (SET) [57]. Through these main mechanisms, the first evaluates the antioxidant capacity to donate hydrogen atom for quenching a radical, while the second evaluates the ability to reduce the oxidation of a species as donors of electrons [60]. Comprehensive antioxidant assessment techniques have been developed over time, usually used in research and industry, to analyze various antioxidant molecules.

The most common techniques used for radical scavenging assays include 2,2-Diphenyl-1-picrylhydrazyl (DPPH) [61], 2,2'-azino-bis (3-ethylbenzothiazoline-6-sulfonic acid) (ABTS) [62], N,N-dimethyl-p-phenylenediamine dihydrochloride (DMPD) [63], Oxygen radical

Table 11.2 Different methods of antioxidant activity.

Bioassay types	Methods
Radical scavenging assays	DPPH, ABTS, DMPD, superoxide anion radical scavenging, nitric oxide radical scavenging assays, peroxynitrite scavenging assays, ORAC, and Total reactive antioxidant potential (TRAP)
Non-radical scavenging assays	Hydrogen peroxide scavenging and singlet oxygen quenching assays
Reducing power assays	FRAP, CUPRAC, Fe^{3+} to Fe^{2+} Ferric ion reducing antioxidant power, Folin–Coicalteu, and phosphomolybdenum method
Cellular assays	Cellular antioxidant activity assay, comet assay, and hemolysis assay
Metal chelating assays	Metal chelating assays
Other assays	Chemiluminescence and β-carotene bleaching assay

Source: Gulcin [57], Hofer et al. [58] and Shahidi and Zhong [59].

absorbance capacity (ORAC) [64], superoxide radicals [65], nitric oxide radicals [66], and peroxynitrite radical assays [67]. ABTS, DPPH, and DMPD techniques represent simple spectrophotometric methodologies, but each presents its challenges. For example, DPPH presents interference with molecules such as carotenes or proteins and is more suited to hydrophilic substances; while ABTS requires a long time for an adequate de-reagent and is pH-sensitive, but could be used in hydrophilic and lipophilic substances; although DMPD presents interference with hydrophobic substances, one of the most important claims is that this class of radicals is not present in biological systems [68]. However, nitric oxide radicals, superoxide radicals, and peroxynitrites are common radicals in biological systems [69]. In the case of ORAC assay, the fluorometric assay is interesting, and therefore colored compounds could be tested, adapted to hydrophilic and lipophilic substances, and related to biological systems [70].

There is another class of oxidizing species besides radicals that could promote the formation of radicals. Some methods evaluate the protective effect of antioxidants against the oxidizing species or their potential transformation as hydrogen peroxide. Hydrogen peroxide scavenging assay is a spectrophotometric technique [71] that evaluates the potential of antioxidant molecules during the presence of H_2O_2 using the maximum absorbance of this key oxidizing molecule, resulting in lower absorbances when antioxidant removes peroxide hydrogen. In addition, variants with the presence of colored compound reactions or fluorometric quantification have also been reported in the technique [72, 73].

There is also some other way, including Ferric reducing antioxidant power (FRAP) [74], CUPric Reducing Antioxidant Capacity (CUPRAC) [75], or Folin–Ciocalteu [76], that can be used to overcome the challenge in the antioxidant capacity of the molecule. This set of methods is based on the ability of antioxidants to donate electrons. In the case of FRAP and CUPRAC (Fe^{3+}/Fe^{2+} and Cu^{2+}/Cu^{+}), both methods generate a spectrophotometrically detectable chromogenic product after reduction, which is both a simple and cheap alternative.

However, FRAP reacts at pH 3.6 allowing the ion solubility involved, the time required varies depending on the antioxidant molecule and is not available for thiol substances [57], while CUPRAC is available for thiol substances and is highly stable at pH 7.0, with low interference with sugars and citric acids and may challenge hydrophilic and lipophilic antioxidants [77].

The Folin-Ciocalteu method works similar but uses Phospho-molybdic-tungstic acid to produce a reduced chromophore, is reliable, stable and has excellent analytical parameters, usually used for the determination of polyphenolic molecules; however, its applicability is not limited to these molecules, it nevertheless has interference with certain metallic ions and amino acid molecules, and is suitable for hydrophilic molecules [78, 79]. The indirect form of antioxidant activity testing is by chelating activity; by oxidative stress atmospheric, the presence of ferrous ion (Fe^{2+}) catalyzes the "Fenton reaction" (Figure 11.5) of the hydroxyl radical [57]. A spectrophotometric study using ferrozine, a red-colored compound with a Fe^{2+} core, was therefore carried out, which, when an antioxidant product traps the key ion species, results in colorless products [80], presenting an alternative method for measuring and understanding the complicated antioxidant properties of various compounds.

It is also important to mention a group of methods that could be used as a chemiluminescent assay, which analyzes the interaction of hydrogen peroxide, a metallic or enzymatic catalyst, and a chemiluminescent reagent, a chemiluminescent reaction with hydrogen peroxide generating a light-measurable compound and competing with an antioxidant molecule for hydrogen peroxide reactions. However, other oxidizing compounds (radicals) could also be replaced by peroxide under essentially the same mechanism [57, 81]. During the adaptation of the assays, combining a wide range of techniques or an in-depth analysis of weaknesses in each process, it is necessary to use different analytical techniques adapted to the resulting requirements, also due to the complexity and synergistic existence of antioxidant compounds.

11.5.2 Cellular Methods

The antioxidant capacity of various microbial compounds could be analyzed from a more sophisticated approach, using a set of biological level analyses, one step closer to the in vivo environment, as a second step in the search for biologically active antioxidants or

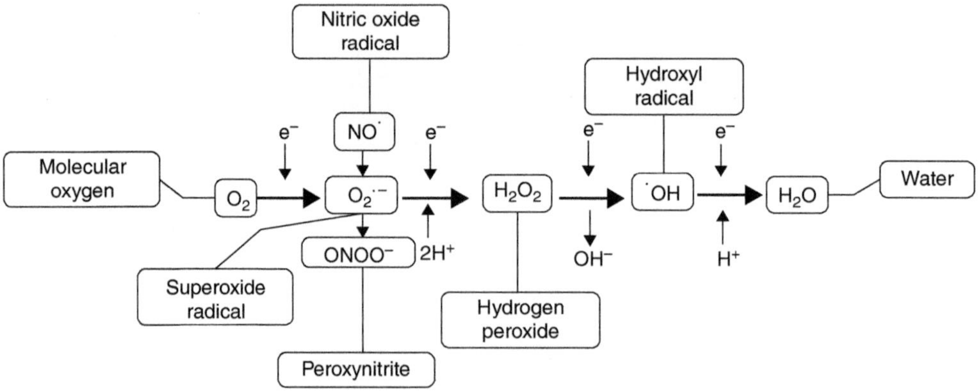

Figure 11.5 Conceptual presentation of the Fenton reaction.

to target their correct use or the challenges that could be faced through their implementation. Cellular antioxidant activity assay (CAAA) is a cellular assay [82] that uses HepG2 cells (human hepatocarcinoma cells), mixed with antioxidant molecules, and 2′,7′-dichlorofluorescein (DCFH-DA). This compound could pass into intracellular space after cellular esterase hydrolysis (DCFH), after which the cell is treated with peroxyl radicals that oxidize the hydrolyzed DCFH and generates the hydrolyzed DCFH. Also generate measurable fluorescence, only antioxidants that could pass through the membrane inhibit peroxyl reactivity [58]. This type of approach may be modified slightly, for instance, with the use of certain cell types such as erythrocytes, which are then washed and fluorescent tested after being properly incubated with antioxidants and controls from a peroxyl-radical source and DCFH-DA [83]. The intracellular-protective effect of antioxidants has been studied in both methods. However, these methods could also be used to study external protection in the biological system, with erythrocytes and antioxidant incubation with a posterior oxidant agent such as radicals or hydrogen peroxide, and to finally measure spectrophotometric hemolysis caused by oxidant reactions to lipids and proteins in blood cell structures [84]. Although this type of test is a result-based technique, this is an excellent biological model because of its mitochondrial absence to ensure scarce interferences with target-biological protection [59]. Comet assay analyzes the protective effects of antioxidants in DNA, cell culture with antioxidants for the incorporation of post-agarose gel, and posterior oxidative treatment. Then, DNA is stained and visualized by fluorescence microscopy with black and white video camera connected to a computer [58]. Another alternative is the use of model organisms as microbial systems for the evaluation of antioxidant activity as it implements a method with the antioxidant mechanism of gene turn-off *E. coli* cells being key genes for survival [85]. The method consists of cells coculture with antioxidant followed by oxidizing treatment and, finally, spectroscopic growth monitoring.

It is important to note that, during in vivo oxidative stress, a number of mechanisms exist that involve oxidation of proliferation and mitigation of chemistry. This mechanism operates simultaneously and is challenging to analyze systematically during in vitro tests. This type of mechanism behind the neutralization or development of oxidizing molecules is primarily an enzyme [86]. In addition to the molecules involved in oxidation reactions, this enhances the measurement target during in vivo study. There has been a generous success in the antioxidant capacity of large-nature molecules.

11.6 Conclusions and Perspectives

Microbial bioactive molecules have become prominent over the last few years due to microorganisms' ability to survive in different extreme habitats. Microbial sources are able to synthesize new bioactive compounds with diverse biological and therapeutic activities that are better or more equal to the bioactivity of already known compounds. Bioactive molecules from microbial sources are promising to treat several chronic degenerative diseases. Microbial production of natural compounds may be carried out and improved by modifying the conditions of culture or modifying gene expression, significant challenges in the industrial production of such compounds. In this chapter, we have discussed some of the chemical, biochemical, and physiological aspects of the microbial production of antioxidants.

References

1 Abd El-Rahman, A.A.E.M., El-Shafei, S.M.A.E.A., Ivanova, E.V. et al. (2014). Cytotoxicity of Trichoderma spp. Cultural filtrate against human cervical and breast cancer cell lines. *Asian Pacific Journal of Cancer Prevention* 15: 7229–7234.

2 Gellen, L.F.A., Silva, D.P., Souza, S.R. et al. (2014). Potencial Farmacoindustrial de Trichoderma harzianum para fins Farmacoterapêuticos. *Biota Amazônia* 4: 91–96.

3 De la Cruz-Quiroz, R., Roussos, S., Rodríguez-Herrera, R. et al. (2018). Growth inhibition of Colletotrichum gloeosporioides and Phytophthora capsici by native Mexican Trichoderma strains. *Karbala International Journal of Modern Science* 4: 237–243.

4 Zhang, S., Xu, B., Zhang, J., and Gan, Y. (2018). Identification of the antifungal activity of Trichoderma longibrachiatum T6 and assessment of bioactive substances in controlling phytopathgens. *Pesticide Biochemistry and Physiology* 147: 59–66.

5 Torres-León, C., Rojas, R., Serna-Cock, L. et al. (2017). Extraction of antioxidants from mango seed kernel: optimization assisted by microwave. *Food and Bioproducts Processing* 105: 188–196.

6 Cipolatti, E.P., Remedi, R.D., dos Santos Sá, C. et al. (2019). Use of agroindustrial byproducts as substrate for production of carotenoids with antioxidant potential by wild yeasts. *Biocatalysis and Agricultural Biotechnology* 20: 101208.

7 Hernández-Almanza, A., Montañez-Sáenz, J., Martínez-Ávila, C. et al. (2014). Carotenoid production by Rhodotorula glutinis YB-252 in solid-state fermentation. *Food Bioscience* 7: 31–36.

8 Torres-León, C., Ramirez-Guzman, N., Ascacio-Valdes, J. et al. (2019). Solid-state fermentation with Aspergillus niger to enhance the phenolic contents and antioxidative activity of Mexican mango seed: a promising source of natural antioxidants. *LWTs* 112: 108236.

9 Chandra, P., Sharma, R.K., and Arora, D.S. (2020). Antioxidant compounds from microbial sources: a review. *Food Research International* 129: 108849.

10 Rizzello, C.G., Lorusso, A., Russo, V. et al. (2017). Improving the antioxidant properties of quinoa flour through fermentation with selected autochthonous lactic acid bacteria. *International Journal of Food Microbiology* 241: 252–261.

11 Gonzalez-Gonzalez, C.R., Machado, J., Correia, S. et al. (2019). Highly proteolytic bacteria from semi-ripened Chiapas cheese elicit angiotensin-I converting enzyme inhibition and antioxidant activity. *LWT* 111: 449–456.

12 Hernández-Almanza, A., Muñiz-Márquez, D.B., De la Rosa, O. et al. (2017). *Microbial Production of Bioactive Pigments, Oligosaccharides, and Peptides*. Elsevier Inc.

13 Babini, E., Tagliazucchi, D., Martini, S. et al. (2017). LC-ESI-QTOF-MS identification of novel antioxidant peptides obtained by enzymatic and microbial hydrolysis of vegetable proteins. *Food Chemistry* 228: 186–196.

14 Sila, A. and Bougatef, A. (2016). Antioxidant peptides from marine by-products: isolation, identification and application in food systems. *A Review. Journal of Functional Foods* 21: 10–26.

15 del Mar Contreras, M., Hernández-Ledesma, B., Amigo, L. et al. (2011). Production of antioxidant hydrolyzates from a whey protein concentrate with thermolysin: optimization by response surface methodology. *LWT – Food Science and Technology* 44: 9–15.

16 Verma, D.K., Patel, A., and Srivastav, P.P. (2018). *Bioprocess Technology in Food and Health: Potential Applications and Emerging Scope*, As part of book series on Innovation in Agricultural Microbiology. Apple Academic Press https://doi.org/10.1201/9781351167888.

17 Hogan, S., Zhang, L., Li, J. et al. (2009). Development of antioxidant rich peptides from milk protein by microbial proteases and analysis of their effects on lipid peroxidation in cooked beef. *Food Chemistry* 117: 438–443.

18 You, L., Regenstein, J.M., and Liu, R.H. (2010). Optimization of hydrolysis conditions for the production of antioxidant peptides from fish gelatin using response surface methodology. *Journal of Food Science* 75: 582–587.

19 Tuli, H.S., Chaudhary, P., and Beniwal, V. (2015). Microbial pigments as natural color sources : current trends and future perspectives. *Journal of Food Science and Technology* 52: 4669–4678.

20 Li, F., Xue, F., and Yu, X. (2017). GC – MS, FTIR and raman analysis of antioxidant components of red pigments from Stemphylium lycopersici. *Current Microbiology* 74: 532–539.

21 Nigam, P.S. and Luke, J.S. (2016). Food additives: production of microbial pigments and their antioxidant properties. *Current Opinion in Food Science* 7: 93–100.

22 Suraiya, S., Siddique, M.P., Lee, J. et al. (2018). Enhancement and characterization of natural pigments produced by Monascus spp using Saccharina japonica as fermentation substrate. *Journal of Applied Phycology* 30 (1): 729–742.

23 Ramesh, C., Vinithkumar, N.V., Kirubagaran, R. et al. (2019). Multifaceted applications of microbial pigments: current knowledge, challenges and future directions for public health implications. *Microorganisms* 7: 186, 1-46.

24 Vendruscolo, F., Marie, R., Bühler, M. et al. (2016). Monascus: a reality on the production and application of microbial pigments. *Applied Biochemistry and Biotechnology* 178: 211–223.

25 Gmoser, R., Ferreira, J.A., Lennartsson, P.R., and Taherzadeh, M.J. (2017). Filamentous ascomycetes fungi as a source of natural pigments. *Fungal Biology and Biotechnology* 17: 1–25.

26 El-Naggar, N.E.A. and El-Ewasy, S.M. (2017). Bioproduction, characterization, anticancer and antioxidant activities of extracellular melanin pigment produced by newly isolated microbial cell factories Streptomyces glaucescens NEAE-H. *Scientific Reports* 7: 1–19.

27 Fiedor, J. and Burda, K. (2014). Potential role of carotenoids as antioxidants in human health and disease. *Nutrients* 6: 466–488.

28 Nosanchuk, J., Stark, R.E., and Casadevall, A. (2015). Fungal melanin: what do we know about structure? *Frontiers in Microbiology* 6: 1463.

29 Yu, X., Huo, L., Liu, H. et al. (2015). Melanin is required for the formation of the multi-cellular conidia in the endophytic fungus Pestalotiopsis microspora. *Microbiological Research* 179: 1–11.

30 Ascacio-Valdés, J.A., Buenrostro-Figueroa, J.J., Aguilera-Carbo, A. et al. (2011). Ellagitannins: biosynthesis, biodegradation and biological properties. *Journal of Medicinal Plant Research* 5: 4696–4703.

31 D'Antuono, I., Bruno, A., Linsalata, V. et al. (2018). Fermented Apulian table olives: effect of selected microbial starters on polyphenols composition, antioxidant activities and bioaccessibility. *Food Chemistry* 248: 137–145.

32 Jeng, K.C., Chen, C.S., Fang, Y.P. et al. (2007). Effect of microbial fermentation on content of statin, GABA, and polyphenols in Pu-erh tea. *Journal of Agricultural and Food Chemistry* 55: 8787–8792.

33 Ascacio-Valdés, J.A., Buenrostro, J.J., De la Cruz, R. et al. (2014). Fungal biodegradation of pomegranate ellagitannins. *Journal of Basic Microbiology* 54: 28–34.

34 Butnariu, M. (2016). Methods of analysis (extraction, separation, identification and quantification) of carotenoids from natural products. *Journal of Ecosystem & Ecography* 6: 1–19.

35 Gomes, T.A., Zanette, C.M., and Spier, M.R. (2020). An overview of cell disruption methods for intracellular biomolecules recovery. *Preparative Biochemistry & Biotechnology* 50 (7): 635–654. 1–20.

36 Lee, A.K., Lewis, D.M., and Ashman, P.J. (2012). Disruption of microalgal cells for the extraction of lipids for biofuels: processes and specific energy requirements. *Biomass and Bioenergy* 46: 89–101.

37 Liu, D., Ding, L., Sun, J. et al. (2016). Yeast cell disruption strategies for recovery of intracellular bio-active compounds — a review. *Innovative Food Science and Emerging Technologies* 36: 181–192.

38 Lebeau, J., Venkatachalam, M., Fouillaud, M. et al. (2017). Production and new extraction method of polyketide red pigments produced by Ascomycetous fungi from terrstrial and marine habitats. *Journal of Fungi* 3: 34.

39 Hernández-Almanza, A., Navarro-Macías, V., Aguilar, O. et al. (2017). Carotenoids extraction from Rhodotorula glutinis cells using various techniques: a comparative study. *Indian Journal of Experimental Biology* 55: 479–484.

40 Ngamwonglumlert, L., Devahastin, S., and Chiewchan, N. (2017). Natural colorants: pigment stability and extraction yield enhancement via utilization of appropriate pretreatment and extraction methods. *Critical Reviews in Food Science and Nutrition* 57: 3243–3259.

41 Papadaki, A., Papapostolou, H., Alexandri, M. et al. (2018). Fumaric acid production using renewable resources from biodiesel and cane sugar production processes. *Environmental Science and Pollution Research* 25: 35960–35970.

42 Kaur, B., Chakraborty, D., Kaur, G., and Kaur, G. (2013). Biotransformation of rice bran to Ferulic acid by Pediococcal Isolates. *Applied Biochemistry and Biotechnology* 170: 854–867.

43 Sun, F., Hu, Y., Yin, X. et al. (2020). Production, purification and biochemical characterization of the microbial protease produced by Lactobacillus fermentum R6 isolated from Harbin dry sausages. *Process Biochemistry* 89: 37–45.

44 Sriram, S.V.A., Sarvamandala, D., Avala, S., and Sharma, M. (2019). Production and characterization of phytocompounds and pigments using Penicillium purpurogenum. *Research Journal of Life Science, Bioinformatics, Pharamceutical and Chemical Science* 5: 307–322.

45 Rehakova, K., Capkova, K., Hrouzek, P. et al. (2019). Microbial photosynthetic and photoprotective pigments in Himalayan soils originating from different elevations and successional stages. *Soil Biology and Biochemistry* 132: 153–164.

46 de Oliveira, C.F.D., da Costa, J.P.V., and Vendruscolo, F. (2019). Maltose syrup residue as the substrate for Monascus pigments production. *Biocatalysis and Agricultural Biotechnology* 18: 101101.

47 Bouhri, Y., Askun, T., Tunca, B. et al. (2020). The orange-red pigment from Penicillium mallochii: pigment production, optimization, and pigment efficacy against Glioblastoma cell lines. *Biocatalysis and Agricultural Biotechnology* 23: 101451.

48 Loh, W.L.C., Huang, K.-C., Ng, H.S., and Lan, J.C.-W. (2020). Exploring the fermentation characteristics of a newly isolated marine bacteria strain, Gordonia terrae TWRH01 for carotenoids production. *Journal of Bioscience and Bioengineering* 130: 187–194.

49 Kot, A.M., Błażejak, S., Kieliszek, M. et al. (2020). Production of lipids and carotenoids by Rhodotorula gracilis ATCC 10788 yeast in a bioreactor using low-cost wastes. *Biocatalysis and Agricultural Biotechnology* 26: 101634.

50 Zhu, L. and Xu, X. (2013). Stimulatory effect of different lignocellulosic materials for phenolic compound production and antioxidant activity from inonotus obliquus in submerged fermentation. *Applied Biochemistry and Biotechnology* 169: 2138–2152.

51 Costa, T.M., Kaufmann, V., Paganelli, C.J. et al. (2019). Kinetic identification of phenolic compounds and potential production of caffeic acid by Ganoderma lipsiense in solid-state fermentation. *Bioprocess and Biosystems Engineering* 42: 1325–1332.

52 Hu, X., Pang, X., Wang, P.G., and Chen, M. (2019). Isolation and characterization of an antioxidant exopolysaccharide produced by Bacillus sp. S-1 from Sichuan Pickles. *Carbohydrate Polymers* 204: 9–16.

53 Giri, S.S., Ryu, E.C., Sukumaran, V., and Park, S.C. (2019). Antioxidant, antibacterial, and anti-adhesive activities of biosurfactants isolated from Bacillus strains. *Microbial Pathogenesis* 132: 66–72.

54 Azmi, W. (2018). Production and antimicrobial, antioxidant and anticancer applications of pyocyanin from isolated Pseudomonas aeruginosa. *Journal of Fram Memory Technology* 1: 1.

55 Murphy, M.P. (2009). How mitochondria produce reactive oxygen species. *Biochemical Journal* 13: 1–13.

56 Davies, J.M.S., Cillard, J., Friguet, B. et al. (2017). The Oxygen Paradox, the French Paradox, and age-related diseases. *GeroScience* 39: 499–550.

57 Gulcin, İ. (2020). Antioxidants and antioxidant methods : an updated overview. *Archives of Toxicology* 94: 651–715.

58 Hofer, T., Eriksen, T.E., Hansen, E. et al. (2011). Cellular and chemical assays for discovery of novel antioxidants in marine organisms. In: *Studies on Experimental Models, Oxidative Stress in Applied Basic Research and Clinical Practice* (eds. S. Basu and L. Wiklund), 637–657. Springer Science + Business Media.

59 Shahidi, F. and Zhong, Y. (2015). Measurement of antioxidant activity. *Journal of Functional Foods* 18: 757–781.

60 Huang, D., Ou, B., and Prior, R.L. (2005). The chemistry behind Antioxidant capacity assays. *Journal of Agricultural and Food Chemistry* 53: 1841–1856.

61 Blois, M.S. (1958). Antioxidant determinations by the use of a stable free radical. *Nature* 26: 1199–1200.

62 Miller, N.J. and Rice-Evans, C.A. (1996). Spectrophotometric determination of antioxidant activity. *Redox Report* 2: 161–171.

63 Fogliano, V., Verde, V., Randazzo, G., and Ritieni, A. (1999). Method for measuring antioxidant activity and its application to monitoring the antioxidant capacity of wines. *Journal of Agricultural and Food Chemistry* 47 (3): 1035–1040.

64 Cao, G., Alessio, H.M., and Cutler, R.G. (1993). Oxygen-radical absorbance capacity assay for antioxidants. *Free Radical Biology and Medicine* 14: 303–311.

65 Valentão, P., Fernandes, E., Carvalho, F. et al. (2002). Studies on the antioxidant activity of Lippia citriodora infusion: scavenging effect on superoxide radical, hydroxyl radical and hypochlorous acid. *Biological and Pharmaceutical Bulletin* 25: 1324–1327.

66 Vanacker, S.A.B.E., Tromp, M.N.J.L., Haenen, G.R.M.M. et al. (1995). Flavonoids as scavengers of nitric oxide radical. *Biochemical and Biophysical Research Communications* 214: 755–759.

67 Kooy, N.W., Royall, J.A., Ischiropoulos, H., and Beckman, J.S. (1994). Peroxynitrite-mediated oxidation of dihydrorhodamine 123. *Free Radical Biology and Medicine* 16: 149–156.

68 Sánchez-Moreno, C. (2002). Methods used to evaluate the free radical scavenging activity in foods and biological systems. *Food Science and Technology International* 36: 703–725.

69 Sánchez, C. (2017). Reactive oxygen species and antioxidant properties from mushrooms. *Synthetic and Systems Biotechnology* 2: 13–22.

70 Wu, X., Beecher, G.R., Holden, J.M. et al. (2004). Lipophilic and hydrophilic antioxidant capacities of common foods in the United States. *Journal of Agricultural and Food Chemistry* 52 (12): 4026–4037.

71 Ruch, R.J., Cheng, S.J., and Klaunig, J.E. (1989). Prevention of cytotoxicity and inhibition of intercellular communication by antioxidant catechins isolated from chinese green tea. *Carcinogenesis* 10 (6): 1003–1008.

72 Mukhopadhyay, D., Dasgupta, P., Sinha Roy, D. et al. (2016). A sensitive in vitro spectrophotometric hydrogen peroxide scavenging assay using 1,10-Phenanthroline. *Free Radicals and Antioxidants* 6 (1): 124–132.

73 Paździoch-Czochra, M. and Wideńska, A. (2002). Spectrofluorimetric determination of hydrogen peroxide scavenging activity. *Analytica Chimica Acta* 452: 177–184.

74 Benzie, I.F.F. and Strain, J.J. (1996). The ferric reducing ability of plasma (FRAP) as a measure of "antioxidant power": the FRAP assay. *Analytical Biochemistry* 239 (1): 70.

75 Apak, R., Güçlü, K., Özyürek, M., and Karademir, S.E. (2004). Novel total antioxidant capacity index for dietary polyphenols and vitamins C and E, using their cupric ion reducing capability in the presence of neocuproine: CUPRAC method. *Journal of Agricultural and Food Chemistry* 52: 7970–7981.

76 Folin, O. and Ciocalteu, V. (1927). On tyrosine and tryptophane determinations in proteins. *The Journal of Biological Chemistry* 73: 627–650.

77 Özyürek, M., Güçlü, K., Tütem, E. et al. (2011). A comprehensive review of CUPRAC methodology. *Analytical Methods* 3: 2439–2453.

78 Sánchez, C. (2017). Bioactives from mushroom and their application. In: *Food Bioactives* (ed. M. Puri), vii–ix. Cham: Springer.

79 Lamuela-Raventós, R.M. (2017). Folin-Ciocalteu method for the measurement of total phenolic content and antioxidant capacity. In: *Measurement of Antioxidant Activity and Capacity: Recent Trends and Applications* (eds. R. Apak, E. Capanoglu and F. Shahidi), 107–115. Wiley.

80 Lopes, G.K.B., Schulman, H.M., and Hermes-Lima, M. (1999). Polyphenol tannic acid inhibits hydroxyl radical formation from Fenton reaction by complexing ferrous ions. *Biochimica et Biophysica Acta – General Subjects* 1472: 142–152.

81 Hirayama, O., Takagi, M., Hukumoto, K., and Katoh, S. (1997). Evaluation of Antioxidant activity by Chemiluminescence. *Analytical Biochemistry* 247: 237–241.

82 Wolfe, K.L. and Rui, H.L. (2007). Cellular antioxidant activity (CAA) assay for assessing antioxidants, foods, and dietary supplements. *Journal of Agricultural and Food Chemistry* 55 (22): 8896.

83 George, A., Ng, C.P., O'Callaghan, M. et al. (2014). in vitro and ex-vivo cellular antioxidant protection and cognitive enhancing effects of an extract of Polygonum minus Huds (LineminusTM) demonstrated in a Barnes Maze animal model for memory and learning. *BMC Complementary and Alternative Medicine* 14: 1–10.

84 Gião, M.S., Leitão, I., Pereira, A. et al. (2010). Plant aqueous extracts: antioxidant capacity via haemolysis and bacteriophage. *Food Control* 21: 633–638.

85 Zhou, D., Wang, D., Yang, L. et al. (2016). A modified and improved assay based on microbial test system (MTS) to evaluate antioxidant activity. *Food Analytical Methods* 9: 895–904.

86 Meng, D., Zhang, P., Zhang, L. et al. (2017). Detection of cellular redox reactions and antioxidant activity assays. *Journal of Functional Foods* 37: 467–479.

12

Microbial Production of Biomethane from Digested Waste and Its Significance

Arun Kumar Pal[1], Vijay Tripathi[1], Prashant Kumar[2], and Pradeep Kumar[3]

[1] *Department of Molecular and Cellular Engineering, Jacob Institute of Biotechnology and Bioengineering, Sam Higginbottom University of Agriculture Technology and Sciences, Prayagraj, Uttar Pradesh, India*
[2] *Department of Civil Engineering, North Eastern Regional Institute of Science and Technology, Nirjuli, Arunachal Pradesh, India*
[3] *Department of Forestry, North Eastern Regional Institute of Science and Technology, Nirjuli, Arunachal Pradesh, India*

12.1 Introduction

The waste generation is increased day by day due to the rapid increase in urbanization and population worldwide [1]. The waste management (or waste disposal) includes the various activities and required actions to manage waste from its inception to its final disposal, which encompasses the collection, transport, treatment and disposal of waste through proper monitoring, and regulation of the whole management process [2]. The management of billions of tons of waste is a severe issue for the environmental safety system because it is the primary source of environmental pollutants that decompose into various hazardous chemicals [3]. Throughout the world waste generation rates are increasing and in 2016 the world cities generated 2.01 billion tonnes of solid waste, amounting to a footprint of 0.74 klogram per person per day, and this may be supposed to increase by 70% from 2016 levels to 3.40 billion tonnes in 2050 due to global population growth and urbanization [4]. The average waste generation worldwide per person per day is 0.74 kg, but that ranges from 0.11 to 4.54 kg [3]. However, developed countries pay a high contribution to the waste generation as compared to developing countries. The Global waste Index indicated that this is an alarming condition, and waste generation will destroy our environment, climate conditions, and our earth planet [5]. Fossil fuels are the only solution to fulfill the energy demands, but that directly affects our climate and environment. Therefore, an alternative of fossil fuels is available to fulfill the higher demand for energy, i.e., "bio-fuel." The biogas is very cheap and environment friendly because it is released from the animal and plant wastes. A few decades ago, various technologies were used to convert the agro and municipal solid wastes into renewable energy. Anaerobic digestion is one of the most used technologies for converting waste into renewable energy [6]. In this technology, wastes are decomposed by anaerobic bacteria during anaerobic respiration in high moisture content and produce biogas [7]. The first-generation biofuel reduces greenhouse gas emission, which relies on the agriculture crops such as corn, sugarcane, sugar

Bioprospecting of Microorganism-Based Industrial Molecules, First Edition.
Edited by Sudhir P. Singh and Santosh Kumar Upadhyay.

beets, soybean, and canola [8], but it displaces the agricultural land [9]. The "first-generation" biofuels appear unsustainable because of the potential stress that their production places on food commodities [10]. However, the production of second- and third-generation biofuel uses the alternative source that cannot affect the agricultural land. Second-generation biofuel uses the substrate as jatropha for biodiesel production, and switchgrass for cellulosic ethanol production and other biomass also use such as plant and crop waste [11]. The third-generation biofuels are produced from algae, sewage sludge, and municipal solid wastes [12]. There is more advantage of having marine algae because it does not require fresh water for their growth. During anaerobic digestion for methane production and fermentation for ethanol production, high water content is requiring [13, 14]. Biomethane is a naturally occurring gas that is produced by anaerobic digestion process from organic matter like dead animal, plant material, manure, and sewage [15]. Biomethane has many environmental benefits because it can reduce the other harmful gasses in the air (methane and nitrous dioxide), which can cause greenhouse effects and climate change [16]. The biomethane production from organic matter can also reduce water pollution and improve the hygiene condition of rural life [17]. In this chapter, we have described the biomethane production from waste materials, processes involved, and their significance.

12.2 Methane

Methane (CH_4) is a colorless and odorless gas, which generally present in nature due to various human activities. Naturally occurring methane is found both below ground and under the seafloor, and is formed by both geological and biological processes. The largest reservoir of methane is under the seafloor in the form of methane clathrates. When methane reaches the surface and then atmosphere, it is known as atmospheric methane [18]. Methane is a hydrocarbon and a very effective greenhouse gas, which is produced by the anaerobic bacterial decomposition of agricultural matter in wetlands. The extraction and processing of oil, natural gas, and coal are the major anthropogenic sources of methane that can release a significant amount of methane in the environment. Other anthropogenic activities such as biomass burning, livestock farming, and waste management are also responsible for methane production. Methane is a source of hydrogen and some organic chemicals, which include methanol, chloroform, carbon tetrachloride, and nitromethane. It yields carbon black during incomplete combustion, which is mainly used as a reinforcing agent in rubber for automobile tires.

12.2.1 Source of Methane

Nowadays, methane gas is a huge problem for the environment because it can contribute to climate change [19]. To save the environment, it is imperative to reduce the emission of methane gas [4]. The source of methane emission is mainly from two ways (i) from human activities (anthropogenic process) such as production and transport of coal, oil, and natural gas and municipal solid waste landfills and (ii) from the breakdown of organic materials (natural process) such as the breakdown of organic materials of agriculture, digestion of food particles by livestock [20]. These processes can introduce the methane gas into the atmosphere. Mostly anthropogenic activities are responsible for introducing 50–65% methane emission worldwide. The critical sector of methane emission are described below:

12.2.1.1 Industry

The energy sector, mainly oil, natural gas, coal, and bioenergy, is the world's second-largest source of methane emission. It is emitted in the atmosphere during the production, processing, storage, transmission, and distribution of natural gas and the production, refinement, transportation, and storage of crude oil. Industrial wastewater treatment plants are also the most important sources of methane emissions (Figure 12.1).

12.2.1.2 Agriculture

The food system of domestic livestock such as cattle, swine, sheep, and goats is one of the largest emitters of methane during the normal digestive process. The gut microbes are involved in the breakdown of plant materials and facilitating to absorb into the animals' bloodstream, and methane will be the by-product of this process. The manure storage in the holding tanks and burning of agricultural wastes are also producing methane gas. The anaerobic condition in the flooded rice paddy field also facilitates microbial methane production. These agricultural processes are the major contributor to CH_4 emissions.

12.2.1.3 Waste

The landfills are the collection of municipal soil wastes, and these waste compounds produce the methane gas through methane-producing bacteria. The methane-producing bacteria decompose the waste and generate methane gas. Landfills are the third-largest source of CH_4 emissions worldwide. Methane is also generated from domestic and industrial wastewater treatment and from composting.

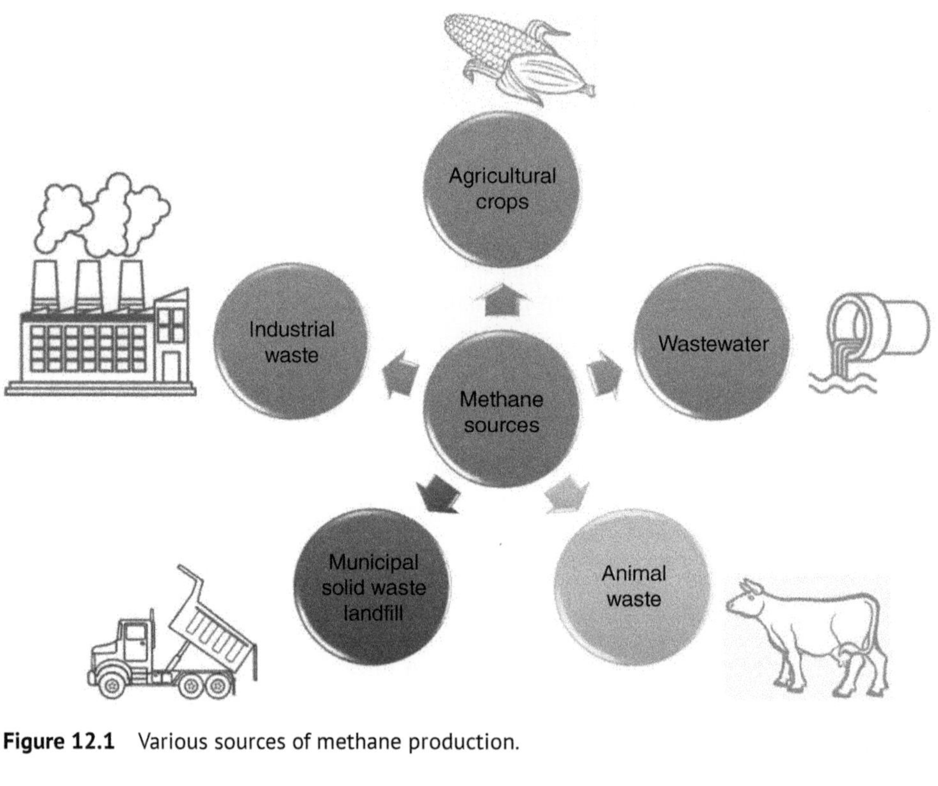

Figure 12.1 Various sources of methane production.

12.2.2 Biomethane

Biomethane is a gas made up of improving the biogas quality by decreasing H_2S, CO_2 moistures, and other gases [21]. Biomethane is the upgraded form of the biogas, which contains the pure methane in a higher percentage. Biomethane is also known as renewable natural gas, but in a vehicle, it is used in a compressed form, known as compressed biomethane gas. This is like a compressed natural gas or natural gas for vehicle. Natural gas is made up of 70–95% methane, and remaining others are hydrocarbons, e.g. butane, propane, ethane, and so on, in trace amounts [22].

According to the Intergovernmental Panel on Climate Change, now the time has come to reduce the greenhouse gas emission level of less than half of 1990 [23]. To initiate toward this goal, European countries produce the six million tons of energy from the biogas that is equivalent to oil in 2007. European Union takes the first position to produce 60% biogas worldwide and likely keep this position stable in the future [24]. Besides this, many European countries are working on biomethane production for some years and have several working plants right now. Most of the countries are in the initial phase of the production and use of biomethane (Table 12.1) [25].

12.3 Types of Waste

Types of waste can be generally classified into four types: biological, domestic, industrial, and agricultural wastes (Table 12.2).

Table 12.1 Use of Biomethane in different European countries. *Source:* Modified from [26].

	Germany	Sweden	UK	France	Netherland	Switzerland
No. of feed in plants (2012)	110	7	2	2	17	13
Total feed in capacity (Nm/h)	66830	2610	160	715	7605	1664
Main market drivers	Government support schemes for energy crop	Vehicle fuel	Heat market	Government support scheme vehicle fuel	Green gas products, government support scheme	Government support scheme vehicle fuel
Dominating feedstock	Energy crops organic waste manure	Sewage sludge organic waste	Organic waste	Sewage sludge organic waste	Organic waste sewage sludge landfill gas extraction	Sewage sludge manure organic waste

Table 12.2 Different source and waste generators. *Source:* Nandan et al. [27] Licensed under CC BY-4.0.

Source	Residential	Industrial	Commercial & institutional	Construction and demolition	Municipal services	Agriculture	Mining
Waste generator	Household activities	Manufacturing unit, power plant, process industries	Hotel, restaurants, Markets, office buildings, schools, hospitals	New construction sites, demolition of existing structures, road repair	Street cleaning, landscaping, parks and other recreational areas, water and wastewater treatment plants	Crops, orchards, vineyards, dairy farms, etc.	Opencast mining, underground mining
Type of waste	Food waste, paper, cardboard, plastic, wood, glass, metal, etc.	Housekeeping wastes, hazardous wastes, ashes, other waste etc.	Biomedical waste, food waste, glass, metals, plastic, paper, etc.	Wood, street concrete, etc.	Tree trimmings, general waste, etc.	Insecticides, pesticides, agricultural waste, etc.	Inert material such as ash

12.3.1 Biological Waste

The waste was created during the healthcare process and contaminated from hazardous biological agents called biological waste. Biological waste includes petri dishes, bandages, surgical wraps, culture tubes, syringes, needles, blood vials, absorbent material, personal protective equipment, and pipette tips. These agents are biological and can create a harmful effect on living organisms. Some biohazardous agents are bacteria, viruses, prion, fungi, chlamydia, parasites, recombinant product, allergens, rickettsiae, infected clinical specimen, and tissue from experimental animals [28]. According to the WHO, these materials should be dumped in a proper place with proper handling; otherwise, they may have a dangerous effect on humans, as well as the environment also [28]. The biological waste source is research centers, chemotherapeutic, animal tissues, veterinary clinics, and human pathological waste (Figure 12.2).

12.3.2 Household Waste

Household waste (HW) is unwanted; useless substances arose in the domestic dwelling, e.g., furniture, toys, clothes, food leftover, used for packaging materials as glass bottle, glass jars, plastic food pots, metal drink, steel food can, paper, textiles, plastic glasses, and so on [29]. The European Commission [30] defines food waste as "waste composed of raw or cooked food materials, and includes food discarded at any time between farm and fork; in households relating to food waste generated before, during or after food preparation such as vegetable peelings, meat trimmings, and spoiled or excess ingredients or prepared food." At present, in worldwide millions of tons of household solid waste was produced; this is very dangerous for human health, as well as for our environment. Furthermore, this is more problematic for urban communities because there is no available place for dumping. This is a very big challenge for the communities to reduce and treat the HW [31].

HW has a positive relevance and it can be exploited as a source of renewable energy. Utilizing HW helps in declining the organic pollutant from the environment and other side

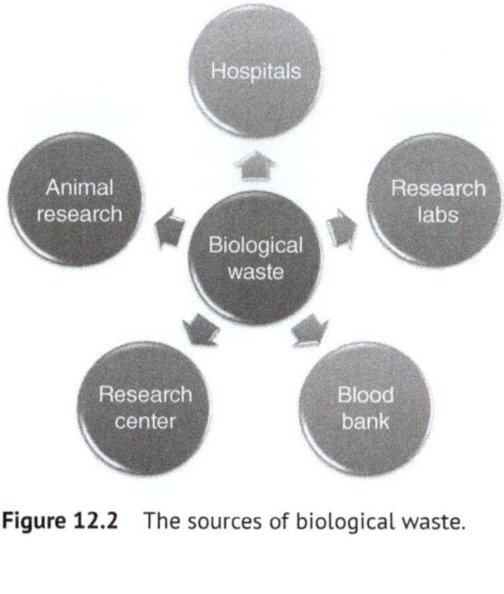

Figure 12.2 The sources of biological waste.

produce the energy in the form of biogas [28]. Waste to energy is a technique that can recover the energy in the form of heat, electricity, or in the form of biogas. This is a pervasive technique that covers different scale and complexity. This work could be done by this technique, e.g., cooking gas in household production, methane production from landfill, gasification in the cement industry, and so on [32].

12.3.3 Agricultural Waste

Agricultural waste is a residual by-product, which is collected during the processing of raw product such as vegetables, fruits, crop, husk, and so on [33]. Presently agricultural waste is becoming huge contaminates for the environment; that is why EPA focused on this more curiously. However, on the other hand, they may also use different beneficial purposes such as energy production, recovery of chemicals, and adsorption of different dye from the environment [34]. For agricultural waste disposition, production of energy from agricultural waste such as rice husk, corncob, crop straw, and crop processing residues is the best method. This method supports the conversion of agricultural waste into clean energy [35]. These wastes mostly include the nonedible portion of plants, such as leaves, corn stover, straw, and so on. Large quantities of crop residues are generated from agricultural activities annually worldwide. In 2006, it was estimated that 75.73 million tons of dry matter were produced from 20 crops in Mexico [36]. Furthermore, this process contributes to the management of agricultural waste, which have not been treated properly, is decomposed, and methane is released to the environment of a noncontrolled manner. Agricultural waste that mostly contains plant materials is the rich source of lignocellulosic biomass.

Lignocellulosic biomass is mainly composed of cellulose, hemicellulose, and lignin, which may vary based on types of plant growth conditions and maturations. Biogas produced from agriculture residues and energy crops cultivated on land unsuitable for feed or food production is regarded as second-generation biofuel, as there is no direct competition for arable land [37, 38].

12.4 Digestion Processes of Organic Wastes

Anaerobic transformation of organic wastes is the process that involves many microorganisms like hydrolytic, acid-forming, acetogenic, and methanogenic bacteria, which produce CO_2 and CH_4 as the main products of the digestion process [39, 40]. Four main steps are involved in the anaerobic digestion process, namely, hydrolysis, acidogenesis, acetogenesis, and methanogenesis [41].

12.4.1 Hydrolysis of Organic Waste

In this process complex, organic matter is transformed into the soluble organic matter like carbohydrates, proteins, and lipids with the help of hydrolytic bacteria. This stage of methanogenesis was carried out by the hydrolytic bacteria producing extracellular enzymes of hydrolases (amylases, proteases, cellulase, and lipases). These hydrolytic enzymes hydrolyzed their respective polymers into smaller molecules and primarily monomeric units, which are then consumed by microbes. During the fermentation of waste containing a high concentration of organic polymers, the hydrolytic activity of relevance to each polymer is of utmost significance, in that polymer hydrolysis may become a rate-limiting step for the production of simpler bacterial substrates to be used in subsequent degradation steps. During the hydrolysis process, approximately 50% of solid waste organic compound undergoes the biodegradation process. The remaining part of the compound remains in the primary state due to the lack of enzymes involved in digestion [42, 43]. The spread of hydrolysis depends on the various biochemical parameters like pH, an enzyme produced, diffusion, and absorption of enzymes, size of waste particles. This step is carried out by the anaerobic group of bacteria belonging to genera *Streptococcus* and *Enterobacter* [41].

Lipases catalyze the hydrolysis of the lipid to long-chain fatty acids, which are further degraded by p-oxidation to produce acetyl Co-A with the help of extracellular lipase producing Clostridia and Micrococci. Further, proteins are hydrolyzed to amino acids with numerous microbes like *Bacteroides, Butyrivibrio, Clostridium, Fusobacterium, Selenomonas,* and *Streptococcus*, etc., producing proteases. Later, amino acids degraded to fatty acids such as acetate, propionate, butyrate, and ammonia, as found in *Clostridium, Peptococcus, Selenomonas, Campylobactor,* and *Bacteroides.* Polysaccharides such as cellulose, starch, and pectin are hydrolyzed by cellulases, amylases, and pectinases. Hexoses and pentoses are generally converted to C2 and C3 intermediates and to reduced electron carriers (e.g. NADH) via common pathways.

12.4.2 Acidogenesis of Hydrolyzed Matter

Acidogenesis is one of the most versatile bioprocesses, employable to produce a multitude of bio-based products. There is conversion of organic matter to organic acids and hydrogen during acidogenesis. Sugars, ammo acids, and fatty acids produced by microbial degradation of biopolymers are successively metabolized by groups of bacteria and are primarily fermented to acetate, propionate, butyrate, lactate, ethanol, carbon dioxide, and hydrogen. Here the water-soluble chemical substances and hydrolyzed product are converted into short-chain organic acids (formic, acetic, propionic, butyric, and pentanoic), alcohol (methanol, and ethanol), aldehydes, carbon dioxide, and hydrogen by the acidifying bacteria. Anaerobic microorganisms get the energy from the breakdown of proteins, amino acids, and peptides. In this stage, microbes involve facultative anaerobes that use oxygen accidentally in the process and create favorable conditions for the development of obligatory anaerobes of following genera: *Pseudomonas, Bacillus, Clostridium, Micrococcus,* and *Flavobacterium* [41, 44].

12.4.3 Acetogenesis

The acidogenesis product is directly converted into acetate and hydrogen with the help of microbial genera, *Syntrophomonas* and *Syntrophobacter, and these products are used by methanogenic microbes* [41, 44]. *Approximately 20% acetate and 4% H_2* are directly produced by acidogenic fermentation of sugar and amino acids. Pentoic acid is converted into propionic acid by *Methanobacterium suboxydans*, and propionic acid is converted into acetic acid by *M. propionicum.* Due to this conversion, hydrogen is released, which shows a toxic effect on microorganisms that carry out this process. The efficiency of biomethane production is illustrated in acetogenesis stage, and approximately 70% of methane arises in the process of acetates reduction [41, 44, 45].

12.4.3.1 Methanogenesis

This is the final stage of methane production by the methanogenic bacteria. They utilize the product (acetic acid, H_2, CO_2, methanol, methylamine, or dimethyl sulfide) of previous phases. Only 30% of the methane produced in this process comes from CO_2 reduction, carried out by autotrophic methane bacteria. During this process, H_2 is used up, which creates good conditions for the development of acid bacteria, giving rise to short-chain organic acids in the acidification phase and, consequently, too low production of H_2 in the acetogenic phase. A consequence of such conversions leads to enrichment of CO_2 gas, as only its insignificant part will be converted into methane [41, 46].

12.5 Conclusions and Perspectives

Various aspects of biogas generation from agricultural wastes are briefly summarised. Utilizing agricultural waste are very economical source of bioenergy and biogas generation. A large number of agricultural wastes like potato peel, bagasse, orange peel, and soy pulp that are used in the production of biogas. Biomethane from the agricultural waste can be one of the most promising energy sources in coming years because it can manage the huge waste generation and meet high energy demands. It is urgent requirement to develop highly efficient method for utilizing the waste products as a source of energy. Moreover, it is required to make a national strategy on the potential use of biogas for the formation of thermal and electrical energy. Government policies for the promotion and motivation to increase the production and development of biofuels should be welcomed.

Acknowledgments

All authors would like to thanks the higher authority of their respective Department and University.

Conflicts of Interest

The authors declare no conflict of interest.

References

1 Vij, D. (2012). Urbanization and solid waste management in India: present practices and future challenges. *Procedia-Social and Behavioral Sciences* 37: 437–447.

2 Ferronato, N. and Torretta, V. (2019). Waste mismanagement in developing countries: a review of global issues. *International Journal of Environmental Research and Public Health* 16: 1060.

3 https://www.worldbank.org/en/topic/urbandevelopment/brief/solid-waste-management.
4 Allen, D.T., Pacsi, A.P., Sullivan, D.W. et al. (2014). Methane emission from prcess equipment at natural gas production sites in the United States: pneumatic controler. *American Chemical Society* 49: 633–640.
5 Mittal, R. and Mittal, C.G. (2013). Impact of population explosion on environment. *National Journal* 1: 1–5.
6 Neves, G.N., Berni, M., Dragone, G., and Mussatto, S.I. (2018). Anaerobic digestion process: technological aspects and recent developments. *International Journal of Environmental Science and Technology* 15: 2033–2046.
7 Anukam, A., Mohammadi, A., Naqvi, M., and Granstrom, K. (2019). A review of the chemistry of anaerobic digestion: methods of accelerating and optimizing process efficiency. *Processes* 7: 504.
8 Hirani, A.H., Javed, N., Asif, M. et al. (2018). *A Review on First- and Second-Generation Biofuel Productions*, 141–154. Springer, Biofuels: Greenhouse Gas Mitigation and Global Warming.
9 Mussgnug, J.H., Klassen, V., Schluter, A., and Kruse, O. (2010). Microalgae as substrates for fermentative biogas production in a combined biorefinery concept. *Journal of Biotechnology* 150: 51–56.
10 Naik, S.N., Goud, V.V., Rout, P.K., and Dalai, A.K. (2010). Production of first and second generation biofuels: a comprehensive review. *Renewable and Sustainable Energy Reviews* 14: 578–597.
11 Nanda, S., Rana, R., Sarangi, P. K., Dalai, A. K., Kozinski, J. A. (2018) A Broad Introduction to First-, Second-, and Third-Generation Biofuels. Springer, Singapore.
12 Rittmann, B.E. (2008). Opportunities for renewable bioenergy using microorganisms. *Biotechnology and Bioengineering* 100: 203–212.
13 Chynoweth, D.P. (2002). *Review of Biomethane from Marine Biomass*, 25. Gainesville: Department of Agricultural and Biological Engineering, University of Florida.
14 Faisal, S., Hafeez, F.Y., Zafar, Y.Y. et al. (2018). A review on nanoparticles as boon for biogas producers – nano fuels and biosensing monitoring. *Applied Sciences* 9: 59.
15 Paolini, V., Petracchini, F., Segreto, M. et al. (2018). Environmental impact of biogas: a short review of current knowledge. *Journal of Environmental Science and Health* 53: 899–906.
16 Ministry of New and Renewable Energy (2019). Government of India (2019) Annual report 2018-2019 in India.
17 Khalil, M.A.K. (1999). Non-Co2 greenhouse gases in the atmosphere. *Annual Review of Energy and the Environment* 24: 645–661.
18 Bryant, M.P. (1979). Microbial methane production: theoretical aspects. *Journal of Animal Science* 48: 193–201.
19 Environmental Defense Fund (2014). https://www.edf.org/climate/methane-other-important-greenhouse-gas.
20 The Obama Administration's Strategy to Reduce Methane Emissions (2015). Methane: An Introduction to Emission Sources and Reduction Strategies. Every CRS Report.com, University of North Texas Libraries Government Documents Department.
21 Mamun, M.R.A. and Torii, S. (2017). Enhancement of methane concentration by removing contaminants from biogas mixtures using combined method of absorption and adsorption. *International Journal of Chemical Engineering* 2017: 9.

22 Koonaphapdeelert, S., Aggarangsi, P., and Moran, J. (2020). *Introduction to Biomethane*, 1–15. Singapore: Springer https://doi.org/10.1007/978-981-13-8307-6.

23 The Intergovernmental panel on climate change (2020). UNEP. https://www.ipcc.ch/ (01 October 2020).

24 Raboni, M. and Urbini, G. (2014). Production and use of biogas in Europe: a survey of current status and perspectives. *An Interdisciplinary Journal of Applied Science* 9: 1324.

25 Busic, A., Kundas, S., Morzak, G. et al. (2018). Recent trends in biodiesel and biogas production. *Food Technology and Biotechnology* 56: 152–173.

26 Scarlat, N., Dallemand, J.F., and Fahl, F. (2018). Biogas: developments and perspectives in Europe. *Renewable Energy 129*: 457–472.

27 Nandan, A., Yadav, B.P., Baksi, S., and Bose, D. (2017). Recent scenario of solid waste management in India. *World Scientific News* 66: 56–74.

28 Cavallini, S., Costanzo, P., Gismondo, M.R. et al. (2014). Handbook on chemical and biological waste management. *American Journal of International Law* 88: 24–75.

29 van der Werf, P., Seabrook, J.A., and Gilliland, J.A. (2018). The quantity of food waste in the garbage stream of southern Ontario, Canada households. *PLoS One* 13 (6): e0198470. https://doi.org/10.1371/journal.pone.0198470.

30 European Commission (2010). Preparatory study on food waste across EU 27, Technical Report 2010–054. http://ec.europa.eu/environment/eussd/pdf/bio_foodwaste_report.pdf.

31 Lastella, G., Testa, C., Cornachhia, G. et al. (2000). Anaerobic digestion of semi-solid organic waste: biogas production and purification. *Energy Conversion and Management* 43: 63–75.

32 Mutz, D., Hengevoss, D., Hugi, C., and Gross, T. (2017). Waste-to-energy options in municipal solid waste management. *International Journal of Environmental* 197: 673–680.

33 Obi, F.O., Ugwuishiwu, B.O., and Nwakaire, J.N. (2016). Agricultural waste concept, generation, utilization and management. *Nigerian Journal of Technogy* 35: 957–964.

34 Chang, H.C. and Li, R.D. (2019). Agricultural waste. *Water Environment Research* 91: 1150–1167.

35 Blikra, E., Martinez-Sanchez, V., and Thomsen, M. (2018). A review of waste management decision support tools and their ability to assess circular biowaste management systems. *Sustainability* 10: 3720.

36 Valdez-Vazquez, I., Acevedo-Benítez, J.A., and Hernández-Santiago, C. (2010). Distribution and potential of bioenergy resources from agricultural activities in Mexico. *Renewable and Sustainable Energy Reviews* 14: 2147–2153.

37 Monlau, F., Latrille, E., Costa, A. et al. (2013). Enhancement of methane production from sunflower oil cakes by dilute acid pretreatment. *Applied Energy* 102: 1105–1113.

38 Chandra, R., Takeuchi, H., and Hasegawa, T. (2012). Methane production from lignocellulosic agricultural crop wastes: a review in context to second generation of biofuel production. *Renewable and Sustainable Energy Reviews* 16: 1462–1476.

39 Demirel, B. and Scherer, P. (2008). The roles of acetotrophic and hydrogenotrophic methanogens during anaerobic conversion of biomass to methane: a review. *Reviews in Environmental Science and Biotechnology* 7: 173–190.

40 Nealson, K.H. (1997). Sediment bacteria: who's there, what are they doing, and what's new? *Annual Review of Earth and Planetary Sciences* 25: 403–434.

41 Shah, A.F., Mahmood, Q., Shah, M.M. et al. (2014). Microbial ecology of anaerobic digesters: the key players of anaerobiosis. *The Scientifc World Journal* 21: 183752.

42 Conrad, R. (1999). Contribution of hydrogen to methane production and control of hydrogen concentrations in methanogenic soils and sediments. *FEMS Microbiology Ecology* 28: 193–202.

43 Parawira, W., Read, J.S., Mattiasson, B., and Björnsson, L. (2008). Energy production from agricultural residues: high methane yields in pilot-scale two-stage anaerobic digestion. *Biomass and Bioenergy* 32: 44–50.

44 Schink, B. (1997). Energetics of syntrophic cooperation in methanogenic degradation. *Microbiology and Molecular Biology Reviews* 61: 262–280.

45 De Bok, F.A.M., Harmsen, H.J.M., and Plugge, C.M. (2005). The first true obligately syntrophic propionate-oxidizing bacterium, *Pelotomaculum schinkii* sp. nov., co-cultured with *Methanospirillum hungatei*, and emended description of the genus Pelotomaculum. *International Journal of Systematic and Evolutionary Microbiology* 55: 1697–1703.

46 Karakashev, D., Batstone, D.J., and Angelidaki, I. (2005). Infuence of environmental conditions on methanogenic compositions in anaerobic biogas reactors. *Applied and Environmental Microbiology* 71: 331–338.

13

Enzymatic Biosynthesis of Carbohydrate Biopolymers and Uses Thereof

Manisha Sharma[1], Jyoti Singh Jadaun[2], Santosh Kumar Upadhyay[3], and Sudhir P. Singh[1]

[1] *Center of Innovative and Applied Bioprocessing (DBT-CIAB), Mohali, Punjab, India*
[2] *Department of Botany, Dayanand Girls Postgraduate College, Kanpur, Uttar Pradesh, India*
[3] *Department of Botany, Panjab University, Chandigarh, India*

13.1 Introduction

Biopolymers are organic macromolecules produced by diverse organisms. They are a versatile group of natural compounds with chemical and functional diversity [1]. Biopolymers may exist either as homopolymers (cellulose, and dextran) or heteropolymers (chitin, and xanthan gum) [2, 3]. Their applications have been established widely from the food to the pharmaceutical industry [4]. This has enhanced researchers' and industries' interest to explore economic industrial processes for the production of biopolymers. These polysaccharides may be produced either chemically or biologically. Due to adverse effect of chemical processes, including alteration in the properties of the product, unwanted side products, high energy consumption, utilization of organic solvents, etc., the biological processes involving microbes and enzymes are preferred over chemical process [5].

The biopolymers of plant origin are cost-effective, as well as involve exploitation of environmental biosphere. This is desirable to develop bioprocesses involving the utilization of the abundant agricultural residual biomass, industrial residues, sea or ocean waste, or by-products for production of value-added biopolymers. Some examples of such feedstocks could be cane molasses, corncob hydrolysate, liquid whey, beverage industry waste, etc. [6].

This chapter focuses on the microbial and enzymatic routes for the production of carbohydrate biopolymers. The biopolymers' unique characteristics are discussed concerning their wide range of applications in different areas, including food, textile, pharmaceutical, medical, etc. The chapter concluded with the need to explore more microbial strains and enzymatic processes for the production of diversified carbohydrate polymers with vast applications.

Bioprospecting of Microorganism-Based Industrial Molecules, First Edition.
Edited by Sudhir P. Singh and Santosh Kumar Upadhyay.

13.2 Dextran

Dextran is a complex homopolysaccharide comprised of a long chain of glucose monomers linked mainly by alpha-1,6 glycosidic bonds, along with alpha-1,4 or alpha-1,2 or alpha-1,3 linked minor branches [7]. Dextran so formed could be of varied molecular weight range from 1.5×10^4 to 2×10^7. In 1930, the first report on the production of dextran from sucrose using *Leuconostoc* species was made by Hucker and Pederson [8]. Later, different bacterial strains (mainly of *Leuconostoc* sp.) were found responsible for dextran production [9]. This polysaccharide is mainly synthesized from sucrose by the microbial phenomenon of various strains belonging to the genera, *Leuconostoc, Lactobacillus,* and *Streptococcus*. Some of the species capable of dextran production are *Leuconostoc mesenteroides, Saccharomyces cerevisiae, Lactobacillus plantarum,* and *Lactobacillus sanfrancisco*. There is hardly any restriction in their use in food processing [7]. Dextran has been produced at a commercial scale by employing *L. mesenteroides* strain [10].

13.2.1 Mechanism of Dextran Production

The enzyme dextransucrase catalyzes transformation of sucrose into dextran involving the glucosyl intermediate. The transfer of glucosyl moieties to the reducing end of a growing glucanosyl chain results in the formation of the polymer, dextran [11]. Different theories explaining the mechanism of dextran biosynthesis have been proposed. As per Robyt et al., two nucleophiles are involved for the synthesis of a chain of α(1,6)-linked glucose residues that attack sucrose and provide two β-glucosyl intermediates by displacing fructose [12, 13]. The other study was performed with 14C-labeled sucrose depicted the addition of glucose to the reducing end of the growing dextran chain, requiring two-site insertion for chain elongation [14–16].

As per the study by Robyt and Eklund, the following steps are involved in the formation of dextran [12, 13]

- The transfer of one glucose moiety to other occurs through the attack of the C-6 OH group of one glucose to C- 1 of the other forming α(1,6)-linkage.
- The nucleophile that gets free from the previous step further attacks on another sucrose molecule and forms enzyme glucosyl intermediate.
- The glucosyl intermediate, containing C-6 OH group, attacks C-1 of the isomaltosyl unit (the growing dextran chain). The alternate transfer of glucosyl and dextranosyl units may happen between the two nucleophiles, when the elongation of dextran chain occurs at the reducing end. The two imidazole groups are believed to be essential for the production of dextran.

13.2.2 Production of Dextran at Industrial Level

The commercial production of dextran has been achieved by culturing *L. mesenteroides* in a medium containing sucrose, peptone as nitrogen source, and minerals and phosphate. The operating conditions for the production include pH 6.5–7, 25–30 °C temperature, sucrose concentration (generally 2%), and fermentation time from 24 to 48 hours [17]. With the continuation of the fermentation process, the pH of the medium falls to 5.0 due to the formation of organic acids. This pH is almost near to the optimum pH required for the maximum catalytic effect of dextransucrase. The dextran so formed is precipitated using alcohol and purified by removing the cell debris. A disadvantage associated with this process is that the cell propagation, enzyme synthesis, and dextran production proceed under the same reaction condition (pH, media composition, oxygen level etc.), which changes during fermentation [11]. Another way is to produce dextran by using the cell-free supernatant containing extracellular secreted dextransucrase to transform sucrose under the optimal conditions [18, 19].

Dextran has been produced from different bacterial strains [20–22], using different agro-industrial substrate [23, 24], with different structural characteristics [25–30].

13.2.3 Applications of Dextran

Uses of dextran have been demonstrated in the manufacturing of blood plasma extenders, anticoagulant therapy as heparin substitutes, and cosmetics [17, 18, 31, 32]. Dextran is used in the development of plasminogen activator, and potent osmotic agents to treat hypovolemia and reduce thrombosis. It is used in the form of lubricants in some eye drops. The solution of clinical grades of dextran, e.g. Dextran 40, Dextran 60 and Dextran 70, have been used as injections for replacing blood loss, substituting plasma, thrombosis prophylaxis. The clinical grade dextran is considered as a safest plasma substitute. The clinical grade dextran is also used for cryopreservation, storage of organs for transplantation, and for carrying vaccines [7]. The requirement of dextran has also been found for the production of sephadex gel beads, utilized for separation of molecules in laboratories [31].

In food industry, dextran is incorporated in the bakery products for the improvement in the softness, texture, and volume of the loaf [33]. It is widely used as additives in ice creams and candles. It is able to stabilize the confectionary products by preventing crystallization, improving retention in moisture, increasing viscosity, and maintaining flavor. It is also used in soft drinks, milk beverages, etc. Dextran has been in demand for its utilization in food preservation applications [7].

The application of dextran has also been explored in improving the quality of silver emulsions in the photography industry. Due to the property of stabilizing the alkali and acids at room temperature, native dextran is applied in wastewater treatment. Being a biodegradable compound, its usage is economical in wastewater management [7]. Furthermore, dextran is used in the matrices for size exclusion chromatography. It can be used as a coating material for iron nanoparticles, enhancing its catalytic properties of the nanoenzymes [34]. It is also used in biosensor applications [35].

13.3 Chitin and Chitosan

The natural polymer, Chitin, is chemically named as poly β-(1→4)-*N*-acetyl-D-glucosamine. It is the second most abundant polymer on the earth crust, after cellulose. It is synthesized by various kinds of living organisms [36]. It exists as ordered crystalline microfibrils and participates in the formation of structural components in the arthropods' exoskeleton and found in the cell wall of fungi and yeasts. Crab and shrimp shells are the main sources for chitin [37]. At the industrial level, acidic and alkaline processing are carried out for the extraction of chitin. The chitin has been generated in a million tonne scale as waste, most of which is utilized to form products like fertilizers, pet foods, and fishmeal [38]. It can be utilized as a matrix for the purification of the enzyme with carbohydrate-binding domains. However, factors like toughness, elasticity, and insolubility in aqueous media limit its application [39–41].

13.3.1 Biological Extraction of Chitin

Chemical extraction of chitin is known to have a negative impact on the physico-chemical properties, intrinsic properties, and economics of the chitin production. Therefore, biological methods of chitin extraction are preferred [37]. Various studies have proven the advantages of biological processes over chemical processes, including a better chitin yield from shrimp shells [42]. The microbial proteolysis led to the formation of high-molecular-weight chitin [43]. It has been found that extraction of chitin using a biological approach involves less time, less energy input, low consumption of solvent, and easy reproducibility. Some of the biological methods explored for the extraction of chitin are discussed below.

13.3.1.1 Biosynthesis of Chitin and Chitosan

The substrate for the production of chitin is glycogen. The mechanism involved in the synthesis of chitin involves the conversion of glycogen to glucose 1-phosphate using phosphorylase enzyme. Then, phosphomutase leads to the formation of glucose-6-P, which is further converted to fructose-6-P by hexokinase. The amination (glutamine to glutamic acid) and acetylation (acetyl CoA to CoA) lead to the conversion of fructose-6-phosphate to N-acetyl glucosamine. The transfer of phosphate from C6 to C1 is catalyzed by phospho-N-acetyl glucosamine mutase, followed by the formation of uridine diphosphate (UDP) N-acetyl glucosamine in the presence of uridine triphosphate (UTP). Finally, chitin synthase catalyzes the formation of chitin from UDP N-acetyl glucosamine. The deacetylation of chitin to chitosan is catalyzed by chitin deacetylase [44, 45].

13.3.1.2 Chitin and Chitosan-Producing Fungi

Many fungal species have been explored for the production of chitin and chitosan, e.g. *Absidia coerulea*, *Absidia glauca*, *Absidia blakesleeana*, *Mucor rouxii*, *Aspergillus niger*, *Phycomyces blakesleeanus*, *Trichoderma reesei*, *Colletotrichum lindemuthianum*, *Gongronella butleri*, *Pleurotus sajo-caju*, *Rhizopus oryzae*, and *Lentinus edodes* [46–48].

13.3.1.3 Enzymatic Deproteinization

The proteases are required for the extraction of chitin. These proteases can be derived from plant, microbes, and animal. Some of the examples of the proteases include alcalase, pepsin, papain, pancreatine, devolvase, and trypsin, which are involved in the chitin isolation by removing proteins from crustacean shells and thus minimizing deacetylation and depolymerization. Mainly bacterial sources are preferred for crude proteases [37]. Marine animals contain proteases, which may be recovered in both active and stable forms for commercial use. The by-products of the fish industry represent about 50% of the seafood harvest in many fish-consuming countries [49]. These largely underutilized materials can be used for protease production and chitin extraction. These steps could decrease the costs of this process, as well as in preserving the environment.

Many reports have demonstrated the application of bacterial proteases in deproteinization. For example, Synowiecki and Al-Khateeb [50] applied enzymatic deproteinization on previously demineralized shrimp waste in order to produce chitin and a nutritionally valuable protein hydrolysate. Alcalase 2.4 L (Novo Nordisk A/S), a serine endopeptidase obtained from *Bacillus licheniformis,* has specificity for terminal hydrophobic amino acids, which leads to the production of non-bitter hydrolysate [37]. The obtained hydrolysate is a good source of essential amino acids in food applications. However, the effectiveness of deproteinization was limited by the presence of residual small peptides and amino acids attached to chitin molecules, which persist after enzymatic hydrolysis [37]. This method allows the isolation of chitin containing about 4% of protein impurities. This level of purity is sufficient for many nonmedical applications of chitin. Gilberg and Stenberg [51] used alcalase 2.41 for chitin, protein hydrolysate, and astaxanthin recovery.

It is noted that the efficiency of enzymatic methods is inferior to chemical methods. The isolated chitin can be treated with an additional NaOH treatment (under milder conditions and for a shorter time) to increase its purity and preservation of the structure of chitin. Manni [52] compared the isolation of chitin from shrimp waste using *Bacillus cereus* SV1 crude alkaline proteases to the use of 1.25 M NaOH. Shrimp shells were demineralized after deproteinization using dilute HCl treatment. The residual protein content was significantly higher in the chitin isolated with the enzymatic deproteinization than that obtained with alkali treatment (10% compared to 6%).

In another study, enzymatic deproteinization was optimized before demineralization [53]. In this study, many microbial proteases were compared on the basis of their efficiency in shrimp shell deproteinization, e.g. *Bacillus mojavensis* A21, *Bacillus subtilis* A26, *B. licheniformis* NH1, *B. licheniformis* MP1, *Vibrio metschnikovii* J1, and *Aspergillus clavatus* ES1. The highest deproteinization degree of about 76% was obtained with *B. mojavensis* A21 proteases. Then, the effect of reaction conditions, mainly enzyme/substrate ratio, temperature, and incubation time, on the deproteinization degree was optimized using response surface methodology to achieve 88% deproteinization under the optimized conditions. Recently, many fish and marine invertebrate-derived alkaline crude proteases have been applied for shrimp shell deproteinization. Mukhin and Novikov [54] studied the possibility of using crustacean waste, both as a substrate and as a source of proteases. The shell proteins were degraded with crude proteases isolated from the hepatopancreas of crab. The objective was to optimize the protein hydrolysate yield. However, even under the optimal conditions, i.e. temperature 50 °C, pH 8.4, enzyme/substrate ratio 6 g/kg, and time 12 hours, the degree of hydrolysis was never higher than 80%. The alkaline proteases from the red scorpionfish *Scorpaena scrofa* were found to deproteinize the shrimp waste up to 85% [53]. Activities of these crude alkaline proteases are related to fish feeding mainly on crustaceans and molluscs, inducing the nature and the specificity of its enzymes. By contrast, when the extraction is carried out by the chemical process, the order of two steps (deproteinization and demineralization) does not have a significant effect on the quality and the yield of the final chitin [55].

However, if enzymatic deproteinization is applied, the minerals presented in the cuticles may decrease the accessibility of the proteases and affect shrimp shell deproteinization efficiency. Thus, demineralization should be performed first.

13.3.1.4 Fermentation

The fermentation process for the deproteinization can be achieved through endogenous microorganisms (called auto-fermentation) or by addition of selected microbial strains. This would reduce the issue of the cost related to the utilization of purified enzymes. A number of microorganisms have been involved in the fermentation of crustacean shells [56]. The methods for fermentation process might be separated into two categories: lactic acid fermentation and non–lactic acid fermentation.

13.3.1.4.1 Lactic Acid Fermentation The lactic acid fermentation of crustacean shells can be carried out using a strain of *Lactobacillus* sp. strain as inoculum that led to the production of lactic acid and proteases. The formation of lactic acid from glucose led to a decrease in pH and ultimately suppresses the growth of spoilage microorganisms. The formation of calcium lactate precipitate after reaction of lactic acid with calcium carbonate could lead to the recovery of lactic acid. Many studies have been reported with the utilization of lactobacillus for the deproteinization of shells [57, 58].

13.3.1.4.2 Non–Lactic Acid Fermentation The bacteria and fungi have been used for the fermentation of crustacean shells, for example *Bacillus* sp. [59–61], *Pseudomonas* sp. [62–64], and *Aspergillus* sp. [65]. The screening of six proteolytic *Bacillus* strains, including *Bacillus pumilus* A1, *B. mojavensis* A21, *B. licheniformis* RP1, *B. cereus* SV1, *B. amyloliquefaciens* An6, and *B. subtilis* A26, was carried out for the fermentation of shrimp waste [61]. *Bacillus cereus* SV1 showed the highest degree of deproteinization. The jaggery broth was utilized for the fermentation of shrimp shells using *B. subtilis*, in which about 84% of the proteins and 72% minerals could be removed [60]. Similarly, *P. aeruginosa* A2 and fungus *A. niger* were utilized for deproteinization and demineralization efficiency of crustacean shells [47, 63]). Basically, during fermentation the release of proteases from bacteria or fungi leads to the deproteinization of shells. Various factors have been noted to affect the fermentation process for the deproteinization and demineralization [58, 62, 66].

13.3.1.5 Enzymatic Deacetylation

Due to disadvantages associated with the chemical deacetylation like consumption of energy, environmental problems generated because of chemical waste discard, etc., enzymatic approach via chitin deacetylase has been favored more for the production of chitosan [67]. The formation of chitosan by hydrolysis of *N*-acetamido bonds of chitin is catalyzed by chitin deacetylase (EC 3.5.1.41). Many fungi have been found to secrete this enzyme, for example, *Mucor rouxii* [68–70], *Absidia coerulea* [71], *Aspergillus nidulans* [72], and two strains of *C. lindemuthianum* [73, 74].

13.3.2 Applications of Chitin and Chitosan

Chitin and chitosan have wide applications due to the biodegradable, biocompatible, and nontoxic properties [75]. Chitin is employed as the raw material for the production of chitin-derived products, such as chitosans, oligosaccharides, and glucosamine [36]. Chitosan is of immense use in the field of nutrition [76]. Chitin and chitosan can be employed for waste-water treatment by adsorbing and fixing the heavy metals. Chitosan has been found effective in coagulating, flocculating, and dehydrating the activated sludge [76, 77]. Chitosan has been employed for the immobilization of microbes or sludge that could be further utilized for the wastewater treatment in extreme condition habitats, e.g. acidic or alkaline water or in the presence of organic solvents. Chitosan has functional, nutritional, and physiological properties and therefore has some significance in the food industry. Chitosan is useful in making stable emulsifiers [78], dietary supplement, and binding agent [79].

Further, chitin and chitosan have immunological, antibacterial, wound-healing [80–82], and hemostatic significance [83, 84]. It is also useful in developing drug delivery system [85, 86]. Chitin is able to retain nutrients in the soil and thus add on to the nitrogen cycle, exhibiting potential in the field of agriculture.

13.4 Xanthan Gum

Xanthan gum is a chain of three to eight monosaccharides with a backbone of repeating subunits [87]. The basic composition of xanthan gum includes D-glucose, D-mannose, and D-glucuronic acid. Usually, the composition of xanthan gum depends on various factors involved in the production process. However, it is found that the polymers produced by *Xanthomonas* genus show more commonly quantitative variation than qualitative [88]. Xanthan gum consists a backbone of linear (1-4) D-glucose linked with a side chain of trisaccharide at C-3 on every other glucose, with glucuronic acid residue linked to a terminal mannose unit and to a second mannose via (1-4) and (1-2) linkages [89]. Even the pyruvic acid moiety and acetyl moiety are linked to the terminal and internal mannose residues, respectively [90]. The synthesis of xanthan gum involves many steps: simple saccharides uptake and their conversion to nucleotide derivatives, pentasaccharide subunits attached to an isopentyl pyrophosphate carrier, polymerization of pentasaccharide repeat units, and secretion. The addition of D-glucose-1-phosphate and D-glucose from uridine diphosphate glucose (UDP-D-glucose) leads to the formation of backbone of xanthan. This is followed by the addition of D-mannose and D-glucuronic acid from guanosine diphosphate mannose (GDP-mannose) and uridine phosphate glucuronic acid (UDPglucuronic acid), respectively. The attachment of O-acetyl groups from acetyl-CoA and pyruvate from phosphoenolpyruvate takes place to the internal mannose residue, and terminal mannose, respectively. Like various polysaccharide-producing bacteria, xanthan production involves activated saccharide as donors and an acceptor molecule [91, 92].

Xanthan gum is approved by FDA for its utilization in food products. The production cost of xanthan gum can be reduced by using feedstock derived from agro-industrial biomass like cheese whey, sugar beet pulp, palm date, sugarcane stalks, peach pulp, molasses, etc. [93–98].

13.4.1 Xanthan Gum Production

Various studies have described the processes and the microorganisms involved in the production of xanthan gum [88, 90, 99–104]. Xanthan gum may be produced through aerobic fermentation at a temperature range from 27 to 30 °C in the presence of organic acids. Commercially, the substrate conversion percent to gum is about 70% via microbial approach. At the commercial level, batch fermentation is preferred for xanthan gum production [105], followed by thermal treatment, removal of cells, product recovery with ethanol, and finally drying and milling of the gum. Different fermentation types have been described as suitable for xanthan gum production, e.g. batch processes [106–110], continuous fermentation [101], and single-stage continuous process [111].

13.4.2 Microbial Production

Xanthomonas belongs to the family of Pseudomonadaceae, is gram-negative, aerobic, short rod-shaped bacteria [112, 113]. Several xanthan-gum-producing strains are included in this genus, e.g. *Xanthomonas campestris* pv. *campestris* 1078, *Xanthomonas arboricola* pv. pruni strain 106 [103], *Xanthomonas arboricola* pv. pruni strain 31 [100]. The microorganisms and individual pathovars contribute to the production, as well as the composition of xanthan gum [102].

13.4.3 Applications of Xanthan Gum

Due to water-soluble and specific rheological properties, xanthan gum is proven to have diverse industrial applications. Xanthan gum plays a crucial role in agriculture to protect barley. Together with fungicides, it acts as an elicitor to prevent *Bipolaris sorokiniana* [114]. Due to emulsifying and thickening properties, xanthan gum is utilized in various products like juice, fruit pulp and powder beverages, chocolates, desserts, jellies, dairy products, margarine, yoghurt, bakery products, frozen foods, sauces, and gravies in the food industry. Xanthan gum has also been utilized in combination with other gums like locust bean gum or guar [115]. Xanthan gum provides texture, viscosity, flavor release, appearance, and water-control properties to the food products. It can also improve the rheological properties of the products [90]. Further, Xanthan gum's role has been demonstrated in enhancing the recovery of the oil [116].

13.5 Bacterial Cellulose

Many bacterial strains have been characterized for the production of glucose polymer, known as bacterial cellulose [117]. Bacterial cellulose exhibits various magnificent properties, e.g. flexibility, good water holding capacity, hydrophilicity, crystallinity, and so on [6]. It has been employed for various medical applications including artificial skin and blood vessel engineering, coverings for wounds and nerve surgery, dura mater material for prosthesis and hemostatic, etc. [5].

13.5.1 Biosynthetic Pathway for Cellulose Production

Many enzymes and complexes of catalytic and regulatory proteins, with the undefined supramolecular structure, are involved in bacterial cellulose synthesis [117]. The mechanism and pathways for the synthesis of uridine diphosphoglucose (UDPGlc) synthesis are relatively well known, but the molecular mechanism of polymerization of glucose into a long unbranched chain is unclear [117]. The elaborated documentation of biochemical reactions for the production of cellulose synthesis by *Acetobacter xylinum* has been deciphered. The polymerization of glucose into b-14 glucan chain, using UDPGlc as the precursor, results into the formation of ribbon-like structure of cellulose chains, followed by their extrusion outside the cell, and self-assembly into fibrils [118–120]. The pentose phosphate cycle, in combination with gluconeogenesis, is involved in bacterial cellulose biosynthesis [121, 122].

13.5.2 Cellulose Precursor

Acetobacter xylinum has a capability to convert about 50% various carbon compounds, like hexoses, glycerol, dihydroxyacetone, pyruvate, and dicarboxylic acids, into cellulose. The pyruvate and dicarboxylic acid enter the Krebs cycle, and decarboxylation leads to the synthesis of hexoses via gluconeogenesis, in a similar manner as with glycerol, dihydroxyacetone, and intermediates of the pentose phosphate cycle (Figure 13.1). UDPGlc is the precursor for cellulose. Phosphorylation of glucose to glucose-6-phosphate (Glc-6-P) by glucokinase, followed by phosphoglucomutase catalyzed isomerization to Glc-a-1-P, and then UDPGlc is formed by UDPGlc pyrophosphorylase [117, 120].

13.5.3 Microbial Source for Cellulose Production

Bacterial cellulose is a nanofiber produced as extracellular polysaccharides. Diverse bacterial strains have been characterized to produce cellulose in media with different carbon sources [6]. The most well-known bacterium related to the production of cellulose is *Komagataeibacter xylinus*, which belongs to acetic acid bacteria (AAB), food-safe

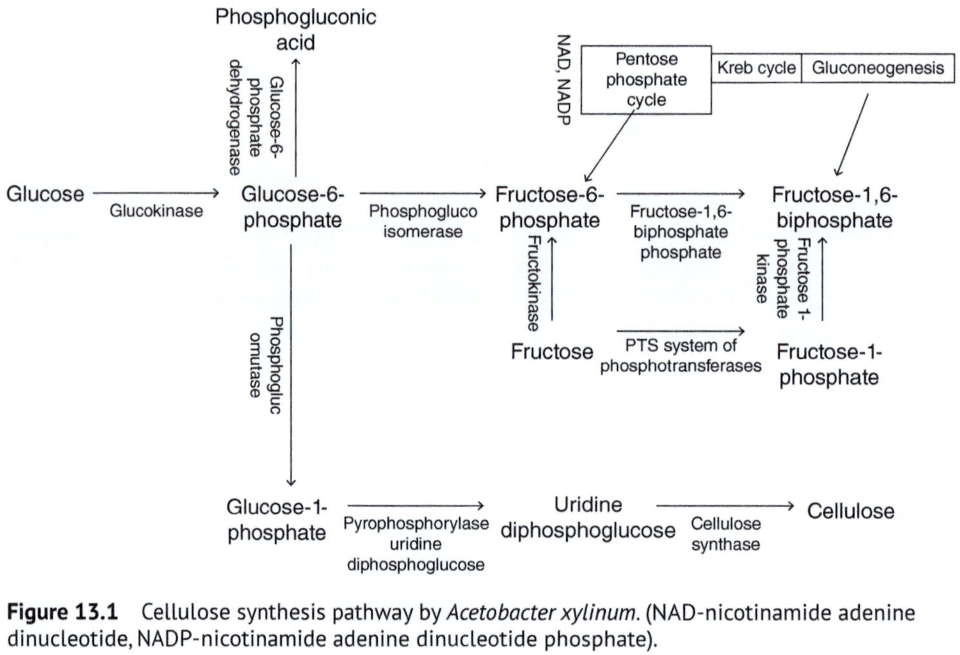

Figure 13.1 Cellulose synthesis pathway by *Acetobacter xylinum*. (NAD-nicotinamide adenine dinucleotide, NADP-nicotinamide adenine dinucleotide phosphate).

microorganisms [123–125]. Some other AAB species involved in the production of cellulose are *Komagataeibacter hansenii*, *Komagataeibacter medellinensis*, *Komagataeibacter nataicola*, *Komagataeibacter oboediens*, *Komagataeibacter rhaeticus*, *Komagataeibacter saccharivorans*, and *Komagataeibacter pomaceti* [124, 126–129]. Some other bacterial genera known for the production of cellulose are *Gluconacetobacter* (formerly *Acetobacter*), *Agrobacterium*, *Aerobacter*, *Achromobacter*, *Azotobacter*, *Rhizobium*, *Sarcina*, and *Salmonella* [130].

The cellulose polymer produced from the bacterial source (C6H10O5)n is different in physical and chemical features from that of plant source [131]. Bacterial cellulose is given preference because of high purity, a higher degree of polymerization, crystallinity, tensile strength, good water holding capacity, etc. [130]. The bacterial cellulose has a microfibril pattern comprised of an interlocked chain of glucan by hydrogen bonds that give a crystalline domain [132].

13.5.4 Applications of Cellulose

With the desirable properties like high purity, a high degree of crystallinity, high density, good shape retention, high water-binding capacity, and higher surface area, bacterial cellulose is useful in various areas including textile, paper, food, and pharmaceutical industry, mining, and refinery [133–135]. Some of the applications of the bacterial cellulose are as follows:

- It can be used as stabilizer and thickener in processed foods. For the first time, bacterial cellulose was utilized in nata de coco in the Philippines. Due to indigestible and gelling properties, bacterial cellulose is considered as a good food base [136]. Another example of its use is in monascus-nata complex, which is a combination of monacolin K and bacterial cellulose [137, 138]. It has also been introduced in beverages in Japan.
- Due to high tensile strength, high porosity, and microfibrillar structure, bacterial cellulose has been employed for various medical applications, including artificial skin and blood vessel engineering, coverings in wounds and nerve surgery, dura mater material for prosthesis and hemostatic, etc. [117].
- Due to reflective, flexible, portable, uniform type of physical and mechanical properties, bacterial cellulose has been considered to be applicable in the electronic paper display [135]. Cellulose has been widely utilized in the process of different forms of paper, like, banknote paper, the Bible paper [139]. The recovery of mineral and oil has somewhat related to microbial cellulose. It has been considered the best material for the sound transducing membrane for maintaining high sonic velocity over a wide range of frequencies [140].

13.6 Levan

Levan is an important homopolymer of fructose, in which D-fructo-furanosyl moieties are linked through β (2-6) linkages with branching chains having β (2-1) linkages, and a D-glucopyranosyl residue is attached at the terminal end. The molecular weight and degree of polymerization (DP) in levan varies according to the sources. Levan synthesized by enzymes generally have lower DP as compared to naturally produced levan by the microbial cell [141]. It is highly soluble in water (completely soluble in hot water) and oil but not in pure organic solvents like alcohols and acetone [142]. In comparison to other biopolymers, levan has exceptionally low intrinsic viscosity, and it has a heat stable polymer as its melting temperature is 225°C [143, 144]. It has numerous industrial applications due to its biodegradable and nontoxic nature. Good biocompatibility and film-forming ability of levan make it a more suitable material for biomedical applications than other synthetic polymers [145]. Levan got more attention in the food industry due to its nondigestible property and prebiotic effect [146, 147].

13.6.1 Levan Producing Organism

Levan is naturally produced by microbes, as well as by plants, but bacterial levan has a molecular weight in the range of 2–100 million Da while for levan of plant origin it ranges from 2000 to 33 000 Da [141]. Microbial levan is produced as an exopolysaccharide, and a number of bacteria are reported for levan production. A few examples are *Zymomonas mobilis* [148], *Gluconobacter* sp. [149], *Bacillus polymyxa* [150], *Bacillus subtilis* (Natto) [151], *Erwinia amylovora*, [152], *Pseudomonas syringae* pv. Phaseolicola [153], *Paenibacillus polymyxa* EJS-3 [154], etc. Other than these organisms, a halophilic bacteria *Halomonas smyrnensis* AAD6T is also reported as a good producer of levan [155]. Different species of *Leuconostoc* are examined for exopolysaccharides production, and among them, levan-producing species are *Leuconostoc citreum* strain BD1707 [150], and *L. mesenteroides* MTCC10508 [156]. *Brachybacterium phenoliresistens* isolated from rhizospheric soil was reported as levan producer [157].

Levan is also reported in some plants, especially temperate grasses, are reported as a good accumulator of levan as storage polysaccharide, which is supposed to provide freezing tolerance to these plants [158]. *Phleum pratense* (timothy), *Dactylis glomerata* (orchardgrass), *Poa secunda* (big bluegrass), *Hordeum vulgare*, and *Triticum aestivum* represent common examples of monocots having levan type fructans [159]. The occurrence of levan is more frequent in monocots but a eudicot, *Pachysandra terminalis*, is also reported for accumulating levan-type fructans [160]. Recently, levan-type fructan is also found in the underground reserve system of *Gomphrena marginata* Seub, which belongs to the dicot family Amaranthaceae [161].

13.6.2 Mechanism for Levan Biosynthesis

Levansucrases (EC 2.4.1.10) enzyme-catalyze the synthesis of levan. This enzyme is also known as sucrose 6-fructosyltransferase (6SFT). It belongs to glycoside hydrolase family 68 (GH-68). It produces levan through execution of three types of chemical reactions: polymerization, transfructosylation, and hydrolysis. Product of reaction is dependent on the nature of accepter molecule. For example, if the accepter is a growing chain of fructan, the transfer of fructosyl residue produces polymer; and if water molecule is available as acceptor, then hydrolysis takes place, and for transfructosylation mono or oligosaccharides are used. Therefore, this enzyme can transfer d-fructosyl residue from sucrose to different mono or oligosaccharides to produce a mixture of fructo-oligosaccharides with free glucose molecules [162]. Levansucrase is reported in many gram-negative bacteria such as *Acetobacter suboxydans*, *Erwinia herbicola*, *Erwinia amylovora*, *Gluconobacter oxydans*, *Gluconobacter japonicas*, *Rhanella aquatilis*, *Zymomonas mobilis*; and several gram-positive bacteria are also identified as, *Bacillus polymyxa*, *Bacillus natto*, *Bacillus amyloliquefaciens*, *Streptococcus mutans*, *Actinomyces viscosus*, *Clostridium acetobutylicum*, *L. mesenteroides* Lm 17, *Clostridium acetobutylicum* and *L. mesenteroides* MTCC10508 [146, 156, 163–165]. Levansucrases are reported as enzymes having signal peptide for extracellular secretion, but the mechanism of their secretion differs in gram-positive and gram-negative strains. Till date, studies have revealed that gram-positive bacteria follow the two-step mechanism for its secretion: in the first step, signal peptide is cleaved, which is followed by the second step of protein folding, involving several external factors such as presence of metal ions (iron or calcium) and pH of the environment [166]. Contrastingly, gram-negative bacteria adapted a signal-peptide-independent pathway for the secretion of levansucrases [163]. In plants, different enzymes are involved in the synthesis of levan-type fructans such as sucrose:sucrose 6-fructosyltransferase (6-SST), and fructan:fructan 6-fructosyl-transferase (6-FFT) [167].

13.6.3 Strategies for Levan Production

In spite of significant industrial applications, there is a lack of a potent system for production of levan at industrial scale. Biotechnological strategies are in trial for achieving this target, one strategy is the search for novel sources having the better efficiency in production, and another may be the transfer of levansucrase gene having good kinetic parameters in homologous or heterologous systems for maximizing the yield of levan [168]. Suitable hosts for heterologous expression include *Escherichia coli* [169, 170]; *S. cerevisiae* [171–173]; *Pichia pastoris* [174]. Another method involves the utilization of different enzymatic sources in vitro conditions either through free enzyme system or with immobilized enzymes [156, 175].

13.6.4 Applications of Levan

Levan being a biopolymer exhibits a wide range of applications in food, medicine, and dairy industries. The property of being less viscous makes it suitable for various food processes [141]. It is known to be present naturally in food products, but only a limited quantity is consumed by humans [176]. Thus, there are various bacterial sources that produce levan, which shows prebiotic effects. Being a nontoxic dietary fiber, it is suitable for improving gut functionality, although the limited resources for levan production limit its extended production [177]. The branched structure of levan restricts the proliferation of tumor cells and also with the increase in levan consumption, growth of lactic acid bacteria is stimulated in the colon that prohibits the carcinogens [178]. Thus, it is known for its antitumor activity in medical industry. Additionally, it exhibits anticlotting property that helps during heart surgeries [179, 180], and healing capability of levan extends its application in dermatology medicines and also, used as carrier matrix for drug delivery that help in fast dissolution of capsules resulting less time in treatment [181]. In addition to various medical applications, levan is widely used in dairy industry as an emulsifier and stabilizer [182]. It also imparts a good flavor, and when levan is hydrolyzed, it acts as sweetener in dairy products. The shear and tensile strength of levan makes it an excellent adhesive [183]. The cross-linked levan can be applied as water-resistant adhesive that gives completion to many water-based adhesives. Levan is also applied in biotechnological industries as a purification system for various biological molecules like egg albumin [184]. In case of nano-technological applications, the levan-blended nanofilms are observed to have good durability with better flexibility and transparency [185]. The improved mechanical properties along with good thermostability of levan nanocomposites enhance its applicability in various industrial aspects.

13.7 Conclusions and Perspectives

Biological approaches involving microbes and biocatalysts have been found to be effective over chemical approach for the production of carbohydrate biopolymers with high yield and purity. This adds a lot in providing platform for the development in food, textile, pharmaceutical, and medical fields. Further research is needed for the development of more biological ways for the production of various polysaccharides with versatile, magnificent characteristics, and wide applications. The utilization of abundant agro-industrial bioresources should be explored as low-cost feedstock for the production of high-value polymers.

Acknowledgments

Authors acknowledge the Department of Biotechnology, Ministry of Science and Technology, Government of India, for all the support. MS acknowledges CSIR-SRF.

References

1 Kaur, T. and Singh, R.P. (2020). Carbohydrate biopolymers: diversity, applications, and challenges bioactive natural products in drug discovery. In: *Bioactive Natural Products in Drug Discovery* (eds. J. Singh, V. Meshram and M. Gupta). Singapore: Springer https://doi.org/10.1007/978-981-15-1394-7_18.
2 Stal, L.J. (2011). Biopolymer. In: *Encyclopedia of Astrobiology* (eds. M. Gargaud et al.). Berlin, Heidelberg: Springer https://doi.org/10.1007/978-3-642-11274-4_188.
3 Hassan, M.E., Bai, J., and Dou, D.-Q. (2019). Biopolymers; definition, classification and applications. *Egyptian Journal of Chemistry* 62 (9): 1725–1737.
4 Verma, S., Manjubala, I., and Kumar, N.U. (2016). Protein and carbohydrate biopolymers for biomedical applications. *International Journal PharmTech, Research* 9 (8): 408–421.
5 Raghunathan, D. (2013). Production of microbial cellulose from the new bacterial strain isolated from temple wash waters. *International Journal of Current Microbiology and Applied Sciences* 2 (12): 275–290.

6 Gorgieva, S. and Trcek, J. (2019). Bacterial cellulose: production, modifification and perspectives in biomedical applications. *Nanomaterials* 9: 1352. https://doi.org/10.3390/nano9101352.

7 Bhavani, A.L. and Nisha, J. (2010). Dextran – the Polysaccharide with versatile uses. *International Journal of Pharma and Bio Sciences* 1 (4): 569–573.

8 Hucker, G.J. and Pederson, C.S. (1930). Studies on coccaceae XVI. Genus Leuconostoc. N.Y. *Agricultural Experiment Station Bulletin* 167: 3–8.

9 Jeans, A., Haynes, W.C., Wilham, C.A. et al. (1954). Characterisation and classification of dextrans from ninety-six strains of bacteria. *Journal of the American Society* 76: 5041.

10 Qader, S.A.U.L., Iqbal, L., Aman, A. et al. (2006). Production of Dextran by newly isolated strains of Leuconostoc mesenteroides PCSIR-4 and PCSIR-9. *Turk. Journal of Biochemistry* 26: 21–26.

11 Naessens, M., Cerdobbel, A., Soetaert, W., and Vandamme, E.J. (2005). Leuconostoc dextransucrase and dextran: production, properties and applications. *Journal of Chemical Technology & Biotechnology* 80: 845–860. https://doi.org/10.1002/jctb.1322.

12 Robyt, J.F. and Eklund, S.H. (1982). Stereochemistry involved in the mechanism of action of dextransucrase in the synthesis of dextran and the formation of acceptor products. *Bioorganic Chemistry* 11: 115–132.

13 Robyt, J.F. (1996). Mechanism and action of glucansucrases. In: *Enzymes for Carbohydrate Engineering* (eds. K.H. Park, J.F. Robyt and Y.D. Choi), 1–21. Amsterdam: Elsevier.

14 Robyt, J.F., Kimble, B.K., and Walseth, T.F. (1974). The mechanism of dextransucrase action: direction of dextran biosynthesis. *Archives of Biochemistry and Biophysics* 165: 634–640.

15 Robyt, J.F. and Martin, P.J. (1983). Mechanism of synthesis of D-glucans by D-glucosyltransferases from Streptococcus mutans 6715. *Carbohydrate Research* 113: 301–315.

16 Luzio, G.A., Parnaik, V.K., and Mayer, R.M. (1983). A D-glucosylated form of dextransucrase: demonstration of partial reactions. *Carbohydrate Research* 121: 269–278.

17 Leathers, T.D. (2002). Dextran. In: *Biopolymers. Vol. 5. Polysaccharides I: Polysaccharides from Prokaryotes* (eds. E.J. Vandamme, S. De Baets and A. Steinbuchel), 299–321. Weinheim: Wiley-VCH.

18 Alsop, R.M. (1983). Industrial production of dextrans. In: *Microbial Polysaccharides* (ed. M.E. Bushell), 1–44. Amsterdam: Elsevier.

19 Robyt, J.F. (1985). Dextran. In: *Encyclopaedia of Polymer Science*, vol. 4 (ed. J.I. Kroschwitz), 753–767. New York: Wiley-VCH.

20 Hu, Y. and Gänzle, M.G. (2018). Effect of temperature on production of oligosaccharides and dextran by Weissella cibaria 10 M. *International Journal of Food Microbiology* 280: 27–34. https://doi.org/10.1016/j.ijfoodmicro.2018.05.003.

21 Bounaix, M.S., Robert, H., Gabriel, V. et al. (2010). Characterization of dextran-producing Weissella strains isolated from sourdoughs and evidence of constitutive dextransucrase expression. *FEMS Microbiology Letters* 311 (1): 18–26. https://doi.org/10.1111/j.1574-6968.2010.02067.x.

22 Vidal, R.F., Martínez, A., Moulis, C. et al. (2011). A novel dextransucrase is produced by Leuconostoc citreum strain B/110-1-2: an isolate used for the industrial production of dextran and dextran derivatives. *Journal of Industrial*

Microbiology and Biotechnology 38 (9): 1499–1506. https://doi.org/10.1007/s10295-010-0936-x.

23 Sharma, M., Sangwan, R.S., Khatkar, B.S., and Singh, S.P. (2019). Alginate-pectin co-encapsulation of dextransucrase and dextranase for oligosaccharide production from sucrose feedstocks. *Bioprocess and Biosystems Engineering* 42 (10): 1681–1693. https://doi.org/10.1007/s00449-019-02164-z.

24 Lata, K., Sharma, M., Patel, S.N. et al. (2018). An integrated bio-process for production of functional biomolecules utilizing raw and by-products from dairy and sugarcane industries. *Bioprocess and Biosystems Engineering* 41 (8): 1121–1131.

25 Wang, C., Chen, S., Zhang, H.B. et al. (2018). Characterization of the inserted mutagenesis dextransucrases from Leuconostoc mesenteroides 0326 to produce hyperbranched dextran. *International Journal of Biological Macromolecules* 112: 584–590. https://doi.org/10.1016/j.ijbiomac.2018.02.001.

26 Fabre, E., Bozonnet, S., Arcache, A. et al. (2005). Role of the two catalytic domains of DSR-E dextransucrase and their involvement in the formation of highly alpha-1,2 branched dextran. *Journal of Bacteriology* 187 (1): 296–303. https://doi.org/10.1128/JB.187.1.296-303.2005.

27 Skory, C.D. and Côté, G.L. (2015). Secreted expression of Leuconostoc mesenteroides glucansucrase in Lactococcus lactis for the production of insoluble glucans. *Applied Microbiology and Biotechnology* 99 (23): 10001–10010. https://doi.org/10.1007/s00253-015-6854-0.

28 Wang, C., Zhang, H.B., Li, M.Q. et al. (2017). Functional analysis of truncated and site-directed mutagenesis dextransucrases to produce different type dextrans. *Enzyme Microbial Technology* 102: 26–34. https://doi.org/10.1016/j.enzmictec.2017.03.011.

29 Brison, Y., Laguerre, S., Lefoulon, F. et al. (2013). Branching pattern of gluco-oligosaccharides and 1.5kDa dextran grafted by the α-1,2 branching sucrase GBD-CD2. *Carbohydrate Polymers* 94 (1): 567–576. https://doi.org/10.1016/j.carbpol.2013.01.064.

30 Claverie, M., Cioci, G., Vuillemin, M. et al. (2020). Processivity of dextransucrases synthesizingvery-high-molar-mass dextran is mediated by sugar-binding pockets in domain V. *Journal of Biological Chemistry* 295 (17): 5602–5613. https://doi.org/10.1074/jbc.RA119.011995.

31 Sutherland, I.W. (1996). *Extracellular Polysaccharides*, 2e. Biotechnology.

32 Kim, D. and Day, D.F. (1994). A New process for the production of clinical dextran by mixed culture fermentation of Lipomyces starkeyi and Leuconostoc mesenteroides. *Enzyme MicrobialTechnology* 16: 844–848.

33 Wilham, C.A., Dimler, R.J., and Senti, F.R. (1959). *Cereal Chemistry* 36: 558–563.

34 Naha, P.C., Liu, Y., Hwang, G. et al. (2019). Dextran-coated iron oxide nanoparticles as biomimetic catalysts for localized and pH-activated biofilm disruption. *ACS Nano* 13 (5): 4960–4971.

35 Song, S., Lu, Y., Li, X. et al. (2017). Optimization of 3D surfaces of dextran with different molecule weights for real-time detection of biomolecular interactions by a QCM biosensor. *Polymers* 9 (9): 409.

36 Rinaudo, M. (2006). Chitin and chitosan: properties and applications. *Progress in Polymer Science* 31: 603–632.

37 Younes, I. and Rinaudo, M. (2015). Chitin and Chitosan preparation from marine sources. Structure, properties and applications. *Marine Drugs* 13: 1133–1174. https://doi.org/10.3390/md13031133.

38 Kim, S.K. (2011). *Chitin, Chitosan, Oligosaccarides and Their Derivatives*, vol. 1. Boca Raton, FL, USA: CRC Press.

39 Bloch, R. and Burger, M.M. (1974). Purifification of wheat germ agglutinin using affinity chromatography on chitin. *Biochemical and Biophysical Research Communications* 58: 13–19.

40 Wolman, F.J., Copello, G.J., Mebert, A.M. et al. (2010). Egg white lysozyme purifification with a chitin–silica-based affinity chromatographic matrix. *European Food Research and Technology* 231: 181–188.

41 Brzezinska, M.S. and Jankiewicz, U. (2012). Production of antifungal chitinase by Aspergillus niger LOCK 62 and its potential role in the biological control. *Current Microbiology* 65: 666–672.

42 Khanafari, A., Marandi, R., and Sanatei, S. (2008). Recovery of chitin and chitosan from shrimp waste by chemical and microbial methods. *Iranian Journal of Environmental Health Science & Engineering* 5: 1–24.

43 Bustos, R.O. and Healy, M.G. (1994). Microbial deproteinization of waste prawn shell. In: *Proceedings of the Second International Symposium on Environmental Biotechnology, Biotechnology'*, vol. 94 (ed. Anon), 15–25. Brighton, UK.

44 Batista, A.C.L., Souza Neto, F.E., and Paiva, W.S. (2018). Review of fungal chitosan: past, present and perspectives in Brazil. *Polimeros* 28 (3): 275–283.

45 Mostafa, M., Abo Elsoud, M.M., and El Kady, E.M. (2019). Current trends in fungal biosynthesis of chitin and chitosan. *Bulletin of the National Research Centre* 43: 59. https://doi.org/10.1186/s42269-019-0105-y.

46 Hu, K.J., Yeung, K.W., Ho, K.P., and Hu, J.L. (1999). Rapid extraction of high-quality chitosan from mycelia of Absidia glauca. *Journal of Food Biochemistry* 23 (2): 187–196.

47 Teng, W.L., Khor, E., Tan, T.K. et al. (2001). Concurrent production of chitin from shrimp shells and fungi. *Carbohydrate Research* 332: 305–316.

48 Chatterjee, S., Adhya, M., Guha, A.K., and Chatterjee, B.P. (2005). Chitosan from Mucor rouxii production and physico-chemical characterization. *Process Biochemistry* 40: 395–400.

49 Rao, M.B., Tanksale, A.M., Ghatge, M.S., and Deshpande, V.V. (1998). Molecular and biotechnological aspects of microbial proteases. *Microbiology and Molecular Biology Reviews* 62: 597–635.

50 Synowiecki, J. and Al-Khateeb, N.A.A.Q. (2000). The recovery of protein hydrolysate during enzymatic isolation of chitin from shrimp Crangon crangon processing discards. *Food Chemistry* 68: 147–152.

51 Gildberg, A. and Stenberg, E. (2001). A new process for advanced utilisation of shrimp waste. *Process Biochemistry* 36: 809–812.

52 Manni, L., Ghorbel-Bellaaj, O., Jellouli, K. et al. (2010). Extraction and characterization of chitin, chitosan, and protein hydrolysates prepared from shrimp waste by treatment with crude protease from Bacillus cereus SV1. *Applied Biochemistry and Biotechnology* 162: 345–357.

53 Younes, I., Ghorbel-Bellaaj, O., Nasri, R. et al. (2012). Chitin and chitosan preparation from shrimp shells using optimized enzymatic deproteinization. *Process Biochemistry* 47: 2032–2039.

54 Mukhin, V.A. and Novikov, V.Y. (2001). Enzymatic hydrolysis of proteins from crustaceans of the Barents Sea. *Applied Biochemistry and Microbiology* 37: 538–542.
55 Kaur, S. and Dhillon, G.S. (2015). Recent trends in biological extraction of chitin from marine shell wastes: a review. *Critical Reviews in Biotechnology* 35: 44–61.
56 Arbia, W., Arbia, L., Adour, L., and Amrane, A. (2013). Chitin extraction from crustacean shells using biological methods – a review. *Food Technology and Biotechnology* 51: 12–25.
57 Rao, M.S., Muñoz, J., and Stevens, W.F. (2000). Critical factors in chitin production by fermentation of shrimp biowaste. *Applied Microbiology and Biotechnology* 54: 808–813.
58 Choorit, W., Patthanamanee, W., and Manurakchinakorn, S. (2008). Use of response surface method for the determination of demineralization efficiency in fermented shrimp shells. *Bioresource Technology* 99: 6168–6173.
59 Yang, J.K., Shih, I.L., Tzeng, Y.M., and Wang, S.L. (2000). Production and purification of protease from a Bacillus subtilis that can deproteinize crustacean wastes. *Enzyme Microbial Technology* 26: 406–413.
60 Sini, T.K., Santhosh, S., and Mathew, P.T. (2007). Study on the production of chitin and chitosan from shrimp shell by using Bacillus subtilis fermentation. *Carbohydrate Research* 342: 2423–2429.
61 Ghorbel-Bellaaj, O., Younes, I., Maalej, H. et al. (2012). Chitin extraction from shrimp shell waste using Bacillus bacteria. *International Journal of Biological Macromolecules* 51: 1196–1201.
62 Oh, K.T., Kim, Y.J., Nguyen, V.N. et al. (2007). Demineralization of crab shell waste by Pseudomonas aeruginosa F722. *Process Biochemistry* 42: 1069–1074.
63 Ghorbel-Bellaaj, O., Jellouli, K., Younes, I. et al. (2011). A solvent-stable metalloprotease produced by Pseudomonas aeruginosa A2 grown on shrimp shell waste and its application in chitin extraction. *Applied Biochemistry and Biotechnology* 164: 410–425.
64 Wang, S.L. and Chio, S.H. (1998). Deproteination of shrimp and crab shell with the protease of Pseudomonas aeruginosa K-1. *Enzyme Microbial Technology* 22: 629–633.
65 Mahmoud, N.S., Ghaly, A.E., and Arab, F. (2007). Unconventional approach for demineralization of deproteinized crustacean shells for chitin production. *Amercian Journal of Biochemistry and Biotechnology* 3: 1–9.
66 Jung, W.J., Kuk, J.H., Kim, K.Y., and Park, R.D. (2005). Demineralization of red crab shell waste by lactic acid fermentation. *Applied Microbiology and Biotechnology* 67: 851–854.
67 Tsigos, I., Martinou, A., Kafetzopoulos, D., and Bouriotis, V. (2000). Chitin deacetylases: new, versatile tools in biotechnology. *Trends in Biotechnology* 18: 305–312.
68 Kafetzopoulos, D., Martinou, A., and Bouriotis, V. (1993). Bioconversion of chitin to chitosan: purification and characterization of chitin deacetylase from Mucor rouxii. *Proceedings of the National Academy of Sciences* 90: 2564–2568.
69 Araki, Y. and Ito, E. (1975). A pathway of chitosan formation in Mucor rouxii: Enzymatic deacetylation of chitin. *European Journal of Biochemistry* 189: 249–253.
70 Martinou, A., Kafetzopoulos, D., and Bouriotis, V. (1993). Isolation of chitin deacetylase from Mucor rouxii by immunoaffinity chromatography. *Journal of Chromatography, A* 644: 35–41.
71 Gao, X.D., Katsumoto, T., and Onodera, K. (1995). Purification and characterization of chitin deacetylase from Absidia coerulea. *Journal of Biochemistry* 117: 257–263.
72 Alfonso, C., Nuero, O.M., Santamaría, F., and Reyes, F. (1995). Purification of a heat stable chitin deacetylase from Aspergillus nidulans and its role in cell wall degradation. *Current Microbiology* 30: 49–54.

73 Tsigos, I. and Bouriotis, V. (1995). Purification and characterization of chitin deacetylase from Colletotrichum lindemuthianum. *Journal of Biological Chemistry* 270: 26286–26291.

74 Tokuyasu, K., Kameyama, M.O., and Hiyashi, K. (1996). Purification and characterization of extracellular chitin deacetylase from Colletotrichum lindemuthianum. *Bioscience, Biotechnology, and Biochemistry* 60: 1598–1603.

75 Johnson, E.L. and Peniston, Q.P. (1982). Utilization of shellfish waste for chitin and chitosan production. In: *Chemistry & Biochemistry of Marine Food Products* (eds. R.E. Martin, G.J. Flick, C.E. Hebard and D.R. Ward), 415. Westport, CT, USA: AVI Publishing Co.

76 Percot, A., Viton, C., and Domard, A. (2003). Characterization of shrimp shell deproteinization. *Biomacromolecules* 4: 1380–1385.

77 Horst, M.N., Walker, A.N., and Klar, E. (1993). The pathway of crustacean chitin synthesis. In: *The Crustacean Integument: Morphology and Biochemistry* (eds. M.N. Horst and J.A. Freeman), 113–149. Boca Raton, FL, USA: CRC Press.

78 Rinaudo, M., Auzely, R., Vallin, C., and Mullagaliev, I. (2005). Specific interactions in modified chitosan systems. *Biomacromolecules* 6: 2396–2407.

79 Desbrieres, J., Martinez, C., and Rinaudo, M. (1996). Hydrophobic derivatives of chitosan: characterization and rheological behaviour. *International Journal of Biological Macromolecules* 19: 21–28.

80 Chang, K.L.B., Tsai, G., Lee, J., and Fu, W.R. (1997). Heterogeneous *N*-deacetylation of chitin in alkaline solution. *Carbohydrate Research* 303: 327–332.

81 Rinaudo, M., Milas, M., and le Dung, P. (1993). Characterization of chitosan. Influence of ionic strength and degree of acetylation on chain expansion. *International Journal of Biological Macromolecules* 15: 281–285.

82 Brugnerotto, J., Desbrieres, J., Roberts, G., and Rinaudo, M. (2001). Characterization of chitosan by steric exclusion chromatography. *Polymer* 42: 9921–9927.

83 Odijk, T. (1979). On the ionic-strength dependence of the intrinsic viscosity of DNA. *Biopolymers* 18: 3111–3113.

84 Mazeau, K., Perez, S., and Rinaudo, M. (2000). Predicted influence of *N*-acetyl group content on the conformational extension of chitin and chitosan chains. *Journal of Carbohydrate Chemistry* 19: 1269–1284.

85 Hirano, S., Yamaguchi, R., Fukui, N., and Iwata, M. (1990). A chitosan oxalate gel: its conversion to an *N*-acetylchitosan gel via a chitosan gel. *Carbohydrate Research* 201: 145–149.

86 Yamaguchi, R., Hirano, S., Arai, Y., and Ito, T. (1978). Chitosan salt gels thermally reversible gelation of chitosan. *Agricultural and Biological Chemistry* 42: 1981–1982.

87 Laws, A., Gu, Y., and Marshall, V. (2001). Biosynthesis, characterization, and design of bacterial exopolysaccharides from lactic acid bacteria. *Biotechnology Advances* 19: 597–625. https://doi.org/10.1016/S0734-9750(01)00084-2.

88 Borges, C. and Vendruscolo, C. (2008). Xanthan Gum: characteristics and operational conditions of production. *Semina: Ciências Biológicas e da Saúde* 29: 171–188.

89 Jansson, P., Kenne, L., and Lindberg, B. (1975). Structure of the extracellular polysaccharide from Xanthomonas campestris. *Carbohydrate Research* 45: 275–282. https://doi.org/10.1016/S0008-6215(00)85885-1.

90 Rosalam, S. and England, R. (2005). Review of xanthan gum production from unmodified starches by Xanthomonas camprestris sp. *Enzyme Microbial Technology* 39: 197–207. https://doi.org/10.1016/j.enzmictec.2005.10.019.

91 Papagianni, M. and Pandey, A. (eds.) (2004). *Concise Encyclopedia of Bioresource Technology*. Binghamton: Haworth Reference Press.

92 Hsu, C. and Lo, Y.M. (2003). Characterization of xanthan gum biosynthesis in a centrifugal, packedbed reactor using metabolic flux analysis. *Process Biochemistry* 38: 1617–1625. https://doi.org/10.1016/S0032-9592(03)00054-2.

93 Gilani, S., Najafpour, G., Heydarzadeh, H., and Zare, H. (2011). Kinetic models for xanthan gum production using Xanthomonas campestris from molasses. *Chemical Industry and Chemical Engineering Quarterly* 17 (2): 179–187.

94 Niknezhad, S.V., Asadollahi, M.A., Zamani, A. et al. (2015). Optimization of xanthan gum production using cheese whey and response surface methodology. *Food Science and Biotechnology* 24 (2): 453–460.

95 Papi, R., Ekateriniadou, L., Beletsiotis, E. et al. (1999). Xanthan gum and ethanol production by Xanthomonas campestris and Zymomonas mobilis from peach pulp. *Biotechnology Letters* 21 (1): 39–43.

96 Salah, R.B., Chaari, K., Besbes, S. et al. (2011). Production of xanthan gum from Xanthomonas campestris NRRL B-1459 by fermentation of date juice palm byproducts (Phoenix dactylifera L.). *Journal of Food Process Engineering* 34 (2): 457–474.

97 Silva, M.F., Fornari, R.C., Mazutti, M.A. et al. (2009). Production and characterization of xantham gum by Xanthomonas campestris using cheese whey as sole carbon source. *Journal of Food Engineering* 90 (1): 119–123.

98 Yoo, S.D. and Harcum, S.W. (1999). Xanthan gum production from waste sugar beet pulp. *Bioresource Technology* 70 (1): 105–109.

99 Palaniraj, A. and Jayaraman, V. (2011). Production, recovery and applications of xanthan gum by Xanthomonas campestris. *Journal of Food Engineering* 106: 1–12. https://doi.org/10.1016/j.jfoodeng.2011.03.035.

100 Antunes, A.E.C., Moreira, A.S., Vendruscolo, J.L., and Vendruscolo, C.T. (2000). Síntese de biopolímero xantana em meios convencionais e alternativos: viscosidade x composição. *Revista Brasileira de Agrociência* 6: 317–322.

101 Zabot, G.L., Silva, M.F., Terra, L.M. et al. (2012). Simulation of the xanthan gum production in continuous fermentation systems. *Biocatalysis and Agricultural Biotechnology* 1: 301–308. https://doi.org/10.1016/j.bcab.2012.08.002.

102 Souza, A.S. and Vendruscolo, C.T. (1999). Produçãoe carac terização dos biopolímeros sintetizados por Xanthomonas campestris pv pruni cepas 24 e 58. *Ciência e Engenharia* 8: 115–123.

103 Borges, C.D., De Paula, R.C.M., Feitosa, J.P.A., and Vendruscolo, C.T. (2009). The influence of thermal treatment and operational conditions on xanthan produced by X. arboricola pv pruni strain 106. *Carbohydrate Polymers* 75: 262–268. https://doi.org/10.1016/j.carbpol.2008.07.013.

104 Rottava, I., Batesini, G., Silva, M.F. et al. (2008). Xanthan gum production and rheological behavior using different strains of Xanthomonas sp. *Carbohydrate Polymers* 77: 65–71.

105 Rosalam, S., Krishnaiah, D., and Bono, A. (2008). Cell free xanthan gum production using continuous recycled packed fibrous-bed bioreactor-membrane. *Malays. Journal of Microbiology* 4: 1–5.

106 Papagianni, M., Psomas, S.K., Batsilas, L. et al. (2001). Xanthan production by Xanthomonas campestris in batch cultures. *Process Biochemistry* 37: 73–80. https://doi.org/10.1016/S0032-9592(01)00174-1.

107 Cacik, F., Dondo, R.G., and Marques, D. (2001). Optimal control of a batch bioreactor for the production of xanthan gum. *Computers & Chemical Engineering* 25: 409–418. https://doi.org/10.1016/S0098-1354(00)00662-1.

108 García-Ochoa, F., Santos, E.V., and Alcon, A. (2004). Chemical structured kinetic model for xanthan production. *Enzyme Microbial Technology* 35: 284–292. https://doi.org/10.1016/j.enzmictec.2003.11.024.

109 Casas, J.A., Santos, V.E., and García-Ochoa, F. (2000). Xanthan gum production under several operational conditions: molecular structure and rheological proper ties. *Enzyme Microbial Technology* 26: 282–291. https://doi.org/10.1016/S0141-0229(99)00160-X.

110 García-Ochoa, F., Santos, E.V., and Alcon, A. (1997). Xanthan gum production in a laboratory aerated stirred tank bioreactor. *Chemical and Biochemical Engineering Quarterly* 11: 69–74.

111 Rogovin, P. and Silman, R.W. (1972). Continuous fermentation to produce xanthan biopolymer: effect of dilution rate. *Biotechnology and Bioengineering* 14: 23–31. https://doi.org/10.1002/bit.260140104.

112 Vendruscolo, C.T., Moreira, A.S., Souza, A.S. et al. (2000). Heteropolysaccharides produced by Xanthomonas campestris pv pruni C24. *Hydrocolloids* (1): 187–191. https://doi.org/10.1016/B978-044450178-3/50022-6.

113 Sharma, A., Gautam, S., and Wadhawan, S. (2014). Xanthomonas. In: *Encyclopedia of Food Microbiology*, 2e (eds. C.A. Batt and M. Tortorello), 811–817. Amsterdam: Academic Press.

114 Antoniazzi, N. and Deschamps, C. (2006). Growth analysis of two barley cultivars after elicitors and fungicides treatment. *Ciência Rural* 36: 1065–1071.

115 Katzbauer, B. (1998). Properties and applications of xanthan gum. *Polymer Degradation* 59: 81–84. https://doi.org/10.1590/S0101-20612001000100018.

116 Lachke, A. (2004). Xanthan: a versatile gum. *Resonance* 9: 25–33. https://doi.org/10.1007/BF02834866.

117 Chawla, P.R., Bajaj, I.B., Survase, S.A., and Singhal, R.S. (2009). Microbial cellulose: fermentative production and applications. *Food Technology and Biotechnology* 47 (2): 107–124.

118 Brown, R.M. Jr. (1987). The biosynthesis of cellulose. *Food Hydrocolloids* 1: 345–351.

119 Delmer, D.P. and Amor, Y. (1995). Cellulose biosynthesis. *Plant Cell* 7: 987–1000.

120 Bielecki, S., Krystynowicz, A., Turkiewicz, M., and Kalinowska, H. (2005). Bacterial Cellulose. In: *Polysaccharides and Polyamides in the Food Industry* (eds. A. Steinbüchel and S.K. Rhee), 31–85. Weinheim, Germany: Wiley-VCH Verlag.

121 Ross, P., Mayer, R., and Benziman, M. (1991). Cellulose biosynthesis and function in bacteria. *Microbiology Reviews* 55: 35–58.

122 Tonouchi, N., Tsuchida, T., Yoshinaga, F. et al. (1996). Characterization of the biosynthetic pathway of cellulose from glucose and fructose in Acetobacter xylinum. *Bioscience, Biotechnology, and Biochemistry* 60: 1377–1379.

123 Trcek, J. and Barja, F. (2015). Updates on quick identifification of acetic acid bacteria with a focus on the 16S–23S rRNA gene internal transcribed spacer and the analysis of cell proteins by MALDI-TOF mass spectrometry. *International Journal of Food Microbiology* 196: 137–144.

124 Škraban, J., Cleenwerck, I., Vandamme, P. et al. (2018). Genome sequences and description of novel exopolysaccharides producing species Komagataeibacter pomaceti sp. nov. and reclassifification of Komagataeibacter kombuchae (Dutta and Gachhui 2007) Yamada et al. 2013 as a later heterotypic synonym of Komagataeibacter. *Systematic and Applied Microbiology* 41: 581–592.

125 Slapšak, N., Cleenwerck, I., de Vos, P., and Trcek, J. (2013). Gluconacetobacter maltaceti sp. nov., a novel vinegar producing acetic acid bacterium. *Systematic and Applied Microbiology* 36: 17–21.

126 Castro, C., Cleenwerck, I., Trcek, J. et al. (2013). Gluconacetobacter medellinensis sp. nov., cellulose- and non-cellulose-producing acetic acid bacteria isolated from vinegar. *International Journal of Systematic and Evolutionary Microbiology* 63: 1119–1125.

127 Castro, C., Zuluaga, R., Álvarez, C. et al. (2012). Bacterial cellulose produced by a new acid-resistant strain of Gluconacetobacter genus. *Carbohydrate Polymers* 89: 1033–1037.

128 Zhang, H., Chen, C., Zhu, C., and Sun, D. (2016). Production of bacterial cellulose by Acetobacter Xylinum: effects of carbon/nitrogen-ratio on cell growth and metabolite production. *Cellulose Chemistry and Technology* 50: 997–1003.

129 Li, J., Chen, G., Zhang, R. et al. (2019). Production of high crystallinity type-I cellulose from Komagataeibacter hansenii JR-02 isolated from Kombucha tea. *Biotechnology and Applied Biochemistry* 66: 108–118.

130 Shoda, M. and Sugano, Y. (2005). Recent advances in bacterial cellulose production. *Bioprocess and Biosystems Engineering* 10: 1–8.

131 Yoshinaga, F., Tonouchi, N., and Watanabe, K. (1997). Research progress in production of bacterial cellulose by aeration and agitation culture and its application as a new industrial material. *Bioscience, Biotechnology, and Biochemistry* 61: 219–224.

132 Mühlethaler, K. (1949). The structure of bacterial cellulose. *Biochimica et Biophysica Acta* 3: 527–535.

133 Czaja, W., Krystynowicz, A., Bielecki, S., and Brown, R.M. Jr. (2006). Microbial cellulose – the natural power to heal wounds. *Biomaterials* 27 (2): 145–151.

134 Legge, R.L. (1990). Microbial cellulose as a specialty chemical. *Biotechnology Advances* 8: 303–319.

135 Shah, J. and Brown, R.M. Jr. (2005). Towards electronic paper displays made from microbial cellulose. *Applied Microbiology and Biotechnology* 66: 352–355.

136 Budhiono, A., Rosidi, B., Taher, H., and Iguchi, M. (1999). Kinetic aspects of bacterial cellulose formation in nata-de-coco culture system. *Carbohydrate Polymers* 40: 137–143.

137 Ng, C.C. and Shyu, Y.T. (2004). Development and production of cholesterol-lowering Monascus-nata comple. *World Journal of Microbiology and Biotechnology* 20: 875–879.

138 Stephens, S. R., Westland, J. A., and Neogi, A. N. (1990). Method of using bacterial cellulose as a dietary fiber component. US patent 4960763.

139 Iguchi, M., Yamanaka, S., and Budhiono, A. (2000). Bacterial cellulose – a masterpiece of nature's arts. *Journal of Material Sciences* 35: 261–270.

140 Iguchi, M., Mitsuhashi, S., Ichimura, K. et al. (1998). Bacterial cellulose-containing molding material having high dynamic strength. US patent 4742164.

141 Srikanth, R., Reddy, C.H.S., Siddartha, G. et al. (2015). Review on production, characterization and applications of microbial levan. *Carbohydrate Polymers* 120: 102–114.

142 Gupta, S., Das, P., Singh, S. et al. (2011). Microbial levari, 683 an ideal prebiotic and immunonutrient in aquaculture. *World Aquaculture* 42 (1): 61.

143 Arvidson, S.A., Rinehart, B.T., and Gadala-Maria, F. (2006). Concentration regimes of solutions of levan polysaccharide from Bacillus spnn. *Carbohydrate Polymers* 65: 144–149.

144 Ullrich, M. (2009). *Bacterial Polysaccharides: Current Innovations and Future Trends.* Horizon Scientific Press.

145 Gomes, T.D., Caridade, S.G., Sousa, M.P. et al. (2018). Adhesive free-standing multilayer films containing sulfated levan for biomedical applications. *Acta Biomaterialia* 69: 183–195.

146 Singh, S.P., Jadaun, J.S., Narnoliya, L.K., and Pandey, A. (2017). Prebiotic oligosaccharides: special focus on fructooligosaccharides, its biosynthesis and bioactivity. *Applied Biochemistry and Biotechnology* 183 (2): 613–635.

147 González-Garcinuño, Á., Tabernero, A., Domínguez, Á. et al. (2018). Levan and levansucrases: polymer, enzyme, micro-organisms and biomedical applications. *Biocatalysis and Biotransformation* 36 (3): 233–244.

148 Bekers, M., Laukevics, J., Upite, D. et al. (2002). Fructooligosaccharide and levan producing activity of Zymomonas mobilis extracellular levansucrase. *Process Biochemistry* 38 (5): 701–706.

149 Jakob, F., Meißner, D., and Vogel, R.F. (2012). Comparison of novel GH 68 levansucrases of levan-overproducing Gluconobacter species. *Acetic Acid Bacteria* 1 (1): e2–e2.

150 Han, J., Xu, X., Gao, C. et al. (2016). Levan-producing Leuconostoc citreum strain BD1707 and its growth in tomato juice supplemented with sucrose. *Applied and Environmental Microbiology* 82 (5): 1383–1390.

151 Shih, I.L., Yu, Y.T., Shieh, C.J., and Hsieh, C.Y. (2005). Selective production and characterization of levan by Bacillus subtilis (Natto) Takahashi. *Journal of Agricultural and Food Chemistry* 53 (21): 8211–8215.

152 Gross, M., Geier, G., Rudolph, K., and Geider, K. (1992). Levan and levansucrase synthesized by the fireblight pathogen Erwinia amylovora. *Physiological and Molecular Plant Pathology* 40 (6): 371–381.

153 Hettwer, U., Gross, M., and Rudolph, K. (1995). Purification and characterization of an extracellular levansucrase from Pseudomonas syringae pv. phaseolicola. *Journal of Bacteriology* 177 (10): 2834–2839.

154 Liu, J., Luo, J., Ye, H., and Zeng, X. (2012). Preparation, antioxidant and antitumor activities in vitro of different derivatives of levan from endophytic bacterium Paenibacillus polymyxa EJS-3. *Food and Chemical Toxicology* 50 (3-4): 767–772.

155 Diken, E., Ozer, T., Arikan, M. et al. (2015). Genomic analysis reveals the biotechnological and industrial potential of levan producing halophilic extremophile, Halomonas smyrnensis AAD6T. *Springerplus* 4 (1): 393.

156 Jadaun, J.S., Narnoliya, L.K., Agarwal, N., and Singh, S.P. (2019). Catalytic biosynthesis of levan and short-chain fructooligosaccharides from sucrose-containing feedstocks by employing the levansucrase from Leuconostoc mesenteroides MTCC10508. *International Journal of Biological Macromolecules* 127: 486–495.

157 Moussa, T.A., Al-Qaysi, S.A., Thabit, Z.A., and Kadhem, S.B. (2017). Microbial levan from Brachybacterium phenoliresistens: characterization and enhancement of production. *Process Biochemistry* 57: 9–15.

158 Livingston, D.P., Hincha, D.K., and Heyer, A.G. (2009). Fructan and its relationship to abiotic stress tolerance in plants. *Cellular and Molecular Life Sciences* 66 (13): 2007–2023.

159 Jun-Zhi, W., Chatterton, N.J., Harrison, P.A. et al. (2002). Characterization of fructan biosyntheis in big bluegrass (Poa secunda). *Journal of Plant Physiology* 159 (7): 705.

160 Van den Ende, W., Coopman, M., Clerens, S. et al. (2011). Unexpected presence of graminan-and levan-type fructans in the evergreen frost-hardy eudicot Pachysandra terminalis (Buxaceae): purification, cloning, and functional analysis of a 6-SST/6-SFT enzyme. *Plant Physiology* 155 (1): 603–614.

161 Joaquim, E.O., Hayashi, A.H., Torres, L. et al. (2018). Chemical structure and localization of levan, the predominant fructan type in underground systems of Gomphrena marginata (Amaranthaceae). *Frontiers in Plant Science* 9: 1745.

162 Li, W., Yu, S., Zhang, T. et al. (2015). Recent novel applications of levansucrases. *Applied Microbiology and Biotechnology* 99 (17): 6959–6969.

163 Öner, E.T., Hernández, L., and Combie, J. (2016). Review of levan polysaccharide: from a century of past experiences to future prospects. *Biotechnology Advances* 34 (5): 827–844.

164 Gao, S., Qi, X., Hart, D.J. et al. (2017). Expression and characterization of levansucrase from Clostridium acetobutylicum. *Journal of Agricultural and Food Chemistry* 65 (4): 867–871.

165 Hövels, M., Kosciow, K., Kniewel, J. et al. (2020). High yield production of levan-type fructans by Gluconobacter japonicus LMG 1417. *International Journal of Biological Macromolecules* 164: 295–303.

166 Szwengiel, A. and Wiesner, M. (2019). Effect of metal ions on levan synthesis efficiency and its parameters by levansucrase from Bacillus subtilis. *International Journal of Biological Macromolecules* 128: 237–243.

167 Ende, W.V., Michiels, A., Roover, J.D., and Laere, A. (2002). Fructan biosynthetic and breakdown enzymes in dicots evolved from different invertases. Expression of fructan genes throughout chicory development. *The Scientific World Journal* 2: 1281–1295.

168 Tajima, K., Tanio, T., Kobayashi, Y. et al. (2000). Cloning and sequencing of the levansucrase gene from Acetobacter xylinum NCI 1005. *DNA Research* 7 (4): 237–242.

169 Kim, C.H., Jang, E.K., Kim, S.H. et al. (2005). Molecular cloning of levan fructotransferase gene from Arthrobacter ureafaciens K2032 and its expression in Escherichia coli for the production of difructose dianhydride IV. *Letters in Applied Microbiology* 40 (3): 228–234.

170 Cai, G., Liu, Y., Li, X., and Lu, J. (2019). New Levan-Type Exopolysaccharide from Bacillus amyloliquefaciens as an Antiadhesive agent against Enterotoxigenic Escherichia coli. *Journal of Agricultural and Food Chemistry* 67 (28): 8029–8034.

171 Trujillo Toledo, L.E., Gómez Riera, R., Banguela Castillo, A. et al. (2004). Catalytical properties of N-glycosylated Gluconacetobacter diazotrophicus levansucrase produced in yeast. *Electronic Journal of Biotechnology* 7 (2): 115–123.

172 Franken, J., Brandt, B.A., Tai, S.L., and Bauer, F.F. (2013). Biosynthesis of levan, a bacterial extracellular polysaccharide, in the yeast Saccharomyces cerevisiae. *PLoS One* 8 (10): e77499.

173 Ko, H., Bae, J.H., Sung, B.H. et al. (2019). Efficient production of levan using a recombinant yeast Saccharomyces cerevisiae hypersecreting a bacterial levansucrase. *Journal of Industrial Microbiology & Biotechnology* 46 (11): 1611–1620.

174 Kang, H.K., Yun, S.I., Lim, T.Y. et al. (2011). Cloning of levansucrase from Leuconostoc mesenteroides and its expression in Pichia pastoris. *Food Science and Biotechnology* 20 (1): 277–281.

175 Bersaneti, G.T., Baldo, C., and Colabone Celligoi, M.A.P. (2019). Immobilization of levansucrase: strategies and biotechnological applications. *Journal of the Chilean Chemical Society* 64 (1): 4377–4381.

176 Marx, S.P., Winkler, S., and Hartmeier, W. (2000). Metabolization of β-(2, 6)-linked fructose-oligosaccharides by different bifidobacteria. *FEMS Microbiology Letters* 182 (1): 163–169.

177 Kang, S.A., Jang, K.H., Seo, J.W. et al. (2009). Levan: applications and perspectives. *Microbial Production of Biopolymers and Polymer Precursors*: 145–162.

178 Basit, A. W. and Ibekwe, V. C. (2015). U.S. Patent No. 9,023,368. Washington, DC: U.S. Patent and Trademark Office.

179 Roberts, E. J. and Garegg, P. J. (1998). Levan derivatives, their preparation, composition and applications including medical and food applications. World Intellectual Property Organization Patent Application No. WO, 98, 03184.

180 Robertson, T.A., Papadimitriou, J.M., Walters, M.N.I., and Wolman, M. (1977). Effects of exposure of murine peritoneal exudate and resident macrophages to high molecular levan: a morphological study. *The Journal of Pathology* 123 (3): 157–164.

181 Han, Y.W. (1990). Microbial levan. *Advances in Applied Microbiology* 35: 171–194. Academic Press.

182 Dal Bello, F., Walter, J., Hertel, C., and Hammes, W.P. (2001). in vitro study of prebiotic properties of levan-type exopolysaccharides from lactobacilli and non-digestible carbohydrates using denaturing gradient gel electrophoresis. *Systematic and Applied Microbiology* 24 (2): 232–237.

183 Barone, J.R. and Medynets, M. (2007). Thermally processed levan polymers. *Carbohydrate Polymers* 69 (3): 554–561.

184 Chung, B., Song, K.B., Kim, C.H., and Rhee, S.K. (1997). Novel polyethylene glycol/levan aqueous two-phase system for protein partitioning. *Biotechnology Techniques* 11 (5): 327–329.

185 Chen, X., Gao, H., and Ploehn, H.J. (2014). Montmorillonite–levan nanocomposites with improved thermal and mechanical properties. *Carbohydrate Polymers* 101: 565–573.

14

Polysaccharides from Marine Microalgal Sources

Ratih Pangestuti[1], Evi Amelia Siahaan[1], Yanuariska Putra[1], and Puji Rahmadi[2]

[1] *Research and Development Division of Marine Bio Industry, Indonesian Institute of Sciences (LIPI), West Nusa Tenggara, Republic of Indonesia*
[2] *Research Center for Oceanography, Indonesian Institute of Sciences (LIPI), Jakarta, Republic of Indonesia*

14.1 Introduction

Algae are a group of aquatic photosynthetic organisms that grow in lakes, ponds, estuaries, rivers, and oceans. Based on the size, algae can be classified as macroalgae (also known as seaweeds) and microalgae. Microalgae are diverse groups (up to 1.000.000 species) of prokaryotic and eukaryotic microorganisms with various cell morphologies [1]. Single-cell microalgae are mostly invisible to the naked eye; they only can be seen in a colony.

Microalgae were able to adapt and tolerate a wide range of stress conditions such as temperatures, salinities, pH values, light intensities, and conditions in reservoirs or oceans. Microalgae can grow alone or in symbiosis with other organisms (i.e. bacteria) [2]. In the marine environment, light, temperature, salinity, pH values, and nutrients are dynamic and diverse for marine creatures, including microalgae [3]. Light intensity, such as the rate of light absorption, must be adjusted to meet algal cellular demands since light intensity variation in the marine environment occurs on short timescales, ranging from seconds to even days. Thus, microalgae must adapt the effective light absorption on the one hand, and effective photoprotective response to light intensities. Therefore, they produce bioactive metabolites to modulate extreme environmental conditions. Bioactive metabolites derived from microalgae such as carotenoids, phycobiliproteins, chlorophylls, functional oils, sterols, vitamins, and polysaccharides have attracted considerable interest not only for scientific but also for industrial communities. Marine microalgae are easy to cultivate in desired (small, medium, and large) and controlled volume. They can be easily engineered and optimized to produce the desired molecules. In addition, these microorganisms grow faster than terrestrial plants with high photosynthetic efficiency and do not need land to grow, needing only seawater with culture medium as a source of nitrogen, phosphorous, and other nutrients [4].

Bioprospecting of Microorganism-Based Industrial Molecules, First Edition.
Edited by Sudhir P. Singh and Santosh Kumar Upadhyay.

Microalgae have so far proved as a renewable yet sustainable source of bioactive compounds in many applications. Marine algae have been known as a source of functional omega-3 fatty acids, which are often reported in the literature to provide health benefit effects [5]. Microalgae also have been reported as potent biofuel sources. In addition to the oil content, polysaccharides from microalgae have increasingly gained interest in recent years. Marine microalgal polysaccharides have been demonstrated to possess several important biological activities such as antiviral, antioxidant, immunomodulatory effects, making them potent biomaterials and bioactive products of increasing importance for a wide range of industrial applications. In this chapter, we focus on the update studies on the polysaccharides found in marine microalgae, their biological roles, and protective activities. This contribution not only presents an overview of the developments and optimization of microalgae culture but also considers the fundamental and innovative green extraction technique to obtain functional polysaccharides from marine microalgae.

14.2 Polysaccharides from Marine Microalgae

Marine microalgae have been known as one of the highest sources of protein and other biochemical molecules, such as carbohydrates, lipids, and fatty acids. Microalgal biochemical composition varies depending on microalgal strains and other factors, including light intensity, photoperiod, temperature, nutrients, pH, salinity, and growth phase. The biochemical compositions of selected species of marine microalgae are shown in Table 14.1.

Table 14.1 Proximate composition of different microalgae species (%, based on dry weight).

Microalgae species	Crude protein	Crude carbohydrates	Crude lipid	References
Spirulina platensis	63.26±0.04	15.17±0.02	7.09±0.03	[6]
Chlorella vulgaris	47.82±0.05	8.08±0.09	13.32±0.07	[6]
Chlorella sp.	6.07±1.14	7.09±0.84	1.82±0.21	[7]
Isochrysis galbana	26.99±0.08	16.98±0.05	17.16±0.04	[6]
Tisochrysis lutea	36.70±1.20	9.40±1.90	22.00±1.60	[8]
Nannochloropsis sp	32.60±3.20	17.60±1.70	11.10±0.90	[9]
Chaetoceros calcitrans	41.60±4.20	8.70±1.20	26.80±5.20	[10]
Chaetoceros sp.	10.5±1.52	1.50±0.09	3.73±0.54	[7]
Nannochloropsis oculata	32.82±3.11	26.13±2.70	13.02±0.91	[10]
Tetraselmis sp	21.70±0.70	14.50±1.60	9.40±0.80	[9]
Tetraselmis chuii	26.30±0.40	11.20±0.20	16.50±0.10	[11]
Cyclotella cryptica	26.00±0.20	35.80±1.20	16.50	[12]
Navicula incerta	50.38±0.03	10.84±0.03	8.76±0.04	[13]
Bellerochea sp	14.2±1.84	3.01±0.39	9.87±0.79	[7]
Rhodomonas sp.	44.9±6.52	8.60±0.48	39.2±3.85	[7]
Thalassiosira sp	73.1±2.60	51.4±5.46	39.6±7.18	[7]

Table 14.2 Sulphated and exopolysaccharides found in marine microalgae.

Microalgae species	Types of polysaccharides	Main neutral sugar	References
Sprulina platensis	Sulphated polysaccharides	GlcA, Gal, Glu, Rha, Ara, Rib	[16]
Navicula sp	Sulphated polysaccharides	Glu, Gal, Rha, Xyl, Man	[17]
Navicula salinarum	Sulphated polysaccharides	Glu, Gal, Man, Rha, Xyl	[18]
Cylindrotheca closterium	Sulphated polysaccharides	Glu, Gal, Man, Rha, Xyl	[18]
Porphyridium cruentum	Exopolysaccharides	Xyl, Gal, Glu, GlcA	[19]
Porphyridium sordidum	Exopolysaccharides	Xyl, Gal, Glu, GlcA	[20]
Porphyridium purpureum	Exopolysaccharides	Xyl, Gal, Glu, GlcA	[20]

Microalgae possess a wide range of polysaccharides, which chemical composition, content, and biological activities vary within microalgal species. As an example, in high luminous intensities, generally, microalgae accumulate carbohydrates, used as energy storage molecules during the respiration phase of photosynthesis. Due to nitrogen deficiencies, the growth of microalgae will decrease, and microalgae enter the stationary phase, and then accumulate lipids and/or carbohydrates. Generally, microalgae produce two types of polysaccharides, the storage polysaccharides in the cytoplasm and secreted/exopolysaccharides. Marine microalgae are one of the most potent sources of nonanimal sulphated polysaccharides. In addition to sulphated polysaccharides, microalgae may excrete extra or exopolysaccharides in their culture. Exopolysaccharides are mainly heteropolysaccharides composed of Glu, Gal, and Xyl but also included significant amounts of other monosaccharides such as Rha, Fuc, and uronic acid [14]. Exopolysaccharides from microalgae are traditionally extracted, fractionated, and purified using solvent precipitation with or a cationic surfactant [15]. Sulphated polysaccharides and exopolysaccharides of selected species of marine microalgae are shown in Table 14.2.

Hot water extraction and hydrolysis with acid are the most common techniques used to isolate polysaccharides from marine microalgae [19]. Acid hydrolysis allows the depolymerizing of microalgal intracellular starch and structural polysaccharides into monomers [20]. Microalgal polysaccharides compositions, chain conformations, and molecular weight may differ due to the extraction methods [21]. Hence, for the efficient recovery of polysaccharides from microalgae with the desired size, structure, and functional property, a suitable method is needed. As an example, temperature, sample-to-solvent ratios, acid ratios, and extraction time must then be adjusted to optimize the extraction process and avoid excessive degradation of monosaccharides. Due to the low extraction yields of hot water extraction, extraction techniques, such as subcritical water extraction, ultrasonic-aided extraction, microwave-assisted extraction, and enzyme-assisted extraction, have been applied in isolation of bioactive compounds from marine microalgae, including polysaccharides.

14.2.1 Subcritical Water Hydrolysis

Subcritical water hydrolysis is an environmentally friendly technique that typically uses water as a solvent. The temperature was set above water's boiling point (100 °C/273 K, 0.1 MPa), but below its critical point (374 °C/647 K, 22.1 MPa) to keep it

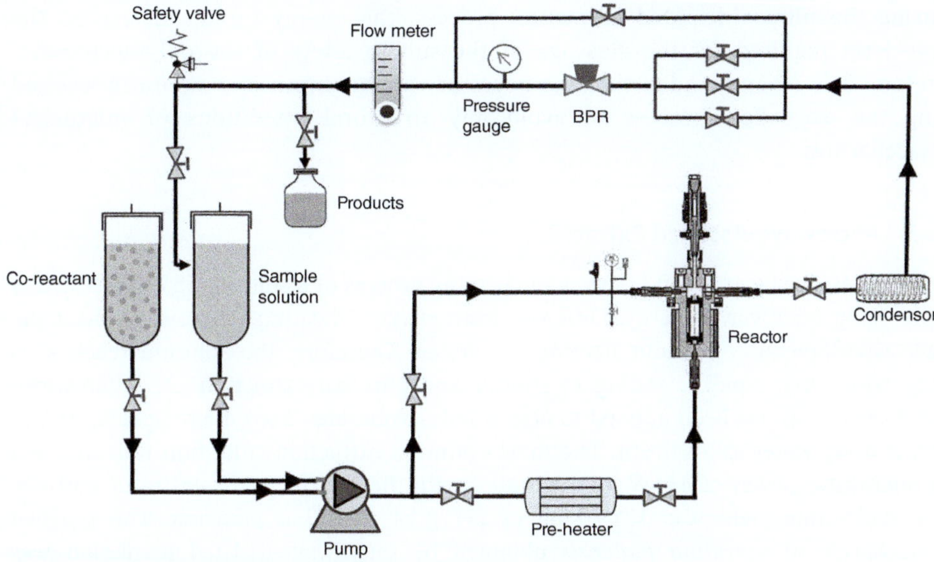

Figure 14.1 Schematic diagram of subcritical water extraction system.

in a liquid state. During these conditions, the chemical and physical properties of the water change dramatically. At this state, the dielectric constant of water drops to 33 at 200 °C from 80 at room temperature. This property makes water able to act effectively as a solvent. The subcritical water apparatus and schematic diagram are presented in Figure 14.1.

Subcritical water hydrolysis has been successfully applied to extract different bioactive molecules from macroalgae [22, 23]. The extraction of crude extract containing polysaccharides from microalgae *Haematococcus pluvialis* has been reported [24]. It was reported that the extraction temperature positively correlated with the extraction yield of polysaccharides and their biological activity.

14.2.2 Ultrasonic-Aided Extraction

Ultrasonic-aided extraction is energy input-assisted extraction methods and has been used to isolate bioactive substances from both terrestrial and marine resources. The principle is based on the waves migrating through medium inducing pressure variations. Kurd et al. (2015) have extracted polysaccharides from *Spirulina platensis* using ultrasound-aided extraction and optimized using response surface methodology (RSM). It was found that optimum extraction conditions were 25 minutes extraction time at 85 °C with ultrasonic power 90 W and the ratio of water to sample of 20 ml/g. *Spirulina platensis* polysaccharides show antioxidant activities [25]. It was also reported elsewhere that ultrasound-assisted extraction was more effective in extracting polysaccharides from *Chlorella* sp. as compared with conventional solvent extraction and fluidized bed extraction [26].

During the ultrasonic-aided extraction process, the energy input can exceed the energy level required for the cleavage of the sulfate esters of the polysaccharides. Therefore, it is recommended that the necessary energy input is temporarily exerted during the extraction process to avoid any structural alterations of microalgal polysaccharides.

14.2.3 Microwave-Assisted Extraction

Microwave-assisted extraction involves the heating process of a solution in contact with a sample using microwave energy. Different from classical heating, microwaves heat the sample simultaneously without heating the vessel. Therefore, the solution reaches its boiling point very rapidly, leading to shorter and effective extraction [3]. Microwave-assisted extraction has been applied to obtain polysaccharides from microalgae *Spirulina platensis* using water as a solvent. The most optimum extraction condition was obtained with microwave power of 434 W for 1 minutes, with the sample to solvent ratio 1:30 w/v. The carbohydrate yield was 127 ± 5 mg of per g of *Spirulina platensis*. The purified polysaccharide of *Spirulina platensis* obtained by microwave-assisted extraction was composed of major glucose and rhamnose [27].

Similar to ultrasound-aided extraction, in the microwave-assisted extraction process, the necessary energy input must be temporarily exerted during the extraction process to avoid any structural alterations of microalgal polysaccharide.

14.2.4 Enzyme-Assisted Extraction

Enzyme-assisted extraction has been reported to improve the extraction efficiency of polysaccharides from terrestrial resources. Some studies showed that enzyme-assisted extraction was an excellent method to extract polysaccharides from macroalgae [28–30]. Charoensiddhi et al. (2016) reported that both enzymes and pH conditions during the enzyme-assisted extraction process affect sugar composition and molecular weight profile of the polysaccharide fractions from brown macroalgae *Ecklonia radiata*. During the enzyme-assisted extraction process, physicochemical conditions of the reaction media, including pH of the protein solution, temperature, and enzyme to sample ratios, must be adjusted in order to optimize enzyme activity. Proteolytic enzymes from different sources, including animals, terrestrial plants, and microbes, can be optimized to obtain polysaccharides from marine microalgae. However, the application of the enzyme-assisted extraction method in polysaccharides extraction from marine microalgae has rarely been reported.

14.3 Optimization of Microalgae Culture Conditions

Microalgae can be grown anywhere and every season because they have a variety of species that have different properties and adaptations. They can grow in seawater, freshwater, open ponds, deserts, etc. There are some important parameters to optimize microalgae growth as well as the quality of the polysaccharides produced, such as light, pH, carbon dioxide, temperature, salinity, and adequate nutrients [31, 32].

Light play an important role in cell growth and polysaccharide production. The type of light source and light intensity is known to be a critical factor for microalgae growth. The impact of light intensity on the growth of microalgae could be classified into several phases, such as limitation phase, saturation phase, and inhibition phase. To enhance biomass production, the saturation of light intensity needs to distribute through the entire cultivation system [33]. Irradiance intensities to produce polysaccharides might have different levels between species of microalgae. The stronger light intensity results in increased biomass and polysaccharides production on the growth of *Spirulina platensis*. The highest light intensity of 192 μmol photons m^2/s gave the maximum amount of the polysaccharides (146.45 ± 16.26 mg/l), while the lower irradiance (96 and 24 μmol photons m^2/s) generated 105.22 ± 18.93 mg/l and 67.69 ± 9.45 mg/l amount of polysaccharides [34]. The optimum growth rate of *Porphyridium cruentum* was achieved at photon flux density 70 μE/m/s with the maximum extracellular polysaccharides production of 0.95 g/l [35]. The simultaneous impact of irradiance and temperature on the carbohydrate concentration of *Pavlova lutheri*. Carbohydrate content in the exponential growth phase increased with irradiance intensity of 150 μmol photons m^2/s, while the concentration of internal carbohydrate content decrease at irradiance 195 μmol photons m^2/s [36]. However, the production of polysaccharide of microalgae strains not only depends upon light exposure but also on other environmental parameters like temperature and nutrient [32].

Temperature is also an important key to promote microalgae growth and target-product production, especially for outdoor cultivation, variation in temperature is highly dependent upon the light intensity and seasonal changes [33]. The temperature impact on microalgae growth and polysaccharides production has been investigated for a variety of microalgae species. For instance, the growth rate of *Spirulina platensis* reached the highest biomass at 28 °C ($672.86 \pm 6.89\%$) with $5.94 \pm 0.55\%$ dry weight ratio of polysaccharides, while the maximum production of polysaccharides was $12.57 \pm 1.06\%$ at 38 °C with $550.66 \pm 5.56\%$ of biomass growth rate [34]. Araujo and Garcia (2005) suggested that temperatures between 20 and 25 °C could be used to enhance the nutritional value of *Chaetoceros* cf. *wighamii* due to higher carbohydrate, lipid, and sufficient protein content under these conditions [37]. In the case of *Dictyosphaerium chlorelloides,* the maximum intracellular polysaccharides are preferentially released at 19.2 °C (2130 ± 16 μg/ml), and extracellular polysaccharides hit the highest level at 25.7 °C (1075 ± 24 μg/ml) [38]. Furthermore, shifting the cultivation temperature of microalgae culture can stimulate sugar content enhancement by 25% [32].

The hydrogen ion concentration is one of the valued variables for microalgae metabolism as it affects not only cell growth but also biochemical composition. In general, the optimal pH for microalgae to accumulate carbohydrates varies based on the group species. Extracellular polysaccharides of *Netrium digitus* attained the maximum yield at pH of 6 ($21.42 \pm 0.28\,g/m^2$) with a combination of high light exposure and nitrogen elevation [39]. The various pH values significantly affected the carbohydrate accumulation on *Chlorella ellipsoidea* and *Dunaliella bardawil*. The carbohydrate content obtained the maximum value at pH 7.5 and pH 9 for *D. bardawil* and *C. ellipsoidea*, respectively. The lowest carbohydrate content was recorded at the acidic pH value (pH 4) for both microalgae [40]. *Skeletonema costatum* also exhibited the highest carbohydrate production at the alkaline pH value (pH 9.4) as much as 17.7%, and in the pH range 6.5–8.5, the average production was only 3.7% [41].

The presence of sodium chloride (NaCl) in the culture medium of microalgae may influence biomass growth and polysaccharide production. For example, the increasing NaCl concentrations in *Spirulina platensis* cultivation simultaneously impede the biomass growth and promote polysaccharide production. The highest polysaccharide yield was found at 0.75 mol/l of NaCl concentration [34]. Mishra and Jha (2009) also reported the extracellular polymeric substances of *Dunaliella salina* increased cooperatively with NaCl concentration. The maximum level (944 mg/l) was attained at 5 M NaCl [42]. The intracellular carbohydrate level in selected species was found to decrease due to salinity increase. It could be the high salinity level triggering the activation of enzymes/pathways that disintegrate stored carbohydrate to form an increased amount of lipid [43].

The basic character of microalgae is phototrophic growth; therefore, they require carbon dioxide to grow, particularly in high biomass cultivation. An adequate carbon dioxide supply is needed related to photosynthesis efficiency that contributes to carbohydrate accumulation as the by-product [32]. Some studies found that the increase in carbon dioxide percentage could improve carbohydrate accumulation. Xia and Gao (2005) reported that carbohydrate content of *Chlorella pyrenoidosa* and *Chlamydomonas reinhardtii* was significantly increased by elevated CO_2 concentration from 3 to 186 μmol/l as a response to photosynthesis enhancement, leading to soluble carbohydrate accumulation [44]. A similar observation was reported when CO_2 concentration was found to enhanced carbohydrate productivity in *Chlorella* sp. [45]. *Chaetoceros gracilis* attained its maximum carbohydrate content accompanied by a higher cell growth at 385 μatm of CO_2 concentration. The elevation of CO_2 to 1050 μatm gradually reduced both cell number and carbohydrate due to the decline of photosynthetic efficiency as the environment becomes more acidic [46, 47]. However, sufficient CO_2 and light energy could be employed under nitrogen starvation conditions to increase the carbohydrate content [48].

Nutrient starvation is the most efficient strategy to stimulate carbohydrate accumulation [32, 49, 50]. It was demonstrated that starch productivity was higher with the deprivation of nitrogen and sulfur in *Tetraselmis subcordiformis* cultivation. Compared with nitrogen deprivation (0 mM KNO_3), starch accumulation under sulfur-limited conditions (0 mM MgSO4) exhibited better performance (0.62 g/l/d). In contrast, starch content was declined at higher nitrogen and sulfur concentration due to cell division in nutrient-sufficient conditions [51]. Dragone et al. (2011) reported that *Chlorella vulgaris* attained the highest starch content (39.8–41.0%) under nitrogen limitation conditions (0 g/l urea) and irradiance level of 70 μmol m^2/s [52]. During essential macronutrient starvation in *Chlorella* cultivation, sulfur limitation and cycloheximide treatment maintained the highest content of starch concentration (up to 60% of dry weight). Despite cycloheximide application has shown the significant increase of starch content, the practical large-scale use of cycloheximide remains unrealistic both environmentally and economically [53]. Another essential macronutrient starvation was demonstrated in *T. subcordiformis* culture by the application of phosphorus-deprived conditions. Starch content immediately increased upon extracellular phosphorus restriction under low and medium initial cell densities. According to a previous study in *T. subcordiformis* cultivation [51], extracellular phosphorus deprivation was less effective than nitrogen and sulfur deprivation in stimulating starch accumulation [54]. Generally, biomass productivity would decrease during nutrient limitation. Hence, two-step cultivation should be employed: a first step is using sufficient nutrient medium to attain an optimum growth rate and biomass concentration, followed by a second step that involves cell cultivation in nutrient deprivation [50, 52].

Overall, understanding the fundamental principle of microalgae carbohydrate metabolism is necessary to develop a more effective strategy to optimize carbohydrate production. The strategy could be enhanced via manipulation of important key factors such as nutrient deprivation, light supply, carbon dioxide supplementation, temperature, pH, and salinity [48].

14.4 Bioactivities and Potential Health Benefits

In this part, the bioactivities of microalgal polysaccharides, including antiviral, antioxidant, anticancer, and immunomodulatory agents, will be discussed.

14.4.1 Antiviral Activity

Virus infections are a worldwide burden on public health and economic stability. There has been a growing concern of numerous viral infectious diseases that have emerged or re-emerged mainly due to an increase in drug resistance. For this reason, the search for antiviral compounds from natural sources has been continuously increasing. Polysaccharides, the natural macromolecules, have been used in traditional Chinese medicine due to their bioactivities [55]. They are widely distributed in marine macro and microalgae. Microalgal polysaccharides, such as galactose, glucose, xylose, rhamnose, and fructose, have been extensively reported to show antiviral activities against several kinds of viruses and retroviruses. The studies on antiviral activity of sulphated polysaccharide from *Porphyridium* sp against HSV (herpes simplex virus) 1, HSV 2, and VZV (varicella zoster virus) have been reported [56, 57]. It is assumed that the mechanisms underlying the antiviral activity of sulphated polysaccharides from *Porphyridium* sp is by preventing adsorption of the virus into the host cells and/or by inhibiting the production of new viral particles inside the host cells [56]. Another investigation of anti-Hepatitis B virus (HBV) activity of extracellular polysaccharide from *Porphyridium cruentum* has been reported by Huang et al. [58]. The extracellular polysaccharide showed a significant inhibitory effect on the HBV replication in vitro. Talyshinsky and coworkers also showed that the polysaccharides from *Porphyridium* sp were effective in inhibiting the replication and transforming ability of retroviruses MuLV (murine leukimia virus) and MuSV-124 (murine sarcoma virus) [59]. The red microalgal polysaccharides from *Porphyridium* could inhibit the infections by blocking the virus penetration into the host cells, or even inhibiting the early stages of virus replication, without any impact on the metabolism of the host cells [60–62].

Another study investigated the antiviral activity of sulphated polysaccharides, naviculan, isolated from *Navicula directa* against HSV 1, HSV 2, and IFV (influenza virus) A and HIV (human immunodeficiency virus). The sugar compositions of this sulphated polysaccharide were fucose, xylose, galactose, mannose, and rhamnose. Naviculan interferes with very early stages of viral replication such as virus binding and penetration into host cells. Naviculan also interferes with cell-to-cell fusion between HIV glycoprotein 160 and CD4 expressing cells [63]. A novel sulphated exopolysaccharide, p-KG03, was isolated from *Gyrodinium impudicum*. It contained a homogenous mixture of galactose conjugated with uronic acids and sulphated groups and found to possess a great inhibitory effect on the EMCV (encephalomyocarditis virus) infection [64, 65]. Moreover, Kim et al. reported that p-KG03 demonstrated a great inhibitory activity on IFV A infections by blocking the cellular internalization of the virus and early stage of virus replication [65, 66]. Numerous studies have already highlighted the impressive activities of microalgal sulphated polysaccharides in preventing virus infections. However, they show different levels of antiviral capacity depending on the molecular weight and the degree of sulphation [57, 67, 68]. Most recently, algal polysaccharide has gained more attention to be an attractive candidate for developing a potential antiviral agent, especially for the treatment of coronavirus [69].

14.4.2 Antioxidant

Reactive oxygen species (ROS) and reactive nitrogen species (RNS) are formed continuously in the human body and play an important role in cell signaling at low concentrations. However, at high concentrations, these reactive species exert oxidative stress that can cause extensive damage to cellular components inducing various diseases, including aging, cancer, and neurodegenerative disease [70, 71]. The antioxidant is a crucial defense against free radical-induced damage [72]. Microalgal polysaccharides are naturally derived compounds that exhibit strong antioxidant activity and can be used to reduce oxidative damage to the human body [71, 73, 74]. In recent years, numerous studies have revealed that microalgal sulphated polysaccharides have shown the highest free-radical scavenging activities on different radicals [17, 75, 76].

It was already demonstrated that polysaccharides isolated from cultured *Isochrysis galbana* exhibit significant scavenging activity against superoxide and hydroxyl radicals. The polysaccharide was rich in uronic acid and sulfate with 15.934 kDa of molecular weight, and contained mannose, glucose, galactose, and rhamnose [75]. The result of this study explained that the strong antioxidant activity was attributed to the higher uronic acid and sulphate content in the polysaccharide [75]. Similar work has been confirmed previously by Chen et al. [77]. This work showed that the crude sulphated exopolysaccharide from *Rhodella reticulata* has a stronger ability than alpha-tocopherol against superoxide anion radical scavenging [77]. Besides the presence of monosaccharides, uronic acids, and sulphate groups, the molecular weight of microalgal polysaccharide was confirmed to have a significant influence on its antioxidant capacity. Several scientific reports have revealed that the antioxidant activity of microalgal polysaccharide can be enhanced by decreasing its molecular weight [73, 78]. Sun and colleagues have obtained undegraded and degraded polysaccharides from marine microalgae *Pavlova viridis* and *Sarcinochrysis marina* Geitler. Polysaccharide samples were observed for their antioxidant capacity using 1,1-diphenyl-2-picrylhydrazyl (DPPH), hydroxyl-radical (OH) scavenging, lipid peroxidation (LPO), inhibition, and the mouse red blood cells (RBCs) hemolysis assay. The presence of uronic acids and sulphate groups was confirmed. They found that the degraded polysaccharides (low molecular weight) with the highest sulphate content exhibited stronger antioxidant activity than the undegraded polysaccharides [78].

Recently, a similar finding was reported by Fimbres-Olivarria et al. [17]. Microalgal sulphated polysaccharides were collected from cultured *Navicula* sp at three different wavelengths, namely WSPN, RSPN, and BSPN. These polysaccharides contained glucose, galactose, rhamnose, xylose, and mannose as their main neutral sugars. The antioxidant activity of WSPN polysaccharide was proved to be higher than others. The high antioxidant result of WSPN was attributed to its highest sulphate content (0.40% w/w) and lowest molecular weight (17 kDa) compared with other polysaccharide samples. Based on this investigation, it could be assumed that microalgal polysaccharides' antioxidant capacity is correlated to their culture conditions. According to Shen et al. [79], three polysaccharides (WL-CPS-1, NaCl-CPS-1, and Glu-CPS-1), purified from *Nostoc flagelliforme* with different culture conditions, had different chemical properties and molecular weight which influenced their antioxidant capacity. The chemical composition of these polysaccharides was composed of nine monosaccharides and one uronic acid with different ratios, with the average molecular weights of 1.02×10^3, 1.12×10^3, 1.33×10^3 kDa, respectively. They all exhibited vigorous scavenging activity on ABTS and hydroxyl radicals and moderate activity on DPPH radical. Glu-CPS-1 showed the highest antioxidant activity. This suggested that different antioxidant of microalgal polysaccharides can be obtained by changing the different cultural conditions of microalgae [73, 79]. Furthermore, the in vivo evaluation of the antioxidant activity of microalgal polysaccharides needs to be extensively studied. These works also suggest that microalgal polysaccharides could be a potential candidate for antioxidant agents.

14.4.3 Anticancer

Cancer is defined as an abnormal multiplication of cells in an uncontrollable manner, which spreads into other tissues causing metastases that are often lethal. Cancer is one of the most threatening diseases and becomes the second primary reason for death globally. Currently, chemoprevention and chemotherapy are the available treatments to cure cancers, but they cause some life-threatening side effects in the human body. These drugs are toxic and affect not only cancer cells but also normal cells [80]. Thus, finding more effective methods and developing the nontoxic, novel therapeutic agents for cancer treatment is still a challenge. Microalgal polysaccharides have been reported to exhibit various bioactivities, including anticancer. Calcium spirulan (Ca-SP), a sulphated polysaccharide from *Spirulina plantesis*, was found to exhibit a remarkable anticancer activity. An in vivo study showed that Ca-SP effectively reduced the lung metastasis of B16-BL6 melanoma cells by inhibiting the tumor invasion of the basement membrane, thus resulting in the blocking of adhesion and migration of tumor cells [81, 82].

A promising compound with anticancer activities, GA3P (d-galactan sulfate, associated with l-(+)-lactic acid), an exopolysaccharide, isolated from *Gymnodinium sp*, inhibited growth of different cancer cell lines, in particular, several colon cancer cell lines (HCC2998, KM-12, HT-29, WiDr, HCT-15, and HCT-116) [83, 84]. Recently, Kamble et al. reported in their investigation of the cancer-preventive effects of sulphated polysaccharides (Cr-SPs) from microalgae *Chlamydomonas reinhardtii*. Cr-SPs have successfully inhibited cell proliferation in MDA-MB-231 breast cancer by disrupting the microtubule network of cancer cells [85]. In an additional investigation, remarkable anticancer activity of sulphated polysaccharide isolated from *Phaeodactylum tricornutum* against HepG2 cells. It is reported that the anticancer effect occurred through induction apoptosis without affecting the cell cycle and mitosis of cancer cells. All these findings suggest the microalgal polysaccharides could be a potential anticancer chemotherapeutic agent.

14.4.4 Immunomodulatory

The polysaccharide is one of the valuable bioactive compounds in microalgae that believe in having potential immunomodulatory properties [86]. Some studies found that the application of microalgae polysaccharide could improve immune response. Sun et al. (2013) reported the polysaccharide from marine microalgae *Sarcinochrysis marina* Geitler could enhance immune response through spleen and thymus index elevation and stimulated lymphocytes propagation [87]. It was demonstrated the isolated extracellular polysaccharides from *Prophyridium cruentum* with different concentrations can promote immune response via macrophage and lymphocytes proliferation and significantly inhibited the growth of the implanted tumor with the inhibition index of 40.5–53.3% [88]. Macrophages play a significant role in infection protection, damage, cell elimination, and immune system enhancement. Chen et al. (2019) found that sulfated polysaccharides extracted from filamentous microalgae, *Tribonema* sp., at a concentration of 25 μg/ml, were remarkably boosting the macrophage viability and its cytokines [80]. Exopolysaccharide derived from *Thraustochytriidae* sp. was revealed to have anticancer activity and promote B cell proliferation that may strengthen the immune system [89]. Pugh et al. (2001) described three high molecular weight polysaccharides isolated from *Spirulina platensis* (immulina), *Aphanizomenon flos-aquae* Immunon), and *Chlorella pyrenoidosa* (Immurella). They have superior macrophage stimulatory compared with commercial polysaccharide immunostimulants, such as schizophyllan, lentinan, krestin, and acemannan [90].

14.5 Conclusions and Perspectives

Marine microalgae are the subject of increasing interest due to their potency as polysaccharides source. Wide range of polysaccharides can be found in microalgae, with chemical composition, content, and biological activities varied within microalgal species. Future studies should be directed toward optimization of microalgal culture to obtain functional polysaccharides, green extraction technique, characterization and identification of microalgal polysaccharides as well as investigation of health benefit effects.

Acknowledgment

This study is supported by National Research Priority (PRN)-Development of Marine Micro and Macro-algae from Indonesia (MALSAI) 2021.

References

1 Rumin, J., Nicolau, E., Gonçalves de Oliveira Junior, R. et al. (2020). Analysis of scientific research driving microalgae market opportunities in Europe. *Marine Drugs* 18 (5): 264.

2 Khan, M.I., Shin, J.H., and Kim, J.D. (2018). The promising future of microalgae: current status, challenges, and optimization of a sustainable and renewable industry for biofuels, feed, and other products. *Microbial Cell Factories* 17 (1): 36.

3 Pangestuti, R., Siahaan, E.A., and Kim, S.-K. (2018). Photoprotective substances derived from marine algae. *Marine Drugs* 16 (11): 399.

4 Vecchi, V., Barera, S., Bassi, R., and Dall'Osto, L. (2020). Potential and challenges of improving photosynthesis in algae. *Plants* 9 (1): 67.

5 Demirbas, A. and Fatih Demirbas, M. (2011). Importance of algae oil as a source of biodiesel. *Energy Conversion and Management* 52 (1): 163–170.

6 Tokuşoglu, Ö. and üUnal, M.K. (2003). Biomass nutrient profiles of three microalgae: spirulina platensis, chlorella vulgaris, and Isochrisis galbana. *Journal of Food Science* 68 (4): 1144–1148.

7 Costard, G.S., Machado, R.R., Barbarino, E. et al. Chemical composition of five marine microalgae that occur on the Brazilian coast. *International Journal of Fisheries and Aquaculture* 4 (9): 191–201.

8 Alkhamis, Y. and Qin, J.G. (2016). Comparison of pigment and proxim ate compositions of Tisochrysis lutea in phototrophic and mixotrophic cultures. *Journal of Applied Phycology* 28 (1): 35–42.

9 Arkronrat, W., Deemark, P., and Oniam, V. (2016). Growth performance and proximate composition of mixed cultures of marine microalgae (Nannochloropsis sp. & Tetraselmis sp.) with monocultures. *Songklanakarin Journal of Science & Technology* 38 (1): 1–5.

10 Banerjee, S., Hew, W.E., Khatoon, H. et al. (2011). Growth and proximate composition of tropical marine Chaetoceros calcitrans and Nannochloropsis oculata cultured outdoors and under laboratory conditions. *African Journal of Biotechnology* 10 (8): 1375–1383.

11 Khatoon, H., Haris, H., Rahman, N.A. et al. (2018). Growth, proximate composition and pigment production of tetraselmis chuii cultured with aquaculture wastewater. *Journal of Ocean University of China* 17 (3): 641–646.

12 Pahl, S.L., Lewis, D.M., Chen, F., and King, K.D. (2010). Growth dynamics and the proximate biochemical composition and fatty acid profile of the heterotrophically grown diatom Cyclotella cryptica. *Journal of Applied Phycology* 22 (2): 165–171.

13 Kang, K.-H., Qian, Z.-J., Ryu, B. et al. (2012). Protective effects of protein hydrolysate from marine microalgae Navicula incerta on ethanol-induced toxicity in HepG2/CYP2E1 cells. *Food Chemistry* 132 (2): 677–685.

14 Michaud, P. (2018). Polysaccharides from microalgae, what's future? *Advances in Biotechnology & Microbiology* 8 (2): 29–30.

15 Marcati, A., Ursu, A.V., Laroche, C. et al. (2014). Extraction and fractionation of polysaccharides and β-phycoerythrin from the microalga Porphyridium cruentum by membrane technology. *Algal Research* 5: 258–263.

16 H, A.E.B., Kf, H.E.B., and Sa, E.-L. (2014). Induction of sulfated polysaccharides in spirulina platensis as response to nitrogen concentration and its biological evaluation. *Journal of Aquaculture Research & Development* 5: 1–8.

17 Fimbres-Olivarria, D., Carvajal-Millan, E., Lopez-Elias, J.A. et al. (2018). Chemical characterization and antioxidant activity of sulfated polysaccharides from Navicula sp. *Food Hydrocolloids* 75: 229–236.

18 Staats, N., Winder, B.D., Stal, L., and Mur, L. (1999). Isolation and characterization of extracellular polysaccharides from the epipelic diatoms Cylindrotheca closterium and Navicula salinarum. *European Journal of Phycology* 34 (2): 161–169.

19 Balti, R., Le Balc'h, R., Brodu, N. et al. (2018). Concentration and purification of Porphyridium cruentum exopolysaccharides by membrane filtration at various cross-flow velocities. *Process Biochemistry* 74: 175–184.

20 Medina-Cabrera, E.V., Rühmann, B., Schmid, J., and Sieber, V. (2020). Characterization and comparison of Porphyridium sordidum and Porphyridium purpureum concerning growth characteristics and polysaccharide production. *Algal Research* 49: 101931.

21 El-Naggar, N.E.-A., Hussein, M.H., Shaaban-Dessuuki, S.A., and Dalal, S.R. (2020). Production, extraction and characterization of Chlorella vulgaris soluble polysaccharides and their applications in AgNPs biosynthesis and biostimulation of plant growth. *Scientific Reports* 10 (1): 3011.

22 de Souza, M.P., Sanchez-Barrios, A., Rizzetti, T.M. et al. (2019). Concepts and trends for extraction and application of microalgae carbohydrates. In: *Microalgae – From Physiology to Application* (ed. M. Vítová), 1–13. IntechOpen.

23 Wijesekara, I., Pangestuti, R., and Kim, S.-K. (2011). Biological activities and potential health benefits of sulfated polysaccharides derived from marine algae. *Carbohydrate Polymers* 84 (1): 14–21.

24 Pangestuti, R., Getachew, A.T., Siahaan, E.A., and Chun, B.-S. (2019). Characterization of functional materials derived from tropical red seaweed Hypnea musciformis produced by subcritical water extraction systems. *Journal of Applied Phycology* 31 (4): 2517–2528.

25 Gereniu, C.R.N., Saravana, P.S., Getachew, A.T., and Chun, B.-S. (2017). Characteristics of functional materials recovered from Solomon Islands red seaweed (Kappaphycus alvarezii) using pressurized hot water extraction. *Journal of Applied Phycology* 29 (3): 1609–1621.

26 Jaime, L., Rodríguez-Meizoso, I., Cifuentes, A. et al. (2010). Pressurized liquids as an alternative process to antioxidant carotenoids' extraction from Haematococcus pluvialis microalgae. *LWT – Food Science and Technology* 43 (1): 105–112.

27 Kurd, F. and Samavati, V. (2015). Water soluble polysaccharides from Spirulina platensis: extraction and in vitro anticancer activity. *International Journal of Biological Macromolecules* 74: 498–506.

28 Zhao, G., Chen, X., Wang, L. et al. (2013). Ultrasound assisted extraction of carbohydrates from microalgae as feedstock for yeast fermentation. *Bioresource Technology* 128: 337–344.

29 Silva, A.d.S.e., de Magalhães, W.T., Moreira, L.M. et al. (2018). Microwave-assisted extraction of polysaccharides from Arthrospira (Spirulina) platensis using the concept of green chemistry. *Algal Research* 35: 178–184.

30 Rhein-Knudsen, N., Ale, M.T., and Meyer, A.S. (2015). Seaweed hydrocolloid production: an update on enzyme assisted extraction and modification technologies. *Marine Drugs* 13 (6): 3340–3359.

31 Dahman, Y., Syed, K., Begum, S. et al. (2019). Biofuels: their characteristics and analysis. In: *Biomass, Biopolymer-Based Materials, and Bioenergy* (eds. D. Verma, E. Fortunati, S. Jain and X. Zhang), 277–325. Woodhead Publishing.

32 Lakatos, G.E., Ranglová, K., Manoel, J.C. et al. (2019). Bioethanol production from microalgae polysaccharides. *Folia Microbiologia (Praha)* 64 (5): 627–644.

33 Yen, H.-W., Hu, I.-C., Chen, C.-Y., and Chang, J.-S. (2014). Design of photobioreactors for algal cultivation. In: *Biofuels from Algae* (eds. A. Pandey, D.-J. Lee, Y. Chisti and C.R. Soccol), 23–45. Amsterdam: Elsevier.

34 Lee, M.-C., Chen, Y.-C., and Peng, T.-C. (2012). Two-stage culture method for optimized polysaccharide production in Spirulina platensis. *Journal of the Science of Food and Agriculture* 92 (7): 1562–1569.

35 You, T. and Barnett, S.M. (2004). Effect of light quality on production of extracellular polysaccharides and growth rate of Porphyridium cruentum. *Biochemical Engineering Journal* 19 (3): 251–258.

36 Carvalho, A.P., Monteiro, C.M., and Malcata, F.X. (2009). Simultaneous effect of irradiance and temperature on biochemical composition of the microalga Pavlova lutheri. *Journal of Applied Phycology* 21 (5): 543–552.

37 de Castro Araújo, S. and Garcia, V.M.T. (2005). Growth and biochemical composition of the diatom Chaetoceros cf. wighamii brightwell under different temperature, salinity and carbon dioxide levels. I. Protein, carbohydrates and lipids. *Aquaculture* 246 (1): 405–412.

38 Kumar, D., Kvíderová, J., Kaštánek, P., and Lukavský, J. (2017). The green alga Dictyosphaerium chlorelloides biomass and polysaccharides production determined using cultivation in crossed gradients of temperature and light. *Engineering in Life Sciences* 17 (9): 1030–1038.

39 Ekelhof, A. and Melkonian, M. (2017). Enhanced extracellular polysaccharide production and growth by microalga Netrium digitus in a porous substrate bioreactor. *Algal Research* 28: 184–191.

40 Khalil, Z.I., Asker, M.M.S., El-Sayed, S., and Kobbia, I.A. (2010). Effect of pH on growth and biochemical responses of Dunaliella bardawil and Chlorella ellipsoidea. *World Journal of Microbiology and Biotechnology* 26 (7): 1225–1231.

41 Taraldsvik, M. and Myklestad, S. (2000). The effect of pH on growth rate, biochemical composition and extracellular carbohydrate production of the marine diatom Skeletonema costatum. *European Journal of Phycology* 35 (2): 189–194.

42 Mishra, A. and Jha, B. (2009). Isolation and characterization of extracellular polymeric substances from micro-algae Dunaliella salina under salt stress. *Bioresource Technology* 100 (13): 3382–3386.

43 Ishika, T., Bahri, P.A., Laird, D.W., and Moheimani, N.R. (2018). The effect of gradual increase in salinity on the biomass productivity and biochemical composition of several marine, halotolerant, and halophilic microalgae. *Journal of Applied Phycology* 30 (3): 1453–1464.

44 Xia, J.-R. and Gao, K.-S. (2005). Impacts of elevated CO_2 concentration on biochemical composition, carbonic anhydrase, and nitrate reductase activity of freshwater green algae. *Journal of Integrative Plant Biology* 47 (6): 668–675.

45 Yadav, G., Dash, S.K., and Sen, R. (2019). A biorefinery for valorization of industrial waste-water and flue gas by microalgae for waste mitigation, carbon-dioxide sequestration and algal biomass production. *Science of the Total Environment* 688: 129–135.

46 Khairy, H.M., Shaltout, N.A., El-Naggar, M.F., and El-Naggar, N.A. (2014). Impact of elevated CO_2 concentrations on the growth and ultrastructure of non-calcifying marine diatom (Chaetoceros gracilis F.Schütt). *Egyptian Journal of Aquatic Research* 40 (3): 243–250.

47 Thornton, D.C.O. (2009). Effect of low pH on carbohydrate production by a marine planktonic diatom (Chaetoceros muelleri). *Research Letters in Ecology* 2009: e105901.

48 Chen, C.-Y., Zhao, X.-Q., Yen, H.-W. et al. (2013). Microalgae-based carbohydrates for biofuel production. *Biochemical Engineering Journal* 78: 1–10.

49 Delattre, C., Pierre, G., Laroche, C., and Michaud, P. (2016). Production, extraction and characterization of microalgal and cyanobacterial exopolysaccharides. *Biotechnology Advances* 34 (7): 1159–1179.

50 de Farias Silva, C.E., Sforza, E., and Bertucco, A. (2018). Stability of carbohydrate production in continuous microalgal cultivation under nitrogen limitation: effect of irradiation regime and intensity on Tetradesmus obliquus. *Journal of Applied Phycology* 30 (1): 261–270.

51 Yao, C., Ai, J., Cao, X. et al. (2012). Enhancing starch production of a marine green microalga Tetraselmis subcordiformis through nutrient limitation. *Bioresource Technology* 118: 438–444.

52 Dragone, G., Fernandes, B.D., Abreu, A.P. et al. (2011). Nutrient limitation as a strategy for increasing starch accumulation in microalgae. *Applied Energy* 88 (10): 3331–3335.

53 Brányiková, I., Maršálková, B., Doucha, J. et al. (2011). Microalgae – novel highly efficient starch producers. *Biotechnology and Bioengineering* 108 (4): 766–776.

54 Yao, C.-H., Ai, J.-N., Cao, X.-P., and Xue, S. (2013). Characterization of cell growth and starch production in the marine green microalga Tetraselmis subcordiformis under extracellular phosphorus-deprived and sequentially phosphorus-replete conditions. *Applied Microbiology and Biotechnology* 97 (13): 6099–6110.

55 Dutot, M., Grassin, D., Salvator, H. et al. (2019). A marine-sourced fucoidan solution inhibits Toll-like-receptor-3-induced cytokine release by human bronchial epithelial cells. *International Journal of Biological Macromolecules* 130: 429–443.

56 Huleihel, M., Ishanu, V., and Tal, J. (2001). Antiviral effect of red microalgal polysaccharides on Herpes simplex and Varicella zoster viruses. *Journal of Applied Phycology* 13 (2): 127–134.

57 Abu-Galiyun, E., Huleihel, M., and Levy-Ontman, O. (2019). Antiviral bioactivity of renewable polysaccharides against Varicella Zoster. *Cell Cycle* 18 (24): 3540–3549.

58 Huang, J., Chen, B., and You, W. (2005). Studies on separation of extracellular polysaccharide from Porphyridium cruentum and its anti-HBV activity in vitro. *Chinese Journal of Marine Drugs* 24: 18–21.

59 Talyshinsky, M.M., Souprun, Y.Y., and Huleihel, M.M. (2002). Antiviral activity of red microalgal polysaccharides against retroviruses. *Cancer Cell International* 2: 1–7.

60 Raposo, M., De Morais, R., and Bernardo de Morais, A. (2013). Bioactivity and applications of sulphated polysaccharides from marine microalgae. *Marine Drugs* 11 (1): 233–252.

61 de Jesus Raposo, M., de Morais, A., and de Morais, R. (2015). Marine polysaccharides from algae with potential biomedical applications. *Marine Drugs* 13 (5): 2967–3028.

62 Prybylski, N., Toucheteau, C., El Alaoui, H. et al. (2020). Bioactive polysaccharides from microalgae. In: *Handbook of Microalgae-Based Processes and Products* (eds. E. Jacob-Lopes, M.M. Manzoni, M.I. Queiroz and L.Q. Zepka), 533–571. Elsevier.

63 Lee, J.-B., Hayashi, K., Hirata, M. et al. (2006). Antiviral sulfated polysaccharide from navicula directa, a diatom collected from deep-sea water in toyama bay. *Biological & Pharmaceutical Bulletin* 29 (10): 2135–2139.

64 Yim, J.H., Kim, S.J., Ahn, S.H. et al. (2004). Antiviral effects of sulfated exopolysaccharide from the marine microalga gyrodinium impudicum strain KG03. *Marine Biotechnology* 6 (1): 17–25.

65 Ahmadi, A., Zorofchian Moghadamtousi, S., Abubakar, S., and Zandi, K. (2015). Antiviral potential of algae polysaccharides isolated from marine sources: a review. *BioMed Research International* 2015: 1–10.

66 Kim, M., Yim, J.H., Kim, S.-Y. et al. (2012). in vitro inhibition of influenza A virus infection by marine microalga-derived sulfated polysaccharide p-KG03. *Antiviral Research* 93 (2): 253–259.

67 Ghosh, T., Chattopadhyay, K., Marschall, M. et al. (2009). Focus on antivirally active sulfated polysaccharides: from structure–activity analysis to clinical evaluation. *Glycobiology* 19 (1): 2–15.

68 Hasui, M., Matsuda, M., Okutani, K., and Shigeta, S. (1995). in vitro antiviral activities of sulfated polysaccharides from a marine microalga (Cochlodinium polykrikoides) against human immunodeficiency virus and other enveloped viruses. *International Journal of Biological Macromolecules* 17 (5): 293–297.

69 Chen, X., Han, W., Wang, G., and Zhao, X. (2020). Application prospect of polysaccharides in the development of anti-novel coronavirus drugs and vaccines. *International Journal of Biological Macromolecules* 164: 331–343.

70 Banskota, A.H., Sperker, S., Stefanova, R. et al. (2019). Antioxidant properties and lipid composition of selected microalgae. *Journal of Applied Phycology* 31 (1): 309–318.

71 Xu, S.-Y., Huang, X., and Cheong, K.-L. (2017). Recent advances in marine algae polysaccharides: isolation, structure, and activities. *Marine Drugs* 15 (12): 388.

72 Sen, S. and Chakraborty, R. (2011). The role of antioxidants in human health. In: *Oxidative Stress: Diagnostics, Prevention, and Therapy*, vol. 1083 (eds. S. Andreescu and M. Hepel), 1–37. Washington, DC: American Chemical Society.

73 Zhong, Q., Wei, B., Wang, S. et al. (2019). The antioxidant activity of polysaccharides derived from marine organisms: an overview. *Marine Drugs* 17 (12): 674.

74 Sousa, W.M., Silva, R.O., Bezerra, F.F. et al. (2016). Sulfated polysaccharide fraction from marine algae Solieria filiformis: structural characterization, gastroprotective and antioxidant effects. *Carbohydrate Polymers* 152: 140–148.

75 Sun, Y., Wang, H., Guo, G. et al. (2014). The isolation and antioxidant activity of polysaccharides from the marine microalgae Isochrysis galbana. *Carbohydrate Polymers* 113: 22–31.

76 Souza, B.W.S., Cerqueira, M.A., Bourbon, A.I. et al. (2012). Chemical characterization and antioxidant activity of sulfated polysaccharide from the red seaweed Gracilaria birdiae. *Food Hydrocolloids* 27 (2): 287–292.

77 Chen, B., You, W., Huang, J. et al. (2010). Isolation and antioxidant property of the extracellular polysaccharide from Rhodella reticulata. *World Journal of Microbiology and Biotechnology* 26 (5): 833–840.

78 Sun, L., Wang, L., Li, J., and Liu, H. (2014). Characterization and antioxidant activities of degraded polysaccharides from two marine Chrysophyta. *Food Chemistry* 160: 1–7.

79 Shen, S., Jia, S., Wu, Y. et al. (2018). Effect of culture conditions on the physicochemical properties and antioxidant activities of polysaccharides from Nostoc flagelliforme. *Carbohydrate Polymers* 198: 426–433.

80 Chen, X., Song, L., Wang, H. et al. (2019). Partial characterization, the immune modulation and anticancer activities of sulfated polysaccharides from filamentous microalgae Tribonema sp. *Molecules* 24 (2): 322.

81 Mishima, T., Murata, J., Toyoshima, M. et al. (1998). Inhibition of tumor invasion and metastasis by calciumspirulan (Ca-SP), a novel sulfated polysaccharide derived from a blue-green alga, Spirulina platensis. *Clinical & Experimental Metastasis* 16 (6): 541–550.

82 Pangestuti, R., Noerdjito, D.R., Siahaan, E.A. et al. (2020). Marine microalgae in food and health applications. In: *Encyclopedia of Marine Biotechnology*, vol. 1 (ed. S.K. Kim), 445–458. Wiley.

83 Umemura, K., Yanase, K., Suzuki, M. et al. (2003). Inhibition of DNA topoisomerases I and II, and growth inhibition of human cancer cell lines by a marine microalgal polysaccharide. *Biochemical Pharmacology* 66 (3): 481–487.

84 Dewi, I.C., Falaise, C., Hellio, C. et al. (2018). Anticancer, antiviral, antibacterial, and antifungal properties in microalgae. In: *Microalgae in Health and Disease Prevention* (eds. I. Levine and J. Fleurence), 235–261. Elsevier.

85 Kamble, P., Cheriyamundath, S., Lopus, M., and Sirisha, V.L. (2018). Chemical characteristics, antioxidant and anticancer potential of sulfated polysaccharides from Chlamydomonas reinhardtii. *Journal of Applied Phycology* 30 (3): 1641–1653.

86 Lordan, S., Ross, R.P., and Stanton, C. (2011). Marine bioactives as functional food ingredients: potential to reduce the incidence of chronic diseases. *Marine Drugs* 9 (6): 1056–1100.

87 Sun, L., Wang, L., Li, J., and Liu, H. (2013). Immunomodulation and antitumor activities of degraded polysaccharide from marine microalgae sarcinochrysis marina geitler. *International Journal Bioautomation* 17 (2): 107–116.

88 Sun, L., Wang, L., and Zhou, Y. (2012). Immunomodulation and antitumor activities of different-molecular-weight polysaccharides from Porphyridium cruentum. *Carbohydrate Polymers* 87 (2): 1206–1210.

89 Park, G.-T., Go, R.-E., Lee, H.-M. et al. (2017). Potential anti-proliferative and immunomodulatory effects of marine microalgal exopolysaccharide on various human cancer cells and lymphocytes in vitro. *Marine Biotechnology* 19 (2): 136–146.

90 Pugh, N., Ross, S.A., ElSohly, H.N. et al. (2001). Isolation of three high molecular weight polysaccharide preparations with potent immunostimulatory activity from Spirulina platensis, aphanizomenon flos-aquae and Chlorella pyrenoidosa. *Planta Medica* 67 (8): 737–742.

15

Microbial Production of Bioplastic

Current Status and Future Prospects

Karishma Seem

Division of Biochemistry, Indian Agricultural Research Institute, New Delhi, India

15.1 Introduction

Fossil hydrocarbon-based plastic has gained widespread use owing to its versatile qualities. However, due to their persistence in nature, plastics have harmful consequences for organisms and ENVIRONMENTAL degradation. Furthermore, synthetic polymers are mostly derived from petroleum, which is a precious nonrenewable resource. Currently, with the rising oil cost and rising renewable energy concerns, the production of plastic is a waste of limited resources. For the last 50 years, global demand and hence the production of plastic rising at an unprecedented pace has led to the accumulation of vast amounts of nondegradable waste materials across our planet. Despite the recycling efforts, extensive use of plastic has led to detrimental effects on our health as well as the environment. It has also made its way to the oceans and has threatened the survival of marine life. Additionally, due to the slow degradation of plastics, the probability of ingestion and accumulation of microplastics in tissues of animals of commercial interest have been reported. These microplastics embedded in the tissue of animals used for human consumption have raised concerns. The primary concern is the toxic and carcinogenic material used to make plastic.

Presently, many countries are becoming aware, thus implementing environmental policies to create a sustainable future, thereby opting for the alternative biodegradable polymer [1]. Bioplastics are natural products, a particular type of biopolymers synthesized by a range of microbes that form a class composed of polyhydroxyalkanoates (PHA), polylactides (PLA), aliphatic polyesters, polysaccharides, and their copolymers and/or blends [2]. Among them, PHA are the most versatile with similar properties as those of plastics. P(3HB-co-3HV) was the first PHA produced by the Monsanto Company under the trade name BIOPOL^{TM}. Currently, Kaneka Corporation produces P(3HB-co-3HH$_x$) under the trade name Aonilex. Other PHA producers are Metabolix (Mirel), Biomer, and Tianjin Green Bioscience Co., Ltd. (Tianjin, China) (Table 15.1).

Bioprospecting of Microorganism-Based Industrial Molecules, First Edition.
Edited by Sudhir P. Singh and Santosh Kumar Upadhyay.

Table 15.1 Major manufacturers of bioplastics worldwide.

Name of the company	Product	Trademark	Strain	Substrate
Biomatera, Canada	PHA resin	Biomater	Non-transgenic soil isolates	Renewable raw material
Biomer, Germany	PHB pellets	Biomer	*Alcaligenes latus*	Sugar
Kaneka Corporation, Japan	PHB-PHHx	Kaneka	*Ralstonia eutropha*	Plant oils
Metabolix, USA	P(3HB-co-4HB)	Mirel	*Ralstonia eutropha*	Glucose
Tianjin GreenBio Materials, China	P(3, 4HB) films	GreenBio	*Escherichia coli*	Glucose + 1,4-butanodiol pellets/foam pellets
Tianan Biologic Material Co. Ltd., Ningbo, China	PHB, PHBV	Enmat	*Ralstonia Entropha*	Glucose + propionate
Danimer Scientific	mcl-PHa	Nodax® PHA	Soil-borne microorganisms	Canola, Palm and Soy oil

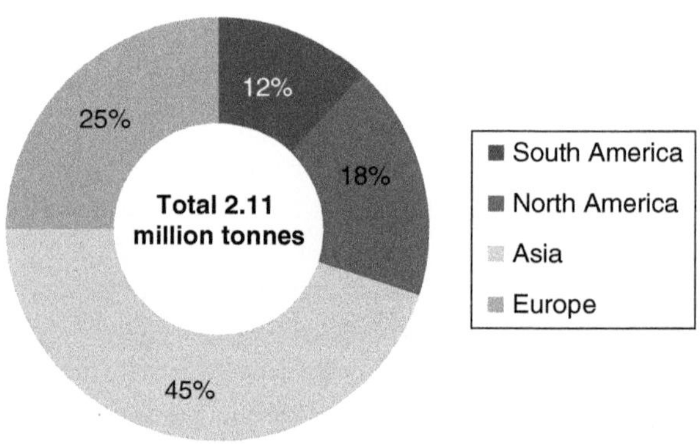

Figure 15.1 Global production capacities of bioplastics. *Source:* Modified from European Bioplastics. Nova institute, 2019.

Currently, bioplastics account for approximately 1% of 335 million tons of plastics produced annually. According to the Institute of Bioplastics and Biocomposites, with the entry of PHA polymer family in the market on a commercial scale, the production capacity of the PHA will increase four times over the next five years [3]. The global generation of PHA from industrial manufacturers reached 2.11 million tons in 2019 (European Bioplastics). As far as global production is concerned, Asia is leading the market in PHA production followed by Europe [4, 5] (see Figure 15.1).

Despite attracting attention as biocompatible and biodegradable, having a minor impact on the environment [6, 7], there are certain obstacles due to which PHA is not able to keep pace with conventional plastics preventing their massive market expansion. Principally, the high production cost of the substrate, use of the aseptic culture of bacteria, and maintenance of sterilization conditions of the medium as well as equipment, and finally, the use of toxic and expensive organic solvents for recovery of PHA from bacterial cells are some of the drawbacks.

This chapter gives an overview of the essential characteristics and biosynthesis of PHA. Various strategies for the utilization of the inexpensive substrate in order to lower the cost and open the route for the biotechnological production of PHA are highlighted. Major challenges in the commercial growth of PHAs and perspectives are also emphasized.

15.2 General Structure of PHA

PHAs are linear polyesters containing monomers of R-hydroxyalkanoic acid (HA) as basic structural units [8]. In this polymer, ester bond is formed between the carboxyl and the hydroxyl group of neighboring monomer. The residual group can consist of different carbon chain lengths. The alkyl side chain could vary; it could be saturated, unsaturated, straight, or branched containing different side groups (see Figure 15.2) [9].

The type of PHA depends on both the structure and length of the main chain, but also on the side chains. PHA can be categorized on the basis of the number of carbon atoms in the monomer as short-chain length [sclPHA] with monomer units varying from C_3 to C_5, medium chain length [mclPHAs] with monomer units varying from C_5 to C_{14}, and long-chain length [lclPHAs] wherein the monomer contains 15 or more carbon [10, 11]. The sclPHA has properties similar to that of conventional plastic; whereas, mclPHAs exhibit resemblance to elastomers. Polyhydroxybutyrate (PHB) is the most well-known example of sclPHA and is of utmost importance as it has excellent thermoplastic properties resembling that of polypropylene polymers [12]. Poly(3-hydroxybutyrate-co-3-hydroxyvalerate), poly (3-hydroxybutyrate-co-3- hydroxyhexanoate), poly (3-hydroxybutyrate-co-3-hydroxypropionate), and poly (3-hydroxybutyrate-co-4-hydroxybutyrate) are four important heteropolymers produced by the polymerization of PHA with other types of monomer.

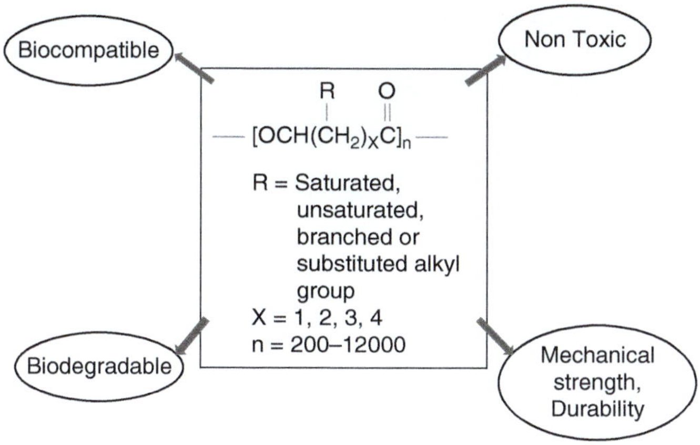

Figure 15.2 General structure and features of PHA. *Source:* Based on Lu, J., Tappel, R.C., and Nomura, C.T. (2009). Mini-review: Biosynthesis of poly(hydroxyalkanoates). *Polymer Review* 49: 226–248.

PHA was first reported in *Bacillus megaterium,* and thereafter a variety of bacterial genera were explored as PHA producers [13]. More than 150 different monomers have been reported to be incorporated into PHA monomer chains, yielding varying characteristic fermentation conditions and distinctive PHA [14].

15.3 Physical Properties

P(3HB) is a highly crystalline thermoplastic due to its stereoregularity. It is optically active and possesses piezoelectricity. However, its widespread application is limited due to its crystalline nature, stiffness, and relatively high melting point (Tm = 175) [15]. The tensile strength of P(3HB) is almost the same as that of petroleum-based polymers, such as polypropylene. However, the elongation to break (5%) is drastically lower than that of polypropylene (400%), which makes them brittle (Table 15.2). The brittleness of PHA can be altered by copolymerization of hydroxybutyrate (HB) and hydroxyl valerate [16] and blending with tough polymers [17]. With the increasing 3HV content, the crystallization rate generally decreases, for instance, 3-hydroxyvalerate PHBV [18].

Glass transition temperature (Tg) of P3HB is 15 whereas that of PHBV is –1. Melting temperature (Tm) of P3HB is 175 while it gets lowered to 145 in case of PHBV. The poor thermal stability of PHB renders it susceptible to thermal degradation.

15.4 Biodegradability of PHA

Environmental degradability of PHA materials is one of its unique characteristics. PHA has been found to be able to degrade naturally by plethora of microbes inhabiting soil [19], seawater [20, 21], and lake water [21]. PHAs are degraded in the environment by the action of microorganisms by using PHA hydrolases, and PHA depolymerizes [22, 23] into water-soluble products innocuous to the environment (see Figure 15.3). There are numerous factors that govern the biodegradability of PHA such as temperature, pH, humidity, microbial activity, composition, substrate specificity, and surface area of the PHA. UV light has been proved to accelerate the degradation process of PHA [24].

Table 15.2 Comparison of properties of PHB polymers with polypropylene.

Physical properties	P(3HB)	PHBV	Polypropylene
Melting temperature (°C)	175	145	176
Glass transition temperature (°C)	75	−1	−50
Crystalline (%)	50–80	56	50
Extension at work (%)	5	50	400
Tensile strength (MPa)	40	35	345
Young's Modulus (GPa)	2	0.7–2.9	1.7

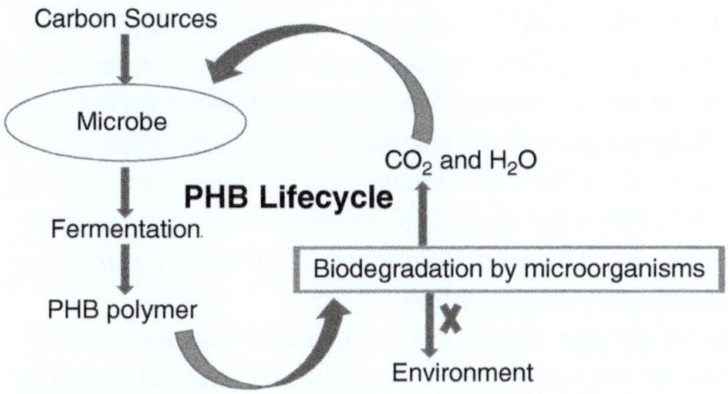

Figure 15.3 Life cycle of PHB.

PHA degradation is performed in microorganisms in two distinct ways. Intracellular degradation occurs under carbon limitation conditions in the cell. Accumulated PHA granules in the cells are hydrolyzed by intracellular PHA (iPHA) depolymerasesas resulting in carbon and energy sources [25].

In extracellular degradation, PHA secreted by PHA-accumulating microbes is utilized as a carbon source. scl-PHA hydrolyzing ability has been demonstrated in many microorganisms. mcl-PHA are difficult to disintegrate by the microorganisms. P(3HB) degradation is extensively studied in many well-characterized microorganisms. A wide variety of PHA degrading microbes have been reported, such as belonging to genera *Pseudomonas* [26], *Comamonas* [27], *Cupriavidus* [28], and *Alcaligenes* [29].

15.5 Biosynthesis of PHA

PHAs are synthesized by a large number of gram-positive and gram-negative bacteria [30]. In spite of a large number of bacteria capable of producing PHA, only a few of them, such as *Cupriavidus necator*, *Azohydromonas lata*, *Pseudomonas* sp., and recombinant *Escherichia coli,* could achieve high yield and productivity at the industrial level.

PHAs are synthesized on exhaustion of essential nutrients such as nitrogen, phosphorus, oxygen, or magnesium in the external medium or under the internal limitation of altering presence and absence of carbon sources [4]. Under these conditions, the substrate is accumulated as intracellular reserve material for carbon and energy [31]. Depending on the type of microorganism used, the PHA production can reach up to 90% of the dry weight of the cell [32]. PHA is known to enhance the robustness of microbes against adverse environmental stress such as high or low temperature, oxidative and osmotic pressure, and ultraviolet irradiation and maintaining cell integrity [33].

The PHA biosynthesis pathway varies considerably among the microbes. So far, 14 pathways have been reported from microbes [34]. The biosynthesis of PHA involves a simple pathway. Microbial biosynthesis of PHA begins with the binding of PHA synthases to the periplasmic membrane of the cell. The first characterized and also the best representative pathway is that of poly(3-hydroxybutyrate) (PHB) synthesis [35, 36]. In general, PHA is synthesized in a three-step reaction beginning with the condensation of 2 molecules of acetyl co A to acetoacetyl co-A. This reaction is catalyzed by the enzyme beta ketothiolase coded by the phaA gene. Acetoacetyl co A is subsequently reduced in the presence of NADPH oxidoreductase (pha B) to 3 hydroxy butyryl. The 3 hydroxy butyryl serves as a precursor monomer to polymerize PHB catalyzed by the enzyme PHB synthase (phaC) [37].

During normal bacterial growth, β-ketothiolase (first enzyme) is inhibited by unbound coenzyme-A (*CoA*) released through the functioning of the Krebs cycle. But when the entry of acetyl-CoA into the Krebs cycle is restricted (during non-carbon nutrient limitation), the surplus acetyl-CoA is channeled into PHB biosynthesis [38].

The composition of PHA depends on the type of microorganism and carbon source used. The ability of microorganisms to produce PHA of varying composition mainly relies on the substrate specificity of PHA synthases. The *PhaC, PhaA, and PhaB* genes are localized on the chromosomes of microorganisms and encode the essential proteins for the production of PHA in the native producers. PhaA encodes a 3-ketothiolase, and phaB encodes an acetoacetyl coenzyme A (acetoacetyl-CoA) reductase.

PHA synthases play a crucial role in PHA biosynthesis and conclude the polymerization of monomer (R)-3-hydroxyacyl-CoA to PHA [39]. PHA synthases are categorized into four classes with respect to their substrate specificities, subunit composition, and structure. Class I, III, and IV polymerize short length (scl) monomers (C_3-C_5), whereas Class II synthases prefer medium chain length (mcl) monomer (C_6-C_{14}) as substrate. Class I (e.g., in *Ralstonia eutropha*) and II (e.g., in *Pseudomonas putida*) synthases are encoded by *Pha* C gene with a molecular mass between 61 kDa and 73 kDa. Class III (e.g., *Allochromatium vinosum*) and class IV (e.g., *Bacillus megaterium*) synthases are composed of two genes PhaC (40.3 kDa) and PhaE (20 or 40 kDa), PhaC (41.5 kDa), and PhaR (22 kDa), respectively [40–42].

15.6 Challenges of Scaling Up of PHA Production on an Industrial Scale

Expensive carbon source is one of the impending factors posing a major hurdle for commercial production and application of PHA, making a microbial synthesis of PHAs 5–10 times more expensive than the petroleum-derived polymers [43]. Currently, research has been carried out for developing low-cost carbon-rich feedstocks to enable economically efficient PHA production. Several industrial/agro-industrial effluents and by-products have been widely used for industrial-scale PHA production [44]. Beet molasses, maltose, corn syrup, cane molasses, malt extract, palm oil mill effluents, whey, or crude glycerol phase from the biodiesel industry have been used as substrates for polymer production. The use of industrial and agricultural by-products could require extensive purification, equaling or even surpassing the energy demand of cost-intensive agricultural feedstocks.

15.6.1 Renewable Sources as Feedstock for PHA Production

15.6.1.1 Food Processing and Agricultural Industries Discharge

An emerging trend in using industrial waste stream still awaits its implication in the industrial-scale PHA production process. The quest for the development of novel, inexpensive, and amply available material is still extensively researched. Utilization of waste streams not only helps the industry to overcome disposal problems but also reduces the production cost. For instance, sawdust hydrolysis by strong acids and enzymes release a considerable amount of fermentable sugars that are utilized by the Burkholderia strains (*Burkholderia cepacia and Burkholderia sacchari*) as a carbon source, and it also contains some growth-limiting components [45]. These growth-limiting substances are removed by the upstream processing strategy designed by the authors, wherein inexpensive lignite is used instead of more expensive charcoal [46].

Steamed soybean wastewater from miso brewing factories is used for the accumulation of PHA by a recombinant *Pseudomonas* strain. Hydrolyzed spray-dried soybean wastewater is used as a carbon and a nitrogen source. This is the first report on the production of poly(3-hydroxybutyrate) and poly(3- hydroxybutyrate-co-4-hydroxybutyrate) from waste [47]. A more recent report of utilization of even petrol-based plastic waste for PHA biosynthesis has been reported. Waste polyethylene pyrolyzed to non-oxygenated PE wax (N-PEW),was used as an unconventional, exotic co-substrate in the cultivation of eubacterial production strain *Cupriavidus necator H16* [30].

B. sacchari produces a blend of 3-hydroxybuytrate (3HB) and 4-hydroxybutyrate (4HB) monomers by utilizing sucrose from the Brazilian sugarcane industry under controlled bioreactor conditions. The strain could be used for major industrial-scale production of PHA copolyester. For economic attainability, integration of high-throughput continuous PHA production facility into existing production lines is required [47]. The integration of municipal wastewater as a potential feedstock for MMC PHA production has not been widely demonstrated. Currently, this method is extensively studied to make it efficient both in sustainability and economic terms. Different types of WWTP sludge were compared as a substrate in different operational conditions [48].

15.6.1.2 Glycerol

Currently, glycerol is gaining attention as a feedstock for PHA production. Glycerol is generated in large quantities as a by-product in the production of biodiesel by transesterification of methanol. Therefore, bacterial fermentation could be an economically viable route to utilize low-value waste glycerol and produce value-added bioproducts. Bacterial fermentation of low-value waste glycerol of PHA has been reported in recent years. *Cupriavidus necator* (*R. eutropha*) using raw glycerol produced 0.84 g/l PHA [49]. In a similar pilot-scale study, Zhu et al. (2010) obtained volumetric productivity of 7.4 g/l P(3HB) [50].

Several bacterial strains such as recombinant *E. coli* [51], *Pseudomonas corrugate,* and *P. oleovorans* [52] have been identified to produce PHA utilizing crude glycerol as a substrate. Co-accumulation of glycogen, 1,3-propanediol (1,3-PDO) along with VFAs is a limiting factor for the yield of PHA and needs to be addressed while using glycerol as a feedstock for sustainable PHA production. However, increasing the selective production of PHA by mixed microbial consortia while leaving 1,3-PDO in the supernatant could overcome the issue [53]. [54]. Recently, efficient PHA production from volatile fatty acids and 1,3-PDO, the main glycerol fermentation effluent, has also been reported [53].

15.6.1.3 Agro-Industrial Oily Wastes

The potential of waste plant oils primarily palm kernel oil, crude palm oil, palm oil, jatropha oil, sludge palm oil, soybean oil as potential substrate for producing PHA has been demonstrated successfully [55–58]. Palm and soybean oil make up a significant fraction of worldwide edible oil production amounting to 73.5 and 56.9 million metric tons, respectively. Other important oils include rapeseed oil and sunflower seed oil with the production of 7.9 and 19.5 million metric tons, respectively [59]. With a tremendous global increase in demand for major vegetable oils, there has been a rise in the widespread problem of disposal of waste generated by these oils [60–62]. *Pseudomonas* sp. is able to yield a myriad of mcl-PHAs with a content of about 43% with various plant oils [63, 64]. Palm oil has been demonstrated to be an excellent carbon source for high yield production of PHAs. Abundant waste edible oils are disposed as industrial waste and need to be managed properly. Waste cooking oil, another option for low-cost sustainable source of plant oil has been demonstrated successfully. *R. eutropha* could produce P(3HB-co-3HV) utilizing used waste rapeseed oil and propanol as a precursor [65, 66]. Both the studies supported the enhanced production of PHA from waste frying oils as they are augmented with carbohydrates and other nutrients. Recently, transformant of *Cupriavidus necator* harboring PHA synthase gene of *Aeromonas caviae* produces 85% of CDW of P(3HB-co-3HHx) copolymer and a biomass concentration of 22.3 g/l utilizing waste cooking oil [67]. *Burkholderia thailandensis* is able to produce 60% of the cell biomass containing P(3HB) and 2.2 g/l of rhamnolipids on used cooking oil [68]. The wild type and engineered strain of *C. necator* produced P(3HB) and P(3HB-*co*-3HHx), respectively, by utilizing waste cooking palm oil (WCO). In extrapolation of the study, fed-batch fermentation strategy was employed to achieve high biomass productivity of 1.9 g/l/h and a high PHA yield of 1.1 g/l/h [58].

15.6.2 Cyanobacteria

Engineering cyanobacteria into commercially important PHA cyanofactories has gained attention among researchers. Cyanobacteria are photosynthetic bacteria that utilize sunlight along with recycling CO_2 for PHB production and, therefore, addressing the development of a sustainable and viable processing environment by counterbalancing the greenhouse gas emissions from industries [69].

An increased yield of 58% was obtained with the thermophilic *Synechococcus* sp. MA19 [70]. *Nostoc muscorum* and *Spirulina platensis* produced PHA when exposed to phosphorus-limiting conditions, wherein PHA production in *N. muscorum* increased to 22.7% of dry weight [71]. Small unicellular organisms, like *Synechocystis* sp., the best-studied cyanobacterium, successfully produced 12.5% PHB dry cell weight in a 2001 horizontal tubular photobioreactor [72]. Generally, nitrogen and phosphorus depletion are the most important factors for increasing the PHB content.

Genetic engineering strategies are extensively being focused on the design of robust cyanobacteria to enhance the PHA production. However, cyanobacterial species like *Synechocystis* sp are preferred over filamentous organisms due to their resistance to shear stress of the pump, thus combating the issues of strain selection [73].

Cyanobacteria-based PHA production faces many research gaps and questions. The non-existence of an economical mass cultivation strategy has hindered utilization of cyanobacteria for commercial PHA production. Notwithstanding, using organic carbon sources will complicate and could easily lead to contaminations and culture crashes.

Exploring new microorganisms for PHA production has led to reports of freshwater purple non-sulfur bacteria strains such as *Rhodospirillum rubrum* [74], *Rhodobacter sphaeroides* [75], and *Rhodopseudomonas palustris* [76]. *Rhodospirillum rubrum* is able to use CO from an inexpensive feedstock, syngas, a fuel gas mixture, consisting mainly of hydrogen and carbon monoxide as carbon and energy source for growth and accumulates PHA (PHB) under anaerobic conditions [77]. Syngas is a waste product from petroleum refining. Further, switchgrass was used as a feedstock for syngas fermentation by *R. rubrum*. The market value for producing syngas-based PHA was determined to be $1.65/kg and biohydrogen as a side-product was commercialized at a value of $2.00/kg [78].

Marine purple non-sulfur bacterium *Rhodovulum sulfidophilum* [79] has been reported to contain active PHA synthases and accumulate PHA up to 50 wt% of its dry weight.

15.6.3 Bacteria from Extreme Niches

Microbes are ubiquitous on Earth, and their diversity and abundance are determined by the diversity of the biogeographical habitat they occupy and the intensity of ecological interaction. It has been claimed that only a tiny fraction of the enormous natural microbial diversity has been satisfactorily characterized and cultured in labs. Their diversity enables them to tolerate extreme environmental conditions such as limited water availability, high salt concentration, extreme hot and cold temperature, and low oxygen levels. These unusual niches are considered extreme environments. Recently, there has been a quest to explore new avenues in PHA bioprospection for finding better PHA producers and spatiotemporal distribution.

15.6.3.1 Halophilic Bacteria

Halophilic microorganisms are natural inhabitants of hypersaline environments representing extreme environments such as salt lakes, salt pans, salt marshes, and saline soils. Halophiles have representatives from Archae, Bacteria, and Eukarya domains [80]. Based on the ability to tolerate high salt concentration, they can be categorized as mild, moderate, and extreme halophiles. Mild halophiles prefer NaCl concentration up to 1–6% (w/v), moderate can survive at salt concentration up to 7–15% (w/v), and lastly extreme halophiles tolerate more than 15% (w/v) salt [81]. Halophiles have evolved strategies and effective mechanisms to confer salt tolerance. One such mechanism that halophiles adopt to survive is by accumulating a high concentration of K^+ salts in equivalent concentration to extracellular NaCl in the environment [82]. The second strategy is the accumulation of osmoprotectants and compatible solutes that act as a stabilizer of biomolecules or stress absorbers [83].

Well-adapted and well-distributed extremely halophilic microorganisms such as archaeal *Halobacterium* sp., Cyanobacteria, and blue-green algae have been the prime focus. Halomonadaceae family consisting of a genus *Halomonas* represents the majority of the species of halophilic bacteria known to accumulate short-chain-length PHA. The *Halomonas* sp. is the most extensively studied genus for bacteria PHA production. They have certain advantages as they are easy to grow and maintain in the lab. Also, the necessity of aseptic conditions is reduced to a minimum. Properties of halophilic bacteria, such as the ability to produce a large amount of PHA and growth at high pH, metabolic versatility, and genetic stability, make them a potential candidate for PHA production. Further, the fragility of halobacteria helps in the simple and rapid recovery of PHA by exposing them to low salt concentration.

Extremely halophilic archaebacteria, *Haloferax mediterranei,* has attracted strong interest in the economical production of poly (3-hydroxybutyrate-co-3 hydroxyvalerate) (PHBV) and poly(3-hydroxybutyrate-*co*-3-hydroxyvalerate-*co*-4-hydroxybutyrate) (PHBV4HB). Moreover, using a simpler substrate such as starch as a carbon source, it has been found to produce 48.6 wt% PHB [84]. *Halobacterium volcanii* is also a promising candidate reported to exhibit PHA accumulating ability from sugarcane bagasse substrate, [85]. Halophilic microorganisms largely belonging to family Halobacteriacea such as *Halococcus, Halorubrum, Natronobacterium, Natronococcus, Halopiger, Haloarcula, Halobiforma, Haloferax*, and *Halobacterium* are capable of synthesizing and accumulating PHA [86].

15.6.3.2 Thermophiles for PHA

Thermophiles flourish at a wide range of temperatures ranging from 41 to 122 °C. They generally belong to the domain archaea. Thermophiles can survive in adverse conditions such as geothermal and volcanic environments, hot springs, and hydrothermal vents in the ocean. Their cellular structures, including protein, are adapted for heat, and enzymes also function at elevated temperatures.

PHA biosynthesis by well-known thermophiles such as *Thermus thermophilus, Chelatococcus* sp., *Aneurinibacillus* sp., *Geobacillus* sp., *Pseudomonas* sp., and *Cladimonas* sp. have been reported to accumulate PHAs in the presence of commercial carbon substrates like glucose [87–90]. *Geobacillus* sp and *Aneurinibacillus* sp were also reported to be efficient PHA producers [89, 91].

15.6.3.3 Psycrophiles for PHA

Microorganisms capable of inhabiting low temperatures are called psycrophiles or cold-adapted organisms. Psycrophiles have developed highly sophisticated adaptations to surmount the constraints inherent to life at low temperatures. Many psycrophiles have been reported to accumulate a high content of PHA. Psycrotrophic bacteria *Pseudomonas extremaustralis* and *Pseudomonas* sp. PAMC 28620 isolated from the pond in the Antarctica and Arctic environment, respectively, were found to produce mclPHA copolymer [92, 93] Further discovery of *Janthinobacterium* and *Pseudomonas* from Antarctic freshwater proved to be potential candidates for PHA production [94].

15.7 Co-synthesis of PHA with Value-Added Products

Regardless of the sincere efforts to alleviate the cost of PHA using inexpensive substrates, its large-scale production is still uneconomical. With the advent of "white technology," CRISPR has revolutionized engineering of microorganisms capable of synthesizing multiple products with economic feasibility. Coproduction would be advantageous, the primary being an optimized biosynthesis of intracellular PHA clubbed together with the extracellular product.

Pseudomonas species was shown to produce PHA (36%, w/w) and rhamnolipids (0.4 g/l) in a single-step fermentation [95]. Since *Pseudomonas* is an opportunistic pathogen, there is a quest for identifying alternate strains. Interestingly, *Thermus thermophilus,* a nonpathogenic, thermophilic, and natural isolate, can produce 0.2 g/l of rhamnolipid and 34.8% (w/w) of mcl-PHA using simple C-sources [87]. Recently, *Burkholderia thailandensis* came out to be a promising strain and could accumulate 7.5 g/l PHA (60%, w/w) and 2.2 g/l rhamnolipids when subjected to cooking oil [68].

Concurrent production of PHA and alginate has been seen in *Pseudomonas mediterranea,* using glycerol. PHA production showed promising results and was reported maximum to be 0.52 g/l (17%, w/w) with concomitant production of 6.93 g/l of extracellular alginate [96]. Alginate is an important component in the food and pharmaceutical industries.

Since carotenoids are expensive as compared with other pigments, their cogeneration with PHA has proven to be successful for the pharma and food industries. *Rhodobacter sphaeroides* was cultivated on oil mill wastewater and reported to coproduce carotenoids and hydrogen (40 mg/l OMW) and PHB (60 mg/l OMW) [97]. Mixed cultures of *E. coli* and *Rhodobacter* by utilizing *Calophyllum inophyllum* oil cake were able to cogenerate biohydrogen and PHA under dark and photofermentation conditions [98]. Recently, *Paracoccus* sp. LL1, a methylotroph, was used to obtain a yield of 7.14 mg/ml of astaxanthin-rich carotenoids and 9.53 g/l PHA using glycerol as a substrate [99]. The downstream processing using methylotrophs is advantageous as it secretes carotenoids in the form of vesicles and is able to utilize a wide range of substrates. However, further optimization of the coproduction parameters and inexpensive carbon sources is required to enhance the yield of valuable products at the industrial level.

15.8 Blends of PHA

Polymer blends emerge as one of the important research areas playing an important role in our lives, with potential applications. Blending is used to create a new polymeric material with different physical properties by physically combining at least two homopolymers. The physical properties of the blend are very much influenced by the behavior of polymer blends that need to be taken into account and the conditions applied. The possibility of blending PHA with biopolymers offers more scope to expand their range of applications. Blending tailors various characteristics such as mechanical properties, morphology, biodegradability, thermal and crystallization behavior to improve the impact strength significantly.

Various PHA blends such as P(HB-HV), PHA-PLA, and P(HB-HO) demonstrating frequently improved ductility and toughness have been developed [100, 101]. PHA blends with natural biodegradable polysaccharides such as cellulose and starch derivatives have attracted attention primarily owing to their sustainability and compatibility with PHB (Table 15.3). The blending of starch substantially reduces the high cost; moreover, it is abundantly available [102]. P(3HB)/starch blend was prepared using a melt compounding method at a ratio of 70 : 30 wt% [103].

P(3HB) with ethyl cellulose (EtC) blends were investigated for the miscibility, crystallization, and melting behavior and phase morphology [104].

Table 15.3 Blends of PHB and their uses.

PHA blends	Applications	References
Chitosan/PHB blend	Food packaging	[102]
PHB/PHBHHx	Biomedical applications	[107]
PHB/starch	Pharmaceutical tablet coating and as carriers for poorly soluble drugs, food packaging	[108, 109]
PHB/PCL	Bone tissue engineering	[110]
PHBV/PLA	Food packaging	[111]
Starch/PHBV	Biofilm carriers for denitrification, food packaging	[112, 113]
PHO/PHB	Biomedical applications	[131]

Thermal stability and compatibility of PLA as compared with P(3HB-*co*-3HV) copolymer has made PLA/PHA copolymer blends an attractive option. In spite of biodegradability of PLA and P(3HB) blend, their commercial application is limited due to brittleness and formation of very large spherulites [105]. Blend microspheres of PHB with cellulose acetate phthalate (CAP) have been prepared in varying compositions to investigate the colon delivery of 5-fluorouracil, an anticancer drug [106].

Blending offers fast and economic performance at a reduced price compared with previously used bottom-up and top-down approaches.

15.9 Applications of PHA

PHA has been widely used for various purposes, including packaging, printing, textile, agriculture, food, medical and pharmaceutical industries, and fine chemicals. PHA has also substituted polypropylene mulch films in agriculture. PHA with antimicrobial activity are used in packaging and storage of food [107, 108], containers, shampoo bottles, disposable items such as razors, pens, bullets, sanitary towels, food additives, cardboards, and paper [109–112]. Their piezoelectric property is harnessed in manufacturing products like keyboard sensors, accelerometer, shock wave sensors, gas lighters, acoustics: microphone, ultrasonic detectors, oscillators, headphones, loudspeaker for ultrasonic therapy and atomization [113].

For agricultural purposes, plants in nurseries are grown in PHA film bags for holding seedlings. PHA has also substituted polypropylene mulch films, herbicides, insecticides, and bacterial inoculants [114–116].

15.9.1 Biomedical Applications

PHAs are receiving more attention in various applications in the biomedical industry. The pioneering in vivo studies were performed by the subcutaneous implantation of poly(3-hydroxyoctanoate-co-3-hydroxyhexanoate), P(3HO-co-3HX) in the mice [117]. Members of the PHA family including PHB, PHBV, P3HB4HB, P4HB, P3HO (poly-(R)-3-hydroxyoctanoate), and PHBHHx have been developed over the years for various biomedical applications [114]. TephaFLEX, a US-based company, absorbable sutures were the first FDA-approved PHA products used in clinical applications. Tepha specializes in manufacturing pericardial patches, artery augments, cardiovascular stents, vascular grafts, heart valves, implants and tablets, sutures, dressings, absorbable microspheres, constructs for local delivery of drugs, dusting powders and prodrugs; it markets P4HB for medical application under the name of PHA4400.55 [118].

The PHA-derived implants/devices do not evoke any in vivo acute immune response at the site of implantation inside the host and are thus considered as biologically safe materials [114]. Researchers have established that PHB- and PHO-based microsphere and tubes were nonimmunogenic in mice models.

15.9.2 Soft Tissue Implants

Highly porous, biodegradable scaffolds with specific cell types are utilized to promote ex-vivo growth in soft tissue engineering. The skin, heart valves, esophagus patches, vascular patches, liver, and nerve tissue guiding are the classic examples of soft tissue engineering.

15.9.3 Esophagus, Pericardial Patches

Extensive esophageal damage is a serious condition associated with a higher incidence of systemic complications, such as renal failure, liver dysfunction, disseminated intravascular coagulation, and hemolysis. The investigation has been carried out on the application of P(3HB-co-3HHx) as a possible material for the construction of artificial esophagus. P(3HB-co3HHx)-based artificial esophagus was implanted in dogs. The study revealed P(3HB-co3HHx) copolymer stimulated the regeneration of the removed esophagus in the dog. The copolymer has higher strength and elasticity; however, the biodegradation of the implanted polymer construct was almost undetected [119].

15.9.4 Heart Valve Tissue Engineering

Because of its desirable mechanical properties, porosity, myocardial stiffness, and surface roughness as compared with innate heart valve tissues, P4HB has been attempted for the fabrication of functional living heart valves. The scl and mcl-PHAs-based scaffold were found successful for the fabrication of heart valves in the past [120]. In further studies, Sodian et al. studied the fabrication of a trileaflet heart valve scaffold. They harvested vascular cells from a carotid arteria, and a jugular vein was seeded in a trileaflet heart valve model made of elastomeric P(3HO). This artificial trileaflet scaffold was reimplanted into a test model. A healthy spectrum of data for the animal after the surgical procedure of reimplantation after 120 days was reported. New fabrication techniques using composite scaffolds of PGA and poly(3-hydroxyoctanoate) [121], P(4HB), and P(3HO) [121] established them as a candidate material for application in heart valve tissue engineering [122]. Silk fibroin-modified P-3HB-co-3HH porus scaffolds were supported by the proliferation of the smooth muscle cells and improved the biocompatibility of the synthesized cells [123].

Recently, MCL Poly(3-hydroxyoctanoate) homopolymer, P(3HO)-based cardiac patches have been seen to meet up with all requirements, exhibiting excellent adhesiveness and proliferation [124].

15.9.5 Nerve Regeneration

Peripheral nervous system injuries and neurodegenerative diseases can lead to a permanent disability of nerve cells and neural tissues to regenerate on their own. Axonal regeneration is considered less than 2 cm in humans. In injuries resulting in longer nerve gaps, nerve grafts are used. However, commonly attempted nerve autografting/allografting techniques for peripheral nerve repair has several limitations like donor site morbidity and loss of function, lack of appropriate donor tissue, uncertain histocompatibility, a limited amount of tissue, and antigenicity [125]. Therefore, nerve conduits comprising of biodegradable natural polyester are a good choice for the regeneration of peripheral nerves.

PHAs are preferred for in vivo application for peripheral nerve regeneration due to their prominent properties such as controllable surface erosion, variability in material properties, lower acidity of degradation products, and longer stability compared with their synthetic counterparts are all of special interest in this field. Ideal nerve conduit needs to have properties like an extremely low inflammatory and immunogenic response, low cytotoxicity, biodegradability, and porous for nutrient exchange [126]. More elastic nerve conduits were developed, employing blends consisting of PDLLA and P(3HB-co-3HHx) for repairing in vivo nerve damage [127]. P(3HB-co-3HHx) conduits produced by microbial fermentation have been shown to repair in vivo nerve gaps of 10 mm gaps in rat sciatic nerves [128]. The identification of PHB, PHBV, P3HBHB, and PHBHHx as heterologous nerve xenografts has brought hope for neural regeneration. PHBV-P(L)-PLGA) with (PHBV-PLGA) has been explored as a nerve conduit for nerve regeneration [129]. Neural stem cells grown on P(3HB-co3HHx)-based nanofiber scaffolds (matrices) were reported to be useful for repairing injury to the central nervous system [130]. Although [P(3HB)] has a low level of inflammatory infiltration, regeneration obtained was not comparable owing to its brittleness and poor mechanical properties. Various modifications of PHA-based scaffolds have been tested in vivo. Enhanced cell proliferation was observed in the composite 25:75 PHO/PHB [poly(3-hydroxyoctanoate)/PHB] scaffold [131]. Nanofibrous membranes of -P(3HB-co4HB), prepared by electrospinning, demonstrated myelinated nerve cell regeneration and high expression of nerve growth factor [132].

15.9.6 Drug Delivery System

PHA can be developed into microcapsules, gels, polymeric micelles, microspheres, films, polymer-linked drugs, porous matrices, and nanoparticles of PHA to deliver the drug in a controlled manner [133]. Polymeric material chosen for drug delivery should not damage the molecular structure of the drug. Ansari and Annuar [2018] conducted a study on MCL-PHA copolymer, i.e. [P(3HO-*co*-3HH$_X$)] as a transporter of glycerol 1, 3-diglycerolate diacrylate (GDGD) depicting their hydrophilicity enabling efficient drug delivery [134]. Furthermore, porphyrin 5,10,15,20-tetrakis(4-hydroxyphenyl)-21H,23H-porphine was encapsulated in nanoparticles formed from PHBV containing 12 and 50% of HV units for anticancer photodynamic therapy [135].

15.10 Conclusions and Perspectives

Biodegradable polymers, the poly(3-hydroxyalkanoates) (PHA) an important polymer family, have grabbed tremendous attention among the researchers in recent years. The material properties can be controlled by tailoring the chemical structure of the polymer. PHAs have many fascinating properties that make them suitable for various industrial and biomedical applications. PHAs have physiochemical properties resembling those of petrochemical polymers and show significant advantages over them. At present, their commercial opportunities are inhibited, mainly due to the stable synthesis of PHA from inexpensive unrelated carbon sources. Since the cost of production of PHAs is still quite high (4–6 USD/kg), current advances in bioprocess engineering to implement an optimized fermentation process, as well as identification of potential superior bacterial strains, are likely to lower the price of PHAs, close to that of a conventional polymer such as PLA and aliphatic polyesters. Isolation, identification, and application of extremophiles need to be pursued intensively for reducing the sterility precautions and thereby improving efficient and sustainable feasibility of the process. Adoption of carbon dioxide as a carbon source could be expected to minimize production costs as well as broaden the industrial application of PHA, for example, in the case of cyanobacterial PHA production.

In addition, the derivation of waste feedstock for the culture of the microorganisms accumulating PHAs is also prompted as a route to decrease the final cost and lead to their greater economic viability and sustainability.

Further progress in genetic engineering aims at knock out of the genes encoding for the enzymes involved in intracellular degradation of PHA, namely depolymerase. Precise genetic modification of *Cupriavidus necator*, by addition of *lac Z, lac I,* and *lac O* genes from *E. coli* had an astonishing effect on its growth on lactose-containing waste material such as cheese and whey [136].

In order to achieve profitable bioprocess, down streaming procedures during the production of PHB could construct a promising PHA producer, paving the way for the introduction of this bioplastic in the high-end markets. Most suitable solvents for enhanced recovery of PHA that are clearly environmentally safer and efficient are being investigated. Further, examination of novel biological lysis methods and cell disruption strategies intend to increase the extraction of high-quality PHA from microbial cells. Also, the solid waste obtained after recovery of PHA non-PHA cell mass (NPCM) poses a constrain in the recovery of PHA that has to be utilized in a sustainable manner. These products are expected to experience a higher demand once the advances mentioned above are in place.

References

1 Alsafadi, D. and Al-Mashaqbeh, O. (2017). A one-stage cultivation process for the production of poly-3-(hydroxybutyrate-co-hydroxyvalerate) from olive mill wastewater by *Haloferax mediterranei*. *New Biotechnology* 34: 47–53.

2 Lambert, S., Sinclair, C.J., and Boxall, A.B.A. (2014). Occurrence, degradation and effects of polymer-based materials in the environment. *Rev Environ Contamination and Toxicolology* 227: 1–53. https://doi.org/10.1007/978-3-319-01327-5_1.

3 Mozejko-Ciesielska, J. and Kiewisz, R. (2016). Bacterial polyhydroxyalkanoates: still fabulous? *Microbiology Research* 192: 271–282.

4 Geueke, B., Groh, K., and Muncke, J. (2018). Food packaging in the circular economy: overview of chemical safety aspects for commonly used materials. *Journal of Cleaner Prouction* 193: 491–505.

5 Imre, B., Garcia, L., Puglia, D., and Vilaplana, F. (2019). Reactive compatibilization of plant polysaccharides and biobased polymers: review on current strategies, expectations and reality. *Carbohydrate Polymers* 209: 20–37.

6 Jiang, X.R. and Chen, G.Q. (2016). Morphology engineering of bacteria for bio-production. *Biotechnology Advances* 34 (4): 435–440.

7 Elhadi, D., Lv, L., Jiang, X.R. et al. (2016). CRISPRi engineering *E. coli* for morphology diversification. *Metabolic Engineering* 38: 358–369.

8 Khanna, S. and Srivastava, A.K. (2005). Recent advances in microbial polyhydroxyalkanoates. *Process Biochemistry* 40: 607–619.

9 Lu, J., Tappel, R.C., and Nomura, C.T. (2009). Mini-review: biosynthesis of poly(hydroxyalkanoates). *Polymer Reviews* 49: 226–248.

10 Li, Z. and Loh, X.J. (2015). Water soluble polyhydroxyalkanoates: future materials for therapeutic applications. *Chemical Society Reviews* 44: 2865–2879.

11 Kai, D., Low, Z.W., Liow, S.S. et al. (2015). Development of lignin supramolecular hydrogels with mechanically responsive and self-healing properties. *ACS Sustainable Chemistry & Engineering* 3: 2160–2169.

12 Soam, A., Singh, A., Singh, R., and Shahi, S. (2012). Optimization of culture conditions for bio polymer producing *Bacillus mycoides* (wss2) bacteria from sewage. *Current Discovery* 1: 27–32.

13 Anderson, A.J. and Dawes, E.A. (1990). Occurrence, metabolism, metabolic role, and industrial uses of bacterial polyhydroxyalkanoates. *Microbiology Review* 54: 450–472.

14 Kim, D.Y., Kim, H.W., Chung, M.G., and Rhee, Y.H. (2007). Biosynthesis, modification, and biodegradation of bacterial medium-chain-length polyhydroxyalkanoates. *Journal of Microbiology* 45: 87–97.

15 Sudesh, K., Abe, H., and Doi, Y. (2000). Synthesis, structure and properties of polyhydroxyalkanoates: biological polyesters. *Progress in Polymer Science* 25 (10): 1503–1555.

16 Yu, L., Dean, K., and Li, L. (2006). Polymer blends and composites from renewable resources. *Progress in Polymer Science* 31 (6): 576–602.

17 Parulekar, Y. and Mohanty, A.K. (2007). Extruded biodegradable cast films from Polyhydroxyalkanoate and thermoplastic starch blends: fabrication and characterization. *Macromolecular Materials and Engineering* 292 (12): 1218–1228.

18 Liu, H., Xie, F., Yu, L. et al. (2009). Thermal processing of starch-based polymers. *Progress in Polymer Science* 34 (12): 1348–1368.

19 Mergaert, J., Webb, A., Anderson, C. et al. (1993). Microbial degradation of poly(3-hydroxybutyrate) and poly(3- hydroxybutyrate-co-3-hydroxyvalerate) in soils. *Appllied Environmental Microbiology* 59: 3233–3238.

20 Doi, Y., Kanesawa, Y., Tanahashi, N., and Kumagai, Y. (1992). Biodegradation of microbial polyesters in the marine environment. *Polymer Degradation and Stability* 36: 173–177.

21 Mergaert, J., Wouters, A., Anderson, C., and Swings, J. (1995). In situ biodegradation of poly(3-hydroxybutyrate) and poly(3-hydroxybutyrate-co-3-hydroxyvalerate) in natural waters. *Canadian Journal of Microbiology* 41 (1): 154–159.

22 Jendrossek, D. and Handrick, R. (2002). Microbial degradation of polyhydroxyalkanoates. *Annual Review of Microbiology* 56: 403–432.

23 Choi, G.G., Kim, H.W., and Rhee, Y.H. (2004). Enzymatic and non-enzymatic degradation of poly (3-hydroxybutyrate-co-3-hydroxyvalerate) copolyesters produced by Alcaligenes sp MT-16. *Journal of Microbiology* 42: 346–352.

24 Shangguan, Y.Y., Wang, Y.W., Wu, Q., and Chen, G.Q. (2006). The mechanical properties and in vitro biodegradation and biocompatibility of UV-treated poly(3-hydroxybutyrate-co-3-hydroxyhexanoate). *Biomaterials* 27: 2349–2357.

25 Luengo, J.M.,·Garcia, B., Sandoval, A. et al. (2003). Bioplastics from microorganisms. *Current Opinion in Microbiology* 6 (3): 251–260.

26 Lusty, C.J. and Doudoroff, M. (1966). Poly-Beta-Hydroxybutyrate depolymerases of *Pseudomonas lemoignei*. *Proceedings of National Academy of Sciences USA* 56: 960–965.

27 Kasuya, K.I., Doi, Y., and Yao, T. (1994). Enzymatic degradation of poly[(R)-3-hydroxybutyrate] by *Comamonas testosteroni* ATSU of soil bacterium. *Polymer Degradation and Stability* 45 (3): 379–386.

28 Jendrossek, D. (2002). Extracellular polyhydroxyalkanoate (PHA) depolymerases: the key enzymes of PHA degradation. Biopolymers Online, 41–77, Wiley-VCH, Weinheim, Germany.

29 Bachmann, B.M. and Seebach, D. (1999). Investigation of the enzymatic cleavage of diastereomeric oligo(3-hydroxybutanoates) containing two to eight HB units. A model for the stereoselectivity of PHB depolymerase from *Alcaligenes faecalis* T_1. *Macromolecules* 32 (6): 1777–1784.

30 Johnston, B., Jiang, G., Hill, D. et al. (2017). The molecular level characterization of biodegradable polymers originated from polyethylene using non-oxygenated polyethylene wax as a carbon source for polyhydroxyalkanoate production. *Bioengineering* 4: 73.

31 Somleva, M.N., Peoples, O.P., and Snell, K.D. (2013). PHA bioplastics, biochemicals, and energy from crops. *Plant Biotechnology Journal* 11: 233–252.

32 Steinbuchel, A. and Lutke-Eversloh, T. (2003). Metabolic engineering and pathway construction for biotechnological production of relevant polyhydroxyalkanoates in microorganisms. *Biochemical Engineering Journal* 16: 81–96.

33 Sedlacek, P., Slaninova, E., Koller, M. et al. (2019). PHA granules help bacterial cells to preserve cell integrity when exposed to sudden osmotic imbalances. *New Biotechnology* 49: 129–136. https://doi.org/10.1016/j.nbt.2018.10.005.

34 Bhubalan, K., Lee, W.H., Loo, C.Y. et al. (2008). Controlled biosynthesis and characterization of poly(3-hydroxybutyrate-co-3-hydroxyvalerateco-3-hydroxyhexanoate) from mixtures of palm kernel oil and 3HV-precursors. *Polymer Degradation and Stability* 93 (1): 17–23.

35 Slater, S.C., Voige, W.H., and Dennis, D.E. (1988). Cloning and expression in *Escherichia coli* of the *Alcaligenes eutrophus* H16 poly-beta-hydroxybutyrate biosynthetic pathway. *Journal of Bacteriology* 170: 4431–4436, 25.

36 Schubert, P., Steinbuchel, A., and Schlegel, H.G. (1988). Cloning of the *Alcaligenes eutrophus* genes for synthesis of poly-beta-hydroxybutyric acid (PHB) and synthesis of PHB in *Escherichia coli*. *Journal of Bacteriology* 170: 5837–5847.

37 Tsuge, T., Yano, K., Imazu, S. et al. (2005). Biosynthesis of polyhydroxyalkanoate (PHA) copolymer from fructose using wild-type and laboratory-evolved PHA synthases. *Macromolecular Bioscience* 5: 112–117.

38 Ratledge, C. and Kristiansen, B. (2001). *Basic Biotechnology*, 2e. Cambridge: Cambridge University Press.

39 Zou, H., Shi, M., Zhang, T. et al. (2017). Natural and engineered polyhydroxyalkanoate (PHA) synthase: key enzyme in biopolyester production. *Applied Microbiology and Biotechnology* 101: 7417–7426.

40 Chen, G.Q., Hajnal, I., Wu, H. et al. (2015). Engineering biosynthesis mechanisms for diversifying polyhydroxyalkanoates. *Trends in Biotechnology* 33 (10): 565–574.

41 Chen, G.Q. and Hajnal, I. (2015). Trends The 'PHAome. *Biotechnology* 33 (10): 559–564.

42 Kumar, P., Ray, S., and Kalia, V.C. (2016). Production of co-polymers of polyhydroxyalkanoates by regulating the hydrolysis of biowastes. *Bioresource Technology* 200: 413–419.

43 Obruca, S., Benesova, P., Marsalek, L., and Marova, I. (2015). Use of lignocellulosic materials for PHA production. *Chemical and Biochemical Engineering Quaterly* 29 (2): 135–144.

44 Valentino, F., Morgan-Sagastume, F., Campanari, S. et al. (2017). Carbon recovery from wastewater through bioconversion into biodegradable polymers. *Biotechnology* 25: 9–23.

45 Kucera, D., Benesova, P., Ladicky, P. et al. (2017). Production of polyhydroxyalkanoates using hydrolyzates of spruce sawdust: comparison of hydrolyzates detoxification by application of overliming, active carbon, and lignite. *Bioengineering* 4: 53.

46 Hokamura, A., Yunoue, Y., Goto, S., and Matsusaki, H. (2017). Biosynthesis of polyhydroxyalkanoate from steamed soybean wastewater by a recombinant strain of *Pseudomonas* sp. *Bioengineering*: 61–63.

47 de Sousa Dias, M.M., Dias, M., Koller, M. et al. (2017). Fed-batch synthesis of poly(3-hydroxybutyrate) and poly(3-hydroxybutyrate-co-4-hydroxybutyrate) from sucrose and 4-hydroxybutyrate precursors by *Burkholderia sacchari* strain DSM 17165. *Bioengineering* 4: 36.

48 Pittmann, T. and Steinmetz, H. (2017). Polyhydroxyalkanoate production on waste water treatment plants: process scheme, operating conditions and potential analysis for german and european municipal waste water treatment plants. *Bioengineering (Basel)* 4 (2): 54. https://doi.org/10.3390/bioengineering4020054.

49 Cavalheiro, J.M., Raposo, R.S., de Almeida, M.C. et al. (2012). Effect of cultivation parameters on the production of poly(3-hydroxybutyrate-co-4-hydroxybutyrate) and poly(3-hydroxybutyrate-4-hydroxybutyrate-3-hydroxyvalerate) by *Cupriavidus necator* using waste glycerol. *Bioresource Technology* 111: 391–397.

50 Zhu, C., Nomura, C.T., Perrotta, J.A. et al. (2010). Production and characterization of poly-3-hydroxybutyrate from biodiesel-glycerol by *Burkholderia cepacia* ATCC 17759. *Biotechnology Progress* 26: 424–430.

51 Almeida, A.D., Nikel, P.I., Giordano, A.M., and Pettinari, M.J. (2007). Effects of granule-associated protein PhaP on glycerol-dependent growth and polymer production in poly(3-hydroxybutyrate)-producing *Escherichia coli*. *Applied and Environmental Microbiology* 73: 7912–7916. https://doi.org/10.1128/AEM.01900-07.

52 Papanikolaou, S., Fakas, S., Fick, M. et al. (2007). Biotechnological valorisation of raw glycerol discharged after bio-diesel (fatty acid methyl esters) manufacturing process: production of 1,3-propanediol, citric acid and single cell oil. *Biomass and Bioenergy* 32: 60–71. https://doi.org/10.1016/j.biombioe.2007.06.007.

53 Burniol-Figols, A., Varrone, C., Le, S.B. et al. (2018). Combined polyhydroxyalkanoates (PHA) and 1, 3-propanediol production from crude glycerol: selective conversion of volatile fatty acids into PHA by mixed microbial consortia. *Water Research* 136: 180–191.

54 Burniol-Figols, A., Varrone, C., Daugaard, A.E. et al. (2018). Polyhydroxyalkanoates (PHA) production from fermented crude glycerol: study on the conversion of 1, 3-propanediol to PHA in mixed microbial consortia. *Water Research* 128: 255–266.

55 Lee, W.H., Loo, C.Y., Nomura, C.T., and Sudesh, K. (2008). Biosynthesis of polyhydroxyalkanoate copolymers from mixtures of plant oils and 3-hydroxyvalerate precursors. *Bioresource Technology* 99 (15): 6844–6851.

56 Sudesh, K., Bhubalan, K., Chuah, J.A. et al. (2011). Synthesis of polyhydroxyalkanoate from palm oil and some new applications. *Applied Microbiolgy Biotechnology* 89 (5): 1373–1386.

57 Rodrigues, P.R., Nunes, J.M.N., Lordelo, L.N., and Druzian, J.I. (2019). Assessment of polyhydroxyalkanoate synthesis in submerged cultivation of *Cupriavidus necator* and *Burkholderia cepacia* strains using soybean as substrate. *Brazilian Journal of Chemical Engineering* 36: 73–83.

58 Thinagaran, L. and Sudesh, K. (2019). Evaluation of sludge palm oil as feedstock and development of efficient method for its utilization to produce polyhydroxyalkanoate. *Waste and Biomass Valorization* 10: 709–720. https://doi.org/10.1007/s12649-017-0078-8.

59 Statista (2019). Worldwide Production Major Vegetable Oils, 2012–2019. https://www.statista.com/statistics/263933/production-of-vegetable-oils-worldwide-since-2000/.

60 Lligadas, G., Ronda, J.C., Galia, M., and Cádiz, V. (2013). Renewable polymeric materials from vegetable oils: a perspective. *Materials Today* 16: 337–343. https://doi.org/10.1016/j.mattod.2013.08.016.

61 Samarth, N.B. and Mahanwar, P.A. (2015). Modified vegetable oil based additives as a future polymeric material. *Open Journal of Organic Polymer Material* 5: 1–22. https://doi.org/10.4236/ojopm.2015.51001.

62 Phang, K. and Lau, S.W. (2017). A survey on the usage of biomass wastes from palm oil mills on sustainable development of oil palm plantations in Sarawak, in IOP Conference Series: Materials Science and Engineering (Miri: IOP Publishing).

63 Song, J.H., Jeon, C.O., Choi, M.H. et al. (2000). Polyhydroxyalkanoate (PHA) production using waste vegetable oil by *Pseudomonas* sp. strain DR2. *Microbiology Biotechnology* 18 (8): 1408–1415.

64 Mozejko, J., Przybyłek, G., and Ciesielski, S. (2011). Waste rapeseed oil as a substrate for medium-chain-length polyhydroxyalkanoates production. *European Journal of Lipid Science and Technology* 113: 1550–1557. https://doi.org/10.1002/ejlt.201100148.

65 Obruca, S., Marova, I., Snajdar, O. et al. (2010). Production of poly(3-hydroxybutyrate-co3-hydroxyvalerate) by *Cupriavidus necator* from waste rapeseed oil using propanol as a precursor of 3-hydroxyvalerate. *Biotechnology Letters* 32: 1925–1932.

66 Verlinden, R.A.J., Hill, D.J., Kenward, M.A. et al. (2011). Production of polyhydroxyalkanoates from waste frying oil by *Cupriavidus necator*. *AMB Express* 1: 3–11.

67 Kamilah, H., Tsuge, T., Yang, T.A., and Sudesh, K. (2012). Waste cooking oil as substrate for biosynthesis of poly(3-hydroxybutyrate) and poly(3-hydroxybutyrate-co-3-hydroxyhexanoate): turning waste into a value-added product. *Malaysian Journal of Microbiology* 9: 51.

68 Kourmentza, C., Costa, J., Azevedo, Z. et al. (2018). *Burkholderia thailandensis* as a microbial cell factory for the bioconversion of used cooking oil to polyhydroxyalkanoates and rhamnolipids. *Bioresource Technology* 247: 829–837.

69 Carpine, R., Du, W., Olivieri, G. et al. (2017). Genetic engineering of *Synechocystis* sp. PCC6803 for poly-β-hydroxybutyrate overproduction. *Algal Research* 25: 117–127.

70 Nishioka, M., Nakai, K., Miyake, M. et al. (2001). Production of the poly-betahydroxyalkanoate by thermophilic cyanobacterium, *Synechococcus* sp. MA19, under phosphate-limited condition. *Biotechnology Letters* 23: 1095–1099.

71 Panda, B., Sharma, L., and Mallick, N. (2005). Poly-β-hydroxybutyrate accumulation in *Nostoc muscorum* and *Spirulina platensis* under phosphate limitation. *Journal of Plant Physiology* 162 (12): 1376–1379. https://doi.org/10.1016/j.jplph.2005.05.002.

72 Panda, B. and Mallick, N. (2007). Enhanced poly-β-hydroxybutyrate accumulation in a unicellular cyanobacterium, *Synechocystis* sp. PCC 6803. *Letters in Applied Microbiology*: 44194–44198. https://doi.org/10.1111/j.1472-765X.2006.02048.x.

73 Troschl, C., Meixner, K., and Drosg, B. (2017). Cyanobacterial PHA production – review of recent advances and a summary of three years' working experience running a pilot plant. *Bioengineering* 4: 26.

74 Brandl, H., Knee, E.J. Jr., Fuller, R.C. et al. (1989). Ability of the phototrophic bacterium *Rhodospirillum rubrum* to produce various poly (beta-hydroxyalkanoates): potential sources for biodegradable polyesters. *International Journal of Biological Macromolecules* 1 (1): 49–55.

75 Brandl, H., Lenz, R.W., Lloyd, R., and Fuller, R.C. (1991). The accumulation of poly(3-hydroxyalkanoates) in *Rhodobacter sphaeroides*. *Archieves of Microbiology* 155 (4): 337–340.

76 Carlozzi, P. and Sacchi, A. (2001). Biomass production and studies on Rhodopseudomonas palustris grown in an outdoor, temperature controlled, underwater tubular photobioreactor. *Journal of Biotechnology* 88 (3): 239–249.

77 Do, Y.S., Smeenk, J., Broer, K.M. et al. (2007). Growth of *Rhodospirillum rubrum* on synthesis gas: conversion of CO to H_2 and poly-β-hydroxyalkanoate. *Biotechnology and Bioengneering* 97: 279–286. https://doi.org/10.1002/bit.21226.

78 Choi, D., Chipman, D.C., Bents, S.C., and Brown, R.C. (2010). A techno-economic analysis of polyhydroxyalkanoate and hydrogen production from syngas fermentation of gasified biomass. *Applied Biochemistry and Biotechnology* 160 (4): 1032–1046. https://doi.org/10.1007/s12010-009-8560-9.

79 Chowdhury, W.Q., Idehara, K., Maeda, I. et al. (1996). Factors affecting polyhydroxybutyrate biosynthesis in the marine photosynthetic bacterium *Rhodopseudomonas* sp strain W-1S. *Applied Biochemistry and Biotechnology* 57 (58): 361–366. https://doi.org/10.1007/978-1-4612-0223-3_31.

80 Quillaguaman, J. (2008). Poly(3-hydroxybutyrate) production by *Halomonas boliviensis* in fed-batch culture. *Applied Microbiology and Biotechnology* 78: 227–232. https://doi.org/10.1007/s00253-007-1297-x.

81 Quillaguaman, J., Guzman, H., Van-Thuoc, D., and Hatti-Kaul, R. (2010). Synthesis and production of polyhydroxyalkanoates by halophiles: current potential and future prospects. *Applied Microbiology and Biotechnology* 85 (6): 1687–1696.

82 Youssef, N.H., Savage-Ashlock, K.N., McCully, A.L. et al. (2014). Trehalose/2-sulfotrehalose biosynthesis and glycine-betaine uptake are widely spread mechanisms for

osmoadaptation in the Halobacteriales. *The ISME Journal* 8 (3): 636–649. https://doi.org/10.1038/ismej.2013.165.

83 Mokashe, N., Chaudhari, B., and Patil, U. (2018). Operative utility of salt-stable proteases of halophilic and halotolerant bacteria in the biotechnology sector. *International Journal of Biological Macromolecules* 117: 493–522. https://doi.org/10.1016/j.ijbiomac.05.217.

84 Don, T.M., Chen, C.W., and Chan, T.H. (2006). Preparation and characterization of poly(hydroxyalkanoate) from the fermentation of Haloferax mediterranei. *Journal of Biomaterials Science Polymer Edition* 17 (12): 1425–1428. https://doi.org/10.1163/156856206778937208.

85 Salgaonkar, B. and Braganca, J. (2017). Utilization of sugarcane bagasse by *Halogeometricum borinquense* strain E3 for biosynthesis of Poly(3-hydroxybutyrate-co-3-hydroxyvalerate). *JM Bioengineering (Basel)* 4 (2).

86 Poli, A., Di Donato, P., Abbamondi, G.R., and Nicolaus, B. (2011). Synthesis, production, and biotechnological applications of exopolysaccharides and polyhydroxyalkanoates by archaea. *Archaea*: 1–13. https://doi.org/10.1155/2011/693253.

87 Pantazaki, A.A., Papaneophytou, C.P., and Lambropoulou, D.A. (2011). Simultaneous polyhydroxyalkanoates and rhamnolipids production by *Thermus thermophilus* HB8. *AMB Express* 1 (1): 17.

88 Ibrahim, M.H.A. and Steinbu, C.A. (2010). Zobellella denitrificans strain MW1, a newly isolated bacterium suitable for poly(3-hydroxybutyrate) production from glycerol. *Journal of Applied Microbiology* 108: 214–225.

89 Xiao, Z., Zhang, Y., Xi, L. et al. (2015). Thermophilic production of polyhydroxyalkanoates by a novel *Aneurinibacillus* strain isolated from Gudao oilfield, China: Thermophilic production of PHA by a novel *Aneurinibacillus* strain. *Journal of Basic Microbiology* 55 (9): 1125–1133. https://doi.org/10.1002/jobm.201400843.

90 Satoh, Y., Tajima, K., Nakamoto, S. et al. (2011). Isolation of a thermotolerant bacterium producing medium-chain-length polyhydroxyalkanoate. *Journal of Applied Microbiology* 111: 811–817.

91 Giedraityte, G. and Kalediene, L. (2015). Purification and characterization of polyhydroxybutyrate produced from thermophilic *Geobacillus* sp. AY 946034 strain. *Chemija* 26 (1): 38–45.

92 Tribelli, P.M., Di Martino, C., Lopez, N.I., and Raiger Iustman, L.J. (2012). Biofilm lifestyle enhances diesel bioremediation and biosurfactant production in the Antarctic polyhydroxyalkanoate producer Pseudomonas extremaustralis. *Biodegradation* 23 (5): 645–651.

93 Sathiyanarayanan, G., Bhatia, S., Song, H. et al. (2017). Production and characterization of medium-chain-length polyhydroxyalkanoate copolymer from Arctic psychrotrophic bacterium *Pseudomonas* sp. PAMC 28620. *International Journal of Macromolecules* 97.

94 Ciesielski, S., Górniak, D., Możejko, J. et al. (2014). The diversity of bacteria isolated from antarctic freshwater reservoirs possessing the ability to produce polyhydroxyalkanoates. *Current Microbiology* 69 (5): 594–603. https://doi.org/10.1007/s00284-014-0629-1.

95 Marsudi, S., Unno, H., and Hori, K. (2008). Palm oil utilization for the simultaneous production of polyhydroxyalkanoates and rhamnolipids by *Pseudomonas aeruginosa*. *Applied Microbiology and Biotechnology* 78: 955–961. https://doi.org/10.1007/s00253-008-1388-3.

96 Licciardello, G., Ferraro, R., Russo, M. et al. (2017). Transcriptome analysis of *Pseudomonas mediterranea* and *P. corrugata* plant pathogens during accumulation of medium-chain-length PHAs by glycerol bioconversion. *New Biotechnology* 37 (Part A): 39–47.

97 Eroglu, I., Tabanoglu, A., Gunduz, U. et al. (2008). Hydrogen production by *Rhodobacter sphaeroides* O.U.001 in a flat plate solar bioreactor. *International Journal of Hydrogen Energy* 33 (2): 531–541.

98 Aru, A., Sandhya, M., and Ponnusami, V. (2014). Biohydrogen and polyhydroxyalkanoate co-production by *Enterobacter aerogenes* and *Rhodobacter sphaeroides* from *Calophyllum inophyllum* oil cake. *Bioresource Technology* 164C: 170–176. https://doi.org/10.1016/j.biortech.2014.04.104.

99 Kumar, P., Jun, H.B., and Kim, B.S. (2018). Co-production of polyhydroxyalkanoates and carotenoids through bioconversion of glycerol by *Paracoccus* sp. strain LL1. *International Journal of Biological Macromolecules* 107: 2552–2558.

100 Basnett, P., Ching, K.Y., Stolz, M. et al. (2013). Novel Poly(3-hydroxyoctanoate)/Poly(3-hydroxybutyrate) blends for medical applications. *Reactive and Functional Polymers* 73 (10): 1340–1348.

101 Akaraonye, E., Keshavarz, T., and Roy, I. (2010). Production of polyhydroxyalkanoates: the future green materials of choice. *Journal of Chemical Technology and Biotechnology* 85 (6): 732–743.

102 Godbole, S., Gote, S., Latkar, M., and Chakrabarti, T. (2003). Preparation and characterization of biodegradable poly-3-hydroxybutyrate–starch blend films. *Bioresource Technology* 86 (1): 33–37.

103 Zhang, L., Deng, X., and Huang, Z. (1997). Miscibility, thermal behaviour and morphological structure of poly(3-hydroxybutyrate) and ethyl cellulose binary blends. *Polymer* 38 (21): 5379–5387.

104 Zhang, M. and Thomas, N. (2009). Preparation and properties of polyhydroxybutyrate blended with different types of starch. *Journal of Applied Polymer Science* 116 https://doi.org/10.1002/app.30991.

105 Zembouai, I., Kaci, M., Bruzaud, S. et al. (2013). A study of morphological, thermal, rheological and barrier properties of Poly (3-hydroxybutyrate-Co-3-Hydroxyvalerate)/polylactide blends prepared by melt mixing. *Polymer Testing* 32 (5): 842–851.

106 Chaturvedi, K., Kulkarni, A., and Aminabhavi, T. (2011). Blend microspheres of poly(3-hydroxybutyrate) and cellulose acetate phthalate for colon delivery of 5-fluorouracil. *Industrial and Engineering Chemistry Research* 50 https://doi.org/10.1021/ie2011005.

107 Rydz, J., Musioł, M., Zawidlak-Węgrzyńska, B., and Sikorska, W. (2018). Present and future of biodegradable polymers for food packaging applications. In: *Biopolymers for Food Design – A Volume in Handbook of Food Bioengineering* (eds. A.M. Grumezescu and A.M. Holban), 431–467.

108 Steinbüchel, A. and Hein, S. (2001). Biochemical and molecular basis of microbial synthesis of polyhydroxyalkanoates in microorganisms. In: *Advances in Biochemical Engineering/Biotechnology* (ed. T. Scheper), 81–123. Berlin Heideberg, Germany: Springer-Verlag.

109 Plackett, D. and Siró, I. (2011). Polyhydroxyalkanoates (PHAs) for food packaging. In: *Multifunctional and Nanoreinforced Polymers for Food Packaging* (ed. J.M. Lagarón), 498–526. Elsevier.

110 Koller, M. (2014). Poly(hydroxyalkanoates) for food packaging: application and attempts towards implementation. *Applied Food Biotechnology* 1 (1): 3–15.

111 Khosravi-Darani, K. and Bucci, D.Z. (2015). Application of poly(hydroxyalkanoate) in food packaging: improvements by nanotechnology. *Chemical Biochemical and Engineering Q* 29 (2): 275–285.

112 Bourbonnais, R. and Marchessault, R.H. (2010). Application of polyhydroxyalkanoate granules for sizing of paper. *Biomacromolecules* (4): 989–993.

113 Sun, J., Shen, J., Chen, S. et al. (2018). Nanofiller reinforced biodegradable PLA/PHA composites: current status and future trends. *Polymers* 10 (5): 505.

114 Chen, G.Q. and Wu, Q. (2005). The application of polyhydroxyalkanoates as tissue engineering materials. *Biomaterials* 26 (33): 6565–6578. https://doi.org/10.1016/j.biomaterials.2005.04.036.

115 Yao, Y.C., Zhan, X.Y., Zhang, J. et al. (2008). A specific drug targeting system based on polyhydroxyalkanoate granule binding protein PhaP fused with targeted cell ligands. *Biomaterials* 29 (36): 4823–4830.

116 Singh, A. and Yakhmi, J. (2017). Polyhydroxyalkanoates: biodegradable plastics and their applications. In: *Handbook of Ecomaterials* (eds. L. Martinez, O. Kharissova and B. Kharisov). Cham: Springer https://doi.org/10.1007/978-3-319-48281-1_84-1.

117 Williams, S.F., Martin, D.P., Horowitz, D.M., and Peoples, O.P. (1999). PHA applications: addressing the price performance issue. *International Journal of Biological Macromolecules* 25 (1-3): 111–121. https://doi.org/10.1016/s0141-8130(99)00022-7.

118 Sasidharan, R. and Bhat, S. (2014). Bacterial Polyhydroxyalkanoates production and its applications. In: *Microbial Bioproducts* (eds. G.S. Bhat and P. Nambisan), 70–96. Directorate of Public Relation and Publication Cochin University of Science and Technology.

119 Cheng, S., Chen, G.Q., Leski, M. et al. (2006). The effect of D, L-β-hydroxybutyric acid on cell death and proliferation in L929 cells. *Biomaterials* 27: 3758.

120 Sodian, R., Hoerstrup, S.P., Sperling, J.S. et al. (2000). Early in vivo experience with tissue-engineered trileaflet heart valves. *Circulation* 102: 22–29.

121 Stock, U.A., Nagashima, M., Khalil, P.N. et al. (2000). Tissue engineered valved conduits in the pulmonary circulation. *Journal of Cardiovascular Surgery* 119: 732–740.

122 Sodian, R., Loebe, M., Hein, A. et al. (2002). Application of stereolithography for scaffold fabrication for tissue engineered heart valves. *ASAIO Journal* 48: 12–16. https://doi.org/10.1097/00002480-200201000-00004.

123 Zhang, Y., Zhou, P., Pan, L.-F. et al. (2007). Growth of human smooth muscle cells on the silk fibroin modified-polyhydroxyalkanoate. *Acta Chimica Sinica* 65: 2935–2940.

124 Bagdadi, A., Safari, M., Dubey, P. et al. (2016). Poly(3-hydroxyoctanoate), a promising new material for cardiac tissue engineering. *Journal of Tissue Engineering and Regenerative Medicine* 12 https://doi.org/10.1002/term.2318.

125 Huang, J., Lu, L., Zhang, J. et al. (2012). Electrical stimulation to conductive scaffold promotes axonal regeneration and remyelination in a rat model of large nerve defect. *PLoS One* 7 (6): e39526. https://doi.org/10.1371/journal.pone.0039526.

126 Muheremu, A. and Ao, Q. (2015). Past, present, and future of nerve conduits in the treatment of peripheral nerve injury. *Biomedical Research International*: 237507. https://doi.org/10.1155/2015/237507.

127 Gao, Y., Kong, L., Zhang, L. et al. (2006). Improvement of mechanical properties of poly(dl-lactide) films by blending of poly(3-hydroxybutyrate-co-3-hydroxyhexanoate). *European Polymer Journal* 42 (4): 764–775. https://doi.org/10.1016/j.eurpolymj.2005.09.028.

128 Bian, Y.Z., Wang, Y., Aibaidoula, G. et al. (2009). Evaluation of poly(3-hydroxybutyrate-co-3-hydoxyhexanoate) conduits for peripheral nerve regeneration. *Biomaterials* 30: 217–225.

129 Yucel, D., Torun Kose, G., and Hasircı, V. (2010). Polyester based nerve guidance conduit design. *Biomaterials* 31 (7): 1596–1603.

130 Xu, X.Y., Li, X.T., Peng, S. et al. (2010). The behaviour of neural stem cells on polyhydroxyalkanoate nanofiber scaffolds. *Biomaterials* 31: 3967–3975. https://doi.org/10.1016/j.biomaterials.2010.01.132.

131 Lizarraga, L., Nigmatullin, R., Taylor, C. et al. (2015). Nerve tissue engineering using blends of poly(3-hydroxyalkanoates) for peripheral nerve regeneration. *Engineering in Life Sciences* 15 (6): 612–621.

132 Zhijiang, C., Yi, X., Haizheng, Y. et al. (2016). Poly(hydroxybutyrate)/cellulose acetate blend nanofiber scaffolds: Preparation, characterization and cytocompatibility. *Materials Science and Engineering: C* 58: 757–767. https://doi.org/10.1016/j.msec.2015.09.048.

133 Orts, W.J., Nobes, G.A., Kawada, J. et al. (2008). Poly (hydroxyalkanoates): biorefinery polymers with a whole range of applications. The work of robert H. Marchessault. *Canadian Journal of Chemistry*: 628–640. https://doi.org/10.1139/v08-050.

134 Ansari, N.F. and Annuar, M.S. (2017). Functionalization of medium-chain-length poly(3-hydroxyalkanoates) as amphiphilic material by graft copolymerization with glycerol 1,3-diglycerolate diacrylate and its mechanism. *Journal of Macromolecular Science* 55: 1–9. https://doi.org/10.1080/10601325.2017.1387490.

135 Pramual, S., Assavanig, A., Bergkvist, M. et al. (2016). Development and characterization of bioderived polyhydroxyalkanoate nanoparticles as a delivery system for hydrophobic pho todynamic therapy agents. *Journal of Materials Science: Materials in Medicine* 27: 40.

136 Povolo, S., Toffano, P., Basaglia, M., and Casella, S. (2010). Polyhydroxyalkanoates production by engineered *Cupriavidus necator* from waste material containing lactose. *Bioresource Technology* 101 (20): 7902–7907. https://doi.org/10.1016/j.biortech.2010.05.029.

16

Microbial Enzymes for the Mineralization of Xenobiotic Compounds

Ankita Chatterjee, Pritha Chakraborty, and Jayanthi Abraham

Microbial Biotechnology Lab, School of Bioscience and Technology, Vellore Institute of Technology, Vellore, Tamil Nadu, India

16.1 Introduction

Filamentous fungi, especially ligninolytic (white-rot) strains, have been studied and used for at least two decades to target specific pollutants in wastes and soils because of their efficient degrading capabilities. Fungi are ubiquitous in nature, present in both aquatic and terrestrial environments, occurring as unicellular yeasts to multicellular and filamentous forms and found as free-living, pathogenic, and symbiotic forms. Fungal mycelia exude powerful extracellular enzymes, and their metabolic activities play a crucial role in bioremediation. *Aspergillus* and other molds are reported to decompose cellulose, hemicellulose, starches, lignin, pectin, and other sugar polymers as well as fat, oil, chitin, and keratin [1]. Ligninolytic fungi, commonly known as white-rot fungus, are able to degrade and mineralize a wide variety of xenobiotic compounds because of their unique enzyme system. Most of these xenobiotic compounds include agricultural wastes (pesticide, organochlorine), pharmaceutical waste (drugs), industrial waste (polychlorinated biphenyls [PCBs], polycyclic aromatic hydrocarbons [PAHs], and synthetic dyes), heavy metals, wood preservatives, and synthetic polymers [2].

Treatment of pollutants from the environment can be carried out using adsorption, ion exchange, nanofiltration, reverse osmosis, distillation, application of phytotechnologies and agricultural wastages, and utilization of microorganisms. Development of replanting programs on a large scale, using symbiotic fungi coupled with the bioremediation techniques based on the use of filamentous fungi or extracellular enzymes, is of great interest in the valorization of polluted soils. Bioremediation is defined as a transformation or degradation of contaminants mediated by organisms (including bacteria, algae, fungi, and plants) into nonhazardous or less-hazardous substances [3]. Unlike organic molecules, metals cannot be biodegraded. But fungi can affect transformation between mobile and

Bioprospecting of Microorganism-Based Industrial Molecules, First Edition.
Edited by Sudhir P. Singh and Santosh Kumar Upadhyay.

immobile forms of metal [4]. Ectomycorrhizal fungi are reported to possess extracellular and intracellular mechanisms, which make them tolerant against high concentrations of heavy metal pollutants. The mechanisms combine the reduction of the absorbed metal in the cytoplasm and immobilization of metallic pollutants outside the cell by secreting ligands or by their retention on the fungal cell wall [5]. Besides bioremedial significance, this phenomenon of fungi confers important environmental roles, including the solubilization of essential metals and associated anionic components such as phosphate, important for plant and microbial productivity.

The fundamental importance of fungi in the environment with regard to decomposition and transformation of both organic and inorganic substrates and resultant cycling of elements is of obvious relevance to the treatment of wastes, while the branching, filamentous mode of growth can allow efficient colonization and exploration, for example, contaminated soil and other solid substrates. This, together with the growing importance of fungi as model systems in a eukaryotic cell and molecular biology, physiology, and biochemistry, provides the rationale for this work.

16.2 Major Pollutants and Their Removal with White-Rot Fungi

White-rot fungi are a group of fungi that are able to biodegrade lignin [6]. The name white-rot derives from the white appearance of the wood attacked by fungi. Taxonomically, white-rot fungi include basidiomycetes (mostly) and a few members of ascomycetes. Recent researches about the capability of white-rot fungus to biodegrade xenobiotic compounds have generated considerable interest in this area of environmental microbiology. White-rot fungi are able to biodegrade pesticides, PCBs, polyaromatic hydrocarbons, dyes, 2,4,6-trinitrotoluene, and several chemicals like azide and cyanides [7].

16.2.1 Pesticides

Pesticides, insecticides, and herbicides are counted as agricultural waste, persistent in the environment resulting in biomagnification through the food chain; toxic effects of several pesticides have been recorded at all levels in the ecosystem [8]. Major classes of pesticides that exert their toxic effect in the environment include organochlorine (most of which are banned), organophosphorus, methylcarbamate, synthetic pyrethroid, and classes of herbicide include chlorophenoxy alkanoates, triazines, and other hydrocarbons [9]. DDT (dichlorodiphenyltrichloroethane), an organochlorine pesticide, is widely used and highly persistent in the environment, which has been banned around the world. Various efficient strains, which are *Phanerochaete chrysosporium*, *Pleurotus ostreatus*, *Phellinus weirii*, and *Polyporus versicolor*, are reported to degrade DDT under ligninolytic conditions. Scientific studies have proved that the biodegradation of DDT can result in DDE (^{14}C-radiolabeled 1,1-dichloro-2,2-bis(4-chlorophenyl)ethene), which is a more toxic and persistent compound but can be mineralized to $^{14}CO_2$ by *P. chrysosporium* [10]. *Phlebia acanthocystis* is able to biodegrade 9.3% of pentachlorophenol (PCP; biocide) to $^{14}CO_2$ within a span of 42 days of interaction [11]. Pure culture of *Phlebia lindtneri* and *Phlebia brevispora* with

DDT (25 lmol/l) showed that 70 and 30% of DDT, respectively, disappeared in a low-nitrogen medium after a 21-day incubation period [12]. *Trametes versicolor* has been reported to degrade fipronil within nine days of interaction with the help of extracellular enzymes produced by the strain and the intracellular cytochrome P450 complex [13]. The involvement of fungal cytochrome P450 complex is found in the degradation of endrin, an insecticide by white-rot fungi, such as *P. acanthocystis* and *P. brevispora* [14]. Clothianidin, belonging to the group of neonicotinoid pesticides, can be degraded to nontoxic metabolite, *N*-(2-chlorothiazol-5-yl-methyl)-*N*′-methylurea, by the application of *Phanerochaete sordida* [15]. Lignin peroxidase (LiP) and manganese peroxidase (MnP) are directly related to the mineralization of these organochlorines via a multistep catalytic process involving oxidation, reduction, and methylation reactions in sequential order to carry out ring cleavage. Further studies showed that hydrolytic and oxidative pathways are also involved in the pesticide degradation model by white-rot fungi. Organochlorine herbicide (2,4,5-T and 2,4-D) biodegradation by white-rot fungi as well as by members of non-white-rot fungi has also been reported by Bhalerao and Puranik [16]. Mineralization of chlorpyrifos, fonofos, terbufos, and atrazine has been demonstrated by *P. chrysosporium*, while the involvement of lignin-modifying enzymes has not been explored [17].

16.2.2 Polychlorinated Biphenyls

PCBs are chlorinated forms of selective biphenyls with extensive use as dielectric fluid, flame retardants, heat-transfer fluid, hydraulic fluids, organic diluents, plasticizers, and solvent extenders in a variety of industries. Other than few anaerobic sediment bacteria, white-rot fungi including *P. chrysosporium*, *P. ostreatus*, *Coriolopsis polyzona*, and *T. versicolor* are reported to remove significant PCBs reported by numerous studies while very few studies approve chlorine substitution mechanism [18, 19]. Few studies have reported the effect of extracellular enzymes and cytochrome P450 monooxygenase in the PCB removal process. Laccase has been found to play an important role in the degradation of PCBs by fungal strains. *Rigidoporus* sp. FMD21 degraded 73% of 2,3,7,8-tetrachlorodibenzo-p-dioxin within 28 days of interaction [20]. In another study, *Pleurotus sajor-caju* LBM105 is revealed to degrade 90% of PCBs within a span of 35 days [21]. Both the abovementioned studies reported the involvement of laccase enzymes secreted by the fungal strains. In case of degradation of PCBs by *P. ostreatus*, synergistic effects of laccase, manganese peroxidases, and lignin peroxidases have been suggested to play an effective role [22].

16.2.3 Polycyclic Aromatic Hydrocarbons

PAHs are composed of four or more benzene rings. Recent research reports that three-ringed homologs like anthracene and phenanthrene can also be included in the definition of PAHs. Natural oil depositions, fossil fuel heating, wood-burning, vehicular transport, waste incarnation, and many other industrial processes are reported sources of PAH formation. Studies have reported PAH mineralization process by white-rot fungi [23]. Studies related with purified cell-free enzyme extracts have revealed the key role of lignin-modifying enzymes in PAH degradation. LiP extracted from different white-rot fungi has shown the ability of PAH oxidation. Veratyl alcohol-mediated PAH oxidation LiP extracted from *Nematoloma forwardii* has been reported [24]. MnP and laccase expressed by white-rot fungi also participate in PAH metabolism. MnP-mediated PAH oxidation by *P. chrysosporium* and *Phanerochaete laevis* produce polar end products without significant quinone accumulation, while MnP from *N. forwardii* oxidize anthracene and pyrene with quinone accumulation [25]. Degradation of PAH (phenanthrene, fluorene, and pyrene) by *Trametes polyzona* was successfully accomplished where MnP and laccase increased the rate of degradation [26]. Besides LiP and MnP, laccase extracted from *T. versicolor* are also reported to degrade PAH with the presence of hydroxybenzotriazole (HBT) and 2,2′-azino-bis-3-ethylbenzothiazoline-6-sulphonic acid (ABTS) [27].

16.2.4 Synthetic Dyes

Synthetic dyes are widely used in textile, biomedical, food, and beverages and plastic industries. Early studies have shown that white-rot fungi are capable of decolorizing a wide variety of synthetic dyes, including azo, triphenylmethane, heterocyclic or polymeric dyes [28]. Although during the early days, dye decolorization studies were concentrated only with *P. chrysosporium*, later on *Pycnoporus sanguineus* has also been reported to degrade azo and triphenylmethane dyes while producing laccase as a sole lignin-modifying enzyme. The study claims that less than 3% dyes can be absorbed by white-rot fungi [29, 30]. The predominant mechanism of dye degradation involves secretion and interaction of lignin-modifying enzymes by the white-rot fungal strains [31]. The increased oxidation potential of the lignin-modifying enzymes aids in effectiveness of white-rot fungi to decolorize textile dyes. Reports proved that crude laccase is capable of decolorizing bromophenol blue, orange II, Congo Red, carmine, reactive black 5, drimarene blue X3LR, acid blue, acid red, and reactive blue [32]. LiP of *P. chrysosporium* is able to decolorize azo, triphenylmethane, and heterocyclic dyes in the presence of H_2O_2 and veratryl alcohol (VA) as co-substrate [33]. Roles of MnP from different white-rot fungi are reported to carry out dye decolorization following a Mn^{2+}-independent mechanism [34]. *Trametes hispida* possesses two laccase isoenzymes, which are also able to carry out dye decolorization.

16.2.5 Synthetic Polymers

Synthetic polymers (plastics) are an undeniable part of modern society. Polyurethane is generated by condensing polyisocyanate and polyol. Extracellular esterase isolated from *Curvularia senegalensis* is different than conventional lignin-modifying enzyme system of white-rot fungi and are able to degrade polyurethane [35]. *P. chrysosporium, Pleurotus sajor caju,* and *P. versicolor* depolymerize polyvinylchloride (PVC) by loss in C–H bonds [36]. Polyvinyl alcohol (PVA) is a synthetic polymer with extensive use in adhesive manufacture and paper/textile coatings. *Pycnoporus cinnabarinus* was reported to carry out oxidative PVA degradation by secreting laccase enzyme. Nylon is also a linear amide-linked polymer, are degradable by *P. chrysosporium* and *T. versicolor* by MnP catalyzed methods [37].

16.2.6 Phenolic Compounds

PCP is most widely used among all the other organopollutants. *P. chrysosporium* is reported to mineralize PCPs up to 50% without any confirmation of lignin-modifying enzymes' involvement [7]. *P. chrysosporium* and *Phanerochaete soridia* are able to grow at 25 ppm concentration of PCP. Rapid and extensive deletion of PCP (64 and 71%, respectively) results in the production of pentachloroanisole (PCA). PCA is a quinone-based intermediate that can be slowly removed by these fungi. Seventy percent mineralization of PCP is reported by cell-free enzyme extraction of *P. chrysosporium,* strongly suggesting the role of extracellular peroxidases and intracellular enzymes [38]. *T. versicolor* is also found to mineralize PCP but at a very low concentration [39]. Degradation of 17 beta estradiol using *T. citrinoviride* AJAC3 secreting MnP and lignin peroxidases have been reported in a recent study [40].

16.2.7 Petroleum Hydrocarbons

Oily sludge is referred to the greasy and thick residues produced from the oil industries during production, refining, and transportation. The two major petroleum hydrocarbon components which are found to be hazardous are aromatic and aliphatic hydrocarbons. Pollution caused by oil hydrocarbons affects the environment and human health adversely. *Polyporus* sp. S133 has been reported in earlier studies to be successful in the degradation of 93% crude oil in liquid cultures when incubated for 60 days [41]. Oxygenases and peroxidases play an important role in the degradation of petroleum hydrocarbons by microorganisms. Heme-thiolate monooxygenases help in degrading oil, chlorinated hydrocarbons, and fuel additives. Several species of *Candida*, such as *Candida maltose, Candida tropicalis, Candida oleophila,* and *Candida apicola,* have been reported to secrete these cytochrome P450 enzymes [42]. The degradation process of petroleum hydrocarbons in the soil is affected by four important factors: properties of the hydrocarbons; structure and metabolism of the microbial cell; environmental conditions; and properties of the soil [43, 44].

A list of organopollutants (with example) present in the environment and which can be degraded by white-rot fungi is presented in Table 16.1.

16.3 Enzyme System of White-Rot Fungi

The wide substrate variety of white-rot fungi can be attributed to the unique enzyme system. Successful mineralization of pollutants by white-rot fungi is considered as a function of lignin-modifying enzymes [45]. Different fungi appear to have the same effect with different combinations of enzymes. The major members of lignin-modifying enzymes are laccase (Lac), LiP, MnP with accessory enzyme like aryl alcohol oxidase, versatile peroxidase, H_2O_2-forming glyoxal oxidase, oxalate decarboxylate (OCD), NAD-dependent formate dehydrogenase (FDH), and P450 monooxygenase are proved to be influenced in the mineralization process [46, 47] (Table 16.2).

16.3.1 Laccase

Laccase is considered as a versatile enzyme, a member of N-glycosylated extracellular blue multicopper oxidases with molecular mass ranging from 57 to 98 kDa [48] Figure 16.1 (a). Laccase is ranked as among the oldest enzymes ever recorded, which was first described by Yoshida in 1883. Distributed among the plants, insects, bacteria, and a wide variety of

Table 16.1 List of persistent organopollutants degraded by white-rot fungi (WRF)

Sl. no	Type of pollutant	Example	WRF strain
1	Polycyclic aromatic hydrocarbons (PAHs)	Anthracene, benzo[α]pyrene, fluorene, naphthalene, phenanthrene, pyrene, biphenylene.	*P. chrysosporiumPhanerochaete sordida,Pleurotus ostreatus,Trametes versicolor,Trametes hirsutus,Kuehneromyces mutabilis.*
2	Chlorinated aromatic compounds	Pentachlorophenol (PCP), trichlorophenol (TCP), bichlorophenols (BCP), polychlorinated biphenyls (PCBs), dioxins, chlorobenzene.	*P. chrysosporium, P. sordida, Lentinula edodes,T. versicolor,P. ostreatus.*
3	Pesticides	Alachlor, aldrin, chlordane, DDT, lindane, atrazine heptachlor.	*P. chrysosporium.*
4	Dyes	Azure B, Congo Red, Orange II, Poly R, Reactive Orange 96, Reactive violet 5, Remazol brilliant blue, R. Tropaeolin.	*Coriolus versicolor, P. chrysosporium, P. ostreatus. T. versicolor, T. trogii.*
5	Nitroaromatics	TNT (2,4,6-trinitrotoluene), 2,4-dinitrotoluene, 2-amino-4,6-dinitrotoluene, 1-chloro-2,4-dinitrobenzene, 2,4-dichloro-1-nitrobenzene.	*P. chrysosporium.*
6	Other compounds	Benzene, toluene, ethylbenzene, *o*-, *m*-,*p*-xylenes (BTEX compounds), linear alkylbenzene sulfonate (LAS), trichloroethylene.	*Pleurotus Xorida,P. chrysosporium,Dichomitus squalens.*

basidiomycetes, laccase performs the reduction of dioxygen to water while oxidizing organic substrate by a one-electron redox process [49]. Broad substrate degradability of laccase is attributed to the unique structure of this enzyme. Laccase contains four copper atoms at three different sites, defined according to spectroscopic properties. The different copper atoms have been separated and identified on the basis of their spectroscopic properties [50]. T1 copper is characterized by having strong absorbance at around 600 nm while T2 copper exhibits very weak absorbance in the visible region. T2 site copper is electron paramagnetic resonance (EPR) active, whereas two copper at T3 site is EPR-silent [51]. The substrate oxidized by T1 copper and the extracted electrons is transferred probably through a strongly conserved His-Cys-His tripeptide motif, to the T2/T3 site, where reduction of molecular oxygen to water takes place [48, 52]. Another form of laccase is also proposed by researchers, which lacks T1 copper and is named as "yellow" laccase and does not show characteristic absorbance at 600 nm. So far, more than 100 different laccases have been extracted and purified from different white-rot fungi, and the diverse biological function of the laccase differs from the corresponding organisms.

Table 16.2 List of major fungal enzymes with their application details in bioremediation of different organopollutants.

Sl. no	Enzyme	Producing fungi	Pollutant	Mechanism	Application
1	Laccase	*P. chrysosporium*, *T. versicolor*, *T. hispida*, *Pycnoporus cinnabarinus*, *Cerrena unicolor*	Ortho and para diphenols, aminophenols, polyphenols, polyamines, lignins, and aryldiamines.	Oxidation, decarboxylation and demethylation of substrate.	Food industry, paper and pulp industry, textile industry, nanotechnology, synthetic chemistry, bioremediation, cosmetics, and so forth.
2	Lignin peroxidase	*P. chrysosporium*, *Chrysonilia sitophila*	Halogenated phenolic compounds, polycyclic aromatic compounds, and other aromatic compounds.	Oxidation of substrate in the presence of co-substrate H_2O_2 and mediators like veratryl alcohol.	Food industry, paper and pulp industry, textile industry, pharmaceutical industry, bioremediation, and so forth.
3	Manganese peroxidase	*P. chrysosporium*, *Nematolona frowardi*, *Phebia radiata*, *Lentinula edodes*	Lignin and other phenolic compounds.	In the presence of Mn2+ and H_2O_2, the co-substrate catalyzes oxidation of Mn^{2+} to Mn^{3+} which results in an Mn3+ chelateoxalate, which in turn oxidizes the phenolic substrates.	Food industry, Paper and pulp industry, textile industry, pharmaceutical industry, bioremediation, and so forth.
4	Versatile peroxidase	*Pleurotuseryngii*	Methoxybenzenes and phenolic aromatic compounds.	The enzyme catalyzes the electron transfer from an oxidizable substrate, with the formation and reduction ofcompound I and compound II intermediates.	Industrial biocatalyst, bioremediation, and so forth.
5	Lipase	*Aspergillus* sp., *Penicillium* sp.	Organic pollutants such as oil spill.	The hydrolysis of triacylglycerols to glycerols and free-fatty acids.	Control of oil spills, detergent production, baking industry, paper and pulp industry, personal care products, and so forth.

(*Continued*)

Table 16.2 (Continued)

Sl. no	Enzyme	Producing fungi	Pollutant	Mechanism	Application
6	Cellulose	*Trichoderma* sp., *Aspergillus* sp.	Cellulosic substance	Hydrolyses the substrate to simple carbohydrates.	Textile manufacturing, detergent production, paper and pulp industry, bioremediation, and so forth.
7	Protease	*Aspergillus* sp., *Penicillium* sp., *Rhizopus* sp., *Mucor* sp., *Humicola* sp.	Proteins	Enzymes that hydrolyze peptide bonds in aqueous environment.	Leather, laundry, biocatalyst, bioremediation, and so forth.

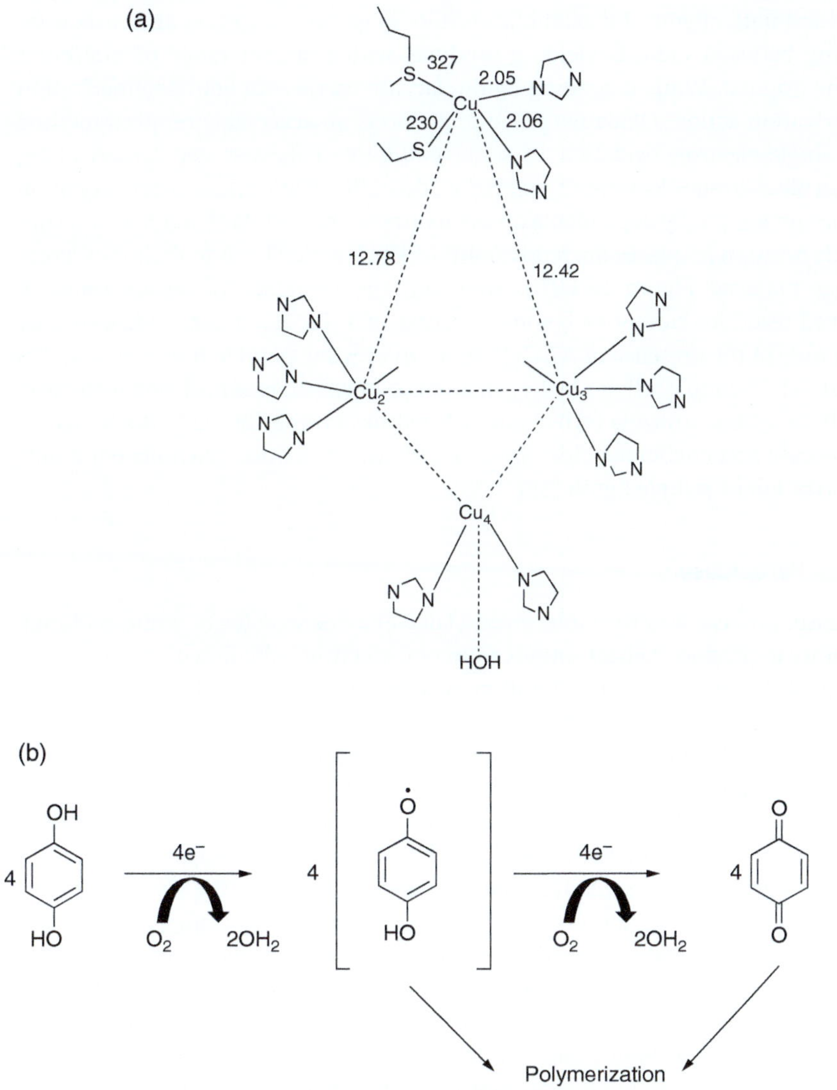

Figure 16.1 Structure of laccase showing the positions of different Cu sites. (a) Structure of laccase showing the positions of different Cu site. (b) Mechanism of laccase to oxidize organopollutants.

Laccase oxidizes many substrates, including phenolic dyes, phenols, chlorophenols, lignin-related diphenyl lignin-modifying enzymes, thane, benzopyrenes, N-substituted *p*-phenylenediamines, organophosphorus, and non-phenolic beta-*O*-lignin model dimer.

16.3.1.1 Mechanisms

Laccase catalyzed degradation of phenolic compounds to radicals by single-electron oxidation. The free radical then reacts further in a sequenced manner, which depends on the nature of the subsequent groups and reaction conditions. Researchers reported that degradation of aromatic, lignin-like phenolics results in carbon-to-carbon and carbon-to-oxygen coupling between radicals yielding products with a higher range of molecular weight than the original. While degrading lignin, laccase carries out both polymerization and depolymerization action. Oxidative polymerization is an acceptable removal method of pollutant. Single-electron oxidation of some pollutants catalyzes depolymerization, which results in alkyl–arene cleavage. C_α-oxidation, demethoxylation, and decarboxylation reactions by laccase are totally dependable on the nature of the pollutant and reaction condition [53]. This reaction increases the redox status of PAH, which leads to their incorporation in cellular material Figure 16.1 (b). However, the influence of co-substrates in laccase-catalyzed reactions cannot be ignored. Laccase of *T. versicolor* can oxidize specific non-phenolics only in the presence of ABTS [54]. An interesting point to note is that in this reaction laccase-ABTS couple follows hydrogen atom abstraction instead of single-electron oxidation. 1-Hydroxybenzotriazole (HBT) and 3,4-hydroxyanthranilic acid also act as co-substrate for laccase and enable the cleavage of a wide variety of non-phenolic pollutants and depolymerization of soluble lignin [55].

16.3.2 Lignin Peroxidase

Lignin peroxidase (LiP) is a water-soluble, glycosylated enzyme secreted by white-rot fungi. LiP has the ability to produce radical cations from one-electron oxidation of non-phenolic aromatic compounds such as VA and 1,4-dimethoxybenzene (DMB). LiP catalysis takes place over an acidic pH range with a maximum value of pH 4.1 and the lower level being dictated by the opposing tendency for protein precipitation [56].

In LiP catalysis, initial oxidation of resting Fe^{III} occurs by H_2O_2, thereby forming oxo-ferryl intermediate followed by reduction of oxo-ferryl intermediate. Further, reduction reaction occurs in order to attain resting ferric oxidation state thus, completing the cycle. This step solely depends on the relative concentration of H_2O_2. In the presence of co-substrate, the end product of oxidation of dimethoxylated aromatics is radical cations. For example, the radical VA^+ breaks down carbon–hydrogen cleavage and produces carbon-centered radical and finally, veratraldehyde. Radical cations, however, are also able to act as one-electron oxidants themselves, creating reactive radical cations in secondary substrates by charge transfer [57] Figure 16.2.

Activated oxygen species like superoxide anion (O_2^-), perhydroxyl radical (HOO^-), H_2O_2, and hydroxyl radical (OH^-) may play a role in the oxidation process carried out by LiP. During the growth period, LiP-producing fungi create an acidic environment that

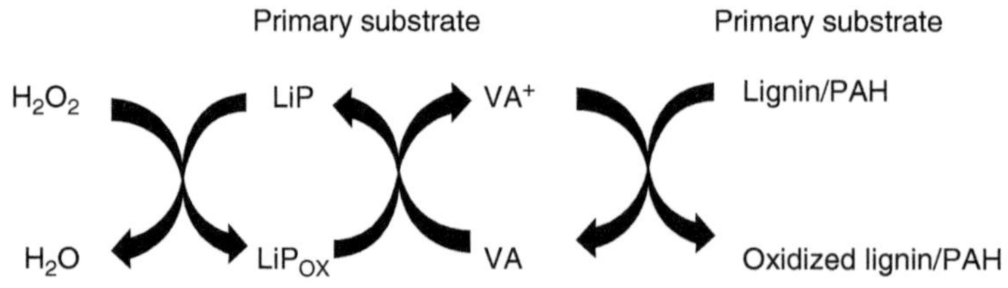

Figure 16.2 Oxidation mechanism of lignin peroxidase.

helps the perhydroxyl radical to initiate the process of the oxidative attack on aromatic structures. It is also able to dismutate with superoxide to provide H_2O_2 to carry out further reactions [58]. The activated oxygen radicals also react with radical cations to bring about aromatic ring opening.

16.3.3 Manganese Peroxidase

MnP is an extracellular glycosylated heme enzyme secreted by a variety of white-rot fungi that uses H_2O_2 as co-substrate to oxidize Mn^{II} and Mn^{III}-chelate, which in turn oxidize phenolic substrate as a more freely diffusible and nonspecific oxidant [59]. The ability of MnP to generate Mn chelate plays a critical role in PAH degradation. MnP belongs to class II of the peroxidase family, with a molecular weight ranging from 32 to 62.4 kDa, optimum temp range of 40–60 °C, and optimum pH range of four to seven. The "heme" moiety is the site of protein oxidation by H_2O_2, resulting in catalytic intermediate (Compound I and Compound II), which is required for the oxidation process [60].

16.3.3.1 Mechanism

The reaction of the native MnP enzyme with H_2O_2 yields Compound I, which contains an oxyferryl heme with a porphyrin cation radical. Two steps of single-electron reduction by Mn^{II} restore the native enzyme via the intermediate Compound II as shown in Figure 16.3.

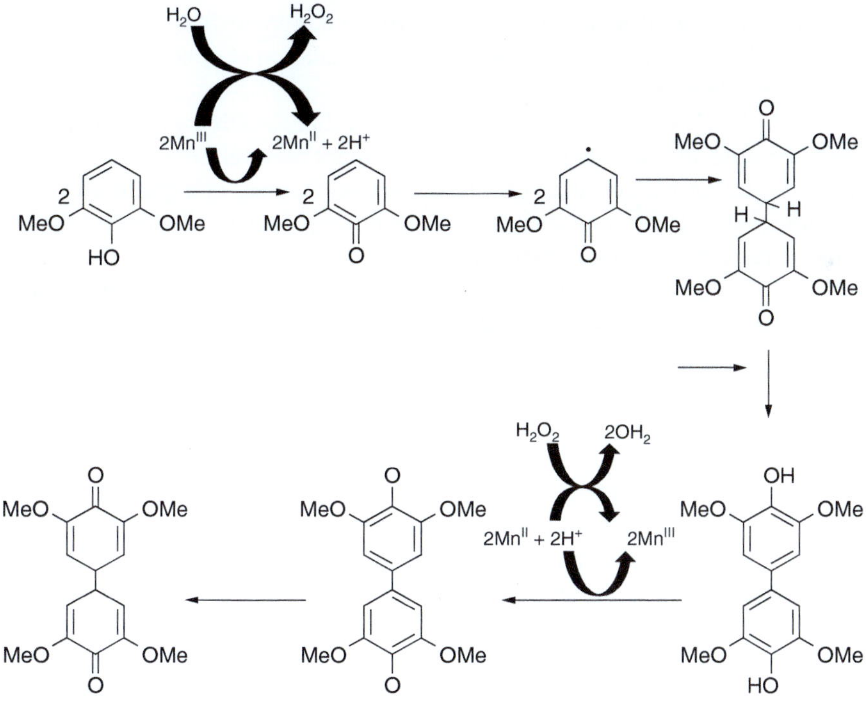

Figure 16.3 Enzymatic oxidation mechanism of MnP peroxidase.

$E + H_2O_2 \rightarrow \text{Compound I} + H_2O$
$\text{Compound I} + Mn^{II} \rightarrow \text{Compound II} + Mn^{III}$
$\text{Compound II} + Mn^{II} \rightarrow E + Mn^{III} + H_2O$
$\text{Compound II} + H_2O_2 \rightarrow \text{Compound III (inactive)}$

Importantly for the catalysis, the supply of H_2O_2 relative to Mn^{II} needs to be poised to ensure that the competing reaction of Compound II with H_2O_2 does not take place since this has the effect of driving the enzyme into a catalytically inactive mode [60]. Malonate, citrate, glyoxalate, and oxalate are essential for chelating and stabilizing Mn^{III} and secreted as secondary metabolites by white-rot fungi along with MnP. Manganese (II) reacts with the oxidized forms of MnP as a monochelated complex, but it is released from the enzyme in its chelated form. In a similar manner to laccase, the range of substrates that can be attacked by MnP is increased in the presence of co-substrates. For example, in the presence of glutathione, VA, which is not normally oxidized by the Mnp-Mn-chelate system, however, can be oxidized to veratraldehyde via thiol radicals generated from the oxidation of glutathione by Mn-chelates [6].

16.3.4 Other Enzymes

Versatile peroxidases are able to oxidize Mn^{2+}, methoxybenzene, phenolic aromatic substances in a likely manner of LiP and MnP. It shows extensive broad substrate specificity (including phenolic and non-phenolic model dimers) and able to oxidize substances even in the absence of manganese, unlike other peroxidases [61]. *Caldariomyces fumago* are reported to produce another form of peroxidase known as chloroperoxidase (CPO), which can oxidize several phenolic substrates in addition to the oxidation of ethanol to aldehyde and oxidation of chloride ions [62].

Extracellular hydrolytic enzymes like amylases, proteases, lipases, DNAses, pullulanases, and xylanases have quite diverse potential usages in pollutant removal from contaminated soil. Cellulases, including extracellular cellulases, hemicellulases, and pectinases, are reported to be secreted at a very low level by white-rot fungi, which are able to convert complex cellulosic material in soil into their simpler form.

Monooxygenase, dioxygenase, lipases, and proteases are also reported to degrade alkanes, steroids, aromatic compounds, fatty acids, and proteins, respectively.

16.4 Molecular Aspect

The ability to degrade a wide variety of organopollutants makes white-rot fungi more acceptable toward pollutant mineralization. Respective genes for encoding lignin-modifying enzymes are discussed in this section. Two isozymes of laccase Lac-I and Lac-II have been identified in *Physisporinus rivulosus, Trametes trogii, Cerrena unicolor,* and *Panus tigrinus* [63]. Two different laccase encoding genes (*Pr-lac1* and *Pr-lac2*) in *Phlebia radiata* are expressed differently and regulated. Pr-lac2 has high sequence similarity with *T. versicolor* laccase [64]. The c-DNA encoding laccase enzyme shows a common sequence of 520–550 amino acid residues, including an N-terminal secretion peptide. In addition to one cysteine and ten histidine residues involved in binding, the four copper atoms found in each laccase molecule are conserved [65]. It is in the copper-binding amino acid residues, and their general distribution in polypeptide chain (two pairs of histidines in the N-terminal domain and other liganding residues all near the C-terminus) shows that laccases are all similar [66].

The *lip* genes encode a mature protein product of 343–345 amino acids prepared by 27–30 residue leader sequence. All the LiP isozymes examined appeared to contain 2.5–3 kDa asparagine-N-linked carbohydrate which includes mannose 6-phosphate and are likely to contain O-linked carbohydrates as well [67]. The coding regions of the *lip* genes contain approximately 60–65% G+C, whereas the 3′ noncoding regions are approximately 44–49% G+C [68]. The gene encoding LG2, encoding H8 isozyme from strain *P. chrysosporium* (OGC101) has the highest G+C codon bias. Since this isozyme accounts for the major fraction of the LiP produced by this strain and this sequence is present as a single-copy gene, a correlation between its high codon bias and high gene expression is likely.

Like LiP, MnP isozymes are also encoded by multiple genes. The coding regions of mnp1 and mnp2 have a nucleotide identity of approximately 70% and an amino acid identity of 88%. The MnP leader sequence (21 amino acids) consists of signal peptides [69]. However, the MnP leader sequence lacks propeptide ending in a dibasic amino acid pair, unlike LiP sequence. The sequences for proximal His, distal His, and distal Arg and sequences surrounding the conserved Asp are similar in LiP and MnP. The MnP proteins lack the proline-rich C-terminal region found in LiP [70]. Whereas MnP1 contains three potential N-glycosylation sites, MnP2 sequence contains four potential N-glycosylation sites. Both isozymes have numerous potential O-glycosylation sites. The coding regions of *mnp* genes have a high G+C content, but their codon bias is lower than those of most *lip* sequences, with the *mnp2* sequence exhibiting less codon bias than the *mnp1* sequence [71].

16.5 Conclusions and Perspectives

Wood-degrading white-rot fungi show remarkable versatility in being able to degrade a wide spectrum of organopollutants. Extracellular enzyme systems of white-rot fungi increase the bioavailability, water solubility, and redox status of PAHs. Implications of the live culture of white-rot fungi are more widespread for bioremediation than using the purified enzyme system. Although, there are certain limitations, including a full understanding of the factors that regulate ligninolytic enzyme systems. However, the development of a technology that employs lignin-degrading fungi to remediate organopollutants is promising.

Acknowledgement

The authors would like to express their gratitude to the management of VIT, Vellore.

Compliance with Ethical Guidelines

The authors have no conflict of interest to declare.

References

1 Rhodes, C.J. (2014). Mycoremediation (bioremediation with fungi) – growing mushrooms to clean the earth. *Chemical Speciation and Bioavailability* 26: 196–198.

2 Zhang, Y.N., Niu, Q., Gu, X. et al. (2019). Recent progress on carbon nanomaterials for the electrochemical detection and removal of environmental pollutants. *Nanoscale* 11: 11992–12014.

3 Rhodes, C.J. (2013). Applications of bioremediation and phytoremediation. *Science Progress* 96: 417–427.

4 Chatterjee, A., Das, R., and Abraham, J. (2020). Bioleaching of heavy metals from spent batteries using *Aspergillus nomius* JAMK1. *International Journal of Environmental Science and Technology* 17: 49–66.

5 Bellion, M., Courbot, M., Jacob, C. et al. (2006). Extracellular and cellular mechanisms sustaining metal tolerance in ectomycorrhizal fungi. *FEMS Microbiology Letters* 254: 173–181.

6 Asgher, M., Bhatti, H.N., Ashraf, M., and Legge, R.L. (2008). Recent developments in biodegradation of industrial pollutants by white rot fungi and their enzyme system. *Biodegradation* 19: 771.

7 Gao, D., Du, L., Yang, J. et al. (2010). A critical review of the application of white rot fungus to environmental pollution control. *Critical Reviews in Biotechnology* 30: 70–77.

8 Uqab, B., Mudasir, S., and Nazir, R. (2016). Review on bioremediation of pesticides. *Journal of Bioremediation and Biodegradion* **7**: 343.

9 Prakash, S., Varadharasan, S., and Abraham, J. (2011). Compatibility studies on *Heterorhabditis indica* ICRI-18 with commonly used pesticides and fungicides at cardamom plantation. *Ecology Environment & Conservation* 17: 563–566.

10 Purnomo, A.S., Mori, T., Kamei, I., and Kondo, R. (2011). Basic studies and applications on bioremediation of DDT: a review. *International Biodeterioration and Biodegradation* 65: 921–930.

11 Xiao, P. and Kondo, R. (2020). Biodegradation and biotransformation of pentachlorophenol by wood-decaying white rot fungus *Phlebia acanthocystis* TMIC34875. *Journal of Wood Science* 66: 1–1.

12 Xiao, P., Mori, T., Kamei, I., and Kondo, R. (2011). A novel metabolic pathway for biodegradation of DDT by the white rot fungi, *Phlebia lindtneri* and *Phlebia brevispora*. *Biodegradation* 22: 859–867.

13 Wolfand, J.M., LeFevre, G.H., and Luthy, R.G. (2016). Metabolization and degradation kinetics of the urban-use pesticide fipronil by white rot fungus *Trametes versicolor*. *Environmental Science: Processes & Impacts* 18: 1256–1265.

14 Xiao, P. and Kondo, R. (2019). Biodegradation and bioconversion of endrin by white rot fungi, *Phlebia acanthocystis* and *Phlebia brevispora*. *Mycoscience* 60: 255–261.

15 Mori, T., Wang, J., Tanaka, Y. et al. (2017). Bioremediation of the neonicotinoid insecticide clothianidin by the white-rot fungus *Phanerochaete sordida*. *Journal of Hazardous Materials* 321: 586–590.

16 Bhalerao, T.S. and Puranik, P.R. (2007). Biodegradation of organochlorine pesticide, endosulfan, by a fungal soil isolate *Aspergillus niger*. *International Biodeterioration and Biodegradation* 59: 315–321.

17 Chirnside, A.E., Ritter, W.F., and Radosevich, M. (2011). Biodegradation of aged residues of atrazine and alachlor in a mix-load site soil by fungal enzymes. *Applied and Environmental Soil Science*: 1–10.

18 Hong, C.Y., Gwak, K.S., Lee, S.Y. et al. (2012). Biodegradation of PCB congeners by white rot fungus, *Ceriporia* sp. ZLY-2010, and analysis of metabolites. *Journal of Environmental Science and Health, Part A: Toxic/Hazardous Substances and Environmental Engineering* 47: 1878–1888.

19 Kamei, I., Kogura, R., and Kondo, R. (2006). Metabolism of 4,4 – dichlorobiphenyl by white-rot fungi *Phanerochaete chrysosporium* and *Phanerochaete* sp. MZ142. *Applied Microbiology and Biotechnology* 72: 566–575.

20 Dao, A.T., Vonck, J., Janssens, T.K. et al. (2019). Screening white-rot fungi for bioremediation potential of 2, 3, 7, 8-tetrachlorodibenzo-p-dioxin. *Industrial Crops and Products* 128: 153–161.

21 Sadañoski, M.A., Benítez, S.F., Fonseca, M.I. et al. (2019). Mycoremediation of high concentrations of polychlorinated biphenyls with *Pleurotus sajor-caju* LBM 105 as an effective and cheap treatment. *Journal of Environmental Chemical Engineering* 7: 103453.

22 Chun, S.C., Muthu, M., Hasan, N. et al. (2019). Mycoremediation of PCBs by *Pleurotus ostreatus*: possibilities and prospects. *Applied Sciences* 9: 4185.

23 Mineki, S., Suzuki, K., Iwata, K. et al. (2015). Degradation of polyaromatic hydrocarbons by fungi isolated from soil in Japan. *Polycyclic Aromatic Compounds* 35: 120–128.

24 Silva, I.S., Grossman, M., and Durrant, L.R. (2009). Degradation of polycyclic aromatic hydrocarbons (2–7 rings) under microaerobic and very-low-oxygen conditions by soil fungi. *International Biodeterioration and Biodegradation* 63: 224–229.

25 Haritash, A.K. and Kaushik, C.P. (2009). Biodegradation aspects of polycyclic aromatic hydrocarbons (PAHs): a review. *Journal of Hazardous Materials* 169: 1–5.

26 Teerapatsakul, C., Pothiratana, C., Chitradon, L., and Thachepan, S. (2016). Biodegradation of polycyclic aromatic hydrocarbons by a thermotolerant white rot fungus *Trametes polyzona* RYNF13. *The Journal of General and Applied Microbiology* 62: 303–312.

27 Anastasi, A., Coppola, T., Prigione, V., and Varese, G.C. (2009). Pyrene degradation and detoxification in soil by a consortium of basidiomycetes isolated from compost: role of laccases and peroxidases. *Journal of Hazardous Materials* 165: 1229–1233.

28 Faraco, V., Pezzella, C., Giardina, P. et al. (2009). Decolourization of textile dyes by the white-rot fungi *Phanerochaete chrysosporium* and *Pleurotus ostreatus*. *Journal of Chemical Technology & Biotechnology: International Research in Process, Environmental & Clean Technology* 84: 414–419.

29 Revankar, M.S. and Lele, S.S. (2007). Synthetic dye decolorization by white rot fungus, *Ganoderma* sp. WR-1. *Bioresource Technology* 8: 775–780.

30 Chakraborty, P. and Abraham, J. (2017). Comparative study on degradation of ciprofloxacin and norfloxacin by *Ganoderma lucidum* JAPC1. *Korean Journal of Chemical Engineering* 34: 1122–1128.

31 Sen, S.K., Raut, S., Bandyopadhyay, P., and Raut, S. (2016). Fungal decolouration and degradation of azo dyes: a review. *Fungal Biology Reviews* 30: 112–133.

32 Yesilada, O., Birhanli, E., and Geckil, H. (2018). Bioremediation and decolorization of textile dyes by white rot fungi and laccase enzymes. In: *Mycoremediation and Environmental Sustainability*, 121–153. Cham: Springer.

33 Chander, M. and Arora, D.S. (2007). Evaluation of some white-rot fungi for their potential to decolourise industrial dyes. *Dyes and Pigments* 72: 192–198.

34 Sakurai, A., Kariminiaae-Hamedaani, H.R., and Sakakibara, M. (2007). Decolorization of synthetic dyes by a new manganese peroxidase-producing white rot fungus. *Dyes and Pigments* 72: 157–162.

35 Schmidt, O. (2006). *Wood and Tree Fungi*. Berlin, Germany: Springer.

36 Hong, C.Y., Gwak, K.S., Lee, S.Y. et al. (2011). *Ceriporia* sp. ZLY-2010 in biodegradation of polychlorinated biphenyls: Extracellular enzymes production and effects of cytochrome P450 monooxygenase. *Journal of Korean Wood Science Technology* 39: 469–480.

37 Tisma, M., Zelic, B., and Vasic-Racki, D. (2010). White-rot fungi in phenols, dyes and other xenobiotics treatment–a brief review. *Croatian Journal of Food Science and Technology* 2: 34–47.

38 Yang, S., Hai, F.I., Nghiem, L.D. et al. (2013). Understanding the factors controlling the removal of trace organic contaminants by white-rot fungi and their lignin modifying enzymes: a critical review. *Bioresource Technology* 141: 97–108.

39 Ergül, F.E., Sargın, S., Öngen, G., and Sukan, F.V. (2009). Dephenolisation of olive mill wastewater using adapted *Trametes versicolor*. *International Biodeterioration and Biodegradation* 63: 1–6.

40 Chatterjee, A. and Abraham, J. (2019). Mycoremediation of 17 β-Estradiol using Trichoderma citrinoviride Strain AJAC3 along with Enzyme studies. *Environmental Progress & Sustainable Energy* 38: 13142.

41 Kristanti, R.A., Hadibarata, T., Toyama, T. et al. (2011). Bioremediation of crude oil by white rot fungi *Polyporus* sp. S133. *Journal of Microbiology and Biotechnology* 21: 995–1000.

42 Barbosa, P., Araújo, P., Oliveira, J. et al. (2019). Metabolic pathways of degradation of malvidin-3-O-monoglucoside by Candida *oleophila*. *International Biodeterioration & Biodegradation* 144: 104768.

43 Varjani, S.J. (2017). Microbial degradation of petroleum hydrocarbons. *Bioresource Technology* 223: 277–286.

44 Al-Hawash, A.B., Dragh, M.A., Li, S. et al. (2018). Principles of microbial degradation of petroleum hydrocarbons in the environment. *The Egyptian Journal of Aquatic Research* 44: 71–76.

45 Wells, A., Teria, M., and Eve, T. (2006). Green oxidations with laccase mediator systems. *Biochemical Society Transactions* 34: 304–308.

46 Giardina, P., Faraco, V., Pezzella, C. et al. (2010). Laccases: a never-ending story. *Cellular and Molecular Life Sciences* 67: 369–385.

47 Taghizadeh, T., Talebian-Kiakalaieh, A., Jahandar, H. et al. (2020). Biodegradation of bisphenol A by the immobilized laccase on some synthesized and modified forms of zeolite Y. *Journal of Hazardous Materials* 386: 121950.

48 Pezzella, C., Lettera, V., Piscitelli, A. et al. (2013). Transcriptional analysis of *Pleurotus ostreatus* laccase genes. *Applied Microbiology and Biotechnology* 97: 705–717.

49 Lettera, V., Piscitelli, A., Leo, G. et al. (2010). Identification of a new member of *Pleurotus ostreatus* laccase family from mature fruiting body. *Fungal Biology* 114: 724–730.

50 Piscitelli, A., Giardina, P., Lettera, V. et al. (2011). Induction and transcriptional regulation of laccases in fungi. *Current Genomics* 12: 104–112.

51 Uzan, E., Nousiainen, P., Balland, V. et al. (2010). High redox potential laccases from the ligninolytic fungi *Pycnoporus coccineus* and *Pycnoporus sanguineus* suitable for white biotechnology: from gene cloning to enzyme characterization and applications. *Journal of Applied Microbiology* 108: 2199–2213.

52 Ghobadi Nejad, Z., Borghei, S.M., and Yaghmaei, S. (2019). Biodegradation of synthetic dye using partially purified and characterized laccase and its proposed mechanism. *International Journal of Environmental Science and Technology* 16: 7805–7816.

53 Dekker, R.F., Barbosa, A.M., Giese, E.C. et al. (2007). Influence of nutrients on enhancing laccase production by *Botryosphaeria rhodina* MAMB-05. *International Microbiology* 10: 177–185.

54 Viswanath, B., Rajesh, B., Janardhan, A. et al. (2014). Fungal laccases and their applications in bioremediation. *Enzyme Research* 2014: 1–21.

55 Ali, M.I., Khalil, N.M., and El-Ghany, M.N. (2012). Biodegradation of some polycyclic aromatic hydrocarbons by *Aspergillus terreus*. *African Journal of Microbiology Research* 6: 3783–3790.

56 Ayodeji, O.F., Uchechukwu, U.N., Benson, C.I. et al. (2017). Lignin peroxidase functionalities and prospective applications. *Microbiology Open* 6: e00394.

57 Furukawa, T., Bello, F.O., and Horsfall, L. (2014). Microbial enzyme systems for lignin degradation and their transcriptonal regulaton. *Frontiers in Biology* 9: 448–471.

58 Luther, G.W. (2010). The role of one- and two-electron transfer reactions in forming thermodynamically unstable intermediates as barriers in multi-electron redox reactions. *Aquatic Geochemistry* 16: 395–420.

59 Santelli, C.M., Pfister, D.H., Lazarus, D. et al. (2010). Promotion of Mn(II) oxidation and remediation of coal mine drainage in passive treatment systems by diverse fungal and bacterial communities. *Applied and Environmental Microbiology* 76: 4871–4875.

60 Hansard, S.P., Easter, H.D., and Voelker, B.M. (2011). Rapid reaction of nanomolar Mn(II) with superoxide radical in seawater and simulated freshwater. *Environmental Science and Technology* 45: 2811–2817.

61 Pogni, R., Teutloff, C., Lendzian, F., and Basosi, R. (2007). Tryptophan radicals as reaction intermediates in versatile peroxidases: multifrequency EPR, ENDOR and density functional theory studies. *Applied Magnetic Resonance* 31: 509–526.

62 Tinoco, R., Verdin, J., and Vazquez-Duhalt, R. (2007). Role of oxidizing mediators and tryptophan 172 in the decoloration of industrial dyes by the versatile peroxidase from *Bjerkandera adusta*. *Journal of Molecular Catalysis B: Enzymatic* 46: 1–7.

63 Lawton, T.J., Sayavedra-Soto, L.A., Arp, D.J., and Rosenzweig, A.C. (2009). Crystal structure of a two-domain multicopper oxidase: implications for the evolution of multicopper blue proteins. *The Journal of Biological Chemistry* 284: 10174–10180.

64 Komori, H., Miyazaki, K., and Higuchi, Y. (2009). X-ray structure of a two-domain type laccase: a missing link in the evolution of multi-copper proteins. *FEBS Letters* 583: 1189–1195.

65 Ye, M., Li, G., Liang, W., and Liu, Y.H. (2010). Molecular cloning and characterization of a novel metagenome-derived multicopper oxidase with alkaline laccase activity and highly soluble expression. *Applied Microbiology and Biotechnology* 87: 1023–1031.

66 Morgenstern, I., Klopman, S., and Hibbett, D.S. (2008). Molecular evolution and diversity of lignin degrading heme peroxidases in the Agaricomycetes. *Journal of Molecular Evolution* 66: 243–257.

67 Janusz, G., Kucharzyk, K.H., Pawlik, A. et al. (2013). Fungal laccase, manganese peroxidase and lignin peroxidase: gene expression and regulation. *Enzyme and Microbial Technology* 52: 1–12.

68 Miki, Y., Ichinose, H., and Wariishi, H. (2010). Molecular characterization of lignin peroxidase from the white-rot basidiomycete *Trametes cervina*: a novel fungal peroxidase. *FEMS Microbiology Letters* 304: 39–46.

69 Morgenstern, I., Robertson, D.L., and Hibbett, D.S. (2010). Characterization of three mnp genes of *Fomitiporia mediterranea* and report of additional class II peroxidases in the order Hymenochaetales. *Applied and Environmental Microbiology* 76: 6431–6440.

70 Nagai, M., Sakamoto, Y., Nakade, K., and Sato, T. (2007). Isolation and characterization of the gene encoding a manganese peroxidase from *Lentinula edodes*. *Mycoscience* 48: 125–130.

71 Sakamoto, Y., Nakade, K., Nagai, M. et al. (2009). Cloning of *Lentinula edodes* lemnp2, a manganese peroxidase that is secreted abundantly in sawdust medium. *Mycoscience* 50: 116–122.

17

Functional Oligosaccharides and Microbial Sources

SA Belorkar

Department of Microbiology and Bioinformatics, Atal Bihari Vajpayee University, Bilaspur, Chattisgarh, India

17.1 Introduction

Microorganisms have been found miraculously capable of various physiological adaptations for their successful survival under normal conditions and in extreme environments of high sugars, salinity, pressure, temperature, and presence of deleterious substances like heavy metals, pesticides, polycyclic aromatic hydrocarbons, and nuclear wastes. Microorganisms adapt and produce novel biomolecules and have been proven to be a significant source for environmental, industrial, and pharmaceutical applications. One such group of novel biomolecules gaining utmost and parallel importance in the food and health industry is the functional food, also known as nutraceuticals. Functional food is an umbrella term coined nowadays for food items, which confer added health benefits to the consumers apart from imparting the standard food deliverables.

The functional foods concept was developed and popularized in Japan, and the oligosaccharides gained much attention as a promising health product for the food and pharmaceutical industry. Early in 1980s, the Japanese government agencies initiated the approval of food items well known and acknowledged for their health benefits with the intention of the betterment of general public health. Figure 17.1 explains the two basic categories of functional foods, namely, conventional and fortified or modified (i.e., artificially added to food products for value enhancements). The conventional functional food includes natural food material like fruits, vegetables, whole-food ingredients, honey, etc. Modified or fortified food is artificially enhanced by the addition of functional food ingredients like vitamins, minerals, probiotics, or fiber, to increase food products' health benefits, e.g. milk, and yogurt, almond, rice, coconut, cashew milk bread, pasta fortified cereal, and eggs [1].

Bioprospecting of Microorganism-Based Industrial Molecules, First Edition.
Edited by Sudhir P. Singh and Santosh Kumar Upadhyay.

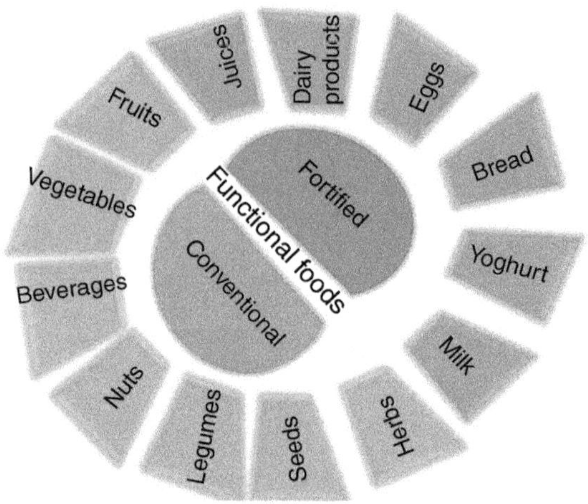

Figure 17.1 Common examples of conventional and fortified FO.

17.1.1 What Are Functional Foods? All You Need to Know

Presently, the health industry is witnessing consumers more inclined toward food items, which confers significant health benefits in addition to their virtue of being low-calorie items. The increased demand for this combinational effect in food is satisfied by dietary carbohydrates that fall under the category of oligosaccharides and offer resistance toward digestion. The rising popularity of the functional oligosaccharides among all other digestion-resistant carbohydrates is bestowed due to the extraordinary benefits conferred to the consumers [2].

Oligosaccharides exhibiting additional health benefits to the consumers are popularly termed as functional oligosaccharides (FO). The oligosaccharides are constituted of 3–10 monosaccharides and share characteristics of intermittent of mono and polysaccharides [3]. The essential characteristic of FO is the osidic bond between the anomeric carbon of the monosaccharides. The hydrolytic human digestive enzymes are unable to hydrolyze this bond, which imparts them the characteristic nondigestible nature [4].

The nondigestible nature makes FO qualify to be used as a prebiotic also. The fortification of food items with FO explicits health benefits like improved gut health and microflora. These FO are fermented by colon inhabitants to release energy, metabolic substrates, lactic acid, and short-chain carboxylic acids. The term prebiotic is conferred due to the growth-promoting effect of FO on the gut microflora or the probiotic organisms.

Some oligosaccharides, gaining recognition under the banner of important functional foods, are cyclodextrins (CD), fructooligosaccharides (FOS), glucooligosaccharides (GOS), galacto-oligosaccharides (GaOS), gentiooligosaccharides (GeOS), isomaltooligosaccharides (IMO), isomaltulose, lactosucrose, mannan oligosaccharides (MOS), maltooligosaccharides (MO), pectin-derived acidic oligosaccharides (pAOS), soybean meal oligosaccharides (SMO), and xylooligosaccharides (XOS) [5].

17.2 Inulin and Oligofructose: The Preliminary Functional Oligosaccharides

Any prebiotic is defined as any food ingredient, which is resistant to human intestinal digestion and confers a favorable growth to the gut microflora and is beneficial for health status of humans. All the information regarding characteristics, nature of FO was based on the studies conducted on the two fundamental FO-inulin or oligofructose. The oligosaccharides (OS) could be obtained after enzymatic hydrolysis of inulin, whereas the oligofructose or fructooligosaccharides is synthesized by fructosyltransferase activity.

17.3 GRAS and FOSHU Status

Being a component of natural food, Chicory inulin and oligofructose have gained affirmation and recognition to acquire the GRAS status (Generally recognized as safe) in the United States. The initial oligosaccharide synthesis was experimented with the use of sucrose as substrate catalyzed by enzyme fructosyltransferase [6]. They have been acknowledged as novel foods and find application in bakeries as sugar substitutes, texture enhancer, foam stabilizer, and used in desserts such as jellies and ice creams; bakery products such as cookies, bread, and pastries; spreads; and infant formulas for increasing the palatability of the food items.

Japan, the country that initiated the use of FO in food items for the betterment of general public health in 1980s, conferred the FOSHU status to oligosaccharides in 1991. It implies "Food for Specified Health Uses." Recent research findings across the globe have acknowledged oligosaccharides, conferring resistance to digestion, as functional foods.

17.4 Conventional and Upcoming Oligosaccharides

After the worldwide acceptance of health benefits of inulin and fructooligosaccharides as functional foods, research was diverted toward finding more such oligosaccharides with different areas of health benefits. Following the two fundamental FO, others that gained recognition were COS, chitooligosaccharides; GOS, galactooligosaccharides; IMO, isomaltooligosaccharides; SOS, soybean oligosaccharides; and XOS, xylooligosaccharides. Substantial level of research has been conducted on these aforesaid oligosaccharides, with well-established health benefits like prebiotic activity, inhibiting microbial adherence, prevention of cancer, improved absorption of minerals, bifidogenic (in vitro and in vivo), laxative effect in constipation, hypocholesterolemic, obesity management, reduction of food intake and weight, antidiabetic effects, improved lipid metabolism, prevention of pathogen proliferation, etc. [7].

Since the path of these oligosaccharides is now well established, they can be termed as conventional oligosaccharides. The present-day research is now targeting a newer source of FO molecules with new health implications. They are classified as upcoming FO, viz., lactosucrose, arabinogalacto-oligosaccharides, arabinoxylo-oligosaccharides,

Conventional FO

- GOS: Galactooligosaccharides
- COS: Chitooligosaccharides
- XOS: Xylooligosaccharides
- IMO: Isomaltooligosaccharides
- SOS: Soybean oligosaccharides

Upcoming FO

- Lactosucrose
- Arabinogalactooligosaccharides
- Arabinoxylooligosaccharides
- Arabinooligosaccharides
- Pectic oligosaccharides
- Galacturonan
- Rhamnogalacturonan
- Mannan oligosaccharides
- Oligodextrans

Figure 17.2 Some important conventional and upcoming functional oligosaccharides.

arabino-oligosaccharides, pectic oligosaccharides, galacturonan, rhamnogalacturonan, mannan oligosaccharides, oligodextrans. Figure 17.2 elaborates some important conventional and novel FO on the basis of their recognition, acceptance, and usage.

17.5 Microbes and Functional Oligosaccharides

The FO, whether conventional or upcoming, can be derived either by the microbial transformation or enzymatic conversion of the suitable substrate, or by extraction from natural sources. The main source of FO are microbial enzymes that convert natural substrates into desired oligosaccharides. The most common enzymes that are involved in various types of FO synthesis are beta-galactosidase, fructosyltransferase, xylanases, etc. Yet another basis of classification is their source, i.e. synthetic or derived FO.

17.6 Arabinoxylo-Oligosaccharides

As earlier stated, these FO are XOS contain variable-length arabinose chain; hence the name derived arabinoxylo-oligosaccharides (AXOS). These FO are comparatively new members of the FO family and descendants of XOS. They differ from the XOS on two grounds: the degree of polymerization, which is from 3 to 67, and the amount of arabinosyl substitution. The key feature to generate this variation is the enzyme action of endoxylanase on arabinoxylans (AX). This enzyme is in the limelight due to its eight glycoside hydrolase (GH) families. There are two types of catalytic mechanisms exhibited by this family – retaining or an inverting. The main category exploited by industries is the retaining endo-xylanases for the production of AXOS. The enzymes GH10, GH11 and GH30 are presently the model biocatalysts used for AXOS production.

17.7 Sources and Properties

Xylan-rich sources are the prospective sources to derive the AXOS. All agro-industrial residues, therefore, possess the potential of being used as its source. Since it is a category of derived FO, therefore, the present knowledge of enzymatic degradation of different types of xylans [8] has been worked out in the past two decades. Being a derived polymer, it gives the opportunity to synthesize tailor-made AXOS conserving the features embedded in the polymers. Enzymatic tools with known specificity are targeted for specific degradation, which is more economically demanding.

The oligomers have positive impact on improvement of bowel habit accompanied with effective fermentation in the colon resulting in its bifidogenic properties [9]. The AXOS is effective against type II diabetes by decreasing post-meal glucose levels and modulates insulin response [10, 11]. The prebiotic effect is mainly dependent on the structure of the AXOS molecule (substitution) and the degree of polymerization [11, 12]. In contrast to humans, the Bifidus group is capable of expressing enzymes required for cleavage of XOS and AXOS [12].

Many reviews have been published focusing on the production of FO using endoxylanases that have a targeted prebiotic effect. The resistance of these natural polysaccharides is of prime importance to the digestive enzymes to execute their growth-promoting effect on gut microflora. A striking characteristic is highlighted that humans are incapable of producing hydrolyzing enzymes like xylanases, xylosidases, or arabinofuranosidases. Thus, OS and substituted OS offer desirable resistance to digestive juices and promote gut health [13].

17.8 Approaches for AXOS Production

As earlier discussed, the selection of enzymes is very critical to decide the composition of the substituted AXOS. The most popular enzyme selected is the main chain-acting endoxylanases [14]. The resultant products are further degraded by the intestinal microbiota resulting in desirable health benefits as an FO [15]. Approaches are also directed toward the production of enzymes in GRAS organisms for future prospects in FO production to facilitate use in functional food manufacturing [16]. Since the most promising source of xylan is agro-industrial wastes, attempts are made for their complete exploitation in FO production. There are various experiments using pretreatment strategies for the breakdown of the polymer at the preliminary level, but Amorim et al. (2019) attempted a direct fermentation using two *Trichoderma* species [17]. This single-step production used brewers' spent grain (BSG) as the substrate. The product was AXOS and reported the degree of polymerization to be in between 2 and 5. Such strategies are new horizons for development of economic, industrial production of AXOS.

Approaches in the form of clinical trials are continuously performed for production and process development to confer the arabinoxylan-oligosaccharides as the next generation FO. Since the xylooligosaccharides and arabinoxylan-oligosaccharides are pentoses, they differ in the mechanism by which the gut flora metabolize them, as the nature of the bonds also present significantly differ from the conventional FO. Clinical trials proved that they are effective to stimulate the growth of *Bifidobacteria* and to some extent of *Lactobacilli* and other probiotic bacteria. These FO have proved their beneficial effect on digestive health, management of blood sugars and lipids, beneficial modification of immune markers, and benefits for laxation. These benefits are registered at a lower dose of concentrations when compared to conventional FO [18].

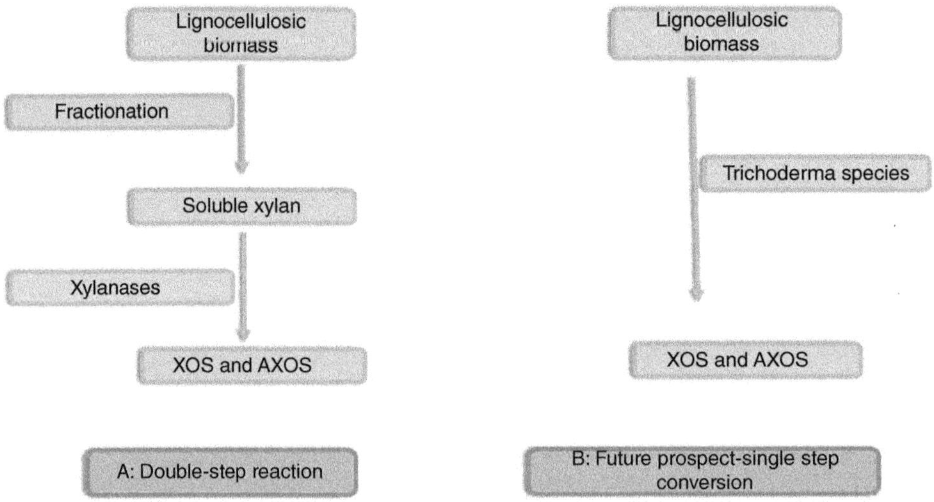

Figure 17.3 The production of AXOS.

The production of AXOS is quite typical with multiple approaches, e.g., chemical method, enzymatic method, and the combination of both [19]. The chemical method involves the hydrolysis of the substrate by using chemicals. This method experiences major drawbacks like contaminated products with traces of chemicals, by-products that are undesirable and toxic, e.g. hydroxymethylfurfural and furfural. The major limitation above all was uncontrolled polymerization, which could affect the product functionalities. Such processes were uneconomical and demanded heavy instrumental support. On the contrary, the enzymatic process provides advantages like economic, no by-products, and is a biodegradable process [20].

Presently, the enzymatic production can be a two-step [21] or a single-step reaction [17], as given in Figure 17.3. The enzymatic method is bestowed upon by virtues like less xylose contamination and toxic by-products. It suffers the drawback of being expensive, and the process is complicated as it depends on both the steps. The microbial sources for enzymes utilized in production are given in Table 17.1.

17.9 Isomaltooligosaccharides

Isomaltooligosaccharides (IMO) arise from polymerization of mainly α-(1,6)-linked glucose residues, rarely α-(1,4) glycosidic bonds are encountered. The resultant is mixture of oligomers with length varying from 2 to 6 monomer. The isomaltooligosaccharides in its structure have glucosyl moiety linked to maltose or isomaltose by α-(1,6) glycosidic bonds. IMO includes oligomers like isomaltose, isomaltotriose, isomaltotetraose, isomaltopentaose, kojibiose, nigerose, panose, and higher branched oligosaccharides ([28–33]. There is yet another term the IMO-like products including branched IMOs, e.g. glucooligosaccharides produced from dextransucrase [34], oligodextran [35]; alditols and IsomaltR [36], isomaltulose orpalatinose (α-D-Glc p-(1 $\rightarrow$ 6)-α-D-Fru f) [37], and cyclic IMOs [38]. The raw material used for its production is starch. Many products marketed use cornstarch as a starting material for IMO synthesis comprising of isomaltose, isomaltotriose, and panose.

Table 17.1 Arabinoxylo-oligosaccharides production and their microbial sources.

Source	Enzyme	References
Trichoderma reesei	Endoxylanases	[17]
Geobacillus stearothermophilus *Rhodothermus marinus*	*Rm*Xyn10A-CM	[22]
Cellvibrio mixtus	endo-1,4-β-xylanase (GH10)	[23]
Trichoderma viride		
	endo-1,4-β-xylanase (GH11).	
Aspergillus aculeatus		
Aspergillus niger		
Thermotoga maritima		
Clostridium thermocellum	GH5_34 (*Ct*Xyl5A)	[24]
Acetivibrio cellulolyticus	(*Ac*GH5)	[25]
Verrucomicrobiae bacterium		
Gonapodya prolifera	(*Vb*GH5)	
	(*Gp*GH5)	
Bacillus subtilis		[26]
Erwinia chrysanthemi	Xylanase A	[27]

17.10 Sources and Properties

The synthetic pathway for IMO involves a multistep enzyme-catalyzed reactions. The natural sources are sugar and honey. They also show occurrence in fermented foods with sugar, as well as products like sake and soybean sauce. These also occur commonly in fermented foods namely rice miso and sake [39]. IMO find wide applications as food additives [10] as a virtue of availability, high stability, and low cost [40].

In contrast, to its counterpart FO, the digestion of IMO initiates in the intestine causing rise in blood sugar level and is fermented by intestinal micro inhabitants [41]. It also shows promotion of growth of bifidogenic microorganisms [9]. The bifidogenic effect is mild in comparison to FO like IOS or FOS [42].

The other properties of IMO are: less sweetness, difficult to crystallize, and less viscousity. The prebiotic property confers improved intestinal health and immunity, mineral absorption, cholesterol regulation, and prevention of dental caries [37, 43]. IMO offer mild sweetness to the consumer, and its production is cost-effective with properties like low viscosity and bulking properties. It prevents dental caries, and could be used as substitute sugars for diabetics [42], or to improve the intestinal flora [41]. IMO find their application in food, pharmaceutical, and cosmetic industries. They are also used as artificial sweeteners for diabetics [10, 43].

17.11 Production of IMO

The synthesis of IMO is an interplay of enzymes and not a single-enzyme catalyzed reactions. The production involves an enzymatic transfer reaction employing a combination of immobilized enzymes. In brief, starch is liquefied using α-amylase (EC 3.2.1.1) and pullulanase (EC 3.2.1.41). Then, the reaction intermediate is treated by β-amylase (hydrolysis) and α-glucosidase (transglucosidation), producing IMO, as shown in Figure 17.4.

Commercially, these FO are produced using starch as a substrate by the process of transglycosylation. Figure 17.4 explains the commercial three-staged process of IMO production and the role of specific enzyme in each catalyzed step of liquefaction, saccharification, and transglycosylation.

17.12 Approaches to Improve IMO Production

Presently, research is being focused on trials with novel approaches using new raw material like banana flour. Novel approach for IMO substrate banana flour has been investigated in tropical Southeast Asia, where fruit wastage is one of the major problem due to overproduction of banana. This approach used banana for IMO synthesis by transglucosidase L with yields of 76.6 ± 2.7. The product of the process was reported to contain 53% isomaltotriose, 21% isomaltotetraose, and 26% maltooligoheptaose, and larger oligomers [44]. Although trials with new substrate yield give a substantial yield of IMO and variable degree of polymerization, researchers have conducted trials for IMO production by using fermented food items. As already known, IMO are naturally encountered in fermented food. Experiments were designed by Cho et al. to study the formation of these prebiotic compounds during *Sauerkraut preparation* [45]. The IMO production was initiated by inoculating psychrotrophic *Leuconostoc citreum* KACC 91035 strain as a source of the

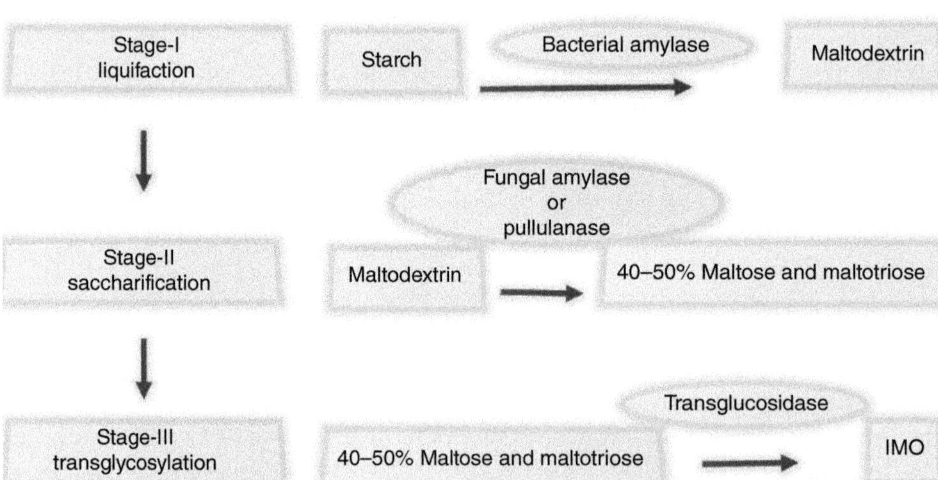

Figure 17.4 Production of IMO.

Table 17.2 Isomaltooligosaccharides production and their microbial sources.

Source	Enzyme	References
Bacillus licheniformis	Cloned amylase	[49]
Bacillus lehensis G1	alkali-stable maltogenic amylase	[50]
Bacillus licheniformis	Amylase	[51]
Bacillus amyloliquefaciens Barley bran *Bacillus naganoensis* *Bacillus licheniformis* (Host) *Aspergillus niger*	α-Amylase β-Amylase Pullulanase α-Transglucosidase	[46]
Bacillus stearothermophilus *Thermotoga maritima*	Maltogenic amylase α-Glucanotransferase	[52]
L. mesenteroides NRRL B-512F	Dextransucrase	[53]

enzyme dextransucrase. This enzyme was found to transfer glucose. Sucrose and maltose were added in Sauerkraut as additional substrates to act as a glucosyl donor and acceptor molecule. This novel approach successfully synthesized IMO by addition of substrates and probiotic lactic acid bacteria, which resulted in value addition of traditional sauerkraut preparation. Table 17.2 provides information about microbial enzymes involved in IMO production.

Another focus area for improved IMO production is the search of potential enzyme sources with promising IMO synthetic enzymes since the commercial production of IMO is a multistage process ending up with low yields. A simple approach of using enzyme mixture for improved and efficient IMO production was developed. The enzyme cocktails was prepared and composed of the pullulanase (*Bacillus licheniformis*), α-amylase (*Bacillus amyloliquefaciens*), β-amylase (barley bran), and α-transglucosidase (*Aspergillus niger*). The stage I liquefaction and stage II transglycosylation were performed simultaneously. The IMOs produced were comprised of isomaltose, isomaltotriose, and panose [46]. With the aim of improvised IMO production, the efficiency of free and immobilized enzymes were also investigated for transformation of sucrose to IMO using immobilized transglucosidase, employing membrane reactor and a modeling mechanistic approach [40, 47, 48].

17.13 Lactosucrose

Various food product regulatory agencies in the world have accorded GRAS (generally recognized as safe) status to lactosucrose. It has also been added to the list of foods for specific health uses (FOSHU) in Japan [54].

Lactosucrose is a trisaccharide composed of galactose, glucose, and fructose exhibiting nonreducing nature. There can be two approaches for lactosucrose synthesis:

a) A galactose when condenses with sucrose molecule
b) A lactose when condenses with a fructose molecule

The synthesis of this FO is dependent on the enzyme β-galactosidase. This enzyme caters the transfer of β-galactosyl group to the C4 hydroxyl group of glucosyl in sucrose. Another pathway uses enzyme β-fructofuranosidase, which transfers the fructose group to the reducing end of the lactose. The pathway of synthesis is given in Figure 17.5.

Naturally, lactosucrose occurs at minimum concentration when the curd is made with milk having sucrose and lactose sugars. Otherwise, lactosucrose is synthesized by enzyme-catalyzed reactions using both sucrose and lactose sugars as substrates.

As lactosucrose is a trisaccharide, it is composed of galactose, glucose, and fructose. The most effective benefit it imparts to the consumer is restored gut health through balanced microbiota. It is the most pronounced prebiotic function of lactosucrose. There is yet another approaches toward the lactosucrose where it is seen as a potential prebiotic awaiting recognition parallel to FOS and IOS [55]. The lactosucrose qualifies the prerequisite of being non-digestible through the digestive tract till it reaches colon and exhibits better laxation without the side effect of diarrhea than counterpart prebiotics [56]. It also has a good water retention, which provides it attributes like improved bowel peristalsis and improved fecal formation and output. As a FO, lactosucrose is low caloric and offers resistance to digestion. Further, it confers health benefits like intestinal mineral absorption. It has now gained popularity as a positively modulating food ingradient for gut microflora [57].

Due to the virtue of water absorption and retention, it finds its application in fermented milk products, such as yogurts or cheese. It is also used as a fat replacer in this kind of product, contributing to a decrease in syneresis and improving some particular characteristics such as consistency and texture, as it was already reported for inulin [58]. It is used as a food additive in baked goods, candies, ice-creams, infant formula, juices, mineral water, pharmaceuticals, and cosmetic formulations [59–61].

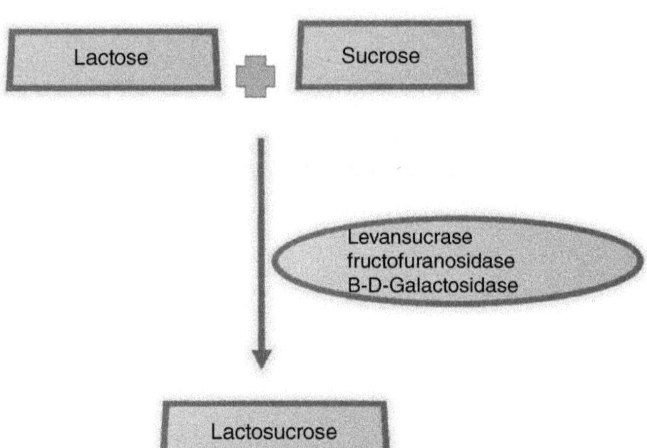

Figure 17.5 Production of lactosucrose.

17.14 Novel Approaches in Lactosucrose Preparation

A synthetic approach was used by Cheul et al. (2009) employing levansucrase of *Zymomonas mobilis* using lactose and sucrose as substrate [62]. The optimization of the production process yielded a conversion of 28.5%. The conversion rate was improvised to 43.2% when the catalysis by levansucrase was conjugated with glucose oxidase. Looking forward to the industrial expectations of the production process, the catalytic enzymes draw exuberate expenses causing hike in the production cost. An economic enzyme usage was attempted by recycling the enzymes via immobilization [63]. Both the enzymes were co-immobilized by sol–gel encapsulation. The so formed system successfully retained 85.39% of their initial activity with extraordinary operational stability after 15 cycles. This process thereby offers a great economical opportunity for food industry as a tool for lactosucrose production. Table 17.3 provides an insight into microbial enzymes involved in lactosucrose production.

Approaches toward finding an economical substrate to lower the production cost were also tried. The agro-industrial by-products tofu whey and cheese whey permeate were experimented to support the enzyme-mediated bioconversion of lactosucrose. The promising productivity of these substrates can provide a new platform for process development [64]. The stability and efficient catalysis are the most desirable qualities an enzyme source must qualify. The coding gene for the enzyme was isolated from *Arthrobacter* sp. 10138 and expressed in *Escherichia coli*. This approach has successfully created an option for the industry to use more potent recombinant enzyme with elevated sustainability [65].

17.15 Xylooligosaccharides

Xylooligosaccharides (XOS) are oligosaccharides of xylose monomers with β-(1,4) linkages. The degree of polymerization is 2–10 monomers. Apart from linear oligomers, branched oligomers are also reported with presence of arabinofuranosyl, glucopyranosyl uronic acid, or its 4-O-methyl derivative, in addition to xylose with consistently different property [47].

Table 17.3 Lactosucrose oligosaccharides production and their microbial sources.

Source	Enzyme	References
LS *Brenneria goodwine*	Levansucrase	[66]
B. circulans	B-galactosidase	[67]
Arthrobacter sp10138	B-fructofuranosidase	[65]
Leuconostoc mesentroids	Levansucrase	[68]
Bacillus licheniformis	Levansucrase	[69]
*Arthrobacter sp*10137	B-fructofuranosidase Glucooxidase (Novonordisk-Immobilised)	[63]
Zymomonas mobilis	Levansucrase	[62]

17.16 Occurrence and Properties

Similar to the other FO, the XOS is found naturally in bamboo shoots, fruits, honey, milk, and vegetables [70]. In the natural form, the oligomers that are predominantly found are xylobiose, xylotriose, and xylo-tetraose [41].

The plant biomass is exuberantly rich in lignin and cellulose with xylan being the major component. The xylan shows its dominance in biological composition of plant biomass, agricultural residues, and proves to be a potent bioresource for XOS [71].

The XOS share some common characteristics with the conventional FO like low calorie with limited sweetness, unobjectionable odor as a food additive, and non-cariogenic. The unique properties of XOS is the stability of the XOS toward pH and temperature variations. In contrast to the most popular conventional oligosaccharides, XOS remains stable and resists digestion in the stomach as a virtue of its stability toward pH 2.5–8 [72]. The XOS products also exhibit striking stability toward temperature up to 100 °C. The temperature stability maintains the nutrient levels to be retained and prevents hydrolysis even though the product undergoes high-temperature variations during pretreatment, extractions, purifications, and storage [73].

The XOS also holds a special feature of strongly promoting bifidobacteria and lactobacilli in pure culture. This enhanced growth of bifidobacteria is health-promoting for the gut. [41]. Interestingly, this bifidogenic property is much stronger than FOS [74].

The raw material used for XOS synthesis is xylan. The synthesis of XOS is executed by hydrolytic reaction of xylan hemicelluloses by β-xylanase enzyme. The xylanases have great biotechnological applications apart from XOS application [75]. The raw materials used for XOS synthesis are agro-industrial wastes rich in xylan. Generally, baggasse, straw, husks corncobs, and hulls are the commonly used sources for xylan. For industrial production of XOS extraction of xylan from the selected raw material is the primary requisite. The XOS synthesis is accomplished in the following steps: extraction of xylan, and enzymatic hydrolysis of xylan as explained in Figure 17.6. The first step utilizes the agro-industrial

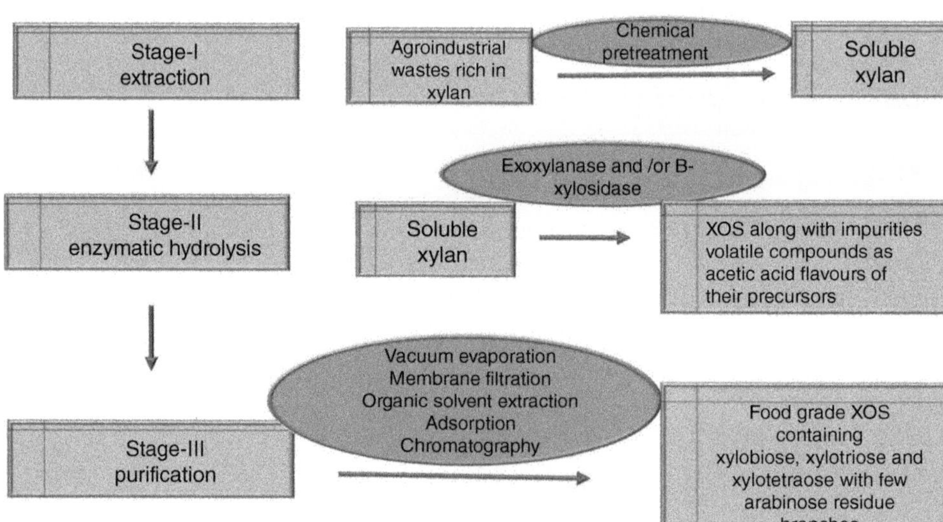

Figure 17.6 Production of XOS.

waste as feedstock for xylan extraction, which are rich in lignocellulosic material. The first step is the treatment of the feedstock with chemical for extraction of xylan to derive a soluble form. The commonly used methods are alkaline hydrolysis, oxidation, precipitation by salts or alcohols to remove lignin or pectic substances. The second step is xylan hydrolysis employing enzymatic treatment. The enzyme commonly used for treatment is exoxylanase and/or xylosidase. A balanced formulation of the enzyme is critical to dominating the unwanted release of xylose that may inhibit the XOS synthesis [70].

On an industrial scale, the production of XOS can be achieved by the chemical approach, enzymatic approach, or combination of the two processes. The products of chemical method exhibit greater polymerization. Contrarily, the enzymatic bioconversions are preferred for the manufacture food-grade XOS and pharmaceutically important XOS on grounds of safety [71].

17.17 Approaches to Improve the Efficiency of XOS

Approaches to improve the efficiency of XOS as FO is targeted toward diverse fields ranging from human trials focused on its prebiotic effect to the screening of novel raw materials for optimization of enzymes required for XOS production. Lin et al. (2016) investigated the prebiotic effect of XOS on human through food supplementation. This research group reported an increase in the fecal *Bifidobacterium* sp., *Lactobacillus* sp. counts [76]. On the contrary, the counts of *Clostridium perfringens* decreased without affecting the total anaerobic microflora counts, providing evidence of improved intestinal microbiota balance due to the consumption of XOS supplemented food. Approaches of the effects of XOS on the growth of probiotic bacteria also proved the prebiotic efficacy of XOS [77]. Similarly, the efficiency of XOS produced by *Streptomyces thermovulgaris* TISTR1948 from corncob was found to be comparable to commercially available XOS when tested on probiotic bacteria like *Lactobacillus casei* TISTR1463, *L. lactis* TISTR1464, and *L. plantarum* TISTR1465. Yet another variation was experimented by using mold-derived enzyme with the commercial enzyme cocktail to yield promising XOS synthesis with prior chemical treatment of the raw material [78]. In order to combat the hindrance caused by macromolecules in the industrial process, microwave-assisted enzymatic hydrolysis was used for XOS production [79]. Table 17.4 provides the information on some microbial sources of enzymes catering production of XOS.

XOS has enormous applications in fields of agriculture, health foods, pharmaceuticals, and feed formulations. The demand of XOS is on a peak in food industry due to its versatile application, but high process cost has limited its use. XOS finds application in food products like confectionaries items, dairy products, soymilk, soft drinks, and sweets. The review of Jain et al. (2015) provides a consolidated knowledge on production, purification, and structural characterization of XOS and their health benefits [71].

17.18 Conclusions and Perspectives

Although much of the experiments are directed toward the investigations of prebiotic properties of FO, yet there are a lot of metabolic implications to be still discovered. The impact of FO on human metabolism is to be targeted by controlled human feeding studies. The modern

Table 17.4 Xylooligosaccharides production and their microbial sources utilizing various substrates.

Source	Raw material	Enzyme	References
Aspergillus fumigatus R1,	wheat husk	Xylanases	[77]
Streptomyces thermovulgaris TISTR1948	Corncob	Thermostable endoxylanase	[80]
Aspergillus fumigatus M51 *Aspergillus fumigatus* U2370 *Trichoderma reesei* CCT 2768	Sugarcane bagasse	Xylanases	[81]
Aspergillus niger 3T5B8 Viscozyme (commercial cocktail)	Brazilian Syrah grape pomace	Xylanases	[78]
Aspergillus oryzae LC1	Sugarcane baggase, wheat straw, and wheat bran	Xylanases	[82]

molecular techniques are to be used as an approach toward the determination of FO effect on the intestinal microbiota. Specific population studies can elaborate on the effect of FO on general health and its correlation with the environment and food habits of the individual.

Efficient bioprocess layout paralleled with economy and environment-friendly approaches are the need of the health industry for designing industrial production of FO. The upcoming fields of research are novel enzyme sources, stable enzymes, innovative and green technology-based production processes, and well-controlled animal and human trials.

References

1 Link, R. (2020). What Are Functional Foods? All You Need to Know. Healthline https://www.healthline.com/nutrition/functional-foods.

2 Belorkar, S.A. (2020). Fungal production of prebiotics Chapter 10. In: *Fungal Biotechnology and Bioengineering* By (ed. A.E.-L. Hesham), 239–254. Switzerland AG: Springer Nature.

3 Weijers, C.A., Franssen, M.C., and Visser, G.M. (2008). Glycosyltransferase-catalyzed synthesis of bioactive oligosaccharides. *Biotechnology Advances* 26 (5): 436–456. https://doi.org/10.1016/j.biotechadv.2008.05.001.

4 Roberfroid, M. and Slavin, J.L. (2000). Non-digestible oligosaccharides. *Critical Reviews in Food Science and Nutrition* 40: 461–480.

5 Belorkar, S.A. and Gupta, A.K. (2016). Oligosaccharides: a boon from nature's desk. *AMB Express* 6: 82.

6 Belorkar, S.A., Gupta, A.K., and Rai, V. (2016). Enhancement of extracellular fructosyltransferase production by *Aspergillus stallus* through batch fermentation. *Journal of Pure and Applied Microbiology.* 10: 649–655.

7 Patel, S. and Goyal, A. (2011). (2011) Functional oligosaccharides: production, properties and applications. *World Journal of Microbiology and Biotechnology* 27: 1119–1128. https://doi.org/10.1007/s11274-010-0558-5.

8 Biely, P., Singh, S., and Puchart, V. (2016). Towards enzymatic breakdown of complex plant xylan structures: state of the art. *Biotechnology Advances* 34 https://doi.org/10.1016/j.biotechadv.2016.09.001.

9 Meyer (2019). Biotechnological Production of Oligosaccharides – Applications in the Food IndustryTathiana Souza Martins Meyer, Ângelo Samir Melim Miguel, Daniel Ernesto Rodríguez Fernández and Gisela Maria Dellamora Ortiz Additional information is available at the end of the chapter doi:10.5772/60934

10 Bharti, S.K., Kumar, A., Krishnan, S. et al. (2015). Mechanism-based antidiabetic activity of Fructo- and isomalto-oligosaccharides: Validation by *in vivo*, *in silico* and *in vitro* interaction potential. *Process Biochemistry* 50: 317–327. https://doi.org/10.1016/j.procbio.2014.10.014.

11 Grootaert, C., Delcour, J.A., Courtin, C.M. et al. (2007). Microbial metabolism and prebiotic potency of arabinoxylan oligosaccharides in the human intestine. *Trends in Food Science and Technology* 18: 64–71.

12 De Vuyst, L., Moens, F., Selak, M. et al. (2014). Summer meeting: growth and physiology of bifidobacteria. *Journal of Applied Microbiology* 116: 477–491.

13 Zhang, T., Yang, Y., Liang, Y. et al. (2018). Beneficial effect of intestinal fermentation of natural polysaccharides. *Nutrients Aug* 10 (8): 1055. https://doi.org/10.3390/nu10081055.

14 Karlsson, E.N., Schmitz, E., Linares-Pastén, J.A., and Adlercreutz, P. (2018). Endo-xylanases as tools for production of substituted xylooligosaccharides with prebiotic properties. *Applied Microbiology and Biotechnology* 102 (21): 9081–9088. https://doi.org/10.1007/s00253-018-9343-4.

15 Berger, K., Falck, P., Linninge, C. et al. (2014). Cereal byproducts have prebiotic potential in mice fed a high-fat diet. *Journal of Agricultural and Food Chemistry* 62: 8169.

16 Linares-Pasten, J.A., Aronsson, A., and Karlsson, E.N. (2018). Structural considerations on the use of Endo-Xylanases for the production of prebiotic Xylooligosaccharides from biomass. *Current Protein & Peptide Science* 19 (1): 48–67. https://doi.org/10.2174/1389203717666160923155209.

17 Amorim, C., Silvério, S.C., and Rodrigues, L.R. (2019). One-step process for producing prebiotic arabino-xylooligosaccharides from brewer's spent grain employing Trichoderma species. *Food Chemistry* 270: 86–94.

18 Saville, B.A. and Saville, S. (2018). Xylooligosaccharidesand arabinoxylanoligosaccharides and their application as prebiotics. *Applied Food Biotechnology* 5 (3): 121–130.

19 Kumar, V. and Satyanarayana, T. (2015). Generation of xylooligosaccharides from microwave irradiated agroresidues using recombinant thermo-alkali-stable endoxylanase of the polyextremophilic bacterium Bacillus halodurans expressed in Pichia pastoris. *Bioresource Technology* 179: 382–389.

20 Mhetras, N., Mapre, V., and Gokhale, D. (2019). Xylooligosaccharides (XOS) as emerging prebiotics: its production from lignocellulosic material. *Advances in Microbiology* 09: 14–20. https://doi.org/10.4236/aim.2019.91002.

21 Chapla, D., Pandit, P., and Shah, A. (2012). Production of xylooligosaccharides from corncob xylan by fungal xylanase and their utilization by probiotics. *Bioresource Technology* 115: 215–221.

22 Mathew, S., Aronsson, A., Karlsson, E.N., and Adlercreutz, P. (2018). Xylo- and arabinoxylooligosaccharides from wheat bran by endoxylanases, utilisation by probiotic bacteria, and structural studies of the enzymes. *Applied Microbiology and Biotechnology* 102: 3105–3120.

23 Dotsenko, G., Meyer, A.S., Canibe, N. et al. (2018). Enzymatic production of wheat and ryegrass derived xylooligosaccharides and evaluation of their in vitro effect on pig gut microbiota. *Biomass Conversion and Biorefinery* 8: 497–507. https://doi.org/10.1007/s13399-017-0298-y.

24 Correia, M.A.S., Mazumder, K., Bras, J.L.A. et al. (2011). Structure and function of an arabinoxylan-specific xylanase. *The Journal of Biological Chemistry* 286 (25): 22510–22520. https://doi.org/10.1074/jbc.M110.217315.

25 Labourel, A., Crouch, L.I., Bras, J.L.A. et al. (2016). The mechanism by which arabinoxylanases can recognize highly decorated xylans. *The Journal of Biological Chemistry* 291 (42): 22149–22159. https://doi.org/10.1074/jbc.M116.743948.

26 St John, F.J., Hurlbert, J.C., Rice, J.D. et al. (2011). Ligand bound structures of a glycosyl hydrolase family 30 glucuronoxylan xylanohydrolase. *Journal of Molecular Biology* 407 (1): 92–109. https://doi.org/10.1016/j.jmb.2011.01.010.

27 Urbanikova, L., Vrsanska, M., Krogh, K. et al. (2011). Structural basis for substrate recognition by *Erwinia chrysanthemi* GH30 glucuronoxylanase. *The FEBS Journal* 278 (12): 2105–2116. https://doi.org/10.1111/j.1742-4658.2011.08127.x.

28 Goffin, D., Delzenne, N., Blecker, C. et al. (2011). Will isomalto-oligosaccharides, a well-established functional food in Asia, break through the European and American market? The status of knowledge on these prebiotics. *Critical Reviews in Food Science and Nutrition* 51: 394–409. https://doi.org/10.1080/10408391003628955.

29 Goffin, D., Wathelet, B., Blecker, C. et al. (2010). Comparison of the glucooligosaccharide profiles produced from maltose by two different transglucosidases from Aspergillus niger. *Biotechnologie, Agronomie, Société et Environnement* 14: 607–616.

30 Lina, B.A.R., Jonker, D., and Kozianowski, G. (2002). Isomaltulose (Palatinose®): a review of biological and toxicological studies. *Food and Chemical Toxicology* 40: 1375–1381. https://doi.org/10.1016/S0278-6915(02)00105-9.

31 Konishi, Y. and Shindo, K. (1997). Production of nigerose, nigerosyl glucose, nigerosylmaltose by Acremonium sp. S4G13. *Bioscience, Biotechnology, and Biochemistry* 61: 439–442.

32 Chaen, H., Nishimoto, T., Nakada, T. et al. (2006). Enzymatic characterization, and gene cloning of a novel maltosyltransferase from an Arthrobacter globiformis strain that produces an alternating α-1,4- and α-1,6-cyclic tetrasaccharide from starch. *Applied and Environmental Microbiology* 72: 1065–1071. https://doi.org/10.1128/AEM.72.2.1065-1071.2006.

33 Fabre, E., Bozonnet, S., Arcache, A. et al. (2005). Role of the two catalytic domains of DSR-E dextransucrase and their involvement in the formation of highly α-1,2 branched dextran. *Journal of Bacteriology* 187: 296–303. https://doi.org/10.1128/JB.187.1.296-303.

34 Leemhuis, H., Pijning, T., Dobruchowska, J.M. et al. (2013). Glucansucrases: three-dimensional structures, reactions, mechanism, α-glucan analysis and their implications in biotechnology and food applications. *Journal of Biotechnology* 163: 250–272. https://doi.org/10.1016/j.jbiotec.2012.06.037.

35 Mountzouris, K.C., Gilmour, S.G., and Rastall, R.A. (2002). Continuous production of oligodextrans via controlled hydrolysis of dextran in an enzyme membrane reactor. *Journal of Food Science* https://doi.org/10.1111/j.1365-2621.2002.tb08720.x.

36 Demuth, K., Jordening, H.J., and Buchholz, K. (2002). Oligosaccharide synthesis by dextransucrase: new unconventional acceptors. *Carbohydrate Research* 337: 1811–1820. https://doi.org/10.1016/S0008-6215(02)00272-0.

37 Moynihan, P.J. (1998). Update on the nomenclature of carbohydrates and their dental effects. *Journal of Dentistry* 26: 209–218. https://doi.org/10.1016/S0300-5712(97)00010-9.

38 Mukai, K., Watanabe, H., Kubota, M. et al. (2006). Purification, characterization, and gene cloning of a novel maltosyltransferase from an Arthrobacter globiformis strain that produces an alternating α-1,4- and α-1,6-cyclic tetrasaccharide from starch. *Applied and Environmental Microbiology* 72: 1065–1071. https://doi.org/10.1128/AEM.72.2.1065-1071.2006.

39 Tungland, B.C. and Meyer, D. (2000). Nondigestible oligo-and polysaccharides (dietary fiber): their physiology and role in human health and food. *Comprehensive Reviews in Food Science and Food Safety* 3: 73–92. https://doi.org/10.1111/j.1541-4337.tb00009.x.

40 Nguyen, T.H. and Haltrich, D. (2013). Microbial production of prebiotic oligosaccharides. In: *Microbial Production of Food Ingredients, Enzymes and Nutraceuticals* (eds. B. McNeil, D. Archer, I. Giavasis and L. Harvey), 494–530. Woodhead Publishing https://doi.org/10.1533/9780857093547.2.494.ch18.

41 Pan, X., Chen, F., Wu, T. et al. (2009). Prebiotic oligosaccharides change the concentrations of short-chain fatty acids and the microbial population of mouse bowel. *Journal of Zhejiang University SCIENCE B* 10: 258–263.

42 Zhang, L., Su, Y., Zheng, Y. et al. (2010). Sandwich structured enzyme membrane reactor for efficient conversion of maltose into isomaltooligosaccharides. *Bioresource Technology* 101: 9144–9149.

43 Florowska, A., Krygier, K., Florowski, T., and Dłużewska, E. (2016). Prebiotics as functional food ingredients preventing diet-related diseases. *Food & Function* 7: 2147–2155. https://doi.org/10.1039/C5FO01459J.

44 Chockchaisawasdee, S. and Poosaran, N. (2013). Production of isomaltooligosaccharides from banana flour. *Journal of the Science of Food and Agriculture* 93: 180–186. https://doi.org/10.1002/jsfa.5747.

45 Cho, S.K., Shin, S.Y., Lee, S.J. et al. (2015). Simple synthesis of isomaltooligosaccharides during *Sauerkraut* fermentation by addition of *Leuconostoc* starter and sugars. *Food Science and Biotechnology* 24: 1443–1446.

46 Niu, D., Qiao, J., Pujun, L. et al. (2016). Highly efficient enzymatic preparation of isomalto-oligosaccharides from starch using an enzyme cocktail. *Electronic Journal of Biotechnology* 26 https://doi.org/10.1016/j.ejbt.2016.12.002.

47 Gibson, G.R. and Rastall, R.A. (2006). *Prebiotics: Development & Application.* Chichester: Wiley.

48 Basu, A., Mutturi, S., and Prapulla, S.G. (2015). Modeling of enzymatic production of isomaltooligosaccharides: a mechanistic approach. *Catalysis Science & Technology* **5**: 2945–2958.

49 Kim, I.C., Cha, J.H., Kim, J.R. et al. (1992). Catalytic properties of the cloned amylase from Bacillus licheniformis. *The Journal of Biological Chemistry* 267: 22108–22114.

50 Abdul Manas, N.H., Pachelles, S., Mahadi, N.M., and Illias, R.M. (2014). The characterisation of an alkali-stable maltogenic amylase from Bacillus lehensis G1 and improved maltooligosaccharide production by hydrolysis suppression. *PLoS One* 9: e106481. https://doi.org/10.1371/journal.pone.0106481.

51 Lee, S.J., Yoo, S.H., Kim, M.J. et al. (1995). Production and characterization of branched oligosaccharides from liquefied starch by the action of Bacillus licheniformis amylase. *Starch-Starke* 47: 127–134. https://doi.org/10.1002/star.19950470403.

52 Lee, H.S., Auh, J.H., Yoon, H.G. et al. (2002). Cooperative action of α-glucanotransferase and maltogenic amylase for an improved process of isomaltooligosaccharide (IMO) production. *Journal of Agricultural and Food Chemistry* 50: 2812–2817. https://doi.org/10.1021/jf011529y.

53 Moulis, C., Vaca Medina, G., Suwannarangsee, S. et al. (2008). One-step synthesis of isomalto-oligosaccharide syrups and dextrans of controlled size using engineered dextransucrase. *Biocatalysis and Biotransformation* 26 (1-2): 141–151. https://doi.org/10.1080/10242420701799493.

54 Alfertsholt-Allen T. (2007). Market developments and industry challenges for lactose and lactose derivatives. IDF Lactose Symposium; 14–16 May; Moscow

55 Gibson, G.R. and Roberfroid, M.B. (2008). Prebiotics: concept, definition, methodologies, criteria and products. In: *A Handbook of Prebiotics* (eds. G.R. Gibson and M.B. Roberfriod), 39–68. CRC Press.

56 Nath, A., Haktanirlar, G., Varga, A. et al. (2018). Review biological activities of lactose-derived prebiotics and symbiotic with probiotics on gastrointestinal system. *Medicina* 54: 18. https://doi.org/10.3390/medicina54020018, www.mdpi.com/journal/medicina.

57 Silverio, S.C., Macedo, E.A., Teixeira, J.A., and Rodrigues, L.R. (2015). Perspectives on the biotechnological production and potential applications of lactosucrose: a review. *Journal of Functional Foods* 19 (Part A): 74–90. https://doi.org/10.1016/j.jff.2015.09.014.

58 Arango, O., Trujillo, A.J., and Castillo, M. (2013). Influence of fat replacement by inulin on rheological properties, kinetics of rennet milk coagulation, and syneresis of milk gels. *Journal of Dairy Science* 96: 1984–1996. https://doi.org/10.3168/jds.2012-5763.

59 Bassarab, S, Bechtold, KP, Fuhrherr, R et al. (2005). Powder comprising new compositions of oligosaccharides and methods for their preparation. US Patent No.; 20050250704

60 Teramoto, F., Rokutan, K., Kawakami, Y. et al. (1996). Effect of 4^G-β-D-galactosylsucrose (lactosucrose) on fecal microflora in patients with chronic inflammatory bowel disease. *Journal of Gastroenterology* 31: 33–39.

61 Nobuaki, I. (1998). Antiathlete's foot composition WO Patent No.; 1998057650

62 Cheul, H.W., Ho, B.S., Hyun, K.M. et al. (2009). Production of lactosucrose from sucrose and lactose by a levansucrase from zymomonas mobilis. *Journal of Microbiology and Biotechnology* 19: 1153–1160. https://doi.org/10.4014/jmb.0901.045.

63 Long, J., Pan, T., Xie, Z. et al. (2019). Effective production of lactosucrose using β-fructofuranosidase and glucose oxidase co-immobilized by sol–gel encapsulation. *Food Science & Nutrition*: 7. https://doi.org/10.1002/fsn3.1195.

64 Corzo-Martínez, M., Luscher, A., Rivas, B. et al. (2015). Valorization of Cheese and Tofu whey through enzymatic synthesis of lactosucrose. *PLoS One* 10: e0139035. https://doi.org/10.1371/journal.pone.0139035.

65 Chen, C., Deng, J., Lv, X. et al. (2020). Biocatalytic synthesis of lactosucrose using a recombinant thermostable β-fructofuranosidase from Arthrobacter sp. 10138. *Bioengineered* 11 (1): 416–427. https://doi.org/10.1080/21655979.2020.1739404.

66 Xu, W., Liu, Q., Yu, S. et al. (2018). Synthesis of lactosucrose using a recombinant levansucrase from Brenneria goodwinii. *Applied Biochemistry and Biotechnology* 186 (2): 292–305.

67 Yin, H., Pijning, T., Meng, X. et al. (2017). Engineering of the Bacillus circulans β-galactosidase product specificity. *Biochemistry* 56 (5): 704–711.

68 Li, W., Yu, S., Zhang, T. et al. (2015). Efficient biosynthesis of lactosucrose from sucrose and lactose by the purified recombinant levansucrase from Leuconostoc mesenteroides B-512 FMC. *Journal of Agricultural and Food Chemistry* 64 (44): 9755–9763.

69 Lu, L., Fu, F., Zhao, R. et al. (2014). A recombinant levansucrase from Bacillus licheniformis 8–37-0-1 catalyzes versatile transfructosylation reactions. *Process Biochemistry* 49 (9): 1503–1510. https://doi.org/10.1016/j.procbio.2014.05.012.

70 Vazquez, M.J., Alonso, J.L., Domınguez, H., and Parajo, J.C. (2000). Xylooligosaccharides: manufacture and applications. *Trends in Food Science and Technology* 11: 387–393.

71 Jain, I., Kumar, V., and Satyanarayana, T. (2015). Xylooligosaccharides: an economical prebiotic from agroresidues and their health benefits. *Indian Journal of Experimental Biology* 53 (3): 131–142.

72 Courtin, C.M., Swennen, K., Verjans, P., and Delcour, J.A. (2009). Heat and pH stability of prebiotic arabinoxylooligosaccharides, xylooligosaccharides and fructooligosaccharides. *Food Chemistry* 112: 831–837.

73 Meyer, T., Ângelo, M., Fernández, D., and Gisela, O. (2015). Biotechnological production of oligosaccharides – applications in the food industry. In: *Food Production and Industry* (ed. A.H.A. Eissa). InTech.

74 Makelainen, H., Forssten, S., Saarinen, M. et al. (2010). Xylo-oligosaccharides enhance the growth of bifidobacteria and Bifidobacterium lactis in a simulated colon model. *Beneficial Microbes* 1 (1): 81–91. https://doi.org/10.3920/BM2009.0025.

75 Singh, A.B. (2019). Production, characteristics, and biotechnological applications of microbial xylanases. *Applied Microbiology and Biotechnology* 103: 8763–8784. https://doi.org/10.1007/s00253-019-10108-6.

76 Lin, S.H., Chou, L.M., Chien, Y.W. et al. (2016). Prebiotic effects of Xylooligosaccharides on the improvement of microbiota balance in human subjects. Gastrointestinal inflammation and repair: role of microbiome, infection, and nutrition. *Gastroenterology Research and Practice*: 5789232. https://doi.org/10.1155/2016/5789232.

77 Jagtap, S., Deshmukh, R.A., Menon, S., and Santanu, D. (2017). Xylooligosaccharides production by crude microbial enzymes from agricultural waste without prior treatment and their potential application as nutraceuticals. *Bioresource Technology* 245 (Pt A): 283–288.

78 Costa, J.R., Tonon, R.V., Gottschalk, L.M. et al. (2019). Enzymatic production of xylooligosaccharides from Brazilian Syrah grape pomace flour: a green alternative to conventional methods for adding value to agricultural by- products. *Journal of the Science of Food and Agriculture* 99 (3): 1250–1257. https://doi.org/10.1002/jsfa.9297.

79 Wang, T.H. and Lu, S. (2013). Production of xylooligosaccharide from wheat bran by microwave assisted enzymatic hydrolysis. *Food Chemistry* 138 (2-3): 1531–1535. https://doi.org/10.1016/j.foodchem.09.124.

80 Boonchuay, P., Techapun, C., Seesuriyachan, P., and Chaiyaso, T. (2014). Production of xylooligosaccharides from corncob using a crude thermostable endo-xylanase from *Streptomyces thermovulgaris* TISTR1948 and prebiotic properties. *Food Science and Biotechnology* 23: 1515–1523. https://doi.org/10.1007/s10068-014-0207-0.

81 Carvalho, A.F., Neto, P.O., Zaghetto, A.P. et al. (2015). Screening of Xylanolytic *Aspergillus fumigatus* for Prebiotic Xylooligosaccharide production using bagasse. *Food Technology and Biotechnology* 53 (4): 428–435. https://doi.org/10.17113/ftb.53.04.15.4160.

82 Bhardwaj, N., Kumar, B., Agarwal, K. et al. (2019). Purification and characterization of a thermo-acid/alkali stable xylanases from Aspergillus oryzae LC1 and its application in Xylo-oligosaccharides production from lignocellulosic agricultural wastes. *International Journal of Biological Macromolecules* 122: 1191–1202. https://doi.org/10.1016/j.ijbiomac.2018.09.070.

18

Algal Biomass and Biofuel Production

Suman Sanju[1], Aditi Thakur[2], Pragati Misra[3], and Pradeep Kumar Shukla[4]

[1] Department of Botany, M.C.M.D.A.V. College, Kangra, Himanchal Pradesh, India
[2] Department of Biotechnology, Harlal institute of Management and Technology, Greater Noida, Uttar Pradesh, India
[3] Department of Molecular and Cellular Engineering, Jacob Institute of Biotechnology and Bioengineering, Sam Higginbottom University of Agriculture, Technology and Sciences, Prayagraj, Uttar Pradesh, India
[4] Department of Biological Sciences, Faculty of Science, Sam Higginbottom University of Agriculture, Technology and Sciences, Prayagraj, Uttar Pradesh, India

18.1 Introduction

Due to sustainable and renewable nature, the demand of fossil fuels is broadening ecumenically owing to the expeditious ever-increase in the population and is expected to grow 40% by 2040. While continuous and over-exhaustive usage of conventional fossil fuels, viz., petroleum, coal, and natural gases, globally, are leading shrinkage and depletion of fossil fuels due to their nonrenewable nature. In integration, due to high carbon percentage, they are potent source of pollution and environmental hazards, thus distressing human health adversely and also inimical to economy due to soaring prices. To combat all these quandaries, the renewable energy sources like hydro, solar, wind, tidal, and biomass have been explored as the potential and promising alternatives. Biomass procured from several resources such as food crops, crop wastes, bio-wastes, and algae are utilized by many developed countries to generate biofuels, i.e. bioethanol and biodiesel at commercial scale [1, 2]. Furthermore, the microalgae predicated fuels are environmentally safe, nonhazardous, sustainable, renewable, and possess inherent integral potential for fixing CO_2, globally. The countries Brazil, United States, France, Sweden, and Germany are pioneers in the biofuel production and consumption.

18.2 Biofuels

Biofuel is a solid, liquid, or gaseous material procured directly or indirectly from biomass feedstock of plant or animal origin. Biomass is the natural material either living or dead can exist as either energy crops, biowastes, and by-products from agricultural and forestry, manure or municipal solid wastes, or any other material containing microbial biomass. Biological biomass can serve as an effective renewable sources of energy and can be

Bioprospecting of Microorganism-Based Industrial Molecules, First Edition.
Edited by Sudhir P. Singh and Santosh Kumar Upadhyay.

replenished conveniently unlike the fossil fuels. The major forms of biofuels include biodiesel, bioethanol, biomethane, biohydrogen, charcoal, and fuelwood, which are found in perpetual engenderment [3], and are considered as key alternative to the conventional fossil fuels due to their cost-efficacy, environment friendly, and renewable nature. Additionally, the high demand, due to incessant usage, of fossil fuels is leading toward the depletion of the fossil fuel reserves. Conjointly, incremented consumption of the fossil fuels has astronomically immense negative impact on human health, as well as on the environment. The extraordinary emission of the greenhouse gases from these fuels is creating imminent threat of global warming. Till now, 35.3 billion tons of CO_2 has been emitted by the fossil fuels at the alarming rate of 29 gigatons/year [4]. In addition, the devastation of flora and fauna habitats for oil resource exploration and their subsequent utilization is provoking a prodigiously adverse impact on the environment. The fossil fuels emit high sulfur levels in the environment and zero oxygen levels compared to the algal biofuels that emit 10–45% oxygen levels with reduced sulfur levels. Further, in context of the ascending petroleum prices, biofuel is commonly advocated as a cost-effective alternative. There are different generations of biofuels depending upon the biomass feedstock, utilized for their production:

18.2.1 First-Generation Biofuels

The first generation of the biofuels generally called as conventional biofuels are procured from the feedstock material of conventional and edible crops [5]. Different types of first-generation biofuels include bioethanol, biodiesel, biobutanol, etc. Several methods, viz., fermentation, transesterification, and distillation, have been utilized for the production of biofuels. The feedstock can be utilized to engender biofuels such as bioethanol through starch fermentation of variety of crops like wheat, corn, potato, barley, sugarcane, sugarbeet, etc. All these victuals crops are most commonly utilized for biofuel engenderment. The oil crops such as rapeseed, palm, coconut, sunflower, and animal fat are utilized as feedstock for biofuel production through chemical extraction, i.e. transesterification to obtain biodiesel [6, 7].

First-generation biofuels are expedient in many ways: (i) they can be coalesced with petroleum-predicated fuels; (ii) they are competent for heat engines like internal combustion engines (ICE) and (iii) are harmonious with dual-fuel automobiles. Conversely, several drawbacks are also associated with first-generation biofuels. Since the crops, utilized for generating first-generation biofuels are food and commercial crops, their utilization increased world food prices and animal feed costs. First-generation biofuels are, therefore, economically inauspicious as they are more sumptuous. Withal, they require a plethora of land to grow adequate quantity of biomass. Moreover, there are several issues associated with the production of first-generation biofuels: (i) competition for food production and food availability, (ii) environment damage due to enhanced land clearing, (iii) biodiversity loss due to habitat devastation, (iv) water depletion, and (v) air pollution [8].

18.2.2 Second-Generation Biofuels

Biomass material for second-generation biofuels includes nonedible, lignocellulosic biomass, municipal solid wastes, manures, agricultural and forest wastes, e.g., jatropha, cassava, and poplar. One of the most noteworthy advantages of second-generation biofuels is the utilization of supplementary and inexpensive feedstock for their procurement. Further, it does not pose any threat to the conventional crop engenderment as in their first-generation counterparts because the plants utilized are inedible and can be grown on the infertile lands where food crops are incompetent to grow [9]. Withal, they generate higher energy yields per acre in contrast to first-generation fuels.

The disadvantage of producing second-generation fuels lies in their extraction process. Since biomaterials for second-generation biofuels are enriched with lignocellulosic biomass, the extraction of biofuel requires the pretreatment step that poses a challenge in the form of technical difficulties and additionally increases the production cost. Lignin reinforces plant cell wall and makes them more vigorous and resistant needs to be broken down first through sundry chemical reactions, followed by hydrolysis of carbohydrate polymers into simple sugar molecules.

18.2.3 Third-Generation Biofuels

Third-generation biofuels include the algal-derived biomass for production of cost efficacious and viable alternative to fossil fuels. Algae are low-cost, high-yielding biomass for biofuel production. It generates 30 times more energy per acre than land crops such as soybeans. Algae can be cultured throughout the year irrespective of the seasons; it is expeditiously growing and consequently has high lipid content than any other biomass feedstock of first and second generation that makes it more promising as biomass for biofuel production. Algal biomass can photosynthetically produce large amount of crude oil, which can be facilely processed into diesel and gasoline and other by-products [10]. Seaweed has high cellulose-to-lignin content ratio compared to other feedstock [11]. However, there is still need to conduct abundant research to minimize the production cost and to make it commercially viable as third-generation biofuel. Reshuffling metabolic circuits and genome editing of microalgae can augment the biofuel yield, in addition to cost depreciation.

The most crucial step in biofuel generation is the selection and processing of biomass as it might be environmentally sustainable and cost-determining step. Microalgae do not entail any kind of pretreatment in contrast to some crops comprising lingocellulosic material. Additionally, microalgae do not exhibit any direct or indirect effect on the productivity and supply of food and other commercial crops and their products/by-products. It can be grown on any type of terrestrial or aquatic ecosystem irrespective of the seasonal conditions. Despite of its prominence in biofuel production, it can withal be utilized in many other industries, as well as in the form of foods, animal feeds, etc. The microalgae are efficient greenhouse gas removers; withal they are considered as the potential source of biofuels. Hence, to compensate the ever-intensifying demand for biofuels, microalgae seem to be the most promising renewable feedstock material for biofuel engenderment [2, 4]. Besides, the application of nanotechnology to increase biofuel production from algae is also being explored [86].

18.3 Algae: The Biomass

Plants have been classified into many divisions predicated on their characteristics. Algae are among diverse division of plant kingdom having prokaryotic and eukaryotic organisms, which populates in various habitats including ponds, rivers, lakes, oceans, wastewater lands, as well as terrestrial habitats, encompassing the extreme habitat conditions like saline water

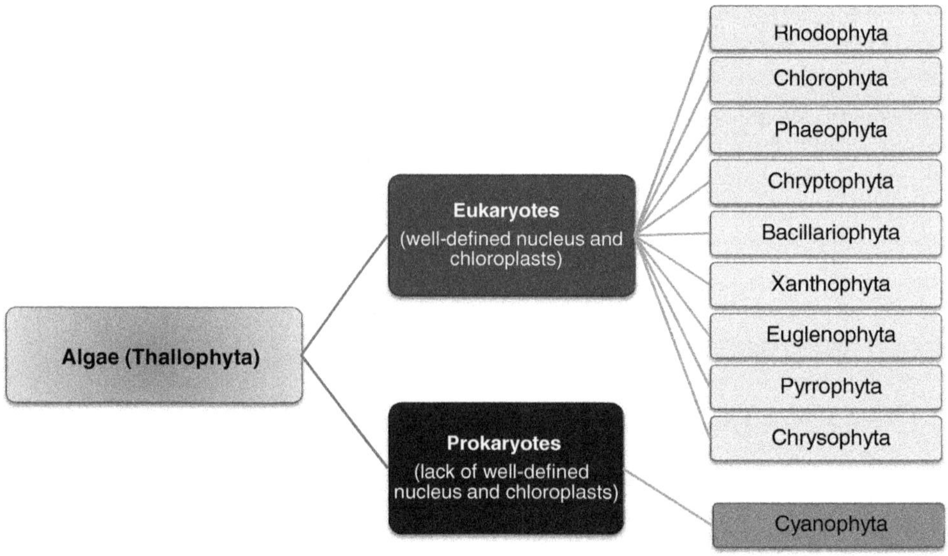

Figure 18.1 Classification of algae.

or inside some other animals, etc. They are either autotrophic or heterotrophic ranging from unicellular such as *Chlorella* to multicellular forms such as giant kelp. One of the most consequential characteristic of algae is to produce biomass utilizing carbon dioxide and light. It is broadly divided into prokaryotic, i.e. Cyanophyceae and eukaryotic algae, e.g. Chlorophyceae and Rhodophyceae.

Algae are further categorized into different divisions predicated upon presence or absence of certain characters such as pigments, flagella, habitat, and size (Figure 18.1). Algae are characterized by the presence of chlorophyll-containing thalloid body; unlike higher plants they have no embryo, are devoid of vasculature, and do not consist of jackets enveloping sex organs. Economically, algae are harnessed for diverse purposes, e.g. bioremediation, engenderment of energy cogeneration, wastewater management, animal fodder, medicinal compounds, and nutraceuticals, as well as CO_2 mitigation.

18.4 Microalgae as Biofuel Biomass

Algae are epitomized into two classes predicated upon the size of organisms: microalgae and macroalgae. Unicellular organisms having miniature size in micrometers are called as microalgae, e.g. *Chlorella*, while the multicellular algae with larger size, quantified in inches are called as macroalgae, e.g. seaweed. Of these two, microalgae are considered the largest unicellular photoautotrophic microorganisms present on the earth. Microalgae are considered as the potential and better source of biofuel compared to other traditional crops utilized for the same purport owing to its distinct characteristics such as prompt multiplication/growth rate, ability to fix greenhouse gases, and higher lipid synthesis capacity [12]. Microalgae have capacity to double its biomass within a period of 24 hours and within few days its growth cycle can also be completed [13]. Consequently, it is considered as 10–20%

more effective in biomass production compared to traditional terrestrial biofuel crops such as soybean, palm oil, etc. [14]. Some microalgae are also reported to double their biomass within three and half hours, during the peak growth phase [15]. The conventional oil content of microalgae ranges between 20 and 50% of dry weight, while it can be as high as 80% in some high oil-yielding strains [16, 17]. About 30000 species of microalgae out of about more than 50000 species are being investigated for biofuel production [18–20]. Withal, the algal mass can be cultured on nonarable waste lands utilizing non-potable saltwater and waste water. The biofuels extracted from microalgae are nonhazardous, environment cordial, and have the vigorous capacity of sequestering global CO_2. To an estimate, 1 kg of algal biomass has been reported to sequester 1.83 kg of CO_2; furthermore, SOx and NOx are additionally utilized by some species as nutrient flow along with CO_2 [21].

Microalgae are photoautotrophic organisms, consequently it consists of high percentage of carbon compounds, which are crucial in biofuel extraction, in pharmaceuticals, cosmetics, and health supplements [22]. CO_2 constitutes 50% of dry weight of algal biomass. Microalgae can provide several variants of renewable biofuels (Figure 18.2). These include methane produced by anaerobic digestion of the microalgal biomass [17]; biodiesel derived

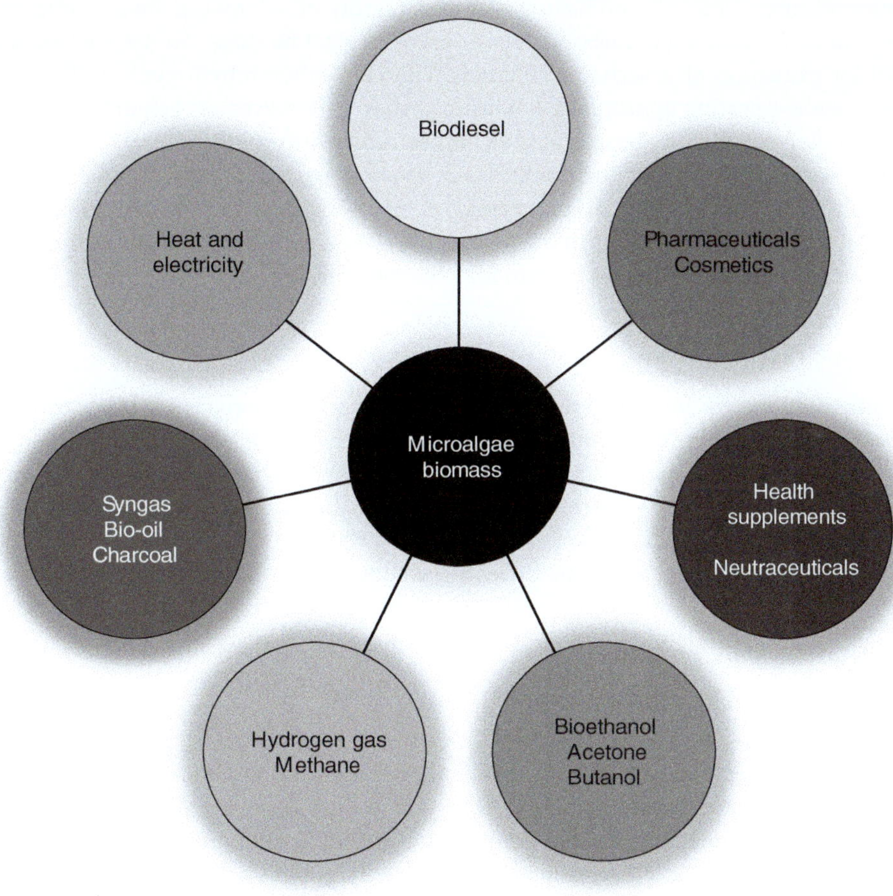

Figure 18.2 Various products of microalgal biomass.

from microalgal lipids [23–27]; bioethanol produced through fermentation, and photobiologically produced bio-hydrogen [28–31].

Microalgae withal synthesize extensive array of biomolecules including pigments, vitamins, proteins, carbohydrates, lipids, antioxidants, and bioactive compounds [8]. Consequently, microalgae are considered as lucrative alternative to biofuel feedstock and are gaining incrementing interest from scientific community, industrialists, and the common public. Microalgae growth enhancement strategies and genome manipulation techniques may be acclimated to amend their potential as a future source of renewable biomaterials. Lipid content, adipose acid composition, growth rate, and cultivation conditions are some of the parameters that need to be considered while screening the microalgal feedstock and to maximize oil yield per acre for biofuel engenderment.

18.5 Microalgae Culture Systems

For the first-rate production of biofuels, the cultivation of microalgae is a consequential factor. Many conventional, as well as improved methods are available to carry out the mass cultivation of microalgae for production of high quantity of microalgal lipids. Different biochemical components, i.e. carbohydrate or oil from algal biomass, are utilized as raw material for extraction of a variety of biofuels. Therefore, the pretreatment of algae are obligatory while selecting appropriate harvesting process. However, no general process is available for harvesting and extracting all algal species and utilizations.

Yield of biofuels and other value-integrated products depends upon the type of cultivation system. Two types of cultivation systems, i.e. open or closed systems are utilized for microalgae culture. A third system, a fusion of a closed and open system kenned as hybrid system, can withal be acclimated to obtain high biomass productivity.

18.5.1 Open Algal Systems

Commonly used system for cultivation of microalgae is open microalgal systems as they have virtuous opportune time to make utilization of the readily available atmospheric CO_2. Diverse configurations of microalgae cultivation systems are expedient for biomass production and for improved phyco-remediation of wastewaters from industries, domestic usage, and agricultural practices; nevertheless, commonly used systems for the industrial microalgal cultivation and research purposes are as follows:

- The raceway pond
- The circular pond tank
- The shallow big pond
- The closed pond

Open-culture systems are usually prudent, more durable, and with high productivity in comparison to the closed systems. For an open system, the selection of location is a paramount factor for the adequate sunlight availability along with other requisites for plentiful algae culture. The open ponds can either be natural or man-made or usually entail natural basins, circular ponds, big/small tanks, and shallow raceway ponds. Conventional culture of *Chlorella* sp. is done in concrete-made circular ponds and is further equipped with rotating arm to ascertain the mixing of culture and to evade the sedimentation of algal biomass. Commonly, the raceway pond entails race tracks or oval channels composed of concrete to circulate essential nutrients and CO_2 routinely to the algal cultures [32].

18.5.2 Closed Algal Systems

Photobioreactors (PBRs) comprise closed system ensuring well-controlled growth conditions for algae culture. PBRs are optimized conferring to the biochemical, as well as physiological properties of the algal species and therefore, sanctions to cultivate any algae species that cannot be cultivated in open systems. Commonly, these bioreactors are designed to enhance the accessibility of light. Several advantages attribute to PBRs, over an open system comprise, better control over culture conditions in terms of optimized pH, temperature, CO_2, mixing, and O_2. PBRs decrease CO_2 losses, apprehend evaporation, sanction to achieve high cell concentration, and microalgae density. PBRs withal provide a more protective environment by averting contamination or incursion of microorganisms. As PBRs resolve many quandaries cognate to the open cultivation, researchers have focused on the design/configuration of PBRs for massive microalgal biomass engenderment [32]. Predicated to the set-up in terms of geometry and architecture, PBRs can be constructed as towers, tanks, or bags. PBRs can be shaped as plates or tubular and engineered from glass or plastic. Tubular PBRs appear to be the utmost appropriate. Nevertheless, airlift PBRs and bubble columns can withal be considered as they cultivate a relatively higher concentration of microalgal biomass product [33]. An auxiliary tank is utilized to disunite the oxygen engendered from the photosynthesis as exorbitant oxygen can urge negative impact on the magnification of microalgae [34]. In spite of the betterments or innovations in the designs of PBRs for enhancing biofuel productivity, bottlenecks are yet to be resolved considering the biofuel productivity and their cost economics.

Among the open and closed systems of algal cultivation, the operational cost and productivity per unit area are the determining factors for selection of cultivation system. The advantages of open system are: (i) the competency to fine-tune atmospheric CO_2, (ii) low cost, (iii) utilization of sunlight reliant to the location of open system. Conversely, the major drawbacks include: (i) contamination by bacteria or other microorganisms, (ii) low light perforation, evaporation loss, and (iii) inhibition to temperature regulation [35, 36].

The PBRs have advantages of meticulous and optimized growth conditions with favorable light perforation that enhance productivity rates. Contamination can withal be obviated. However, the production cost increases due to infrastructure, operation, and maintenance [37]. Consequently, the open and closed systems were combined and a third setup called hybrid system was established.

18.5.3 Hybrid Algal Systems

The hybrid system setup integrates both open and closed systems. They are cost-efficacious and can be utilized for cultivation of voluminous algal biomass [38]. Besides, they also overcome many constraints of open systems and high initial and operating cost linked with closed systems. In the hybrid system, the microalgae are primarily cultured in closed and controlled PBR systems and then transferred to open system for the upsurge in the biomass yield [39]. This system can engender high biomass along with high nutrient abstraction. Hybrid system offers promising alternatives for astronomically immense algal cultivation toward biofuel production.

18.6 Microalgae Harvesting

Harvesting of algae is the collection of algal cells from cultivation medium deprived of losing its water content. Harvesting of microalgae contributes 20–30% of the total biomass production cost [40]. Harvesting procedures that are commonly used comprise of ultrafiltration, sedimentation, centrifugation, sonication, precipitation, and sometimes flocculation or the combination of flotation–flocculation. Flocculation increases the particle size by aggregation of microalgal cells leading to better centrifugation recovery [40]. Four main harvesting method studied by Weismann and Goebel are (i) micro-straining, (ii) belt filtering, (iii) flotation with float collection, and (iv) sedimentation. These methods separate the algal biomass on the basis of size and density. Due to the simple mechanical designing and the accessibility in large unit sizes, micro-strainers is an attractive method. In comparison to the conventional filtrations, ultracentrifugation and membrane micro-filtrations are more adequate for frangible cells and small-scale production process. Otherwise, dehydration can be done afore harvesting. Dehydration is the practice of abstracting the water content from algal cells to get dry mass [41].

18.7 Processing and Extraction of Components

After the disunion of algal biomass, it must be processed expeditiously to evade spoilage within a few hours, especially in humid climate. Generally, dehydration of biomass is applied to enhance its shelf life and the yield of final product. After drying of the biomass, several cell disruption methods predicated upon the mechanical and nonmechanical methods can be utilized. Mechanical methods comprise of cell homogenizers, bead mills, autoclave, spray drying, ultrasounds, enzyme reactions, acid base assays, freezing, osmotic shocking, etc. Fatty acids and lipids need to be extracted for the engenderment of biodiesel from the microalgal biomass. Solvent extraction is customarily done directly from the lyophilized biomass for the expeditious and adequate extraction of lipids from the microalgal biomass. About 98% of the quantitative extraction of fatty acids is possible by utilizing the solvents such as hexane, ethanol (96%), or ethanol:hexane amalgamation (96%). Research studies have already shown that compared to the conventional methods, the utilization of ultrasound/multimode microwave irradiation worked best and incremented the extraction yield [42].

18.8 Biofuel Conversion Processes

The transformation of microalgae biomass to different energy resources is carried out by different processes called as conversion processes. The variants of conversion processes are:

1) Chemical reactions:
   - Transesterification
2) Biochemical conversion
   - Fermentation
   - Anaerobic digestion
   - Photobiological H_2 production
3) Thermochemical conversion
   - Gasification
   - Pyrolysis
   - Liquefaction
4) Direct combustion
   - Power generation

18.8.1 Transesterification

Transesterification is the chemical conversion of microalgal biomass to biodiesel and glycerol as by-product. There are two methods of transesterification- i) conventional method where lipid extraction is performed through esterification algal biomass, and the second method is ii) in situ direct single-step method without lipid extraction. Albeit direct transesterification is energy, cost, and time efficacious, but the yield of biodiesel is higher in conventional method. Transesterification is performed on triacylglycerols (TAG), the biomolecules comprising of three long variable chains of fatty acids (saturated or unsaturated) annexed to a glycerol molecule through ester bond formation. Transesterification is a reversible process where these bonds are broken down and it comprises of three steps. First, the TAG are converted to diacylglycerols (DAG), diacylglycerols are converted into monoacylglycerols (MAG) and determinately converted into esters (biodiesel) and glycerol as the by-product (Figure 18.3).

Acid or base catalyst is utilized to enhance the reaction rate of transesterification, during the manufacturing process. Alkali catalyst, i.e. NaOH and KOH, is generally utilized for the industrial process, and this alkali catalyst transesterification is the most conventionally used method for biodiesel production [43]. Monounsaturated fatty acids are the desirable fatty acids as they elevate the cloud point and viscosity of biodiesel and are favored for

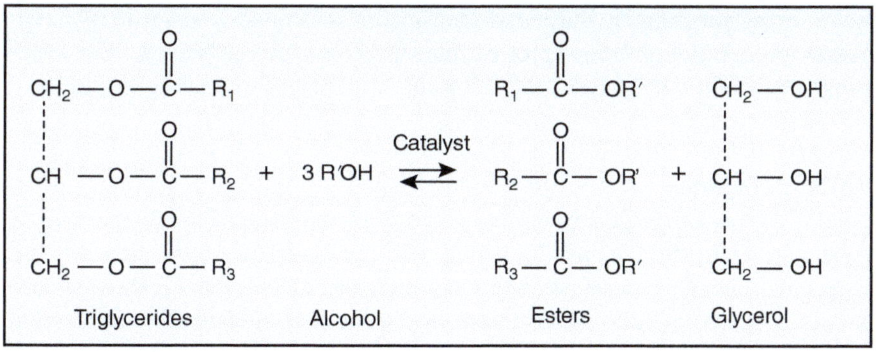

Figure 18.3 Transesterification reaction of triglycerides (overall reaction). *Source:* Mata et al. [32] / with permission of Elsevier.

biodiesel production compared to the polyunsaturated fatty acids as they tend to oxidized easily. Consequently, palmitic acid (*n*-hexadecanoic acid C16:0), stearic acid (*n*-octadecanoic acid C18:0), oleic acid (*cis*-9-octadecenoic acid C18:1Δ^{9}), linoleic acid (*cis,cis*-9,12-octadecadienoic acid C18:2$\Delta^{9,12}$), and linolenic acid (*all-cis*-9,12,15-octadecatrienoic acid C18:3$\Delta^{9,12,15}$) are the major fatty acids used for biodiesel production [44]. The properties of biodiesel such as viscosity, melting point, and heat of combustion of biodiesel are directly cognate to the length of the fatty acid chain and inversely linked to unsaturation [45]. However, increasing unsaturation withal upsurges autoxidation and lubricity [46].

18.8.2 Biochemical Methods

Biochemical conversion involves the hydrolysis of cell wall of algae-utilizing microorganisms and/or enzymes to generate biofuel. These processes are slow and less energy intensive compared to the other processes. These processes require pretreatment of biomass and include fermentation, anaerobic digestion, and photo-biological H_2 production.

18.8.2.1 Fermentation

Fermentation is the anaerobic digestion of carbohydrates such as sugar, starch and so on into biofuel, i.e. bioethanol and biogas utilizing microorganisms. Microalgae should have high starch content for efficient fermentation. This is an enzymatic process where the hydrolysis of starch converts it into simple sugars that are then fermented to bioethanol with the help of yeast through fermentation. Before fermentation, pre-treatment of microalgae is achieved through different processes like hydrolytic enzymatic conversion or ultrasonification to disintegrate the cell wall to relinquish carbohydrates [47, 48]. Biogas consists of methane (55–75%) and CO_2 (25–45%), and its production involves two main processes: acetogenesis and methanogenesis. In acetogenesis, the carbohydrates are converted into acetate through oxidation and then converted to methane and CO_2 during methanogenesis [35]. C:N ratio in the feedstock is the consequential factor as low C:N ratio leads to generation of ammonia, which act as the inhibitor of the process. Various other factors like degradation of algae by bacteria, temperature, pH, etc. additionally influence the biogas production. The lower C:N ratio can be maintained by co-digestion of waste papers and sewage sludge along with the biomass [36, 49]. Biogas yield can be enhanced up to 56% through microwave pretreatment via altering its cell wall structure [50].

Bioethanol is extracted from hydrolysis of hydrocarbons through yeast [51]. For efficient bioethanol production, different pretreatment processes are critical like hot water wash, milling, enzymatic hydrolysis, liquefaction, and saccharification or alginate extraction [52, 53].

18.8.2.2 Anaerobic Digestion

Anaerobic digestion involves degeneration of algal organic biomass, through enzymes or microorganisms, into biomethane and CO_2 along with traces of H_2S. It proceeds through three processes: hydrolysis, acetogenesis, and methanogenesis. Hydrolysis is the rate-constraining step in this process. The pretreatment in the form of mechanical and thermal pretreatment is done to enhance the enzymatic action [54]. Primarily, the process relies on the hydraulic retention time (HRT), digester structure (batch or perpetual), system microorganism colony, and specific temperature settings, in particular cases of either thermophilic (49–57 °C) or mesophilic (30–38 °C) microbes [55]. The biomethane generation potential of algae is highly reliant on its biochemical composition so that the integral minerals compensate the nutrient requisite of the anaerobic microorganisms [56].

18.8.3 Thermochemical Conversions

Thermochemical processes are habituated to convert microalgal biomass into biofuels. These processes include gasification, pyrolysis, and hydrothermal liquefaction (HTL). These processes can convert total microalgal biomass into biofuels unlike other processes where either carbohydrates or lipids or both are converted into biofuels.

18.8.3.1 Gasification

Gasification process implicates partial oxidation of microalgal biomass in a gasifier without combustion under controlled quantity of oxygen, air, or steam at 700–1000 °C to produce syngas and the by-products char, tar, and ash. Syngas (synthesis gas) is an amalgamation of different gases mainly H_2, CO, CO_2, and CH_4. Syngas can be reformed catalytically into liquid fuel through Fischer-Tropsch process, which converts carbon monoxide and hydrogen into long chain hydrocarbons. Commonly used gasification reactors in the industry are of three main types: fixed bed, fluidized bed, and moving bed. Reliant upon the type of reactor's design, there may be several advantages and disadvantages, eventually affecting product quality and operating costs. There are two broad categories of gasification: (i) conventional gasification is performed either utilizing fixed bed or fluidized bed at high temperature, i.e. 800–1000 °C, and (ii) supercritical water gasification (SCWG) is done at 375–550 °C temperature range and a pressure range between 22.1 and 36 MPa and can be aided by catalyst. As compared to conventional gasification or electrolysis, supercritical water gasification is more cost efficacious if operated at moderately low temperatures for syngas engenderment [57]. The water-gas shift reaction is enhanced by the integration of alkali such as hydroxides of sodium, potassium, and calcium during gasification, and tar and char formation is suppressed [58].

18.8.3.2 Pyrolysis

Pyrolysis involves thermochemical conversion of microalgae biomass to produce syngas, bio-oils, and charred solid residues under extreme heat and pressure, but in the absence of oxygen. Conventional pyrolysis operates at temperature range of 400–600 °C, microwave pyrolysis activates up to 800 °C, and catalytic pyrolysis at 300 °C at atmospheric pressure [41]. The rate of heating and the residence time of the reaction are the paramount parameters for pyrolysis process. There are two methods of pyrolysis, i.e. fast pyrolysis or flash pyrolysis and conventional or slow pyrolysis. The fast and flash pyrolysis enhances crude bio-oil yield [59], whereas biochar production and the raised production of pyrolytic gases such as methane and CO_2 are associated with conventional pyrolysis [47]. The composition of the bio-oils, gases, and chars are greatly affected by the residence time of pyrolysis reaction. Pyrolysis is envisaged as the most endorsed conversion process due to its ability to produce high ash content from algal biomass, but the produced oil still has some complications with its acidity, viscosity, and stability [35]. The microwave-enhanced pyrolysis has been recognized as an expeditious, efficient method for bio-oil engenderment [60].

18.8.3.3 Liquefaction

Liquefaction is the physicochemical conversion of microalgal biomass to bio-oil at high temperature (250–500 °C), at high pressure (5–20 bar), with or without catalysts, and in the presence of a solvent. Water, acetone, or alcohol can be used as typical solvents for converting microalgal biomass into slurry feed up to 5–50% mass fraction [41, 61]. Liquefaction can be either direct or indirect. In the direct method, biomass is directly converted into liquid fuel, and it produces lesser amount of char, while in the indirect method the biomass is first converted into gas and then to liquid [62]. In the hydrothermal liquefaction, to engender biocrude, gas, and char, the algal slurries are exposed to temperature between 300–400 °C and 40–200 bar pressure (10–73, 8–20, and 0.2–0.5%, respectively) [63]. Nine to ninety-seven percent of the oil yield can be obtained from hydrothermal liquefaction process [64, 65], which is higher than the pyrolysis bio-oil [66].

18.8.4 Direct Combustion

In this process, the chemical energy of the microalgal biomass is transformed into heat or electricity by burning the algal biomass in a furnace, boiler, or steam in the presence of oxygen provided the moisture content of microalgae is less than 50% of its dry weight [67]. The temperature of about 1000 °C is required to produce hot gases. Pretreatment such as grinding and drying of biomass is required afore combustion process that can maximize energy production [35]. Consequently, electricity can efficiently be generated through curing of algal biomass and coal.

18.9 Microalgal Biofuels

18.9.1 Biodiesel

Algal biomass consists of the three major macromolecules: carbohydrates, proteins, and lipids/natural oils. Lipids are the long chain fatty acids with high-value energy material, which are synthesized by microalgae more efficiently by converting sunlight energy. Lipids, constitute the bulk of the natural oil synthesized by microalgae, are considered as the appropriate feedstock for biodiesel production. Most of the microalgae species exhibit high lipids contents, i.e. 50–70% and may reach to 80% in case of the microalga *Botryococcus braunii* that can accumulate up to 80% of oil in it biomass [68]. Hence, it is preferred for production of biodiesel. The composition of microalgal fatty acids has a paramount effect on the fuel properties of biodiesel. Microalgae (with high oil content) have capacity to generate 136 900 l/ha/year algal oil, which in turn can produce 121 104 kg/ha/year biodiesel [69, 70]. Thus, microalgae especially cyanobacteria, chlorophytes, and diatoms are profoundly adept at composing lipids and growing expeditiously. Despite being minimization in emission of sulfur and particulate matter, biodiesel has proportional engine performance compared to the petroleum-diesel fuel [71, 72]. It is a biodegradable, nontoxic alternative fuel obtained from the renewable sources [73]. Biodiesel is a coalescence of fatty acid alkyl esters, derived from the ester exchange reaction, i.e. trans-esterification of animal and vegetable oils. These lipids are composed of 90–98% by weight of triglycerides with an iota of mono and di-glycerides. In juxtaposition of this, free fatty acids (1–5%), phospholipids, tocopherols, sulfur compounds, traces of water, and phosphatides are withal present [74]. Biodiesel, obtained from microalgae, is an exceptional substitute to the perpetual diesel crisis; however, the strains of high oil content and magnification rate need to be culled for the proficient generation of biodiesel [75]. A scheme for microalgal biodiesel value chain stages is shown in Figure 18.4.

18.9.2 Bioethanol

Engenderment of ethanol from the algal biomass is in the exploratory stage. Bioethanol can supersede the fossil-derived petrol, but the more research is anticipated to investigate the astronomically immense scale production. Presently, the bioethanol is produced from the fermenting sugars. Through the developing advancements, the cellulose

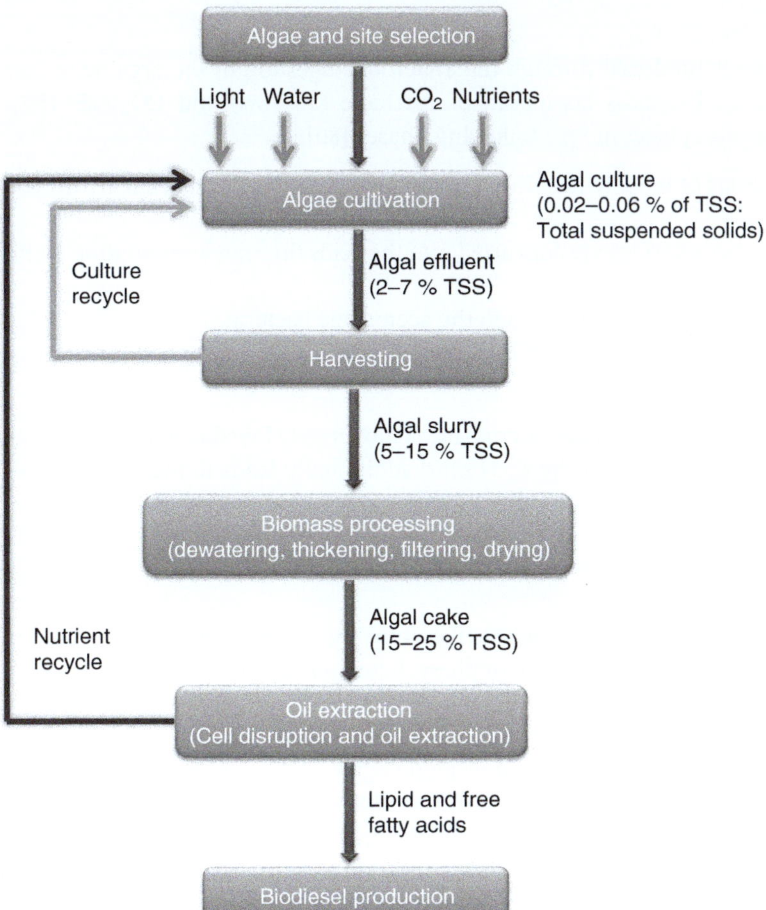

Figure 18.4 Microalgal biodiesel value chain stages. *Source:* Mata et al. [32] / with permission of Elsevier.

and hemicelluloses are hydrolyzed into sugars, which ease the generation of ethanol from the major part of the dry algal biomass [76]. Microalgae biomass is a better substitute in comparison to the woody biomass for ethanol engenderment due to the following reasons [77]:

- Processing becomes easy due to the absence of lignin in microalgae.
- Cellular composition is not involute, so biomass can be rarely processed.
- Sufficient amplitude of polysaccharides is present in microalgae that can be converted into sugar.
- Microalgae can be facilely engineered genetically to engender ethanol.

Microalgae amass higher concentrations of a variety of carbohydrates, viz., cellulose, starch, glycogen, agar, etc., which can be conveniently transformed to fermentable sugars for bioethanol engenderment [32]. Phaeophyceae, owing to its high sugar content, is considered as the most opportune feedstock for bioethanol [78, 79].

18.9.3 Biogas

Biogas or biomethane is produced through the anaerobic digestion of the organic matter (involute bioploymers). Biogas is composed of methane (55–75%) and CO_2 (25–45%). Formation of biogas takes places in four following phases [80]:

- Hydrolysis: Hydrolysis of biopolymers into the monosaccharides through the hydrolytic bacteria.
- Acidogenesis: Monosaccharides are converted into the acids through fermentation facilitated by fermentive bacteria.
- Acetogenesis: Formation of acetate through the acetogenic bacteria.
- Methanogenesis: Acetate is converted into methane and CO_2 through the action of methanogens.

Engenderment of biogas through the microalgae is moderately low due to the sensitivity of algal cells to the degradation and the C: N ratio additionally leads to the formation of ammonia, which act as an inhibitor.

18.9.4 Bio-Oil and Bio-Syngas

When the algal biomass is processed at high temperature and in the absence of oxygen, the products are engendered in three phases: vapor phase, liquid phase, and solid phase. Intricate mixture in the liquid phase is kenned as bio-oil. Bio-oil is the primary product achieved through the pyrolysis. For power generation through external (steam cycles, organic Rankine cycles, and sterling engines) and internal combustion engine (diesel engine and gas turbine engine), bio-oil is found to be appropriate [81–83]. Recent research studies have revealed that the bio-oil obtained from microalgal biomass is of higher quality than bio-oil from wood.

In gasification process, the organic substrate is converted to syngas by chemical biotransformation. Syngas is utilized as an intermediate for the engenderment of various biofuels or can be directly utilized in turbines and engines. High temperature (80–200 °C) and the water content less than 20% in biomass are indispensable for the gasification [84].

18.9.5 Biohydrogen

Hydrogen is a consequential fuel, with an extensive array of applications in fuel cells, e.g. liquefaction of coal and in the refining of heavy oils. Microalgae, under anaerobic condition, engender hydrogen, using sunlight and water. Presently, the hydrogen is produced by bio-sources denotes through the steam reformation, photo-fermentation and photolysis of water catalyzed by many photosynthetic algae [85].

18.10 Conclusions and Perspectives

Improved biofuel production strategies are the prerequisite of a moment due to the perpetual exhaustion of the nonrenewable energy sources. Environmental pollution is withal an ecumenical concern. Biodiesel production from algae needs massive scale cultivation and harvesting systems. But the challenge is to minimize the cost per unit area. Ergo at astronomically immense scale, the optimum magnification conditions, and the nurturing environment should be needed to lookout thoroughly. The processes for the biofuel production from microalgae can become more paramount and cost efficacious if the extraction of another consequential biomolecule is integrated with wastewater remediation process and the sequestration of CO_2 from the fuel gas emissions. Research efforts should fixate on the economical designing of PBRs, as well as the integration of downstream along with upstream processes of algal biomass for a more efficient and pragmatic energy recovery. Techno-economic analysis should be done for the opportune implementation of the future strategies to make the algal biofuels commercially available and viable for a better future.

References

1. Harun, R., Danquah, M.K., and Forde, G.M. (2010). Microalgal biomass as a fermentation feedstock for bioethanol production. *Journal of Chemical Technology and Biotechnology* 85: 199–203.
2. Ho, S.H., Chen, W.M., and Chang, G. (2010). *Scenedesmus obliquus* CNW-N as a potential candidate for CO2 mitigation and biodiesel production. *Bioresource Technology* 101 (22): 8725–8730.
3. Saladini, F., Patrizi, N., Pulselli, F.M. et al. (2016). Guidelines for energy evaluation of first, second and third generation biofuels. *Renewable and Sustainable Energy Reviews* 66: 221–227.
4. Paul Abishek, M., Patel, J., and Prem Rajan, A. (2014). Algae oil: a sustainable renewable fuel of future. *Biotechnology Research International* 272: 814.
5. Ullah, K., Ahmad, A., Sharma, V.K. et al. (2015). Assessing the potential of algal biomass opportunities for bioenergy industry: a review. *Fuel* 143: 414–423.
6. Lee, R. and Lavoie, J.M. (2013). From first to third –generation biofuels: challenges of producing a commodity from a biomass of increasing complexty. *Animal Frontiers* 3: 6–11.
7. Maity, J.P., Bundschuh, J., Chen, C.Y., and Bhattacharya, P. (2014). Microalgae for third generation biofuel production, mitigation of greenhouse gas emissions and wastewater treatment: Present and future perspectives – a mini review. *Energy* 78: 104–113.

8 Brennan, L. and Owende, P. (2010). Biofuels from microalgae: a review of the technologies for production processing and extractions of biofuels and co-products. *Renewable and Sustainable Energy Reviews* 14: 557–577.

9 Aro, E.M. (2016). From first generation biofuels to advanced solar biofuels. *Ambio* 45: 24–31.

10 Gaurav, N., Sivasankari, S., Kiran, G.S. et al. (2017). Utilization of bioresources for sustainable biofuels: a review. *Renewable and Sustainable Energy Reviews* 73: 205–214.

11 Yanagisawa, M., Nakamura, K., Ariga, O., and Nakasaki, K. (2011). Production of high concentrations of bioethanol from seaweeds that contain easily hydrolysable polysaccharides. *Process Biochemistry* 46: 2111–2116.

12 Alam, F., Date, A., Rasjidin, R. et al. (2012). Biofuel from algae- Is it a viable alternative? *Procedia Engineering* 49: 221–227.

13 Chisti, Y. (2008). Biodiesel from microalgae beats bioethanol. *Trends in Biotechnology* 26: 126–131.

14 Narala, R.R., Garg, S., Sharma, K., and Schenk, P.M. (2016). Comparison of microalgae cultivation in photobioreactor, open raceway pond, and a two-stage hybrid system. *Frontiers in Energy Research* 4 (29): 1–10.

15 Chisti, Y. (2007). Biodiesel from microalgae. *Biotechnology Advances* 25: 294–306.

16 Metting, F.B. (1996). Biodiversity and application of microalgae. *Journal of Industrial Microbiology* 17: 477–489.

17 Spolaore, P., Joannis-Cassan, C., Duran, E., and Isambert, A. (2006). Commercial applications of microalgae. *Journal of Bioscience and Bioengineering* 101 (2): 87–96.

18 Surendhiran, D. and Vijay, M. (2012). Microalgal biodiesel: a comprehensive review on the potential and alternative Biofuel. *Research Journal of Chemical Sciences* 2 (11): 71–82.

19 Richmond, A. and Hu, Q. (2013). *Handbook of Microalgal Culture: Applied Phycology and Biotechnology*, 2e. Wiley https://doi.org/10.1002/9781118567166.

20 Rajkumar, R., Yaakob, Z., and Takriff, M.S. (2014). Potential of the micro and macro algae for biofuel production: a brief review. *BioResources* 9: 1606–1633.

21 Gendy, T.S. and El-Temtamy, S.A. (2013). Commercialization potential aspects of microalgae for biofuel production: an overview. *Egyptian Journal of Petroleum* 22: 43–51.

22 Das, P., Lei, W., Aziz, S.S., and Obbard, J. (2011). Enhanced algae growth in both phototrophic and mixotrophic culture under blue light. *Bioresource Technology* 102: 3883–3887.

23 Roessler, P.G., Brown, L.M., Himmel, M.E. et al. (1994). Genetic engineering approaches for enhanced production of biodiesel fuel from microalgae. In: *Enzymatic Conversion of Biomass for Fuels Production*, ACS Symposium Series 566 (eds. M.E. Himmel, J.O. Baker and R.P. Overend), 256–270. ACS Publications.

24 Senousy, H.H., Beaks, G.W., and Hack, E. (2004). Phylogenetic placement of *Botryococcus braunii* (trebouxiophyceae) and *Botryococcus sudeticus* isolate utex 2629 (chlorophyceae). *Journal of Phycology* 40: 412–423.

25 Sheehan, J., Dunahay, T., Benemann, J., and Roessler, P.A. (1998). *Look Back at the U.S. Departmentof Energy's Aquatic Species Program-Biodiesel from Algae*, 18. Golden, CO, USA: National Renewable Energy Laboratory.

26 Banerjee, A., Sharma, R., Chisti, Y., and Banerjee, U.C. (2002). Botryococus Braunii: a renewable source of hydrocarbons and other chemicals. *Critical Reviews in Biotechnology* 22: 245–279.

27 Gravirilescu, M. and Chisti, Y. (2005). Biotechnology- a sustainable alternative for chemical industry. *Biotechnology Advances* 23: 471–499.

28 Ghirardi, M.L., Zhang, L., Lee, J.W. et al. (2000). Microalgae: a green source of renewable H2. *Trends in Biotechnology* 18 (12): 506–511.

29 Akkerman, I., Janssen, M., Rocha, J., and Wijffels, R.H. (2002). Photobiological hydrogen production, Photochemical efficiency and bioreacter design. *International Journal of Hydrogen Energy* 27 (11): 1195–1208.

30 Melis, A. (2002). Green alga hydrogen production: progress, challenges and prospectes. *International Journal of Hydrogen Energy* 127 (11-12): 1217–1228.

31 Kapdan, I.K. and Kargi, F. (2006). Bio-hydrogen production from waste materials. *Enzyme and Microbial Technology* 38: 569–582.

32 Mata, T.M., Martins, A.A., and Caetano, N.S. (2010). Microalgae for biodiesel production and other applications: a review. *Renewable and Sustainable Energy Reviews* 14: 217–232.

33 Ugwu, C.U., Aoyagi, H., and Uchiyama, H. (2008). Photobioreactors for mass cultivation of algae. *Bioresource Technology* 99: 4021–4028.

34 Rawat, I., Ranjith Kumar, R., Mutanda, T., and Bux, F. (2013). Biodiesel from microalgae: a critical evaluation from laboratory to large scale production. *Applied Energy* 103: 444–467.

35 Adeniyi, O.M., Azimov, U., and Burluka, A. (2018). Algae biofuel: current status and future applications. *Renewable and Sustainable Energy Reviews* 90: 316–335.

36 Suganya, T., Varman, M., Masjuki, H.H., and Renganathan, S. (2016). Macroalgae and microalgae as a potential source for commercial applications along with biofuels production: a biorefinery approach. *Renewable and Sustainable Energy Reviews* 55: 909–941.

37 Leite, G.B., Abdelaziz, A.E.M., and Hallenbeck, P.C. (2013). Bioresource technology algal biofuels: challenges and opportunities. *Bioresource Technology* 145: 134–141.

38 Fon Sing, S., Isdepsky, A., Borowitzka, M., and Moheimani, N.R. (2011). Production of biofuels from microalgae. *Mitigation and Adaptation Strategies for Global Change* 18: 47–72.

39 Schenk, P.M., Thomas-Hall, S.R., Stephens, E. et al. (2008). Second generation biofuels: high-efficiency microalgae for biodiesel production. *Bioenergy Research* 1: 20–43.

40 Grima, E.M., Belarbi, E.H., Fernández, F.A. et al. (2003). Recovery of microalgal biomass and metabolites: process options and economics. *Biotechnology Advances* 20: 491–515.

41 Chen, C.L., Chang, J.S., and Lee, D.J. (2015). Dewatering and drying methods for Microalgae. *Drying Technology* 33: 443–454.

42 Cravotto, G., Boffa, L., Mantegna, S. et al. (2008). Improved extraction of vegetable oils under high-intensity ultrasound and/or microwaves. *Ultrasonics Sonochemistry* 15 (5): 898–902.

43 Razzak, S.A., Hossain, M.M., Lucky, R.A. et al. (2013). Integrated CO2 capture, wastewater treatment and biofuel production by microalgae culturing – a review. *Renewable and Sustainable Energy Reviews* 27: 622–653.

44 Lee, A.F., Bennet, J.A., Manayil, J.C., and Wilson, K. (2014). Heterogenous catalysis for sustainable biodiesel production *via* esterification and transesterification. *Chemical Society Reviews* 43: 7887–7916.

45 Knothe, G. (2012). Fuel properties of highly polyunsaturated fatty acid methyl esters. Prediction of fuel properties of algal biodiesel. *Energy & Fuels* 26: 5265–5273.

46 Dunn, R. and Bagby, M. (1995). Low-temperature properties of triglyceride-based diesel fuels: transesterified methyl esters and petroleum middle distillate/ester blends. *Journal of the American Oil Chemists' Society* 72: 895–904.

47 Suali, E. and Sarbatly, R. (2012). Conversion of microalgae to biofuel. *Renewable and Sustainable Energy Reviews* 16 (6): 4316–4342.

48 Naveena, B., Armshaw, P., and Pembroke, J.T. (2015). Ultrasonic intensication as a tool for enhanced microbial biofuel yields. *Biotechnology for Biofuels* 8: 140.

49 Hidaka, T., Inoue, K., Suzuki, Y., and Tsumori, J. (2014). Growth and anaerobic digestion characteristics of microalgae cultivated using various types of sewage. *Bioresource Technology* 170: 83–89.

50 Passos, F., Solé, M., García, J., and Ferrer, I. (2013). Biogas production from microalgae grown in wastewater: effect of microwave pretreatment. *Applied Energy* 108: 168–175.

51 Laurens, L.M.L., Nagle, N., Davis, R. et al. (2015). Acid-catalyzed algal biomass pretreatment for integrated lipid and carbohydrate-based biofuels production. *Green Chemistry* 17: 1145–1158.

52 Fasahati, P., Woo, H., and Liu, J. (2015). Industrial -scale bioethanol production from brown algae: effects of pretreatment processes on plant economics. *Applied Energy* 1139: 175–187.

53 Teixeira, A.C.R., Sodré, J.R., Guarieiro, L.L.N. et al. (2016). *A Review on Second and Third Generation Bioethanol Production*, 1–7. Warrendale, PA, USA: SAE *International*.

54 Rodriguez, C., Alaswad, A., Mooney, J. et al. (2015). Pretreatment techniques used for anaerobic digestion of algae. *Fuel Processing Technology* 138: 765–779.

55 Fasahati, P., Saffron, C.M., Woo, H.C., and Liu, J.J. (2017). Potential of brown algae for sustainable electricity production through anaerobic digestion. *Energy Conversion and Management* 135: 297–307.

56 Sialve, B., Bernet, N., and Bernard, O. (2009). Anaerobic digestion of microalgae as a necessary step to make microalgal biodiesel sustainable. *Biotechnology Advances* 27: 409–416.

57 Ni, M., Leung, D.Y.C., Leung, M.K.H., and Sumathy, K. (2006). An overview of hydrogen production from biomass. *Fuel Processing Technology* 87: 461–472.

58 Muangrat, R., Onwudili, J.A., and Williams, P.T. (2010). Infuence of alkali catalysts on the production of hydrogen-rich gas from the hydrothermal gasification of food processing waste. *Applied Catalysis B: Environmental* 100: 440–449.

59 Chiaramonti, D., Prussi, M., Buf, M. et al. (2017). Review and experimental study on pyrolysis and hydrothermal liquefaction of microalgae for biofuel production. *Applied Energy* 185: 963–972.

60 Campanella, A., Rachel, M., Harold, M. et al. (2012). Thermolysis of microalgae and duckweed in a CO_2-swept fixed-bed reactor: Bio-oil yield and compositional effects. *Bioresource Technology* 109: 154–162.

61 Raheem, A., Azlina, W.A.K.G.W., Yap, Y.H.T. et al. (2015). Thermochemical conversion of microalgal biomass for biofuel production. *Renewable and Sustainable Energy Reviews* 49: 990–999.

62 Rustamov, V.R., Abdullayev, K.M., and Samedov, E.A. (1998). Biomass conversion to liquid fuel by two- stage thermochemical cycle. *Energy Conversion and Management* 39 (9): 869–875.

63 Brand, S., Hardi, F., Kim, J., and Suh, D.J. (2014). Effect of heating rate on biomass liquefaction: differences between subcritical water and supercritical ethanol. *Energy* 68: 420e–427e.

64 Dote, Y., Inoue, S., Ogi, T., and Yokoyama, S. (1998). Distribution of nitrogen to oil products from liquefaction of amino acids. *Bioresource Technology* 64 (2): 157–160.

65 Zou, S., Wu, Y., Yang, M. et al. (2010). Bio-oil production from sub-and supercritical water liquefaction of microalgae Dunaliella tertiolecta and related properties. *Energy and Environment Sciences* 3: 1073–1078.

66 Duan, P. and Savage, P.E. (2010). Hydrothermal liquefication of a microalga with heterogenous catalysis. *Industrial and Engineering Chemistry Research* 50 (1): 52–61.

67 Goyal, H.B., Seal, D., and Saxena, R.C. (2008). Bio-fuels from thermochemical conversion of renewable resources: a review. *Renewable and Sustainable Energy Reviews* 12: 504–517.

68 Zheng, G., Wang, Y., and Qin, J. (2012). Microalgal motility measurement microfluidic chip for toxicity assessment of heavy metals. *Analytical and Bioanalytical Chemistry* 404: 3061–3069.

69 Deng, X., Li, Y., and Fei, X. (2009). Microalgae: a promising feedstock for biodiesel. *African Journal of Microbiology Research* 3 (13): 1008–1014.

70 Wang, K., Brown, R.C., Homsy, S. et al. (2013). Fast pyrolysis of microalgae remnants in a fluidized bed reactor for bio-oil and biochar production. *Bioresource Technology* 127: 494–499.

71 Miao, X.L. and Wu, Q. (2006). Biodiesel production from heterotrophic microalgal oil. *Bioresource Technology* 97 (6): 841–846.

72 Scragg, A.H., Morrison, J., and Shales, S.W. (2003). The use of a fuel containing *Chlorella vulgaris* in a diesel engine. *Enzyme and Microbial Technology* 33: 884–889.

73 Hossain, A., Salleh, A., Boyce, A. et al. (2008). Biodiesel fuel production from microalgae as renewable energy. *American Journal of Biochemistry and Biotechnology* 4: 250–254.

74 Bozbas, K. (2008). Biodiesel as an alternative motor fuel: production and policies in the European Union. *Renewable and Sustainable Energy Reviews* 12: 542–552.

75 Xiao-li, L., Thomas Kiran, M., Ling, T. et al. (2017). A novel growth method for diatom algae in aquaculture wastewater for natural food development and nutrient removal. *Water Science and Technology* 75 (12): 2777–2783.

76 Hamelinck, C.N., van Hooijdonk, G., and Faaij, A.P.C. (2005). Ethanol from lignocellulosic biomass: techno-economic performance in short-, middle- and long-term. *Biomass & Bioenergy* 28 (4): 384–410.

77 Deng, M.D. and Coleman, J.R. (1999). Ethanol synthesis by genetic engineering in cyanobacteria. *Applied and Environmental Microbiology* 65 (2): 523–528.

78 Lee, O.K. and Lee, E.Y. (2016). Sustainable production of bioethanol from renewable brown algae biomass September. *Biomass and Bioenergy* 92: 70–75.

79 Slade, R. and Bauen, A. (2013). Micro-algae cultivation for biofuels: cost, energy balance, environmental impacts and future prospects. *Biomass and Bioenergy* 53: 29–38.

80 Oncel, S.S. (2013). Microalgae for a macroenergy world. *Renewable and Sustainable Energy Reviews* 26: 241–264.

81 Czernik, S. and Bridgwater, A.V. (2004). Overview of the application of biomass fast pyrolysis oil. *Energy & Fuels* 18: 590–598.

82 Bridgwater, A.V., Meier, D., and Radlein, D. (1999). An overview of fast pyrolysis of biomass. *Organic Geochemistry* 30: 1479–1493.
83 Chiarmonti, D., Osama, A., and Solantastua, Y. (2007). Power generation using fast pyrolysis liquids from biomass. *Renewable and Sustainable Energy Reviews* 11: 1056–1086.
84 Ghasemi, Y., Rasoul-Amini, S., Naseri, A.T. et al. (2012). Microalgae biofuel potentials (review). *Applied Biochemistry and Microbiology* 48: 126–144.
85 Ran, C.Q., Chen, Z.A., Zhang, W. et al. (2006). Characterization of photo-biological hydrogen production by several marine green microalgae. *Wuhan Ligong Daxue Xuebao* 28 (2): 258–263.
86 Mandotra S.K., Kumar R., Upadhyay S.K., Ramteke P.W. (2018) Nanotechnology: A New Tool for Biofuel Production. In: Srivastava N., Srivastava M., Pandey H., Mishra P., Ramteke P. (eds) Green Nanotechnology for Biofuel Production. Biofuel and Biorefinery Technologies, vol 5. Springer, Cham. https://doi.org/10.1007/978-3-319-75052-1_2.

19

Microbial Source of Insect-Toxic Proteins

Tripti Yadav and Geetanjali Mishra

Ladybird Research Laboratory, Department of Zoology, University of Lucknow, Lucknow, Uttar Pradesh, India

19.1 Introduction

Insects are the most diverse groups occupying almost every environmental niche. More than 1 million insect species have been identified globally [1]. Scientists believe that there are many more unidentified ones, with estimates ranging from 3 to 30 million. This wide-ranging class Insecta harbors a great diversity. Of this diverse order, around one percent of insects are known to cause severe economic damage to humans and their domesticated animals and plants. Losses to crop pests globally are believed to range somewhere between 20 and 40%; these estimates are for five major crops and 137 pests and pathogens infesting them [2]. Each year, plant diseases cost the global economy around $220 billion, and invasive insects around US$70 billion [3].

Since time immemorial, humans have waged war against this one percent of total insect population with the search for better and more innovative, eco-friendly, self-sustaining methods of pest management continuing till date [4]. Generations of pesticides have been developed and given up owing to their broad range of toxicity, bioaccumulation, development of resistance, and overall negative impact on ecosystem and humans. Each generation of pesticides has done better than the previous ones, but we are still far from the utopian pesticide. In the meantime, more eco-friendly pest management options such as biological control, entomopathogens, antifeedants, semiochemicals, insect hormones, etc., are available [4]. The return to cultural practices has also been considered an important means of pest management. Among biological control practices is the use of microbes as natural enemies and potential pest control.

Symbiotic as well as pathogenic microorganisms have adapted specifically to insects as host systems [5–7]. The interactions between microbes and insect communities go back a long way. Since insects inhabit a range of environments, there are chances that they are colonized with different microorganisms. This microbial association of insects not only accounts for basic nutritional deficiencies and biochemical needs, but also plays an

Bioprospecting of Microorganism-Based Industrial Molecules, First Edition.
Edited by Sudhir P. Singh and Santosh Kumar Upadhyay.

essential role in sexual compatibility among mates [8–10]. Besides these, over 100 bacteria, 800 fungi, and more than 1000 entomopathogenic viruses have been isolated and identified from different insect orders [11–13]. These entomopathogenic microbes have developed a piece of chemical machinery to disarm their target host by producing a wide range of toxins. These microbial sources of insect-toxic compounds have drawn the attention of microbiologist toward the production of biopesticides that are safer and eco-friendly alternative to chemical pesticides to control the pest population [14]. In addition, with the advancement in technology and development of new analytical methods including UHPLC–DAD and UHPLC–MS-based techniques, imaging mass spectrometry (IMS) and high-resolution NMR, the identification and characterization of these toxins have become easier and expeditious. Additionally, with the help of genetic engineering techniques, we can produce microbial biopesticides with increased virulence and broad host range along with their tolerance to physical and chemical conditions [15, 16].

In this chapter, we have discussed the major microbial sources, i.e. fungi, bacteria, and viruses, which are known to produce insect-toxic proteins and their development of biopesticides that are safer and cost-effective alternative to chemical pesticides.

19.2 Fungi

Fungi share a diverse history of coevolutionary relationship with animals, especially arthropods. They have evolved a formidable range of chemical defensive strategies as protection against the mycophagous organisms including mammal and insects along with parasitic fungi. Besides plants, fungi are also known for being an extensive producer of toxic compounds mainly in the form of peptidic secondary metabolites. These secondary metabolites are generally the intermediate products of primary metabolism and arise mainly from five metabolic sources including amino acids, polyketide biosynthetic pathway (from acetyl coenzyme A), shikimic acid pathway (for the biosynthesis of aromatic amino acids), mevalonic acid pathway (from acetyl coenzyme A), and polysaccharides and peptidopolysaccharides which along with their biosynthesis have been reviewed by Griffin [17]. Besides secondary metabolites, fungi are also known to produce proteins with different defensive functions including lectin acting as toxic deterrent, protein inhibitors, ribotoxins, cell lysing proteins, and others enzymes such as proteases, oxidases, and phospholipases.

The fungal infection starts with the attachment of their dispersive forms, i.e., blastophore or conidia, onto the host cuticle, they do so by using a variety of hydrolytic enzymes including proteases, lipases, and chitinases. They enter into the hemocoel where other factors promote their germination and growth across the host's surface [18, 19].

The modern exploration in this field started with the study of Agostino Bassi who was working on the white muscardine disease of silkworm and reported for the first time that a microorganism (later known as *Beauveria bassiana* (Balsamo) Vuillemin) could cause infectious disease in insects, which led to the first germ theory of diseases [20, 21]. Among the entomotoxic fungi, *Metarhizium anisopliae* (Metsch) and *B. bassiana* are the most extensively studied species for their toxicity against different insect orders (Table 19.1); Lepidoptera, Coleoptera, Orthoptera, and Hemiptera [22–26]. Besides

Table 19.1 Entomopathogenic microbes with their target hosts.

Species	Target host	References
Fungi		
Beauveria bassiana	Aphididae spp.	[123, 124]
	Western flower thrips, *Frankliniella occidentalis* (Pergande) (Thysanoptera)	[125–127]
	Tarnished plant bug, *Lygus lineolaris* (Palisot de Beauvois) (Hemiptera)	[128]
	Psyllidae spp.	[129, 130]
	Curculionoidea spp.	[131]
	Pseudococcidae spp.	[132–134]
	Glasshouse whitefly, *Trialeurodes vaporariorum* Westwood (Hemiptera)	[135, 136]
	Diamondback moth, *Plutella xylostella* (L.) (Lepidoptera)	[137, 138]
	Greater Waxmoth, *Galleria mellonella* (L.), (Lepidopetra)	[139–141]
	Colorado Potato beetle, *Leptinotarsa decemlineata* (Say) (Coleoptera)	[142]
	Blister beetles, *Lytta nuttali* Say (Coleoptera)	[143]
	European corn worm, *Ostrinia nubilalis* (Hübner) (Lepidoptera)	[144–146]
Metarhizium anisopliae	Mediterranean fruit fly, *Ceratitis capitate* (Wiedemann) (Diptera)	[24, 25, 147]
	Sugar-beet weevil, *Cleonus punctiventris* Germar (Coleoptera)	[148, 149]
	Black vine weevil, *Otiorhynchus ovatus* (L.) (Coleoptera)	[150, 151]
	Bean flower thrips, *Megalurothrips sjostedti* Trybom, 1908 (Thysanoptera)	[152–154]
	Turnip moth, *Agrotis segetum* Schiff. (Lepidoptera)	[155]
	Pine chafer, *Polyphylla fullo* (L.) (Coleoptera)	[156]
Verticillium lecanii	Apididae spp.; Chrysanthemum Aphid, *Macrosiphoniella sanborni* (Gillette, 1908); Green peach aphid, *Myzus persicae* (Sulzer)	[157–159]
Hirsutella thompsonii	Citrus mites and other acarine pests, Citrus rust mite, *Phyllocoptruta oleivora* (Ashmead)	[160–162]
Entomophthora thaxteriana Petch; *Entomophthora sphaerosperma*	Aphids and Spider mites	[163, 164]
Lecanicillium muscarium	Silverleaf whitefly, *Bemisia tabaci* (Genn.)(Hemiptera); *Rhynchota* spp.; *Ricaniidae* spp.; *T. vaporariorum*	[165–168]

(*Continued*)

Table 19.1 (Continued)

Species	Target host	References
Isaria spp.	*T. vaporariorum;* Asian citrus psyllid, *Diaphorina citri* Kuwayama (Hemiptera); Citrus aphid, *Toxoptera citricida* Kirkaldy; Glassy-winged sharpshooter, *Homalodisca vitripennis* (Germar)(Homoptera); Root weevil, *Diaprepes abbreviates* (L.)(Coleoptera), *B. tabaci*	[169–173]
Beauveria medogensis	Cotton bollworm, *Helicoverpa armigera* (Hübner) (Lepidoptera); Yellow mealworm beetle, *Tenebrio molitor* L. (Coleoptera)	[174]
Beauveria asiatica	Coleoptera	[175]
Beauveria sinensis	Lepidoptera	[176]
Beauveria lii	Epilachna beetle, *Henosepilachna vigintioctopunctata* Fabr.	[177]
Bacteria		
Bacillus thuringiensis subsp. thuringiensis	Lepidoptera	[14, 178–182]
Bacillus thuringiensis subsp. kurstaki	Lepidoptera	
Bacillus thuringiensis subsp. israelensis	Diptera	
Bacillus thuringiensis subsp. tenebrionis	Coleoptera	
Bacillus thuringiensis subsp. aizawai	*G. mellonella* (Lepidopetra)	
Bacillus thuringiensis subsp. san diego	Certain Coleopterans	
B. laterosporus	Coleoptera spp., Diptera spp., Lepidoptera spp., and other invertebrates	
Bacillus papillae	Scarabaeid beetles	
Bacillus sphaericus	Mosquitoes	
Pseudomona aeruginosa	Major Insect taxa	
Pseudomonas fluorescens	Aphidoidea spp.; Isoptera spp.; *Epilachna Vigintioctopunctata* (Fabricius) (Coleoptera); *Cnaphalocrocis* *Medinalis* Guenée (Lepidoptera)	[183, 184]
Pseudomona protegens	Lepidoptera spp.; Diptera spp.; *Drosophila melanogaster*	[185]
P. chlororaphis	Lepidopteran spp.	[186]
P. entomophila	*D. melanogaster*	[187]
P. taiwanensis	*D. melanogaster*	[188]
Lysinibacillus sphaericus	*Culex* spp.; *Anopheles* spp.; *Aedes* spp.; *Simuliidae* spp.; *Nematocera* spp.	[189–191]

Table 19.1 (Continued)

Species	Target host	References
Brevibacillus laterosporus	Diptera spp.; *P. xylostella* (L.)(Lepidoptera); *Leptinotarsa decemlineata* (Say) (Coleoptera)	[192, 193]
Photorhabdus luminescens *P. asymbiotica*	*Manduca sexta* (L.); *L. decemlineata*; *B. tabaci*	[103, 104, 194]
Xenorhabdus nematophila	*P. xylostella*	[103, 104]
Serratia entomophila *S. proteamaculans*	Grass grub, *Costelytra zealandica* (White)(Coleoptera)	[195–197]
S. marcescens	Orthoptera spp.; Coleoptera spp.; Hymenoptera spp.; Lepidoptera spp.; Diptera spp.; Apple maggot fly, *Rhagoletis pomonella* (Walsh)(Diptera)	[198–200]
Virus		
Baculovirus		[201, 202]
Heliothis Nucleopolyhedroviruses (NPV)	*Heliothis sp.* (Lepidoptera)	
Spodoptera frugiperda NPV	Fall armyworm, *Spodoptera frugiperda* (JE Smith) (Lepidoptera)	[203–206]
NPV	Tree sawfly, *Neodiprion sertifer* (Geoffroy) (Hymenoptera)	[207, 208]
	Cabbage looper, *Trichoplusia ni* (High Five) (Lepidoptera)	[209, 210]
	Cotton pest, *Spodoptera litura* (Fabricius) (Lepidoptera)	[211, 212]
	Rhinoceros beetle, *Oryctes rhinoceros* (L.)(Coleoptera)	[213]
AcMNPV	Alfalfa Looper, *Autographa californica* (Speyer) (Lepidoptera); Beet armyworm, *Spodoptera exigua* (Hübner)(Lepidoptera)	[214, 215]
HearNPV	*H. armigera; Heliothis assulta* Guenée (Lepidoptera)	[216, 217]
HezeSNPV	Corn earworm, *Helicoverpa zea* (Boddie)(Lepidoptera)	[218, 219]
MabrNPV	Cabbage moth, *Mamestra brassicae* L. (Lepidoptera); *H. armigera*; *S. exigua*	[220–222]
LdMNPV	Gypsymoth, *Lymantria dispar* L. (Lepidoptera)	[223–225]
cytoplasmic-polyhedrosis virus (CPV)	Spruce budworm, *Choristoneura fumiferana* (Clemens) (Lepidoptera)	[226, 227]
CPV	Forest caterpillar, *Malacosoma disstria* (L.) (Lepidoptera)	[228, 229]
Granulovirus CpGV,	Codling moth, *Cydia pomonella* (L.)(Lepidoptera); Large cabbage whitefly, *Pieris brassicae* (L.) (Lepidoptera)	[230, 231]
AdorGV	Smaller tea tortrix, *Adoxophyes oranai* (FvR) (Lepidoptera)	[232, 233]

(Continued)

Table 19.1 (Continued)

Species	Target host	References
CrleGV	False codling moth, *Cryptophlebia leucotreta* Meyr. (Lepidoptera)	[234, 235]
PhopGV	Potato tuber moth, *Phthorimaea operculella* Zeller (Lepidoptera)	[236, 237]
PiraGV	Cabbage white butterfly, *Pieris rapae* L. (Lepidoptera)	[238, 239]
PlxyGV	*P. xylostella*	[240, 241]
Phthorimaea operculella GV	Guatemalan potato moth, *Tecia solanivora* Povolny (Lepidoptera)	[242, 243]
Aphodius tasmaniae entomopoxvirus (EPV)	Coleoptera	[244, 245]
Dermolepida albohirtum EPV	Coleoptera	[244]
Heliothis armigera EPV	Lepidoptera	[246]
Locusta migratoria EPV	Orthoptera	[244]
Operophtera brumata EPV	Lepidoptera	[246]
Aedes aegypti EPV	Diptera	[246]
Cypovirus	Lepidoptera sp. *Dendrolimus punctatus* Walker; *D. kikuchii* Matsumura; *D. spectabilis* Butler	[247–249]

these, *Paecilomyces fumosoroseus* (Wize) Brown and Smith, *Tolypocladium* spp., *Verticillium lecanii* (Zimm.), and *Hirsutella* spp. are also known to have entomotoxic effects. *B. bassiana* is reported to produce entomotoxins such as beauverolide, beauveriolides, bassianin, bassianolide, bassiacridin, oosporein, isarolides, and insect-toxic protein Bb70p. In addition, destruxins and cytochalasins have been isolated from *M. anisopliae*-infected insects.

Beauverolide, a cyclic tetradepsipeptide, has been isolated from different entomopathogenic fungus including *B. bassiana*, *Beauveria brongniartii*, and *Isaria* spp. The stereostructure and bioactivity of beauverolide from *B. bassiana* and *P. fumosoroseus* have been studied earlier [27–30]. Jegorov et al. (1994) isolated and identified beauverolide L and La from entomopathogenic fungi, *Beauveria tenella* and *P. fumosoroseus,* respectively [31]. Later, Vilcinskas and Gotz (1999) investigated the immunomodulatory activity of beauverolide L and cyclosporine A (another insecticidal depsipeptide isolated from *Tolypocladium niveum*) on *Galleria mellonella* (L.). Although, the study reported that beauverolide L and cyclosporin A did not cause any mortality in *G. mellonella* larvae, but it did induce strong cellular and humoral responses when injected with sublethal concentrations coated on particles [32]. Besides its insecticidal properties, cyclosporin A is also used as an immune response-suppressing agent after organ transplantations in humans. Another cyclic depsipeptide, beauveriolide I and II, structurally similar to beauverolide H and I, is reported to have mild insecticidal activity on *Spodoptera litura* (Fabricius) and *Collosobruchus chinensis* (L.) [33].

Beauvericin, a cationophoric cyclic hexadepsipeptide ($C_{46}H_{59}N_3O_9$; MW: 797.99 g/mol) isolated from *B. bassiana*, *P. fumosoroseus,* and *Fusarium* spp. (phytopathogenic) [34, 35], is reported to increase membrane permeability to Na^+ and K^+ ions [36]. Insecticidal activity of beauvericin has been reported earlier on fall armyworm, *Spodoptera frugiperda* (JE Smith) [37, 38], wheat aphid, *Schizaphis graminum* (Rondani) [39, 40], Colorado potato beetle, *Leptinotarsa decemlineata* (Say) [41], blue blowfly, *Calliphora erythrocephala* Meigen [42], and yellow fever mosquito, *Aedes aegypti* L. [43]. Its antibacterial activity against humans, animals, and plant pathogenic bacteria [44–46] and antiviral activity against human immunodeficiency virus type-1 integrase [47] have also been investigated. Besides these, its in vitro genotoxic, estrogenic, cytotoxic, and immune-toxic activities have been reviewed in detail by Mallebrera et al. (2018) and Caloni et al. (2020) [34, 35].

More recently, another insect-toxin protein Bb70p (MW: 35.5 kDa) was isolated from *B. bassiana* 70 using ammonium sulfate precipitation, ion-exchange chromatography, and gel filtration [48]. Intra-hemocoelic injections of Bb70 in *G. mellonella* resulted in high mortality and activation of phenol oxidase cascade [48].

Destruxins, another group of cyclic hexadepsipeptides, composed of five amino acids and a variable α-hydroxy acid (characterizing the destruxin types A-F) has been isolated from *Metarhizium aniplosae*. Till date, 39 destruxin derivatives have been isolated, characterized, and reviewed from different fungal cultures and new destruxins are still being investigated [52]. Studies on *G. mellonella* [49] and *Manduca sexta* (L.) [50] reported that the infection by destruxins induced tetanic paralysis in larvae by depolarizing the membranes and activating calcium channels. Sowjanya Sree et al. [51] in their study on insecticidal activity of destruxin in *S. litura* also reported that the destruxin exposure (topical or ingestion) elicited oxidative stress in *S. litura* larvae. Besides insecticidal activity, destruxins are reported to have a broad range of biological activities including cytotoxic, antiviral, immunosuppressant, phytotoxic, antimicrobial, antitrypanosomal, and antiproliferative activities [52–54] making them a good candidate for development of modern drugs for cancer therapies or as mycopesticides.

Ribotoxins, a group of extracellular ribonucleases (~150 amino acids), produced by several fungal species including *Aspergillus*, *Hirsutella thompsonii* Fisher, and *M. anisopliae*, are known for their extraordinarily specific and toxic ribonucleolytic activity [55, 56]. The target of ribotoxins, i.e. sarcin/ricin loop, is localized in the larger subunit of ribosome. The ribotoxin specifically cleaves single phosphodiester bond of the sarcin/ricin loop, which is enough for inactivating its function. *H. thompsonii,* an entomopathogenic fungus, produces ribotoxin, hirsutellin A, whose insecticidal activity has opened a new view as a potential biological control agent of pests [57, 58]. Another member of the ribotoxin family includes α-sarcin and restrictocin from *Aspergillus giganteus* and *Aspergillus restrictus,* respectively. Olombrada et al. (2014) compared the toxicity of ribotoxins α-sarcin and hirsutellin A (HtA) against *G. mellonella* larvae and insect cell lines *Trichoplusia ni* (High Five) and *Spodoptera frugiperda* (Sf9). Both were found to be equally toxic and resulted in decreased cell viability and protein synthesis inhibition (in vivo) [56]. More recently, another ribotoxin, anisoplin, structurally similar to hirsutellin A has been isolated and characterized from *M. anisopliae* [59] and still being investigated for its biological activities.

Lectins, proteins of nonimmune origin, have been isolated from fungi belonging to phyla Ascomycota and Basidiomycota. It consists of a domain that binds specifically and reversibly to carbohydrate (glycan) structure. The entomotoxic activity has been earlier investigated for *Xerocomus chrysenteron* lectin (XCL; 15kDa) on aphid, *Acrythosiphon pisum* and *Drosophila melanogaster*, by addition to their artificial diet. In addition, Hamshou et al. [60] studied the insecticidal activity of *Sclerotinia sclerotiorum* agglutinin, a fungal lectin from the sclerotes, on *A. pisum*, demonstrating high mortality when provided with different concentrations of this fungal lectin in artificial diet. Hogervorst et al. [61] studied the effect of snowdrop lectin (of plant origin) on *Chrysoperla carnea, Coccinella septempunctata,* and *Adalia bipunctata*. The study revealed that the lectin binds to terminal mannose residues on glycoproteins in the gut of aphid larvae and other insects [62–64].

19.3 Bacteria

Since the discovery of first insecticidal bacteria, dated back to the 1960s, entomopathogenic bacteria have received considerable attention because of their utilization, not only in agriculture as a biocontrol agent of insect pests but also as insect vectors of human diseases. Over the past century, many species of bacterial pathogens have been isolated from insects, and by far most extensively studied species is *Bacillus thuringiensis* [65–67].

B. thuringiensis, endospore-forming gram-positive bacteria, is characterized by the presence of insect-toxic protein (δ-endotoxins) crystal within the sporulating cell cytoplasm. Besides endotoxins, *B. thuringiensis* is also reported to produce several other virulence factors such as α- and β-exotoxins, hemolysins, enterotoxins, phospholipases, proteases, chitinases, S-Layer protein, parasporins, secreted insecticidal protein (Sip), and vegetative insecticidal proteins (Vip) [66, 68–70]. Over 70 subspecies of *B. thuringiensis* are described, and most of them are reported to be insecticidal to lepidopteran [71–73], coleopteran [74, 75], hymenopteran [76–78], or dipteran [79, 80] insect extending to other groups, including protozoa [81], nematodes [82], and mites [81].

δ-endotoxins include crystal (Cry) and cytolytic (Cyt) toxic proteins that belong to the pore-forming class of bacterial toxin. More than 500 Cry genes have been classified into 67 groups (Cry1- Cry67) based on their primary amino acid sequence [83, 84]. These toxins are secreted as water-soluble proteins that undergo conformational changes in order to insert and to translocate across the cell membranes of their host. The Cry and Cyt endotoxins are solubilized and are activated by proteases present in the insect's midgut lumen. The activated Cry toxins (60 kDa) consists of three functional domains, domain I, a seven a-helices bundle (involved in membrane insertion), and domains II and III composed of two b-sheet (for receptor interactions). In contrast, Cyt (25 kDa) protein is composed only of a single domain [85]. The Cry toxins specifically bind to protein receptors present in the microvilli of midgut epithelial cells, whereas Cyt is the hemolytic toxins that directly interact with membrane lipids, inserting into the membrane and forming pores.

In addition, vegetative insecticidal proteins (Vip) secreted during the vegetative growth of certain *B. thuringiensis* strains [86, 87] consists of Vip1, Vip2, Vip3, and Vip4, where first three are reported to be insecticidal; no such activity has been recorded till date for Vip4. It is believed that Vip1 component of Vip1/Vip2 binary toxin binds to insect's midgut membrane receptors whereas the Vip2 component enters the cell preventing microfilament formation through its ADP-ribosyltransferase activity against actin [88]. In addition, the mode of action of Vip3 is reported to resemble that of Cry proteins; in terms of proteolytic activation, binding to insect's midgut epithelial membrane, and pore formation [88–90]. Vip1 and Vip2 have been reported to be toxic to coleopterans (*Holotrichia parallela, Holotrichia oblita,* and *Anomala corpulenta*) [90, 91] and homopterans [92] whereas Vip3a reported being toxic against a broad range of lepidopteran insects (*Spodoptera exigua, Agrotis ipsilon, Helicoverpa punctigera, Chilo partellus*, and *Diatraea saccharalis*) [72, 87, 93]. Owing to the rich toxin collection, transgenic plants expressing *B. thuringiensis* are being employed for controlling insect pest population as an alternative for chemical pesticides over the past decades.

Another insecticidal bacterium that has been commercialized over the past decade is *Lysinibacillus sphaericus* (=*Bacillus sphaericus* Neide), many isolates of which have proven to have larvicidal activity against mosquitoes (*Anopheles* spp. and *Culex* spp.) [94–97]. It is almost 50 years since the first report of insect pathogenic strain of *L. sphaericus* by Kellen et al. (1965). Meanwhile, there have been significant advances in the understanding of this bacterium, including its taxonomy and mechanisms of pathogenic stains activity [94]. Recently, Yu et al. (2017) studied the effect of sublethal doses of *L. sphaericus* on two continuous generations of *Anopheles dirus* and recorded the gravidity, oviposition, pupation, hatch, and eclosion. The study reported a significant reduction in oviposition without any effect on gravidity, pupation, hatch, and eclosion [97].

The entomopathogenic strains of *L. sphaericus* are reported to produce a range of insect-toxic proteins like sphaericolysins, vegetative mosquitocidal toxins, and binary toxin [98]. Nishiwaki et al. (2007) studied the insecticidal activity of sphaericolysin (53 kDa) isolated from bacterial pathogen (A3-2 strain) of *Myrmeleon bore* (Tjeder 1941) larvae on German cockroach, *Blattela germanica* L. and the common cutworm *S. litura* and reported greater sphaericolysin-induced damage in *B. germanica* as compared with *S. litura* in vivo [99].

The mosquitocidal toxins (Mtx) and the binary toxin (Bin) produced by *L. sphaericus* during vegetative growth and sporulation, respectively, are reported to have strong larvicidal activity against *Anopheles* and *Culex* mosquitos. To date, four different types of Mtx toxins have been isolated and categorized into two toxin families, i.e. Mtx1 family consisting of Mtx1 (100 kDa) toxic protein and Mtx2 family consisting of Mtx2 (31.8 kDa), Mtx3 (35.8 kDa) toxic proteins. More recently, Mtx4 has been added to the group but has yet not been characterized for its toxic properties [98]. In addition, studies have reported that Mtx1 toxin has a synergistic activity, when combined with Cyt1Aa, against *Culex quinquefasciatus* Say [100]. This synergy may raise a possibility for the production of bacterial insecticides with enhanced insecticidal/mosquitocidal activity.

Like Vip1/Vip2 binary toxin of *B. thuringiensis*, BinA (42 kDa) and BinB (51 kDa) also act together as a binary toxin in which BinB is responsible for receptor binding, and BinA acts as a toxic component. The toxins acts by first solubilizing in alkaline medium of insect's gut, followed by its activation by proteolytic cleavage by midgut proteinases converted into their active form. Also, the maximum potency of the toxin has been evaluated when both the proteins are present in equimolar (1:1M) ratio [101]. More recently, its mosquitocidal activity in malarial mosquito, *Anopheles gambiae* Ag55 cell lines, has been demonstrated by Riaz et al. [102].

Besides these many other bacterial pathogens such as *Photorhabdus* and *Xenorhabdus* that exist in symbiotic association with entomopathogenic nematodes, *Heterorhabditis* and *Steinernema* genera, respectively, are reported to produce insect-toxic proteins. A high molecular weight insecticidal protein Mcf (Makes caterpillars floppy; Mcf1 and Mcf2) produced by *Photorhabdus luminescens* is known to induce apoptosis in insect's midgut epithelium and hemocytes. Other high molecular weight toxin complexes (Tc) with broad range of insecticidal activity have also been isolated from *Photorhabdus* and *Xenorhabdus*. The toxins have been grouped into different complexes Tca, Tcb, Tcc, and Tcd [103, 104]. *Photorhabdus luminescens* strain W14 contains toxins (toxins A and B) that are known to have insecticidal activity. Liu et al. (2003) demonstrated the entomotoxicity of toxin A (283 kDa) protein isolated from *P. luminescens* strain W14 expressed in transgenic *Arabidopsis thaliana* against European corn borer (*Ostrinia nubilalis*), southern corn rootworm (*Diabrotica undecimpunctata howardi*), and tobacco hornworm (*M. sexta*) [194]. Besides these, many more insect-toxic proteins from have been isolated (Table 19.1), characterized, and commercialized as an alternative to chemical pesticides and new discoveries in this field are opening new routes for insect pest control.

19.4 Virus

More than 800 entomopathogenic viruses have been isolated till date, mostly from Lepidoptera followed by Hymenoptera, Coleoptera, Orthoptera, and Diptera [11]. The viral pathogenesis begins with the ingestion of polyhedra (infective stage) by the host. The polyhedra dissolve in the gut, and the virion is released, which later invades the midgut epithelial cells. The virus replicates within these infected cells and are able to produce hundreds of polyhedra and thousands of granules per cell. The infection subsequently invades the hemocoel and rest of the host cells, resulting in the death of the host. Winstanley and Crook (1993) have summarized the replication of *Cydia pomonella* granulosis virus in cell cultures [105].

The different entomopathogenic viruses belong to Baculoviridae, Reoviridae, Poxviridae, Iridoviridae, Picornoviridae, Parvoviridae, and Rhabdoviridae families [106], of which baculoviruses and granulosis viruses have received the maximum attention because of their host specificity and inherent safety to humans and nontarget organisms [107, 108]. Till date, several nuclear polyhedrosis viruses have been commercially developed and registered, some of which include Nuclear polyhedrosis virus for *Heliothis zea* (Boddie) (Cotton ballworm), *Heliothis virescens* F. (Tobacco budworm), *Orgyia pseudotsugata* (McDunnough) (Douglas fir tussock moth), *Lymantria dispar* L. (Gypsy moth), *Neodiprion sertifer* (Geoffroy) (European pine sawfly), *N. lecontei* (Fitch)(Red-headed pine sawfly) and *N. abietis* (Harris) (Balsam fir sawfly), *Mamestra brassicae* L. (Cabbage moth), *Spodoptera exigua* (Hübner) (Beet armyworm), *Anagrapha* (*Syngrapha*) *falcifera* (Kirby) (Celery looper), *Anticarsia gemmatalis* Hübner (Velvetbean caterpillar), and *Autographa californica* (Speyer) (Alfalfa looper) [14, 109] (Table 19.1).

Based on the chemical nature of the genome present and the online database of International Committee on Taxonomy of Viruses (ICTVdB) and the Eighth Report of the International Committee on Taxonomy of Viruses [110], entomopathogenic viruses can be grouped into two, i.e. DNA or RNA virus. The DNA virus includes members of family Ascoviridae (dsDNA), Baculoviridae (dsDNA), Parvoviridae (ssDNA), Poxviridae (dsDNA), Nudiviruses (dsDNA), and Polydnaviridae (dsDNA) while the RNA virus includes family Dicistroviridae (ssRNA), Nodaviruses (ssRNA), Tetraviridae (ssRNA), and Reoviridae (ssRNA).

Ascoviruses from the family Ascoviridae are known to infect Lepidopteran larvae and pupae and are transmitted by parasitoids during oviposition. Baculoviridae have been isolated from Lepidoptera, Diptera, and Hymenoptera. It is further classified into four genera: Alpha-, Beta-, Gamma-, and Deltabaculovirus. Alphabaculovirus and betabaculovirus include all lepidopteran nuclear polyhedrosis viruses, whereas gammabaculovirus and deltabaculovirus are specific to hymenoptera and dipetaran baculoviruses, respectively [106, 111]. Other family of DNA viruses includes Parvoviridae, which consists of insect-specific genus Densovirus (ssDNA). These are single-stranded DNA viruses, which have been isolated from different insect orders, including Lepidoptera, Orthoptera, Hemiptera, Dictyoptera, and Diptera. In addition, Entomopoxvirinae of the family Poxviridae have reported infecting Coleoptera, Lepidoptera, and Diptera.

Salivary Gland Hypertrophy Viruses (SGHVs) (family Hytrosaviridae) [112, 113] comprise unclassified members of entomopathogenic dsDNA viruses that have been reported from three genera of Diptera: the housefly *Musca domestica* L. (MdSGHV), the tsetse fly *Glossina pallidipes* (Austen)(GpSGHV), and the narcissus bulb fly *Merodon equestris* (MeSGHV) [114, 115].

Nudiviruses forming a highly diverse group of dsDNA have been reported from the coconut beetle, *Oryctes rhinoceros* (L.) and family Polydnaviridae contains insect Polydnaviruses which have been reported from hymenopteran insects [116, 117].

The family Dicistroviridae consisting of RNA viruses have been reported to infect a wide range of arthropods, especially insect orders Diptera, Hemiptera, Hymenoptera, Lepidoptera, and Orthoptera whereas members of family Tetraviridae exclusively infect larvae of Lepidoptera [106, 118]. Various other insect-infecting viruses are enlisted in the table (Table 19.1) along with their target host.

Using recombinant baculovirus technology, we can enhance the insecticidal activity of viruses, making them potential economic substitutes [15, 16]. Earlier studies have reported that *Buthus eupeus* insect toxin-1, the *B. thuringiensis* ssp. *kurstaki* HD-73 delta-endotoxin, the *M. sexta* diuretic hormone, *Androctonus australis* Hector neurotoxin, the *Heliothis virescens* juvenile hormone esterase, the *Pyrenophora tritici* TxP-I toxin, T-urf 13 genes, and Dol m V genes have been inserted into baculoviruses [119–122] for the purpose of developing viral pesticides.

19.5 Conclusions and Perspectives

With the existence of considerable diversity among entomopathogenic populations in nature and the wide range of secondary metabolites produced, they can be considered in the selection of isolates for commercialization and successful production of biopesticides. However, we are still scratching the surface of the biochemical potential of the microbial world, but chemical investigations will certainly expand the list of new bioactive secondary metabolites in the near future. As our understanding of the multiple fungal, bacterial, and viral entomopathogens increases, so does the potential for their employment in new and interesting ways.

References

1 Stork, N.E. (2018). How many species of insects and other terrestrial arthropods are there on earth? *Annual Review of Entomology* 63 (1): 31–45.

2 Savary, S., Willocquet, L., Pethybridge, S.J. et al. (2019). The global burden of pathogens and pests on major food crops. *Nature Ecology and Evolution* 3 (3): 430–439.

3 FAO (2018). FAO – News Article: New standards to curb the global spread of plant pests and diseases. FAO.

4 Omkar (2016). *Ecofriendly Pest Management for Food Security*. Academic Press an imprint of Elsevier Inc., USA, pp. 762.

5 Biedermann, P.H. and Rohlfs, M. (2017). Evolutionary feedbacks between insect sociality and microbial management. *Current Opinion in Insect Science* 22: 92–100.

6 Ollerton, J. (2006). "Biological barter": patterns of specialization compared across different mutualisms. In: *Plant-Pollinator Interactions: From Specialization to Generalization* (eds. J. Ollerton and N. Waser), 411–435. University of Chicago Press.

7 Douglas, A.E. (1998). Nutritional interactions in insect-microbial symbioses: aphids and their symbiotic bacteria *Buchnera*. *Annual Review of Entomology* 43: 17–37.

8 Archetti, M., Scheuring, I., Hoffman, M. et al. (2011). Economic game theory for mutualism and cooperation. *Ecology Letters* 14 (12): 1300–1312.

9 Biedermann, P.H.W. and Kaltenpoth, M. (2014). New synthesis: the chemistry of partner choice in insect-microbe mutualisms. *Journal of Chemical Ecology* 40 (2): 99.

10 Zhang, M.M., Poulsen, M., and Currie, C.R. (2007). Symbiont recognition of mutualistic bacteria by *Acromyrmex* leaf-cutting ants. *ISME Journal* 1 (4): 313–320.

11 Martignoni, M. and Iwai, P. (1981). A catalogue of viral diseases of insects, mites and ticks. In: *Microbial Control of Pests and Plant Diseases 1970-1980* (ed. H.D. Burges), 897–911. London: Academic Press.

12 Glare, T.R., Jurat-Fuentes, J.L., and O'Callaghan, M. (2017). *Basic and Applied Research: Entomopathogenic Bacteria*. Elsevier Inc.

13 Aung, O.M., Soytong, K., and Hyde, K.D. (2008). Diversity of entomopathogenic fungi in rainforests of Chiang Mai Province, Thailand. *Fungal Diversity* 30: 15–22.

14 Koul, O. (2011). Microbial biopesticides: opportunities and challenges. *CAB Reviews: Perspectives in Agriculture, Veterinary Science, Nutrition and Natural Resources* 6 (056): 1–26.

15 Beas-Catena, A., Sánchez-Mirón, A., García-Camacho, F. et al. (2014). Baculovirus biopesticides: an overview. *Journal of Animal and Plant Sciences* 24 (2): 362–373.

16 Gramkow, A.W., Perecmanis, S., Sousa, R.L.B. et al. (2010). Insecticidal activity of two proteases against *Spodoptera frugiperda* larvae infected with recombinant baculoviruses. *Virology Journal* 7 (1): 1–10.

17 Griffin, D. (1994). *Fungal Physiology*. New York: Wiley-Liss.

18 Mannino, M.C., Huarte-Bonnet, C., Davyt-Colo, B., and Pedrini, N. (2019). Is the insect cuticle the only entry gate for fungal infection? Insights into alternative modes of action of entomopathogenic fungi. *Journal of Fungi* 5 (2).

19 Koca, A.S., Altinok, H.H., and Altinok, M.A. (2019). Modes of action of entomopathogenic fungi. *Current Trends in Natural Sciences* 8: 117–124.

20 Bassi, A. (1836). *Del mal del segno e di altre malattie dei bachi da seta*. Parte seconda: Practica, Tipografia Orcesi, Lodi.

21 Bassi, A. (1835). *Del mal del segno, calcinaccio o moscardino, malattia che affigge i bachi da seta, e sul modo di liberarne le bigattaje anche le piui infestate*. Parte prima: della teoria, Tipografia Orcesi, Lodi.

22 Ekesi, S., Maniania, N.K., and Lux, S.A. (2002). Mortality in three African tephritid fruit fly puparia and adults caused by the entomopathogenic fungi, *Metarhizium anisopliae* and *Beauveria bassiana*. *Biocontrol Science and Technology* 12 (1): 7–17.

23 Ortiz-Urquiza, A., Riveiro-Miranda, L., Santiago-Alvarez, C., and Quesada-Moraga, E. (2010). Insect-toxic secreted proteins and virulence of the entomopathogenic fungus *Beauveria bassiana*. *Journal of Invertebrate Pathology* 105 (3): 270–278.

24 Ortiz-Urquiza, A., Keyhani, N.O., and Quesada-Moraga, E. (2013). Culture conditions affect virulence and production of insect toxic proteins in the entomopathogenic fungus *Metarhizium anisopliae*. *Biocontrol Science and Technology* 23 (10): 1199–1212.

25 Quesada-Moraga, E., Ruiz-GarcíA, A., and Santiago-ÁLvarez, C. (2006). Laboratory evaluation of entomopathogenic fungi *Beauveria bassiana* and *Metarhizium anisopliae* against puparia and adults of *Ceratitis capitata* (Diptera: Tephritidae). *Journal of Economic Entomology* 99 (6): 1955–1966.

26 Samuels, R.I., Coracini, D.L.A., Martins Dos Santos, C.A., and Gava, C.A.T. (2002). Infection of *Blissus antillus* (Hemiptera: Lygaeidae) eggs by the entomopathogenic fungi *Metarhizium anisopliae* and *Beauveria bassiana*. *Biological Control* 23 (3): 269–273.

27 Elsworth, J.F. and Grove, J.F. (1980). Cyclodepsipeptides from *Beauveria bassiana*. Part 2. Beauverolides A to F and their relationship to isarolide. *Journal of the Chemical Society, Perkin Transactions* 1: 1795–1799.

28 Elsworth, J.F. and Grove, J.F. (1977). Cyclodepsipeptides from *Beauveria bassiana* bals. Part 1. Beauverolides H and I. *Journal of the Chemical Society, Perkin Transactions* 1 (3): 270–273.

29 Grove, J.F. (1980). Cyclodepsipeptides from *Beauveria bassiana*. Part 3. The isolation of beauverolides Ba, Ca, Ja, and Ka. *Journal of the Chemical Society. Perkin Transactions* 1: 2878–2880.

30 Isogai, A., Kanaoka, M., Matsuda, H. et al. (1978). Structure of a new cyclodepsipeptide, beauverilide a from *Beauveria bassiana*. *Agricultural and Biological Chemistry* 42 (9): 1797–1798.

31 Jegorov, A., Sedmera, P., Maťha, V. et al. (1994). Beauverolides L and La from *Beauveria tenella* and *Paecilomyces fumosoroseus*. *Phytochemistry* 37 (5): 1301–1303.

32 Vilcinskas, A. and Gotz, P. (1999). Parasitic fungi and their interactions with the insect immune system. *Advances in Parasitology* 43: 267–313.

33 Mochizuki, K., Ohmori, K., Tamura, H. et al. (1993). The structures of bioactive cyclodepsipeptides, beauveriolides I and II, metabolites of entomopathogenic fungi *Beauveria* sp. *Bulletin of the Chemical Society of Japan* 66 (10): 3041–3046.

34 Mallebrera, B., Prosperini, A., Font, G., and Ruiz, M.J. (2018). in vitro mechanisms of Beauvericin toxicity: a review. *Food and Chemical Toxicology* 111: 537–545.

35 Caloni, F., Fossati, P., Anadón, A., and Bertero, A. (2020). Beauvericin: the beauty and the beast. *Environmental Toxicology and Pharmacology* 75: 103349.

36 Steinrauf, L.K. (1985). Beauvericin and the other enniatins. *Metal Ions in Biological Systems* 19: 139–171.

37 Fornelli, F., Minervini, F., and Logrieco, A. (2004). Cytotoxicity of fungal metabolites to lepidopteran (*Spodoptera frugiperda*) cell line (SF-9). *Journal of Invertebrate Pathology* 85 (2): 74–79.

38 Calo, L., Fornelli, F., Nenna, S. et al. (2003). Beauvericin cytotoxicity to the invertebrate cell line SF-9. *Journal of Applied Genetics* 44 (4): 515–520.

39 Ganassi, S., Moretti, A., Stornelli, C. et al. (2001). Effect of *Fusarium*, *Paecilomyces* and *Trichoderma* formulations against aphid *Schizaphis graminum*. *Mycopathologia* 151 (3): 131–138.

40 Ganassi, S., Moretti, A., Maria Bonvicini Pagliai, A. et al. (2002). Effects of beauvericin on *Schizaphis graminum* (Aphididae). *Journal of Invertebrate Pathology* 80 (2): 90–96.

41 Gupta, S., Krasnoff, S.B., Underwood, N.L. et al. (1991). Isolation of beauvericin as an insect toxin from *Fusarium semitectum* and *Fusarium moniliforme* var. *subglutinans*. *Mycopathologia* 115 (3): 185–189.

42 Grove, J.F. and Pople, M. (1980). The insecticidal activity of beauvericin and the enniatin complex. *Mycopathologia* 70 (2): 103–105.

43 Daniel, J.F.S., Silva, A.A., Nakagawa, D.H. et al. (2017). Larvicidal activity of *Beauveria bassiana* extracts against *Aedes aegypti* and identification of beauvericins. *Journal of the Brazilian Chemical Society* 28 (6): 1003–1013.

44 Meca, G., Sospedra, I., Soriano, J.M. et al. (2010). Antibacterial effect of the bioactive compound beauvericin produced by *Fusarium proliferatum* on solid medium of wheat. *Toxicon* 56 (3): 349–354.

45 Zhang, H., Ruan, C., Bai, X. et al. (2016). Isolation and identification of the antimicrobial agent beauvericin from the endophytic *Fusarium oxysporum* 5-19 with NMR and ESI-MS/MS. *BioMed Research International* 2016: 1–4.

46 Castlebury, L.A., Sutherland, J.B., Tanner, L.A. et al. (1999). Use of a bioassay to evaluate the toxicity of beauvericin to bacteria. *World Journal of Microbiology and Biotechnology* 15 (1): 131–133.

47 Shin, C.G., An, D.G., Song, H.H., and Lee, C. (2009). Beauvericin and enniatins H, i and MK1688 are new potent inhibitors of human immunodeficiency virus type-1 integrase. *Journal of Antibiotics* 62 (12): 687–690.

48 Khan, S., Nadir, S., Lihua, G. et al. (2016). Identification and characterization of an insect toxin protein, Bb70p, from the entomopathogenic fungus, *Beauveria bassiana*, using *Galleria mellonella* as a model system. *Journal of Invertebrate Pathology* 133: 87–94.

49 Hubert, M., Cherton, J.C., Vey, A., and Lange, C. (1999). Metabolites of A and E destruxins in *Galleria mellonella* larvae using negative-ion fast-atom bombardment mass spectrometry. *Rapid Communications in Mass Spectrometry* 13 (3): 179–184.

50 Samuels, R.I., Reynolds, S.E., and Charnley, A.K. (1988). Calcium channel activation of insect muscle by destruxins, insecticidal compounds produced by the entomopathogenic fungus *Metarhizium anisopliae*. *Comparative Biochemistry and Physiology Part C, Comparative* 90 (2): 403–412.

51 Sowjanya Sree, K., Padmaja, V., and Murthy, Y.L.N. (2008). Insecticidal activity of destruxin, a mycotoxin from *Metarhizium anisopliae* (Hypocreales), against *Spodoptera litura* (Lepidoptera: Noctuidae) larval stages. *Pest Management Science* 64 (2): 119–125.

52 Liu, B.L. and Tzeng, Y.M. (2012). Development and applications of destruxins: a review. *Biotechnology Advances* 30 (6): 1242–1254.

53 Pedras, M.S.C., Irina Zaharia, L.I., and Ward, D.E. (2002). The destruxins: Synthesis, biosynthesis, biotransformation, and biological activity. *Phytochemistry* 59 (6): 579–596.

54 Wang, X., Gong, X., Li, P. et al. (2018). Structural diversity and biological activities of cyclic depsipeptides from fungi. *Molecules* 23 (1).

55 Olombrada, M., Lázaro-Gorines, R., López-Rodríguez, J.C. et al. (2017). Fungal ribotoxins: a review of potential biotechnological applications. *Toxins* 9 (2): 1–21.

56 Olombrada, M., Martínez-Del-Pozo, Á., Medina, P. et al. (2014). Fungal ribotoxins: natural protein-based weapons against insects. *Toxicon* 83: 69–74.

57 Herrero-Galán, E., García-Ortega, L., Olombrada, M. et al. (2013). Hirsutellin A: a paradigmatic example of the insecticidal function of fungal ribotoxins. *Insects* 4 (3): 339–356.

58 Mazet, I. and Vey, A. (1995). Hirsutellin A, a toxic protein produced in vitro by *Hirsutella thompsonii*. *Microbiology* 141 (6): 1343–1348.

59 Olombrada, M., Medina, P., Budia, F. et al. (2017). Characterization of a new toxin from the entomopathogenic fungus *Metarhizium anisopliae*: the ribotoxin anisoplin. *Biological Chemistry* 398 (1): 135–142.

60 Hamshou, M., Smagghe, G., Shahidi-Noghabi, S. et al. (2010). Insecticidal properties of Sclerotinia sclerotiorum agglutinin and its interaction with insect tissues and cells. *Insect Biochemistry and Molecular Biology* 40 (12): 883–890.

61 Hogervorst, P.A.M., Ferry, N., Gatehouse, A.M.R. et al. (2006). Direct effects of snowdrop lectin (GNA) on larvae of three aphid predators and fate of GNA after ingestion. *Journal of Insect Physiology* 52 (6): 614–624.

62 Upadhyay, S.K., Mishra, M., Singh, H. et al. (2010a). Interaction of Allium sativum leaf agglutinin (ASAL) with midgut BBMV proteins and its stability in Helicoverpa armigera. *Proteomics* 10: 4431–4440.

63 Upadhyay, S.K., Saurabh, S., Rai, P. et al. (2010b). SUMO fusion facilitates expression and purification of garlic lectin but modifies some of its properties. *Journal of Biotechnology* 146: 1–8.

64 Upadhyay, S.K. and Singh, P.K. (2012). Receptors of garlic (Allium sativum) lectins and their role in insecticidal action. *The Protein Journal* 31 (6): 439–446.

65 Cory, J.S. and Franklin, M.T. (2012). Evolution and the microbial control of insects. *Evolutionary Applications* 5 (5): 455–469.

66 Neto, J.M.W.D., Wanderley, M.C.D.A., da Silva, T.A.F. et al. (2020). *Bacillus thuringiensis* endotoxin production: a systematic review of the past 10 years. *World Journal of Microbiology and Biotechnology* 36 (9).

67 Jisha, V.N., Smitha, R.B., and Benjamin, S. (2013). An overview on the crystal Toxins from *Bacillus thuringiensis*. *Advances in Microbiology* 03 (05): 462–472.

68 Chakroun, M., Banyuls, N., Bel, Y. et al. (2016). Bacterial Vegetative Insecticidal Proteins (Vip) from Entomopathogenic Bacteria. *Microbiology and Molecular Biology Reviews* 80 (2): 329–350.

69 Donovan, W.P., Engleman, J.T., Donovan, J.C. et al. (2006). Discovery and characterization of Sip1A: a novel secreted protein from *Bacillus thuringiensis* with activity against coleopteran larvae. *Applied Microbiology and Biotechnology* 72 (4): 713–719.

70 Mizuki, E., Park, Y.S., Saitoh, H. et al. (2000). Parasporin, a human leukemic cell-recognizing parasporal protein of *Bacillus thuringiensis*. *Clinical and Diagnostic Laboratory Immunology* 7 (4): 625–634.

71 Upadhyay, S.K. and Singh, P.K. (2011). Role of alkaline phosphatase in insecticidal action of Cry1Ac against Helicoverpa armigera larvae. *Biotechnology Letters* 33: 2027–2036.

72 Chakrabarty, S., Jin, M., Wu, C. et al. (2020). *Bacillus thuringiensis* vegetative insecticidal protein family Vip3A and mode of action against pest Lepidoptera. *Pest Management Science* 76 (5): 1612–1617.

73 Daquila, B.V., Scudeler, E.L., Dossi, F.C.A. et al. (2019). Action of *Bacillus thuringiensis* (Bacillales: Bacillaceae) in the midgut of the sugarcane borer *Diatraea saccharalis* (Fabricius, 1794) (Lepidoptera: Crambidae). *Ecotoxicology and Environmental Safety* 184: 109642.

74 Domínguez-Arrizabalaga, M., Villanueva, M., Escriche, B. et al. (2020). Insecticidal activity of *Bacillus thuringiensis* proteins against coleopteran pests. *Toxins* 12 (7): 1–29.

75 Domínguez-Arrizabalaga, M., Villanueva, M., Fernandez, A.B., and Caballero, P. (2019). A strain of *Bacillus thuringiensis* containing a novel cry7Aa2 gene that is toxic to *Leptinotarsa decemlineata* (Say) (Coleoptera: Chrysomelidae). *Insects* 10 (9).

76 Porcar, M., Gómez, F., Gruppe, A. et al. (2008). Hymenopteran specificity of *Bacillus thuringiensis* strain PS86Q3. *Biological Control* 45 (3): 427–432.

77 Upadhyay, S.K., Singh, S., Chandrashekar, K. et al. (2012). Compatibility of garlic (Allium sativum L.) leaf agglutinin and Cry1Ac δ-endotoxin for gene pyramiding. *Applied Microbiology and Biotechnology* 93: 2365–2375.

78 Wang, D.S., He, Y.R., Guo, X.L., and Luo, Y.L. (2012). Acute toxicities and sublethal effects of some conventional insecticides on *Trichogramma chilonis* (Hymenoptera: Trichogrammatidae). *Journal of Economic Entomology* 105 (4): 1157–1163.

79 Shishir, M.A., Akter, A., Bodiuzzaman, M. et al. (2015). Novel toxicity of *Bacillus thuringiensis* strains against the melon fruit fly, *Bactrocera cucurbitae* (Diptera: Tephritidae). *Biocontrol Science* 20 (2): 115–123.

80 Ben-Dov, E. (2014). *Bacillus thuringiensis* subsp. *israelensis* and Its dipteran-specific toxins. *Toxins* 6 (4): 1222–1243.

81 Veloorvalappil Narayanan, J., Robinson Babysarojam, S., Prakasan, P. et al. (2018). Crude *Bacillus thuringiensis* pellets efficiently combats *Eutetranychus orientalis*, the spider mite. *International Journal of Pest Management* 64 (3): 243–251.

82 Wang, P., Zhang, C., Guo, M. et al. (2014). Complete genome sequence of *Bacillus thuringiensis* YBT-1518, a typical strain with high toxicity to nematodes. *Journal of Biotechnology* 171 (1): 1–2.

83 Crickmore, N., Baum, J., Bravo, A. et al. (2014). *Bacillus thuringiensis* toxin nomenclature. http://www.BtnomenclatureInfo/.

84 Crickmore, N. (2017). *Bacillus thuringiensis* toxin classification. In: *Bacillus Thuringiensis and Lysinibacillus Sphaericus: Characterization and use in the Field of Biocontrol* (eds. L. Fiuza, R. Polanczyk and N. Crickmore), 41–52. Springer International Publishing.

85 De Maagd, R.A., Bravo, A., and Crickmore, N. (2001). How *Bacillus thuringiensis* has evolved specific toxins to colonize the insect world. *Trends in Genetics* 17 (4): 193–199.

86 Osman, G., Assaeedi, A., Osman, Y. et al. (2013). Purification and characterization of *Bacillus thuringiensis* vegetative insecticidal toxin protein(s). *Letters in Applied Microbiology* 57 (4): 310–316.

87 Estruch, J.J., Warren, G.W., Mullins, M.A. et al. (1996). Vip3A, a novel *Bacillus thuringiensis* vegetative insecticidal protein with a wide spectrum of activities against lepidopteran insects. *Proceedings of the National Academy of Sciences of the United States of America* 93 (11): 5389–5394.

88 Palma, L., Muñoz, D., Berry, C. et al. (2014). *Bacillus thuringiensis* toxins: An overview of their biocidal activity. *Toxins* 6 (12): 3296–3325.

89 Güney, E., Adıgüzel, A., Demirbağ, Z., and Sezen, K. (2019). *Bacillus thuringiensis kurstaki* strains produce vegetative insecticidal proteins (Vip 3) with high potential. *Egyptian Journal of Biological Pest Control* 29 (1).

90 Geng, J., Jiang, J., Shu, C. et al. (2019). *Bacillus thuringiensis* vip1 functions as a receptor of vip2 toxin for binary insecticidal activity against *Holotrichia parallela*. *Toxins* 11 (8).

91 Bi, Y., Zhang, Y., Shu, C. et al. (2015). Genomic sequencing identifies novel *Bacillus thuringiensis* Vip1/Vip2 binary and Cry8 toxins that have high toxicity to Scarabaeoidea larvae. *Applied Microbiology and Biotechnology* 99 (2): 753–760.

92 Sattar, S., Biswas, P.K., Hossain, M.A. et al. (2008). Search for Vegetative Insecticidal Proteins (VIPs) from local isolates of *Bacillus thuringiensis* effective against lepidopteran and homopteran insect pests. *Journal of Biopesticides* 1 (2): 216–222.

93 Liao, C., Heckel, D.G., and Akhurst, R. (2002). Toxicity of *Bacillus thuringiensis* insecticidal proteins for *Helicoverpa armigera* and *Helicoverpa punctigera* (Lepidoptera: Noctuidae), major pests of cotton. *Journal of Invertebrate Pathology* 80 (1): 55–63.

94 Kellen, W.R., Clark, T.B., Lindegren, J.E. et al. (1965). *Bacillus sphaericus* Neide as a pathogen of mosquitoes. *Journal of Invertebrate Pathology* 7 (4): 442–448.

95 Davidson, E.W. and Yousten, A.A. (1990). The Mosquito Larval Toxin of *Bacillus sphaericus*. In: *Bacterial Control of Mosquitoes & Black Flies* (eds. H. de Barjac and D.J. Sutherland), 237–255. Netherlands: Springer.

96 Priest, F.G. (1992). Biological control of mosquitoes and other biting flies by *Bacillus sphaericus* and *Bacillus thuringiensis*. *Journal of Applied Bacteriology* 72 (5): 357–369.

97 Yu, S., Ji, C., Zhu, X. et al. (2017). Impact of *Bacillus sphaericus* exposure on *Anopheles dirus*'s fecundity and resistance development. *Parasitology Research* 116 (3): 859–864.

98 Berry, C. (2012). The bacterium, *Lysinibacillus sphaericus*, as an insect pathogen. *Journal of Invertebrate Pathology* 109 (1): 1–10.

99 Nishiwaki, H., Nakashima, K., Ishida, C. et al. (2007). Cloning, functional characterization, and mode of action of a novel insecticidal pore-forming toxin, sphaericolysin, produced by *Bacillus sphaericus*. *Applied and Environmental Microbiology* 73 (10): 3404–3411.

100 Wirth, M.C., Berry, C., Walton, W.E., and Federici, B.A. (2014). Mtx toxins from *Lysinibacillus sphaericus* enhance mosquitocidal cry-toxin activity and suppress cry-resistance in *Culex quinquefasciatus*. *Journal of Invertebrate Pathology* 115 (1): 62–67.

101 Oei, C., Hindley, J., and Berry, C. (1990). An analysis of the genes encoding the 51.4- and 41.9-kDa toxins of *Bacillus sphaericus* 2297 by deletion mutagenesis: the construction of fusion proteins. *FEMS Microbiology Letters* 72 (3): 265–273.

102 Riaz, M.A., Adang, M.J., Hua, G. et al. (2020). Identification of *Lysinibacillus sphaericus* Binary toxin binding proteins in a malarial mosquito cell line by proteomics: a novel approach towards improving mosquito control. *Journal of Proteomics* 227: 103918.

103 Ffrench-Constant, R.H. and Bowen, D.J. (2000). Novel insecticidal toxins from nematode-symbiotic bacteria. *Cellular and Molecular Life Sciences* 57 (5): 828–833.

104 Bowen, D., Blackburn, M., Rocheleau, T.A. et al. (1999). Insecticidal toxins from the bacterium *Photorhabdus luminescens* : gene cloning and toxin histopathology. *Pesticide Science* 55 (6): 666–668.

105 Winstanley, D. and Crook, N.E. (1993). Replication of *Cydia pomonella* granulosis virus in cell cultures. *Journal of General Virology* 74 (8): 1599–1609.

106 Valicente, F.H. (2019). Entomopathogenic viruses. In: *Natural Enemies of Insect Pests in Neotropical Agroecosystems: Biological Control and Functional Biodiversity* (eds. B. Souza, L. Vázquez and R. Marucci), 137–150. Springer International Publishing.

107 Herniou, E.A., Arif, B.M., Becnel, J.J. et al. (2011). Baculoviridae. In: *Virus Taxonomy: Ninth Report of the International Committee on Taxonomy of Viruses* (eds. A.M.Q. King, M.J. Adams, E.B. Carstens and E.J. Lefkowitz), 163–174. Elsevier.

108 Cuartas, P., Barrera, G., Barreto, E., and Villamizar, L. (2014). Characterisation of a Colombian granulovirus (Baculoviridae: Betabaculovirus) isolated from *Spodoptera frugiperda* (Lepidoptera: Noctuidae) larvae. *Biocontrol Science and Technology* 24 (11): 1265–1285.

109 Miller, L.K., Lingg, A.J., and Bulla, L.A. (1983). Bacterial, viral, and fungal insecticides. *Science* 219 (4585): 715–721.

110 Fauquet, C., Mayo, M., Maniloff, J., and Desselberger, U. (2005). *Virus taxonomy: VIIIth report of the International Committee on Taxonomy of Viruses.*

111 Lacey, L.A. (2012). *Manual of Techniques in Invertebrate Pathology*. Academic Press, pp 504.

112 Lietze, V.U., Abd-Alla, A.M.M., Vreysen, M.J.B. et al. (2011). Salivary gland hypertrophy viruses: a novel group of insect pathogenic viruses. *Annual Review of Entomology* 56: 63–80.

113 Abd-Alla, A.M.M., Vlak, J.M., Bergoin, M. et al. (2009). Hytrosaviridae: a proposal for classification and nomenclature of a new insect virus family. *Archives of Virology* 154 (6): 909–918.

114 Garcia-Maruniak, A., Abd-Alla, A.M.M., Salem, T.Z. et al. (2009). Two viruses that cause salivary gland hypertrophy in *Glossina pallidipes* and *Musca domestica* are related and form a distinct phylogenetic clade. *Journal of General Virology* 90 (2): 334–346.

115 Demirbas-Uzel, G., Kariithi, H.M., Parker, A.G. et al. (2018). Susceptibility of tsetse species to *Glossina pallidipes* salivary gland hypertrophy virus (GpSGHV). *Frontiers in Microbiology* 9: 701.

116 Burand, J.P. (1998). Nudiviruses. In: *The Insect Viruses* (eds. L.K. Miller and L.A. Ball), 69–90. Springer.

117 Wang, Y.J., Burand, J.P., and Jehle, J.A. (2007). Nudivirus genomics: diversity and classification. *Virologica Sinica* 22 (2): 128–136.

118 Susevich, M.L., Metz, G.E., Marti, G.A., and Echeverría, M.G. (2017). Dicistroviridae: a new viral purification technique. *Revista Argentina de Microbiologia* 49 (4): 311–314.

119 Arora, R. and Shera, P.S. (2014). Genetic improvement of biocontrol agents for sustainable pest management. In: *Basic and Applied Aspects of Biopesticides*, vol. 9788132218 (ed. K. Sahayaraj), 255–285. India: Springer.

120 Haase, S., Ferrelli, L., Luis, M., and Romanowski, V. (2013). Genetic engineering of Baculoviruses. In: *Current Issues in Molecular Virology – Viral Genetics and Biotechnological Applications* (ed. V. Romanowski), 79–101. IntechOpen.

121 Kroemer, J.A., Bonning, B.C., Harrison, R.L. et al. (2015). Expression, delivery and function of insecticidal proteins expressed by recombinant baculoviruses. *Viruses* 7: 422–455.

122 Kamita, S.G., Kang, K.D., Inceoglu, A.B., and Hammock, B.D. (2017). Genetically modified baculoviruses for pest insect control. In: *Reference Module in Life Sciences* (eds. L.I. Gilbert and S.S. Gill), 331–369. Elsevier.

123 Feng, M.G. and Johnson, J.B. (1990). Relative virulence of six isolates of *Beauveria bassiana* on *Diuraphis noxia* (Homoptera: Aphididae). *Environmental Entomology* 19 (6): 785–790.

124 Pacheco, J.C., Poltronieri, A.S., Porsani, M.V. et al. (2017). Entomopathogenic potential of fungi isolated from intertidal environments against the cabbage aphid *Brevicoryne brassicae* (Hemiptera: aphididae). *Biocontrol Science and Technology* 27 (4): 496–509.

125 Jacobson, R.J., Chandler, D., Fenlon, J., and Russell, K.M. (2001). Compatibility of *Beauveria bassiana* (Balsamo) Vuillemin with *Amblyseius cucumeris* Oudemans (Acarina: Phytoseiidae) to control *Frankliniella occidentalis* Pergande (Thysanoptera: Thripidae) on cucumber plants. *Biocontrol Science and Technology* 11 (3): 391–400.

126 Ludwig, S.W. and Oetting, R.D. (2003). Efficacy of *Beauveria bassiana* plus insect attractants for enhanced control of *Frankliniella occidentalis* (Thysanoptera: Thripidae). *Florida Entomologist* 85 (1): 270–272.

127 Lee, S.J., Kim, S., Kim, J.C. et al. (2017). Entomopathogenic *Beauveria bassiana* granules to control soil-dwelling stage of western flower thrips, *Frankliniella occidentalis* (Thysanoptera: Thripidae). *BioControl* 62 (5): 639–648.

128 Al-Mazra'Awi, M.S., Shipp, L., Broadbent, B., and Kevan, P. (2006). Biological control of *Lygus lineolaris* (Hemiptera: Miridae) and *Frankliniella occidentalis* (Thysanoptera: Thripidae) by *Bombus impatiens* (Hymenoptera: Apidae) vectored *Beauveria bassiana* in greenhouse sweet pepper. *Biological Control* 37 (1): 89–97.

129 Majeed, M.Z., Fiaz, M., Ma, C.S., and Afzal, M. (2017). Entomopathogenicity of three muscardine fungi, *Beauveria bassiana*, *Isaria fumosorosea* and *Metarhizium anisopliae*, against the asian citrus psyllid, *Diaphorina citri kuwayama* (Hemiptera: Psyllidae). *Egyptian Journal of Biological Pest Control* 27 (2): 211–216.

130 Alves, L.F.A., Formentini, M.A., Fanti, A.L.P. et al. (2013). Susceptibility of *Gyropsylla spegazziniana* (Lizer & Trelles) (Hemiptera: Psyllidae) to *Beauveria bassiana* (Bals.) Vuill. *Arquivos do Instituto Biológico* 80 (3): 363–366.

131 Khun, K.K., Wilson, B.A.L., Stevens, M.M. et al. (2020). Integration of entomopathogenic fungi into IPM programs: studies involving weevils (coleoptera: Curculionoidea) affecting horticultural crops. *Insects* 11 (10): 1–36.

132 Amutha, M. and Gulsar Banu, J. (2017). Variation in Mycosis of Entomopathogenic Fungi on Mealybug, *Paracoccus marginatus* (Homoptera: Pseudococcidae). *Proceedings of the National Academy of Sciences India Section B – Biological Sciences* 87 (2): 343–349.

133 Sharma, L., Gonçalves, F., Oliveira, I. et al. (2018). Insect-associated fungi from naturally mycosed vine mealybug *Planococcus ficus* (Signoret) (Hemiptera: Pseudococcidae). *Biocontrol Science and Technology* 28 (2): 122–141.

134 Mohamed, G.S. (2016). Virulence of entomopathogenic fungi against the vine mealy bug, *Planococcus ficus* (Signoret) (Hemiptera: Pseudococcidae). *Egyptian Journal of Biological Pest Control* 26 (1): 47–51.

135 Feng, M.G., Chen, B., and Ying, S.H. (2004). Trials of *Beauveria bassiana*, *Paecilomyces fumosoroseus* and imidacloprid for management of *Trialeurodes vaporariorum* (Homoptera: Aleyrodidae) on greenhouse grown lettuce. *Biocontrol Science and Technology* 14 (6): 531–544.

136 Quesada-Moraga, E., Maranhao, E.A.A., Valverde-García, P., and Santiago-Álvarez, C. (2006). Selection of *Beauveria bassiana* isolates for control of the whiteflies *Bemisia* *tabaci* and *Trialeurodes vaporariorum* on the basis of their virulence, thermal requirements, and toxicogenic activity. *Biological Control* 36 (3): 274–287.

137 Selman, B.J., Dayer, M.S., and Hasan, M. (1997). Pathogenicity of *Beauveria bassiana* (Bals.) Vuill. to the larvae of diamondback moth, *Plutella xylostella* L. (Lep., Yponomeutidae). *Journal of Applied Entomology* 121 (1): 47–49.

138 Ibrahim, Y.B. and Low, W. (1993). Potential of mass-production and field efficacy of isolates of the entomopathogenic fungi *Beauveria bassiana* and *Paecilomyces fumosoroseus* against *Plutella xylostella*. *International Journal of Pest Management* 39 (3): 288–292.

139 Wojda, I., Kowalski, P., and Jakubowicz, T. (2009). Humoral immune response of *Galleria mellonella* larvae after infection by *Beauveria bassiana* under optimal and heat-shock conditions. *Journal of Insect Physiology* 55 (6): 525–531.

140 Samšiňáková, A., Mišíková, S., and Leopold, J. (1971). Action of enzymatic systems of *Beauveria bassiana* on the cuticle of the greater wax moth larvae (*Galleria mellonella*). *Journal of Invertebrate Pathology* 18 (3): 322–330.

141 Vilcinskas, A. and Maťha, V. (1997). Effect of the entomopathogenic fungus *Beauveria bassiana* on the humoral immune response of *Galleria mellonella* larvae (Lepidoptera: Pyralidae). *European Journal of Entomology* 94 (4): 461–472.

142 Miranpuri, G.S. and Khachatourians, G.G. (1995). Comparative virulence of different isolates of *Beauveria bassiana* and *Verticillium lecanii* against Colorado potato beetle, *Leptinotarsa decemlineata* (Say). *Journal of Insect Science* 8 (2): 160–166.

143 Miranpuri, G.S. and Khachatourians, G.G. (1994). Pathogenicity of *Beauveria bassiana* (Bals.) Vuill. and *Verticillium lecanii* (Zimm.) toward blister beetle, *Lytta nuttali* Say (Col., Meloidae). *Journal of Applied Entomology* 118 (1–5): 103–110.

144 Bing, L.A. and Lewis, L.C. (1991). Suppression of *Ostrinia nubilalis* (Hübner) (Lepidoptera: Pyralidae) by endophytic *Beauveria bassiana* (Balsamo) Vuillemin. *Environmental Entomology* 20 (6): 1207–1211.

145 Bing, L.A. and Lewis, L.C. (1993). Occurrence of the entomopathogen *Beauveria bassiana* (Balsamo) Vuillemin in different tillage regimes and in *Zea mays* L. and virulence towards *Ostrinia nubilalis* (Hübner). *Agriculture, Ecosystems and Environment* 45 (1–2): 147–156.

146 Feng, Z., Carruthers, R.I., Roberts, D.W., and Robson, D.S. (1985). Age-specific dose-mortality effects of *Beauveria bassiana* (Deuteromycotina: Hyphomycetes) on the European corn borer, *Ostrinia nubilalis* (Lepidoptera: Pyralidae). *Journal of Invertebrate Pathology* 46 (3): 259–264.

147 Boudjelida, H. and Soltani, N. (2011). Pathogenicity of *Metarhizium anisopliae* (Metsch) on *Ceratitis capitata* L. (Diptera: Tephritidae). *Annals of Biological Research* 2 (2): 104–110.

148 Fabrice, C.E.S., Tonussi, R.L., Orlandelli, R.C. et al. (2013). Compatibility of entomopathogenic fungus *Metarhizium anisopliae* (Metschnikoff) Sorokin with fungicide thiophanate-methyl assessed by germination speed parameter. *Journal of Food, Agriculture and Environment* 11 (1): 368–372.

149 Roberts, D.W. and St. Leger, R.J. (2004). *Metarhizium* spp., cosmopolitan insect-pathogenic fungi: Mycological aspects. *Advances in Applied Microbiology* 54: 1–70.

150 Vainio, A. and Hokkanen, H.M.T. (1993). The potential of entomopathogenic fungi and nematodes against *Otiorhynchus ovatus* L. and *O. dubius* Ström (Col., Curculionidae) in the field. *Journal of Applied Entomology* 115 (1–5): 379–387.

151 Bruck, D.J. (2007). Efficacy of *Metarhizium anisopliae* as a curative application for Black Vine Weevil (*Otiorhynchus sulcatus*) infesting container-grown nursery crops. *Journal of Environmental Horticulture* 25 (3).

152 Ekesi, S. and Maniania, N.K. (2000). Susceptibility of *Megalurothrips sjostedti* developmental stages to *Metarhizium anisopliae* and the effects of infection on feeding, adult fecundity, egg fertility and longevity. *Entomologia Experimentalis et Applicata* 94 (3): 229–236.

153 Pereira, A., Casals, P., Salazar, A.M., and Gerding, M. (2011). virulencia y efectos pre-letales en la reproducción de *Metarhizium anisopliae* var. *anisopliae* en *Pseudococcus viburni* (Hemiptera: Pseudococcidae). *Chilean Journal of Agricultural Research* 71 (4): 554–559.

154 Diabate, S., Deletre, E., Murungi, L.K. et al. (2019). Behavioural responses of bean flower thrips (*Megalurothrips sjostedti*) to vegetative and floral volatiles from different cowpea cultivars. *Chemoecology* 29 (2): 73–88.

155 Thomsen, L. and Eilenberg, J. (2000). Time-concentration mortality of *Pieris brassicae* (Lepidoptera: Pieridae) and *Agrotis segetum* (Lepidoptera: Noctuidae) larvae from different destruxins. *Environmental Entomology* 29 (5): 1041–1047.

156 Erler, F. and Ates, A.O. (2015). Potential of two entomopathogenic fungi, *Beauveria bassiana* and *Metarhizium anisopliae* (Coleoptera: Scarabaeidae), as biological control agents against the june beetle. *Journal of Insect Science* 15 (1): 44.

157 Milner, R.J. (1997). Prospects for biopesticides for aphid control. *Entomophaga* 42 (1–2): 227–239.

158 Fournier, V. and Brodeur, J. (2000). Dose-response susceptibility of pest aphids (Homoptera: Aphididae) and their control on hydroponically grown lettuce with the entomopathogenic Fungus *Verticillium lecanii*, Azadirachtin, and Insecticidal Soap. *Environmental Entomology* 29 (3): 568–578.

159 Kim, J.J., Lee, M.H., Yoon, C.S. et al. (2001). Control of cotton aphid and greenhouse whitefly with a fungal pathogen. *Biological Control of Greenhouse Pests*: 8–15.

160 Villalon, B. and Dean, H.A. (1974). *Hirsutella thompsonii* a fungal parasite of the citrus rust mite *Phyllocoptruta oleivora* in the Rio Grande Valley of Texas. *Entomophaga* 19 (4): 431–436.

161 McCoy, C.W., Selhime, A.G., Kanavel, R.F., and Hill, A.J. (1971). Suppression of citrus rust mite populations with application of fragmented mycelia of *Hirsutella thompsonii*. *Journal of Invertebrate Pathology* 17 (2): 270–276.

162 McCoy, C.W. and Kanavel, R.F. (1969). Isolation of *Hirsutella thompsonii* from the citrus rust mite, *Phyllocoptruta oleivora*, and its cultivation on various synthetic media. *Journal of Invertebrate Pathology* 14 (3): 386–390.

163 Wilding, N. (1971). Discharge of Conidia of *Entomophthora thaxteriana* Petch from the Pea Aphid *Acyrthosiphon pisum* Harris. *Journal of General Microbiology* 69 (3): 417–422.

164 Dean, G.J.W. and Wilding, N. (1973). Infection of cereal aphids by the fungus *Entomophthora*. *Annals of Applied Biology* 74 (2): 133–138.

165 Askary, H. and Yarmand, H. (2007). Development of the entomopathogenic hyphomycete *Lecanicillium muscarium* (Hyphomycetes: Moniliales) on various hosts. *European Journal of Entomology* 104 (1): 67–72.

166 Güçlü, S., Ak, K., Eken, C. et al. (2010). Pathogenicity of *Lecanicillium muscarium* against *Ricania simulans*. *Bulletin of Insectology* 63 (2): 243–246.

167 Cuthbertson, A.G.S., Blackburn, L.F., Northing, P. et al. (2008). Further compatibility tests of the entomopathogenic fungus *Lecanicillium muscarium* with conventional insecticide products for control of sweetpotato whitefly, *Bemisia tabaci* on poinsettia plants. *Insect Science* 15 (4): 355–360.

168 Cuthbertson, A.G.S. and Walters, K.F.A. (2005). Pathogenicity of the entomopathogenic fungus, *Lecanicillium muscarium*, against the sweetpotato whitefly *Bemisia tabaci* under laboratory and glasshouse conditions. *Mycopathologia* 160 (4): 315–319.

169 Scorsetti, A.C., Humber, R.A., De Gregorio, C., and López Lastra, C.C. (2008). New records of entomopathogenic fungi infecting *Bemisia tabaci* and *Trialeurodes vaporariorum*, pests of horticultural crops, in Argentina. *BioControl* 53 (5): 787–796.

170 Zimmermann, G. (2008). The entomopathogenic fungi *Isaria farinosa* (formerly *Paecilomyces farinosus*) and the *Isaria fumosorosea* species complex (formerly *Paecilomyces fumosoroseus*): biology, ecology and use in biological control. *Biocontrol Science and Technology* 18 (9): 865–901.

171 D'Alessandro, C.P., Padin, S., Urrutia, M.I., and López Lastra, C.C. (2011). Interaction of fungicides with the entomopathogenic fungus *Isaria fumosorosea*. *Biocontrol Science and Technology* 21 (2): 189–197.

172 Pick, D.A., Avery, P.B., Hunter, W.B. et al. (2012). Effect of *Isaria fumosorosea* (Hypocreales: Cordycipitaceae) and *Lysiphlebus testaceipes*, (Hymenoptera: Braconidae) on the Brown Citrus aphid: Preliminary assessment of a compatibility study. *Florida Entomologist* 95 (3): 764–766.

173 Hunter, W.B., Avery, P.B., Pick, D., and Powell, C.A. (2011). Broad spectrum potential of *Isaria fumosorosea* against insect pests of citrus. *Florida Entomologist* 94 (4): 1051–1054.

174 Imoulan, A., Wu, H.J., Lu, W.L. et al. (2016). *Beauveria medogensis* sp. nov., a new fungus of the entomopathogenic genus from China. *Journal of Invertebrate Pathology* 139: 74–81.

175 Rehner, S.A., Minnis, A.M., Sung, G.H. et al. (2011). Phylogeny and systematics of the anamorphic, entomopathogenic genus *Beauveria*. *Mycologia* 103 (5): 1055–1073.

176 Chen, M.J., Huang, B., Li, Z.Z., and Spatafora, J.W. (2013). Morphological and genetic characterisation of *Beauveria sinensis* sp. nov. from China. *Mycotaxon* 124: 301–308.

177 Zhang, S.L., He, L.M., Chen, X., and Huang, B. (2012). *Beauveria lii* sp. nov. isolated from *Henosepilachna vigintioctopunctata*. *Mycotaxon* 121: 199–206.

178 Bravo, A., Likitvivatanavong, S., Gill, S.S., and Soberón, M. (2011). *Bacillus thuringiensis*: a story of a successful bioinsecticide. *Insect Biochemistry and Molecular Biology* 41 (7): 423–431.

179 Chattopadhyay, A., Bhatnagar, N.B., and Bhatnagar, R. (2004). Bacterial insecticidal toxins. *Critical Reviews in Microbiology* 30 (1): 33–54.

180 El-Bendary, M.A. (2006). *Bacillus thuringiensis* and *Bacillus sphaericus* biopesticides production. *Journal of Basic Microbiology* 46 (2): 158–170.

181 Bravo, A., Gill, S.S., and Soberón, M. (2007). Mode of action of *Bacillus thuringiensis* Cry and Cyt toxins and their potential for insect control. *Toxicon* 49 (4): 423–435.

182 Soberón, M., Fernández, L.E., Pérez, C. et al. (2007). Mode of action of mosquitocidal *Bacillus thuringiensis* toxins. *Toxicon* 49 (5): 597–600.

183 Ffrench-Constant, R.H., Dowling, A., and Waterfield, N.R. (2007). Insecticidal toxins from *Photorhabdus* bacteria and their potential use in agriculture. *Toxicon* 49 (4): 436–451.

184 Rangel, L.I., Henkels, M.D., Shaffer, B.T. et al. (2016). Characterization of toxin complex gene clusters and insect toxicity of bacteria representing four subgroups of *Pseudomonas fluorescens*. *PLoS One* 11 (8).

185 Loper, J.E., Henkels, M.D., Rangel, L.I. et al. (2016). Rhizoxin analogs, orfamide A and chitinase production contribute to the toxicity of *Pseudomonas protegens* strain Pf-5 to *Drosophila melanogaster*. *Environmental Microbiology* 18 (10): 3509–3521.

186 Dangar, T.K. (2008). Infectivity and ecology of *Pseudomonas* spp. from natural epizootics in the rice leaf folder, *Cnaphalocrocis medinalis* (Lepidoptera: Pyralidae) in India. *Biocontrol Science and Technology* 18 (3): 241–253.

187 Mulet, M., Gomila, M., Lemaitre, B. et al. (2012). Taxonomic characterisation of *Pseudomonas* strain L48 and formal proposal of *Pseudomonas entomophila* sp. nov. *Systematic and Applied Microbiology* 35 (3): 145–149.

188 Chen, W.J., Hsieh, F.C., Hsu, F.C. et al. (2014). Characterization of an Insecticidal Toxin and Pathogenicity of *Pseudomonas taiwanensis* against Insects. *PLoS Pathogens* 10 (8).

189 Lacey, L.A. (2017). Microbial control of medically important mosquitoes in tropical climates. In: *Microbial Control of Insect and Mite Pests: From Theory to Practice* (ed. L.A. Lacey), 409–430. Academic Press.

190 Su, T. (2016). Resistance and Its Management to Microbial and Insect Growth Regulator Larvicides in Mosquitoes. Insecticides Resistance.

191 Becker, N. and Lüthy, P. (2017). Mosquito control with Entomopathogenic Bacteria in Europe. In: *Microbial Control of Insect and Mite Pests: From Theory to Practice* (ed. L.A. Lacey), 379–392. Academic Press.

192 Echeverri-Molina, D. and Santolamazza-Carbone, S. (2010). Toxicity of synthetic and biological insecticides against adults of the Eucalyptus snout-beetle *Gonipterus scutellatus* Gyllenhal (Coleoptera: Curculionidae). *Journal of Pest Science* 83 (3): 297–305.

193 Du Rand, N. and Laing, M.D. (2011). Determination of insecticidal toxicity of three species of entomopathogenic spore-forming bacterial isolates against *Tenebrio molitor* L. (Coleoptera: Tenebrionidae). *African Journal of Microbiology Research* 5 (16): 2222–2228.

194 Liu, D., Burton, S., Glancy, T. et al. (2003). Insect resistance conferred by 283-kDa *Photorhabdus luminescens* protein TcdA in *Arabidopsis thaliana*. *Nature Biotechnology* 21 (10): 1222–1228.

195 Grimont, P.A.D., Jackson, T.A., Ageron, E., and Noonan, M.J. (1988). *Serratia entomophila* sp. nov. associated with amber disease in the New Zealand grass grub *Costelytra zealandica*. *International Journal of Systematic Bacteriology* 38 (1): 1–6.

196 Johnson, V.W., Pearson, J., and Jackson, T.A. (2001). Formulation of *Serratia entomophila* for biological control of grass grub. *New Zealand Plant Protection* 54.

197 Hurst, M.R.H., Glare, T.R., and Jackson, T.A. (2004). Cloning *Serratia entomophila* antifeeding genes – a putative defective prophage active against the grass grub *Costelytra zealandica*. *Journal of Bacteriology* 186 (15): 5116–5128.

198 Flyg, C., Kenne, K., and Boman, H.G. (1980). Insect pathogenic properties of *Serratia marcescens*: phage-resistant mutants with a decreased resistance to Cecropia immunity and a decreased virulence to *Drosophila*. *Journal of General Microbiology* 120 (1).

199 Wu, D., Li, P., Zhou, J. et al. (2016). Identification of a toxic serralysin family protease with unique thermostable property from *S. marcescens* FS14. *International Journal of Biological Macromolecules* 93: 98–106.

200 Fritsch, M.J., Trunk, K., Diniz, J.A. et al. (2013). Proteomic identification of novel secreted antibacterial toxins of the *Serratia marcescens* Type VI secretion system. *Molecular and Cellular Proteomics* 12 (10): 2735–2749.

201 Chen, X., Ijkel, W.F.J., Tarchini, R. et al. (2001). The sequence of the *Helicoverpa armigera* single nucleocapsid nucleopolyhedrovirus genome. *Journal of General Virology* 82 (1): 241–257.

202 Long, G., Westenberg, M., Wang, H. et al. (2006). Function, oligomerization and N-linked glycosylation of the *Helicoverpa armigera* single nucleopolyhedrovirus envelope fusion protein. *Journal of General Virology* 87 (4): 839–846.

203 Escribano, A., Williams, T., Goulson, D. et al. (1999). Selection of a nucleopolyhedrovirus for control of *Spodoptera frugiperda* (Lepidoptera: Noctuidae): structural, genetic, and biological comparison of four isolates from the Americas. *Journal of Economic Entomology* 92 (5).

204 Bateman, M.L., Day, R.K., Luke, B. et al. (2018). Assessment of potential biopesticide options for managing fall armyworm (*Spodoptera frugiperda*) in Africa. *Journal of Applied Entomology* 142 (9): 805–819.

205 Raghunandan, B., Patel, N., Dave, H., and Mehta, D. (2019). Natural occurrence of nucleopolyhedrovirus infecting fall armyworm, *Spodoptera frugiperda* (J. E. Smith) (Lepidoptera : Noctuidae) in Gujarat, India. *Journal of entomology and zoology studies* 7 (2): 1040–1043.

206 Rios-Velasco, C., Gallegos-Morales, G., Berlanga-Reyes, D. et al. (2012). Mortality and production of occlusion bodies in *Spodoptera frugiperda* larvae (Lepidoptera: Noctuidae) treated with nucleopolyhedrovirus. *Florida Entomologist* 95 (3): 752–757.

207 Yaman, M., Nalçacioğlu, R., and Demirbağ, Z. (2001). Viral control of the European Pine Sawfly, *Neodiprion sertifer* (Geoffroy) in Turkey. *Turkish Journal of Biology* 25 (4).

208 Podgwaite, J.D., Rush, P., Hall, D., and Walton, G.S. (1984). Efficacy of the *Neodiprion sertifer* (Hymenoptera: Diprionidae) Nucleopolyhedrosis Virus (Baculovirus) Product, Neochek-S. *Journal of Economic Entomology* 77 (2).

209 Willis, L.G., Siepp, R., Stewart, T.M. et al. (2005). Sequence analysis of the complete genome of *Trichoplusia ni* single nucleopolyhedrovirus and the identification of a baculoviral photolyase gene. *Virology* 338 (2): 209–226.

210 Fuxa, J.R., Richter, A.R., Ameen, A.O., and Hammock, B.D. (2002). Vertical transmission of TnSNPV, TnCPV, AcMNPV, and possibly recombinant NPV in *Trichoplusia ni*. *Journal of Invertebrate Pathology* 79 (1).

211 Kumari, V. and Singh, N.P. (2009). *Spodoptera litura* nuclear polyhedrosis virus (NPV-S) as a component in Integrated Pest Management (IPM) of *Spodoptera litura* (Fab.) on cabbage. *Journal of Biopesticides* 2 (1): 84–86.

212 Senthil Nathan, S. and Kalaivani, K. (2005). Efficacy of nucleopolyhedrovirus and azadirachtin on *Spodoptera litura* Fabricius (Lepidoptera: Noctuidae). *Biological Control* 34 (1): 93–98.

213 Payne, C.C. (1974). The isolation and characterization of a virus from *Oryctes rhinoceros*. *Journal of General Virology* 25 (1): 105–116.

214 Dáder, B., Aguirre, E., Caballero, P., and Medina, P. (2020). Synergy of lepidopteran nucleopolyhedroviruses AcMNPV and SpliNPV with insecticides. *Insects* 11 (5): 316.

215 Karabörklü, S., Azizoglu, U., and Azizoglu, Z.B. (2018). Recombinant entomopathogenic agents: a review of biotechnological approaches to pest insect control. *World Journal of Microbiology and Biotechnology* 34 (1): 14.

216 Moore, S.D., Pittaway, T., Bouwer, G., and Fourie, J.G. (2004). Evaluation of *Helicoverpa armigera* nucleopolyhedrovirus (HearNPV) for control of *Helicoverpa armigera* (Lepidoptera: Noctuidae) on citrus in South Africa. *Biocontrol Science and Technology* 14 (3): 239–250.

217 Arrizubieta, M., Simón, O., Torres-Vila, L.M. et al. (2016). Insecticidal efficacy and persistence of a co-occluded binary mixture of *Helicoverpa armigera* nucleopolyhedrovirus (HearNPV) variants in protected and field-grown tomato crops on the Iberian Peninsula. *Pest Management Science* 72 (4): 660–670.

218 Rowley, D.L., Popham, H.J.R., and Harrison, R.L. (2011). Genetic variation and virulence of nucleopolyhedroviruses isolated worldwide from the heliothine pests *Helicoverpa armigera, Helicoverpa zea*, and *Heliothis virescens*. *Journal of Invertebrate Pathology* 107 (2): 112–126.

219 Lacey, L.A. (2016). Entomopathogens used as microbial control agents. In: *Microbial Control of Insect and Mite Pests: From Theory to Practice* (ed. L.A. Lacey), 1–10. Academic Press.

220 Bae, S. (2008). Characterization of *Mamestra brassicae* Nucleopolyhedrovirus (MabrNPV)-K1 Isolated in Korea. *International Journal of Industrial Entomology* 17 (1).

221 Bae, S. (2009). Pathogenicity and Production of *Mamestra brassicae* Nucleopolyhedrovirus (MabrNPV)-K1. *International Journal of Industrial Entomology* 19 (2).

222 Mukawa, S. and Goto, C. (2010). *Mamestra brassicae* nucleopolyhedrovirus infection and enhancing effect of proteins derived from *Xestia c-nigrum* granulovirus in larvae of mamestra brassicae and *Helicoverpa armigera* (Lepidoptera: Noctuidae) on cabbage. *Journal of Economic Entomology* 103 (2): 257–264.

223 Ilyinykh, A.V. and Polenogova, O.V. (2012). Demonstration of the remote effect of baculovirus vertical transmission, with gypsy moth *Lymantria dispar* L. (Lepidoptera, Lymantriidae) as an example. *Zhurnal Obshcheĭ Biologii* 73 (5): 389–395.

224 Harrison, R.L., Keena, M.A., and Rowley, D.L. (2014). Classification, genetic variation and pathogenicity of *Lymantria dispar* nucleopolyhedrovirus isolates from Asia, Europe, and North America. *Journal of Invertebrate Pathology* 116 (1): 27–35.

225 Bakhvalov, S.A., Martemyanov, V.V., Bakhvalova, V.N., and Morozova, O.V. (2012). Detection of the nuclear polyhedrosis Virus DNA in samples from eggs and Caterpillars at different stages of the Gypsy Moth *Lymantria Dispar* (L.) population dynamics. *Voprosy Virusologii* 57 (4): 35–37.

226 Bird, F.T. (1966). The development of a cytoplasmic polyhedrosis virus in the spruce budworm, *Choristoneura fumiferana* (Clemens). *Canadian Journal of Microbiology* 12 (2): 337–339.

227 Kemp, E.M., Woodward, D.T., and Cory, J.S. (2011). Detection of single and mixed covert baculovirus infections in eastern spruce budworm, *Choristoneura fumiferana* populations. *Journal of Invertebrate Pathology* 107 (3): 202–205.

228 Hayashi, Y. (1970). Occluded and free virions in midgut cells of *Malacosoma disstria* infected with cytoplasmic-polyhedrosis virus (CPV). *Journal of Invertebrate Pathology* 16 (3): 442–450.

229 Donaghue, T.P. and Hayashi, Y. (1972). Cytoplasmic polyhedrosis virus (CPV) of *Malacosoma disstria*: RNA polymerase activity in purified free viroin. *Canadian Journal of Microbiology* 18 (2): 207–210.

230 Stará, J. and Kocourek, F. (2018). Evaluation of efficacy of *Cydia pomonella* granulovirus (CpGV) to control the codling moth (*Cydia pomonella L.*, Lep.: Tortricidae) in field trials. *Plant Protection Science* 39 (4): 117–125.

231 Luque, T., Finch, R., Crook, N. et al. (2001). The complete sequence of the *Cydia pomonella* granulovirus genome. *Journal of General Virology* 82 (10): 2531–2547.

232 Nakai, M., Harrison, R.L., Uchida, H. et al. (2015). Isolation of an *Adoxophyes orana* granulovirus (AdorGV) occlusion body morphology mutant: biological activity, genome sequence and relationship to other isolates of AdorGV. *Journal of General Virology* 96 (4): 904–914.

233 Tsuruta, K., Wennmann, J.T., Kunimi, Y. et al. (2018). Morphological properties of the occlusion body of *Adoxophyes orana* granulovirus. *Journal of Invertebrate Pathology* 154: 58–64.

234 Opoku-debrah, J., Moore, S., Hill, M., and Knox, C. (2013). Characterisation of novel CrleGV isolates for false codling moth control – lessons learnt from codling moth resistance to CpGV. *Cabdirectorg* 90: 155–159.

235 Opoku-Debrah, J. K. (2011). Studies on existing and new isolates of Cryptophlebia Leucotreta (CrleGV) on Thaumatotibia leucotreta populations from a range of geographic regions in South Africa. 203.

236 Schmitz, H. (2018). Evaluation of the biological activity of granulovirus isolates from *Tuta absoluta* (TuabGV) and *Phthorimaea operculella* (PhopGV) in its primary and secondary host.

237 Léry, X., Villamizar, L., Espinel-Correal, C. et al. (2008). Analysis of several Colombian *Phthorimaea operculella* granulovirus isolated from *Tecia solanivora*: detection of a new variable region in the PhopGV genome. *IOBC/wprs Bulletin* 31: 83.

238 Wen, R., Ou, R., and Chen, B. (2007). Identification, transcriptional and phylogenetic analysis of the DNA polymerase gene of *Pieris rapae* granulovirus. *Virus Genes* 34 (3): 351–358.

239 Li, P., Wu, L., and Chen, G. (2019). Identification of a novel bipartite nuclear localization signal in the DNA polymerase of the betabaculovirus *Pieris rapae* granulovirus. *Archives of Virology* 164 (3): 839–845.

240 Abdulkadir, F., Knox, C., Marsberg, T. et al. (2015). Genetic and biological characterisation of a novel *Plutella xylostella* granulovirus, PlxyGV-SA. *BioControl* 60 (4): 507–515.

241 Abdulkadir, F., Marsberg, T., Knox, C.M. et al. (2013). Morphological and genetic characterization of a South African *Plutella xylostella* Granulovirus (PlxyGV) isolate. *African Entomology: Journal of the Entomological Society of Southern Africa* 21 (1): 168–171.

242 Espinel-Correal, C., Léry, X., Villamizar, L. et al. (2010). Genetic and biological analysis of colombian *Phthorimaea operculella* granulovirus isolated from *Tecia solanivora* (Lepidoptera: Gelechiidae). *Applied and Environmental Microbiology* 76 (22): 7617–7625.

243 Sporleder, M., Kroschel, J., Huber, J., and Lagnaoui, A. (2005). An improved method to determine the biological activity (LC50) of the granulovirus PoGV in its host *Phthorimaea operculella*. *Entomologia Experimentalis et Applicata* 116 (3): 191–197.

244 King, L.A., Wilkinson, N., Miller, D.P., and Marlow, S.A. (1998). Entomopoxviruses. In: *The Insect Viruses*, 1–29. US: Springer.

245 Goodwin, R.H. and Filshie, B.K. (1975). Morphology and development of entomopoxviruses from two Australian scarab beetle larvae (Coleoptera: Scarabaeidae). *Journal of Invertebrate Pathology* 25 (1): 35–46.

246 Crnov, R. and Dall, D.J. (1999). Virions of *Heliothis armigera* entomopoxvirus contain a homologue of the vaccinia VP8 major core protein. *Virus Genes* 19 (1): 23–31.

247 Zhou, Y., Qin, T., Xiao, Y. et al. (2014). Genomic and biological characterization of a new cypovirus isolated from *Dendrolimus punctatus*. *PLoS One* 9 (11).

248 Belloncik, S. and Mori, H. (1998). Cypoviruses. In: *The Insect Viruses* (eds. L.K. Miller and L.A. Ball), 337–369. US: Springer.

249 Graham, R.I., Rao, S., Sait, S.M. et al. (2007). Characterisation and partial sequence analysis of two novel cypoviruses isolated from the winter moth *Operophtera brumata* (Lepidoptera: Geometridae). *Virus Genes* 35 (2): 463–471.

20

Recent Trends in Conventional and Nonconventional Bioprocessing

Saswata Goswami[1], *Keyur Raval*[2], *Anjana*[1], *and Priyanka Bhat*[2]

[1] *Center of Innovative and Applied Bioprocessing, Mohali, Punjab, India*
[2] *Department of Chemical Engineering, National Institute of Technology Karnataka, Mangalore, Karnataka, India*

20.1 Advances in Conventional Bioprocessing

Since the beginning of bioprocessing industries, two bioreactor configurations have been widely used: the shake flasks or conical flasks and the stirred-tank bioreactors. These two have been conventionally used for bioprocess development, optimization, and scale-up. Shaken cultures have been used in all biotechnology laboratories, but its significance was made visible to the scientific community very recently [1]. The characterization of shaken cultures helps scientists and engineers for the improvement of productivity, process scale-up, or scale-down. The primary issue with shaking bioreactors is the difficulty of scale-up, especially for large-volume products. Therefore, stirred-tank bioreactors have always been the preferred choice for large-scale operations. The disadvantages of shaken cultures, such as reproducibility and longer bioprocessing time, have been addressed largely. However, one needs to understand which parameters would overcome these disadvantages. For example, how one can make a bioprocess more efficient and fast using shaken cultures. Aeration and mixing are two hurdles to overcome for improvement of any bioprocess performance. It is known that oxygen has the least solubility of all the medium components of any aerobic bioprocess, and therefore, it is supplied continuously using air. Figure 20.1 explains the dependency of the microbial growth rate on oxygen concentration in the medium.

In region A of Figure 20.1, the growth rate of a given microorganism is dependent on the concentration of dissolved oxygen in the medium. The growth rate is independent of dissolved oxygen concentration in region B. Ideally, the bioprocess should operate in region B. However, in most of the shaken cultures, the conditions shown in region A prevail, in which the growth rate of the cultivated cells is dependent on the oxygen concentration because of less oxygen transfer from the air into the medium. For shaken cultures, the air present in the headspace provides oxygen to the liquid medium being shaken in the conical flask. The cells would grow faster if the rate of transfer of oxygen from the headspace to the

Bioprospecting of Microorganism-Based Industrial Molecules, First Edition.
Edited by Sudhir P. Singh and Santosh Kumar Upadhyay.

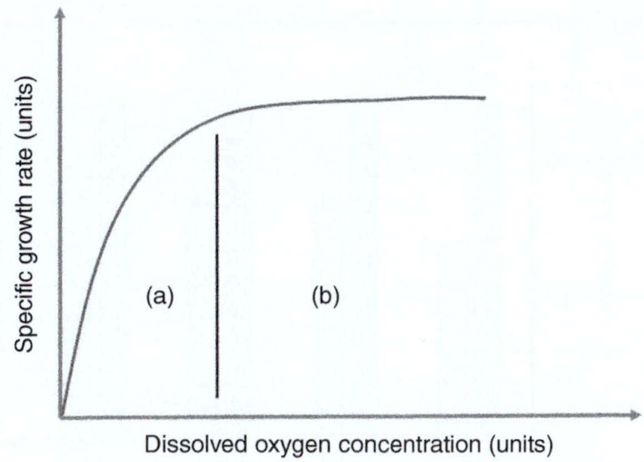

Figure 20.1 Dependency of the growth rate of aerobic microorganism on the dissolved oxygen levels in the medium. (a) represents oxygen dependent growth rate, and region (b) represents oxygen independent growth rate.

liquid medium is improved. The oxygen transfer rate in the shaken cultures depends on the following parameters. Meier et al. [2] provided a generalized equation of the oxygen transfer rate in shaken cultures.

$$OTR_{\max} = 3.72 \cdot 10^{-7} \cdot Osmol^{0.05} \cdot n^{\left(1.18 - \frac{osmol}{10.1}\right)} \cdot V_L^{-0.74} \cdot d^{1.88} \cdot d_0^{0.33} \cdot p_R \cdot y_{O_2}^{*}$$

Where, OTR_{max}=maximum oxygen transfer rate (mol/L/h), $Osmol$= osmolality of the medium (Osmol/kg), n=shaking frequency (rpm), V_L=medium filling volume inside the flask (ml), d=maximum diameter of the conical flask (mm), d_0=shaking diameter of the shaker table (cm), p_R=pressure inside the bioreactor (bar), $y_{O_2}^{*}$ =oxygen mole fraction in the gas phase.

If the effect of osmolality is neglected, then the medium volume inside the flask and diameter of the flask has a profound effect on the OTR, followed by the shaking diameter and the revolution speed of the shaker. This means, if one is not able to increase the revolution speed or shaking diameter of the shaker, then also, oxygen transfer would significantly improve by reducing the medium's volume and increasing the flask size. This technique would improve oxygen availability to the cells grown in the flask and hence would improve the growth rate, which in turn would reduce the time of bioprocessing. Figure 20.2 shows the application of this equation for microbial cultures for dye degradation and recombinant enzyme production [3].

Figure 20.2a shows the effect of reducing the filling volume inside the conical flask on dye degradation.

Figure 20.2b shows how the process time is reduced considerably by reducing the medium filling volume and increasing the rotational speed of the shaker.

Another point of consideration is the use of shake flask closures, which are mostly overlooked. The flask closure has an effect on the supply of air into the headspace of the flask. A correct closure is very much necessary for a continuous supply of air into the headspace. Many times, aluminum foils are used to close the flask opening. However, this reduces air supply to the headspace, which in turn reduces the availability of oxygen to the cells as well as the removal of carbon dioxide from the headspace. Moreover, the foil does not filter the

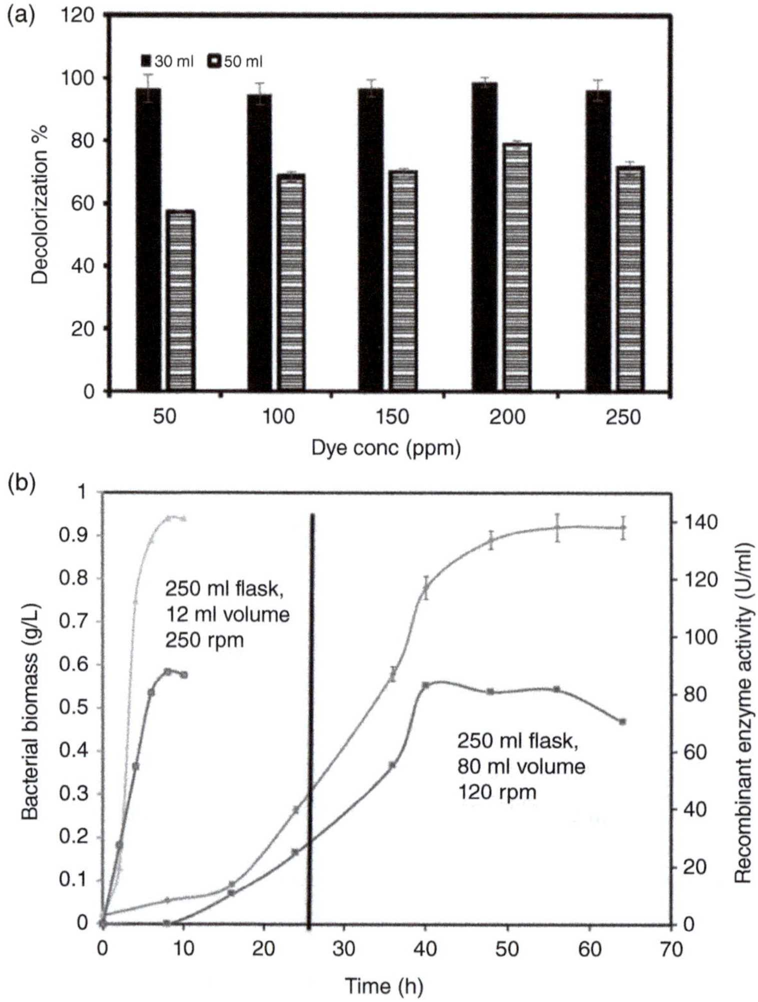

Figure 20.2 (a) Degradation of reactive black 5 dye in shake flask operated at 110 rpm with different medium filling volumes, (b) effect of rotational speed and medium filling volume on recombinant enzyme production in *E. coli* using LB medium.

air and hence increases the chances of contamination. It is recommended to use standard cotton closures or commercially available flask closures to maintain reproducibility and continuous air supply to the culture [4]. One crucial aspect of the shaken cultures is the relatively homogeneous distribution of the momentum and reduced shear stress generation. The flask wall transfers momentum to the fluid being mixed in the flask. Thus, the ratio of momentum transfer area to the liquid volume being handled is relatively higher, which reduces the velocity gradient and hence substantially reduces the shear stress generation. The shear stress generation in the shake flask cultures is about 10 times less than that of the stirred-tank bioreactor for the same specific power consumption. This positive effect of reduced shear stresses has been observed in the cultivation of plant and animal cell cultures, which are shear sensitive [5]. Although mixing is reported to be homogenous in shake flask, care must be taken off the out-of-phase phenomenon, which is typical for viscous cultures. In the out-of-phase phenomenon, the medium inside the flask is not well mixed. This usually happens if the culture viscosity is more or the ratio of shaking diameter to flask diameter is too small. Buechs et al. provided a dimensionless Phase number (Ph) to identify the out-of-phase condition in the shaken cultures. They concluded that given operating conditions should provide good mixing conditions if $Ph>1.26$ [6]. The characterization of the shaken bioreactors also paved the way for scale-up of shaken cultures to the 2000 Liter scale. This is discussed later in the chapter.

20.1.1 The Stirred-Tank Bioreactor Systems

The stirred-tank bioreactors have been widely used at the small, pilot, and large scale. The system has been the benchmark for decades for large-scale bioproduct production. The cells growing in the bioreactor require a homogenous environment for good growth. The homogeneity is maintained by mixing, and impellers are the source of mixing in the stirred-tank bioreactor. Therefore, impeller designs have been researched extensively because mixing governs all the transport aspects (mass, momentum, and heat) inside the bioreactor and helps in providing a homogenous nutrient-rich environment to the cells. We know that there are mainly three kinds of impellers being used extensively for bioprocessing, which are radial flow, axial flow, and mixed flow impellers. The mixed flow and radial impellers provide better dispersion of air inside the bioreactor as compared with the axial impellers. Moreover, axial impeller generates an excessive longitudinal force on the shaft of the impeller at a large scale, which increases the maintenance of a bioreactor mixing assembly. Therefore, radial and mixed flow impellers have been the preferred choice [7]. The Scaba (6SRGT) and hydrofoil impellers are the most widely used impellers for large-scale operations related to gas–liquid dispersion. These two impellers are relatively recent developments in the field of impeller design for gas–liquid dispersion. The advantage of these impellers is efficient mixing using very low power consumption [8]. A dimensionless power number related to the degree of mixing by a dimensionless Reynolds number is well reported for different impellers. The Power number becomes independent of the Reynolds number in the fully developed flow. The value of Power number to achieve fully developed flow depends on the ability of the impeller to provide homogenous distribution of the momentum and hence its geometry. The Rushton turbine requires Power number 5.2 to achieve homogenous mixing. On the other hand, the Scaba (SRGT6) impellers require a power number of 2.1 to achieve homogenous mixing or fully developed flow. Since the Power number indicates how much power is required to achieve homogenous mixing, less Power number impeller will use less power and hence less electricity. This substantially reduces the capital cost and running cost of a bioreactor. Therefore, all the large-scale stirred bioreactors use the low power number impellers such as Scaba, Pitched blade, or hydrofoil impellers. Another interesting point is Scaba impellers generate fully developed flow at almost one-tenth of the Reynolds number as compared with the Rushton turbine [9]. As one can imagine, the power consumption or degree of mixing is more near the impeller region, as compared with that of the rest of the fluid. The maximum degree of mixing is observed within one diameter of the impeller. The power consumption in this region is called the maximum power consumption, which is different from the average power consumption per unit liquid volume. The ratio of average power consumption to the maximum power consumption measures the inhomogeneity of mixing. The higher the ratio, higher the inhomogeneous environment for the microbial cells. This maximum power consumption causes the shear stress being experienced by the microbial culture being cultivated and is measured by the Kolmogorov eddy length given by the following equation:

$$\gamma = \left(\frac{\mu_L}{\rho_L}\right)^{3/4} \cdot \varepsilon^{-1/4}$$

Where, γ = Kolmogorov length scale (m), μL = viscosity of the liquid medium (PaS), ρL = density of the liquid medium, and ε = energy dissipation rate near impeller region which corresponds to the maximum power consumption (watts/kg).

Thus, a high value of the maximum power consumption generates eddies of smaller size and generates more shear stress to the cells. This indicates that impellers that generate less power number at fully developed flow would also generate less shear stress. There is one very important point to consider at this juncture. For aeration, the air supply is usually provided just below the impeller by a sparger for better dispersion of air inside the bioreactor. The impeller region is exposed to an air–liquid mixture, which reduces the density of fluid element being used for the transfer of momentum from the impeller to the rest of the liquid. This reduces the power consumption at a given operating condition, thereby the degree of mixing is reduced, which generates inhomogeneity in the bioreactor. If power consumption is reduced drastically, then a condition called "flooding" happens, in which the impeller region is flooded with air bubbles. Therefore, two kinds of power consumption are reported in the literature. The ungassed power consumption (in the absence of air supply) and the gassed power consumption (in the presence of air supply). The value of gassed power consumption depends on the airflow rate, type of the impeller, the diameter of the impeller, and the rotational speed of the impeller [7]. The ratio of gassed to ungassed power consumption indicates how efficient a given impeller is in dispersing the air/gas. Ideally, a ratio close to 1 is desired; however, inefficient impeller design reduces this ratio to below 0.7. As discussed above, this phenomenon is called flooding, and it should be avoided at any scale of operation. Flooding substantially reduces all transport processes inside the bioreactor and causes an inhomogeneous environment. The Scaba or BT6 impellers proved to be the most efficient radial flow impellers as they are not flooded at low revolutions and high aeration rate [10]. Nienow et al. [7] provided the following equations for operating conditions required to overcome flooding,

$$Fl = k \cdot \left(\frac{D_i}{D_T}\right)^{3.5} \cdot Fr, \text{where}, F_l = \frac{Q_G}{n.D_i^3}, F_r = \frac{n^2 D_i}{g}$$

The value of k = 30 for the Rushton impeller and 170 for the Scaba (BT6) impeller. This shows that Scaba impellers can provide much better gas–liquid dispersion as compared with Rushton turbines at high airflow rates. Similarly, the following correlation [7] is used to predict operating conditions in which complete dispersion will be achieved,

$$F_l = k_{CD}\left(F_r\right)^{0.5}\left(\frac{D_i}{D_T}\right)^{0.5}$$

Where, $k_{CD} = 0.2$ for Rushton turbine and 0.4 for Scaba impellers.

These findings indicate that Scaba or SRGT impellers provide very efficient gas–liquid dispersion as compared with Rushton or pitched blade turbines. Wide-blade Hydrofoil impellers are also an excellent alternative to Rushton turbines.

20.2 Nonconventional Bioprocessing

The commercial era of bioprocessing started with the development of bioprocesses for antibiotics and vaccine production about five decades back. This era of large-volume products was the most prominent until the 1990s. The development of monoclonal antibodies, therapeutic proteins, and other animal cell culture technology started a new era of low-volume high-value products. The commercial-scale volume of such products is only a few thousand liters as compared with hundreds of kiloliters of large-volume low-value products, e.g. 2000 Liter scale is already a commercial scale for low-volume large-value products such as monoclonal antibodies. These low-volume large-value products are being produced using animal, insect, or plant cell cultures, which are susceptible to shear stresses. Moreover, it was needed to produce various types of such low-volume high-value products to reach the market faster. These two bottlenecks prompted researchers to develop a new system of bioreactors, which are called as disposable or single-use bioreactors. Eibl et al. provided the following table, which explains the market factors that are driving the use of single-use technology in the biopharmaceutical sector (Table 20.1).

Shaking bioreactors were first used as single-use bioreactors [12]; however, it was the Wave bioreactors that were first to be introduced to the market. The Wave bioreactors provided the flexibility of product change over as well as high titers of the products because shear stress generation was lower than the stirred-tank bioreactor [13]. The success of Wave bioreactors opened a plethora of options for single-use bioreactors. However, only three categories of bioreactors were successfully scaled-up to the 2000 Liter scale. These three categories based on the mixing pattern of the medium are the wave bioreactors, stirred-tank bioreactors, and shaking bioreactors.

20.2.1 Wave Bioreactors

Vertical movement of the wave bioreactor's platform induces a wave in the sterile plastic bag containing culture medium and cells. This facilitates surface renewal of the culture medium and provides mixing. The important parameters which govern the mixing inside the wave-type bioreactor are rocking rate, rocking angle, vibration frequency, temperature, and aeration rate [14]. BioWave®, Wave Bioreactor™, BIOSTAT®CultiBagRM, and CELL-tainer® are some successful examples of mixed wave bioreactors. The CELL-tainer® combines rocking and horizontal displacement to provide a three-dimensional mixing pattern, which is reported to provide very high oxygen transfer rates up to 700/h and better carbon

Table 20.1 Factors affecting the development of single-use technology.

Market factors	Advantages	Current limitations
Emphasis on production costs	Reduced capital costs for plant construction and commissioning	Leachable and extractable
Flexible, multiproduct manufacturing facilities	Reduced risk for product cross-contamination in a multiproduct facility	Prior investment in fixed equipment
Biosimilar	Rapid changeover	Scales limited by current 2000-liter cell culture bioreactor capacity
		A limited number of vendors
Multiple, smaller manufacturing plants collocated with markets	Lower utility costs due to reduced need for SIP	The high cost of disposables
The increasing number of low-volume biopharmaceutical products	Reduced need for cleaning validation	Lack of universal standards for vendors
		Solid waste disposal

Source: Based on Eibl et al. [11].

dioxide exchange than in the wave-mixed bioreactors previously mentioned. Because of high oxygen transfer rates, the CELL-tainer® is also suitable for microbial high-cell density cultures. The wave bioreactors have been a widely used platform for animal cell culture-related applications. The mixing and transport processes are characterized for wave bioreactors. However, the mixing patterns vary from one scale to another scale, and flow behavior is largely nonuniform due to the wave motion of the fluid [13]. The bag geometry plays a profound role in the fluid motion inside the wave-type bioreactors. This generates numerous possibilities of flow patterns inside the bag. The resonance phenomenon induced by the wave motion also provides a heterogeneous environment for the growing culture inside the wave motion-type bioreactors.

20.2.2 Orbital Shaken Bioreactors

Orbitally shaken vessels, especially "Erlenmeyer" flasks or conical flasks with nominal volumes up to 5 l, are widely used for the cultivation of microbial hosts. The fluid flow characterization over the past decade has made them attractive alternatives to conventional stirred-tank bioreactors. Some successful examples include Tubespin bioreactors for small scale and shaken bioreactors or helical bioreactors [11, 13, 15]. These bioreactors are well characterized in terms of mixing, heat, and mass transfer, which makes them suitable for scale-up. The flow patterns are homogenous at a different scale, which makes the prediction of the bioprocess profile easier. As discussed previously in Section 20.1, the parameters that affect the fluid flow behavior in conical shaking bioreactors are similar to the ones observed for cylindrical shaken bioreactors and only the intensity of parameters on the mixing changes. The following equation [16] provides the critical shaking frequency above which the shaking bioreactor should be operated for homogenous mixing.

$$N_c = \frac{1}{d^2} \cdot \sqrt{0.28 \cdot V_L \cdot g}$$

Where, N_c is critical shaking frequency (rps) above which homogenous mixing is observed, d is cylindrical vessel diameter (m), V_L is filling volume inside the vessel (m^3), and g is the gravitational acceleration. This equation is derived using bioreactors sizes ranging from 10 to 2000 l and filling volumes ranging from 10 to 60%. A well-defined fluid flow profile helps in getting reproducible results in any bioprocess, which also helps to achieve GMP requirements. Mammalian and insect cells were first reported to grow in a 50 l single-use shaking bioreactor by Liu et al. [12], followed by the plant cell culture cultivation by Raval et al. [17]. The k_La values in the range of 7 to 10 1/h are reasonably good for mammalian cell cultivation. The k_La values between 1 and 30 1/h are reported for shaking bioreactors up to 2000 l scale bioreactors [18]. The helical track cylindrical shaking bioreactors are reported to provide further improvement of the oxygen transfer rate. The helical track inserted in the periphery of the cylindrical vessel developed a rotating flow pattern due to orbital shaking. The fluid velocity created due to the helical track allowed the liquid to lift itself toward the oxygen-rich headspace of the bioreactor. This typically resulted in the increased interfacial area available for oxygen transfer and improved oxygen transfer rate to the tune of 5- to 10-folds depending upon the operating conditions [18]. Even in a 1500-l helical track vessel with a working volume of 1000 l, a surprisingly high k_La value of 10/h was obtained at a relatively low shaking speed of 39 rpm and an axial displacement radius of 10 cm. The new developments in shaking technology combined with the development of efficient optical sensors, shaken bioreactor systems at scales beyond 2000 l are expected to be introduced in the future.

20.2.3 Stirred Tank Bioreactors

The disposable stirred-tank bioreactor assembly resembles their stainless-steel counterparts to a large extent. This technology entered the biopharmaceutical market in 2006. Since the working principle is similar to standard stirred-tank reactors, the technology was quickly adopted by the companies thinking process transfer to the large-scale facility would be efficient. The disposable stirred-tank bioreactors up to 1000 l (HyClone) and 2000 l (Xcellerex) culture volume provide specially designed disposable bag, which includes optical sensors, assembly tubes, filters, spargers, ports, and sampling probes. The bag is integrated into a permanent support vessel (stainless steel). HyClone has an integrated pitch blade impeller with an angled shaft connected to the top. These bioreactors have been routinely applied for animal cell culture cultivation. However, the mixing patterns in disposable stirred-tank bioreactors are very different from their stainless-steel counterparts due to different design and aeration patterns [11]. The baffles are removed from the reactor to reduce shear stress generation, especially for mammalian cell cultivation. Still, there are different zones of mixing in which different levels of shear stresses are observed. The Kolmogorov length scale measured at different operating conditions using different volumes of cultivations bags in BIOSTAT showed that the length scale in the stirred zone and rotating stirrer zone is different by the magnitude of order one. The authors demonstrated that for all scales, the rotating stirrer volume was approximately 7% of the total volume [11]. Thus, a mammalian cell will experience different levels of shear stress in different regions. The CHO cells have a size in the range of 10 to 20 μm. If the cells encounter eddies size less than their diameter, then they would experience the shear stress. It is reported that for CHO cells, the power input values up to $10 W/m^3$ should not damage the cells. If power input exceeds $10 W/m^3$, then the cells may get damaged near the stirrer region. The disposable stirred-tank bioreactor assemblies, such as BIOSTAT and others, have an option to provide baffles like perturbations in the stainless-steel support vessel. The disposable bag shape alters, which provides efficient mixing but at the cost of high shear rates. However, this arrangement provides the advantage to cultivate high cell density microbial cells. Interestingly, most of the stirred-tank bioreactors are equipped with either Rushton impellers, pitched blade impellers, or marine impellers. On the contrary, there are more efficient impeller options available, as discussed in Section 20.1.

Table 20.2 summarizes the differences between the three types of single-use bioreactors discussed so far. Do note that the medium volume and medium physical properties affect fluid flow pattern in all the three kinds of systems.

The biopharmaceutical industries control the product quality using the Quality by design (QbD) approach. This method uses process-specific critical quality attributes (CQA), such as product purity, product properties, glycosylation pattern, etc. These CQAs depend on the key process parameters (KPP), such as mixing efficiency, homogenous environment concerning shear stress, oxygen transfer rate, feeding strategy, etc. Therefore, any bioreactor that provides minimum variations in KPP from a small scale to a large scale would also provide better product quality [11].

Table 20.2 Comparison of the bioreactors depending on the three different flow patterns.

	Wave motion	Shaking motion	Stirred motion
Parameters influencing flow pattern	Rocking angle, rocking rate, height, and width of culture bag	Bioreactor geometry, shaking diameter, revolution speed	The geometry of the bioreactor, revolution speed, air bubbles, type of impeller, number of the impeller, mounting of the impeller
Type of aeration	Headspace	Headspace	Air bubbles
Fluid flow pattern	Not well defined	Well defined	Not well defined
Ease of scale-up from lab scale to 2000 l scale	Moderate to difficult	Easy to moderate	Moderate to difficult
Shear stress variations inside the bioreactor	Heterogenous pattern	Homogeneous pattern	Heterogenous pattern
Process intensification using a perfusion system	Possible	Possible	Possible

20.3 Brief Note on the Recent Trends in Downstream Bioprocessing

Single-use technology is used for upstream bioprocesses development for the past two decades because of the numerous benefits as compared with stainless steel-based conventional technologies. The technology is now widely adopted in downstream bioprocessing as well. The complex unit operations such as chromatography, filtration, tangential flow filtration (TFF), etc., are replaced by single-use platforms. The single-use downstream bioprocessing offers flexibility, robustness, and improved productivity. As single-use technologies have grown in importance and acceptance; there has been a dramatic improvement in production efficiencies around the turn of the twenty-first century [11, 19, 20].

The sustainable, efficient, and cost-effective processes are the need of an hour in biological industries. Bioprocessing has the characteristic of high labor cost, material cost, and extensive demand experimentation. All bioprocessing industries are developing a functional first-generation process to move toward goals such as capacity building, robustness, consistency, efficiency, manufacturing flexibility, and sustainability Bioprocess intensification (BI) is a method to comply with such requirements [21]. There are several ways through which the manufacturing of biological products can be intensified. The term "bioprocess intensification" refers to systems for producing more product per cell, volume, time footprint, or cost. This has become the need of an hour because of these two emerging priorities: cost control and process efficiency. Research and developments in disciplines such as process simplification, synthetic biology, continuous bioprocessing, and the use of digital techniques for optimizing bioprocess development and manufacturing enabled BI. Furthermore, the developments that are aiding in BI include process automation, prefabricated facility, and improved monitoring. The improvement in the high yield cell lines, optimized culture media, and expansion systems have provided many more opportunities for adoption and development of continuous bioprocessing. Cell engineering significantly increased cell-specific production rates in the last 10 years [22].

The BI is a subordinate to the next generation of "factory of the future" initiatives. BI can be enabled by such corollary technologies as digital or continuous biomanufacturing.

The BI can be divided into two main areas:

1) Equipment intensification, which involves the use of novel reactors, mass transfer units, and heat exchangers.
2) Process intensification, which involves the use of new separation strategies, phase transition, integration of reaction and separation, techniques using alternative energy sources, and new control method.

Table 20.3 describes a few successful examples of BI.

20.4 Perfusion Culture for Bioprocess Intensification

The cost-effective perfusion systems have been used at a commercial scale in a single-use format for BI. It includes perfusion technology adoption like alternating tangential flow (ATF)-based systems that pump and filter to retain the cells within the bioreactors with

Table 20.3 Overview of the various applications of intensifying processes which can be used in bioprocesses.

Equipment	Typical applications	Results of the study
Higee Bioreactor (HBR)	Production of Polyhydroxyalkanoate (PHA) through the fermentation process	Higher biomass concentration, mass transfer capability, and PHA yield
Spinning Cloth Disc Reactor	Enzyme immobilization	Rapid mixing and improved mass transfer rates
Static mixer	Multiphase reactions	Higher mass transfer
Micro-reactors	Manufacturing fine chemicals and drugs	Improved diffusion
Monolithic reactors	1) Enzymatic oxidation of glucose 2) Formation of Gluconic acid by gluconobacter suboxydans 3) Production of Hydrogen by Clostridium butyricum	
Heat exchange reactors	Fast exothermic reactions, azo coupling, hydrogenation, halogenation, oxidation, amination, sulfonation, and alkylation	Energy-efficient, safer processing, and better economics
Electrically enhanced reactions	Droplets can break up into micron sizes, Hydrolytic splitting of esters into free fatty acids and glycerol	Improvement in heat and mass transfer
Continuous stirred-tank reactor	Bioreactions, Hazardous reactions like fine chemical reactions, Crystallization, general pharmaceutical reactions	Can be used in series, easy to clean, improved temperature control than batch STR
Fixed-bed bioreactor using micro-cellular polymer-immobilized cells	Production of α-amylase in porous polymeric polyHIPE by immobilized Bacillus subtilis	Matrix clogging
METHODS	**Typical applications**	**Results of the study**
Enzymes immobilization	a) Glucose isomerase (on an inorganic carrier) – producing high fructose corn syrup, b) penicillin G acylase (covalently attached to polyacrylate) – producing semisynthetic penicillins, c) lactase (on an ion-exchange resin) – producing low-lactose milk, d) TL lipase (on silica) for fat modification	Enhancement in the activity, stability, and selectivity, obtaining adequate productivity
Cells immobilization	Immobilized yeast cells used for beer fermentation	High productivity
Reactive crystallization	Fermentation coupled with crystallization-based in-situ product removal for the efficient calcium lactate production	Higher average productivity

(Continued)

Table 20.3 (Continued)

Equipment	Typical applications	Results of the study
Membrane absorption	Recovery and purification of biofuel compounds such as isopropanol, ethanol, n-propanol, n-butanol, iso-butanol, n-pentanol, etc.	Economic recovery
Liquid–liquid extraction	In-situ product removal of phenol from P. Putida S12TPL fermentations Recovery of fermentative carboxylic acids	Enhanced yields and productivity
Membrane absorption	In-situ Product Recovery of n-Butanol In-situ product removal of phenol from P. Putida S12TPL fermentations	Improved yield, productivity, and economical purification
Membrane extraction **Anion exchanger-based**	Microbial production of propionic acid with *Propionibacterium freudenreichii*	Enhanced yield and productivity, less substrate consumption, and decrease in the number of downstream processing steps
Adsorption–desorption	In-situ recovery of 3-hydroxypropionaldehyde (3HPA) during biotransformation of glycerol by Lactobacillus reuteri	2 times higher productivity of 3HPA

Source: Vaghari et al. [21]. Licensed under CC BY-4.0.

continuous outflow of culture media [22]. Perfusion system uses various means such as centrifugation, gravity settling, external loop flow-through filters, and pumping through internal filters to separate cells from the media [23]. Other methods for perfusion generally involve the host cells to be inherently adherent to micro-carrier particles that are suspended in the cell culture, magnetic beads, hollow capillary, flat plates, sponge-like matrices, other membranes, and fiber-based bioreactors. The cells adhere to these substrates during cell culture. Perfusion provides integration with purification and complete end-to-end continuous production of monoclonal antibodies in bulk with automated control and real-time release [23]. According to the methods and equipment used, the perfusion mode requires less culture media as compared with fed-batch mode. Product-related quality problems are common with fed-batch processing as the product will remain exposed to stressful conditions (mixing and heating) and by-products (e.g. lactic acid and ammonia) that can intoxicate the cells. With perfusion, there is a continuous removal of contaminants [22]. The perfusion media is combined with cell retention through an external hollow tangential flow filter (TFF). The ATF and the automated TFF use a continuous media feed stream to the reactor. An ultrafiltration membrane (usually 30–50 KDa) is connected to the reactor so that the cells and products are retained within the reactor to remove the undesired metabolites in the permeate. This concept of BI is so efficient that the volumetric productivity of a 2000 l disposable bioreactor based on perfusion technology is more than that of a 10 000 l stainless steel reactor operated in a fed-batch mode that too with significantly lower capital cost [23]. Optimized culture media composition in batch, fed-batch, perfusion-enhanced batch, and perfusion-based continuous operations can vary greatly. Intensified perfusion processes can dramatically increase productivity in terms of both time and footprints [24].

Digital approaches, e.g. "4.0" in biopharma are showing promise to become major contributors to BI. The rapid improvement in process control systems, which include supervisory control, data acquisition, BI data sources, transmission, storage, and governance. The two important factors to consider in BI are bioreactor type and process control. Microfluidics is aiding cell line development and gene editing [25].

20.5 Conclusions and Perspectives

The bioprocess technology has evolved in leaps and bounds in the past century. The traditional stainless-steel units have been replaced by the single use units in upstream and downstream bioprocessing. The chapter describes the evolution in conventional bioprocessing to the non-conventional bioprocessing, i.e., single use technology. The sustainability of single use technology is explained in brief. The advantages and disadvantages ofconventional and no-conventional bioprocessing are demonstrated with examples. The future of the non-conventional technology lies in the multifaceted approachof biology, engineering, and electronics. The artificial intelligence would pave the way for the increased efficiency of the bioprocessing technology.

References

1 Büchs, J. (2001). Introduction to advantages and problems of shaken cultures. *Biochemical Engineering Journal*. Elsevier 7: 91–98.

2 Meier, K., Klöckner, W., Bonhage, B. et al. (2016). Correlation for the maximum oxygen transfer capacity in shake flasks for a wide range of operating conditions and for different culture media. *Biochemical Engineering Journal* 109: 228–235.

3 Raval, K., Thaira, H., Bhat, P., and Raval, R. (2019). Improving productivity of shaken cultures. In: *BESCON 2019* (ed. S. Srivastava). IIT Madras.

4 Mrotzek, C., Anderlei, T., Henzler, H.J., and Büchs, J. (2001). Mass transfer resistance of sterile plugs in shaking bioreactors. *Biochemical Engineering Journal*. Elsevier 7: 107–112.

5 BÜCHS, J. and ZOELS, B. (2001). Evaluation of maximum to specific power consumption ratio in shaking bioreactors. *Journal of Chemical Engineering of Japan* 34 (5): 647–653.

6 Büchs, J., Maier, U., Milbradt, C., and Zoels, B. (2000). Power consumption in shaking flasks on rotary shaking machines: II. Nondimensional description of specific power consumption and flow regimes in unbaffled flasks at elevated liquid viscosity. *Biotechnology and Bioengineering*.

7 Nienow, A.W. (2015). Aeration in Biotechnology. In: *Kirk-Othmer Encyclopedia of Chemical Technology* (ed. C. Ley), 1–23. Hoboken, NJ, USA: John Wiley & Sons, Inc.

8 Pinelli, D., Bakker, A., Myers, K.J. et al. (2003). Some features of a novel gas dispersion impeller in a dual-impeller configuration. *Chemical Engineering Research and Design* 81 (4): 448–454.

9 Myers, K.J., Thomas, A.J., Bakker, A., and Reeder, M.F. (1999). Performance of a gas dispersion impeller with vertically asymmetric blades. *Chemical Engineering Research and Design* 77 (8): 728–730.

10 Bujalski, W., Amanullah, A., and Schmitz R (1996). A comparison of torque fluctuations and gassed power characteristics for a range of radial flow impellers in water. In: Institution of chemical engineers symposium series. Hemsphere publishing corporation; 1996. p. 315–26.

11 Eibl, D., Eibl, R., and Adams, T. (2014). *Disposable bioreactors II*, 1e (eds. D. Eibl and R. Eibl), 292. Berlin and Heidelberg: Springer-Verlag.

12 Liu, C.M. and Hong, L.N. (2001 Mar 1). Development of a shaking bioreactor system for animal cell cultures. *Biochemical Engineering Journal* 7 (2): 121–125.

13 Eibl, R. and Eibl, D. (eds.) (2010). *Disposable Bioreactors*, Advances in Biochemical Engineering/Biotechnology, vol. 115. SpringerLink.

14 Singh, V. (1999). Disposable bioreactor for cell culture using wave-induced agitation. *Cytotechnology* 30 (1–3): 149–158.

15 Klöckner, W., Diederichs, S., and Büchs, J. (2014). Orbitally shaken single-use bioreactors. In: *Advances in biochemical engineering/biotechnology* (eds. R. Eibl and D. Eibl), 45–60. Netherlands: Springer.

16 Klöckner, W., Tissot, S., Wurm, F., and Büchs, J. (2012 Jun 15). Power input correlation to characterize the hydrodynamics of cylindrical orbitally shaken bioreactors. *Biochemical Engineering Journal* 65: 63–69.

17 Raval, K., Liu, C.-M., and Büchs, J. (2006). Large-scale disposable shaking bioreactors : a promising choice. *Bioprocess International* 4 (1): 46–50.

18 Zhang, X., Stettler, M., de Sanctis, D. et al. (2010). Use of orbital shaken disposable bioreactors for mammalian cell cultures from the milliliter-scale to the 1,000-liter scale. In: *Advances in Biochemical Engineering/Biotechnology* (eds. R. Eibl and D. Eibl), 33–53. Berlin, Heidelberg: Springer.

19 Haigney, S. (2015). Using single-use technologies in downstream processing. *Bioprocess International*.

20 Junne, S. and Neubauer, P. (2018). How scalable and suitable are single-use bioreactors? *Current Opinion in Biotechnology* 53: 240–247.

21 Vaghari, H., Eskandari, M., Berenjian, A., and Jafarizadeh, H. (2015). Process intensification for production and recovery of biological products. *American Journal of Biochemistry and Biotechnology* 11 (1): 37.

22 Langer, B.E.S. and Rader, R.A. (2014). Continuous bioprocessing and perfusion: wider adoption coming as bioprocessing matures. *Bioprocessing Journal* 13 (1): 50–55.

23 Pollard, D. and Kistler, C. (2017). Disposable bioreactors. In: *Current Developments in Biotechnology and Bioengineering* (eds. C. Larroche, M. Sanroman, G. Du and A. Pandey), 353–379. Elsevier Inc.

24 Godawat, R., Konstantinov, K., Rohani, M., and Warikoo, V. (2015). End-to-end integrated fully continuous production of recombinant monoclonal antibodies. *Journal of Biotechnology*: 1–7.

25 Whitford, B. (2020). Bioprocess intensification : aspirations and achievements. *Bioprocess Technology*.

Index

Bioprospecting of Microorganism-Based Industrial Molecules, First Edition.
Edited by Sudhir P. Singh and Santosh Kumar Upadhyay.

c

q

r